# Retinal Degeneration

## Clinical and Laboratory Applications

# Retinal Degeneration

## Clinical and Laboratory Applications

Edited by

## Joe G. Hollyfield and Robert E. Anderson

Cullen Eye Institute
Baylor College of Medicine
Houston, Texas

and

## Matthew M. LaVail

Beckman Vision Center
University of California, San Francisco
San Francisco, California

Springer Science+Business Media, LLC

Library of Congress Cataloging-in-Publication Data

Retinal degeneration : clinical and laboratory applications / edited
  by Joe G. Hollyfield and Robert E. Anderson, Matthew M. LaVail.
        p.   cm.
    "Proceedings of an International Symposium on Retinal
  Degeneration, held September 15-20, 1992, in Costa Smeralda,
  Sardinia"--T.p. verso.
    "Satellite meeting of the 10th International Congress of Eye
  Research"--Pref.
    Includes bibliographical references and index.
    ISBN 978-1-4613-6294-4     ISBN 978-1-4615-2974-3 (eBook)
    DOI 10.1007/978-1-4615-2974-3
    1. Retinal degeneration--Congresses.   I. Hollyfield, Joe G.
  II. Anderson, Robert E. (Robert Eugene)   III. LaVail, Matthew M.
  IV. International Symposium on Retinal Degeneration (1992 :
  Sardinia, Italy)  V. International Congress of Eye Research (10th :
  1992 : Sardinia, Italy)
    [DNLM: 1. Retinal Degeneration--congresses.   2. Disease Models,
  Animal--congresses.   WW 270 R43824 1992]
  RE661.D3R476   1993
  617.7'3--dc20
  DNLM/DLC
  for Library of Congress                                93-37104
                                                              CIP

Proceedings of an International Symposium on Retinal Degeneration, held September 15–20, 1992, in Costa Smeralda, Sardinia

©1993 Springer Science+Business Media New York
Originally published by Plenum Press, New York in 1993
Softcover reprint of the hardcover 1st edition 1993

Dedicated to

**Nicola Orzalesi, M.D.**

Professor and Chairman of Ophthalmology, Institute of Biomedical Sciences, San Paolo Hospital, University of Milan, Milan, Italy, for his early insight into the origin of phagosomes in the pigment epithelium, his lifetime contributions to basic and clinical visual science, and his efforts in organizing the Sardinia Symposium on Retinal Degeneration.

# PREFACE

During the last few years, an explosion of information has come from human genetics and molecular and cell biological studies as to the genetic basis for a number of forms of inherited retinal degenerations. These disorders have plagued mankind for millennia because they take from otherwise healthy individuals the precious gift of sight. The fundamental advances in recent years have identified a number of genes involved in the groups of diseases which hopefully will lead to discoveries that may, in the not too distant future, allow the prevention and possible cure of some of these blinding eye disorders. To foster a forum for discussions of studies on degenerative retinal disorders, we convened a symposium on retinal degenerations in 1984, at the VIth International Congress of Eye Research Meeting, held in Alicante, Spain. Because of the success of this meeting and the subsequent publication, we have since organized a series of biennial satellite meetings on retinal degenerations for the ISER congresses held in Nagoya, Japan (1986), San Francisco (1988) and Helsinki (1990). Each of these satellite symposium on retinal degenerations was accompanied by a published proceedings volume.

This volume is the fifth in this series and contains the proceedings of the Sardinia Symposium on Retinal Degeneration held September 15-20, 1992, as a satellite meeting of the 10th International Congress of Eye Research. Most of the participants of this meeting have contributed chapters for this book, as did a number of individuals who work in this area but were unable to attend the symposium. The symposium was held north of Olbia at Porto Cervo on the Costa Smeralda. This stunning site provided an idyllic atmosphere that contributed enormously to the comradery and interactions which took place at this meeting. We are grateful to Professor Nicola Orzalesi, Chairman of Ophthalmology at the Institute of Biomedical Sciences, San Paolo Hospital, University of Milan, Milan, Italy, for his outstanding efforts in attention to all of the local details. We are also grateful to Professor Orzalesi's colleagues who assisted in the execution of this symposium: Professor Stefano Miglior, Dr. Chiara Pierrottet, Dr. Alessandro Porta, Professor Antonina Serra, Dr. Maurizio Fossarello and Mrs. Ann Orzalesi.

The meeting received extensive financial support from a number of organizations. We are happy to thank and acknowledge important contributions from the National Retinitis Pigmentosa Foundation Fighting Blindness, Inc., Baltimore, Maryland, USA; The National Institutes of Health, Bethesda, Maryland, USA; The Regional Government of Sardinia; Merck, Sharp and Dohme, Italy; Alcon Pharmaceuticals, Italy; Allergan, Italy; ACR Angelini; Frau Medica; Kabi Pharmacia; Sifi; Fidia; Surgitek; Opticon; Rodenstock, Italy; Louisiana State University Medical Center Foundation, New Orleans, USA; Inverni della Beffa; Sinax; and Esse Emme.

We thank May Lin Bell for her help in corresponding with all of the participants during the organization of this meeting, for preparing the meeting program and for her help in organizing the manuscripts for this volume. Special thanks go to Mary E. Rayborn for all her efforts in proofreading the final manuscript.

Joe G. Hollyfield
Robert E. Anderson
Matthew M. LaVail

## THE EDITORS

**Joe G. Hollyfield, Ph.D.,** is a Professor of Ophthalmology and Neuroscience at Baylor College of Medicine, Houston, Texas. He received his Ph.D. degree in Zoology (1966) from the University of Texas at Austin and was a postdoctoral fellow at the Hubrecht Laboratory in Utrecht, The Netherlands. He was appointed Assistant Professor of Anatomy assigned to Ophthalmology at Columbia University College of Physicians and Surgeons in New York City in 1969 and was promoted to Associate Professor in 1975. In 1977, he moved to the Cullen Eye Institute, Baylor College of Medicine, and was promoted to Professor in 1979. He has been Director of the Retinitis Pigmentosa Research Center in the Cullen Eye Institute since 1978.

Dr. Hollyfield has published extensively in the area of cell and developmental biology of the retina and pigment epithelium in both normal and retinal degenerative tissues. He has edited six books, five on retinal degeneration and one on the structure of the eye.

Dr. Hollyfield has received the Marjorie W. Margolin Prize (1981) and the Sam and Bertha Brochstein Award (1985) from the Retina Research Foundation, the Olga Keith Wiess Distinguished Scholars' Award (1981), a Senior Scientific Investigator Award (1988) from Research to Prevent Blindness, Inc., an Award for Outstanding Contributions to Vision Research from the Alcon Research Institute (1987), and the Distinguished Alumnus Award (1991) from Hendrix College, Conway, Arkansas. He has previously served on the editorial boards of *Vision Research* and *Survey of Ophthalmology*. He is currently Editor-in-Chief of *Experimental Eye Research*. He has received grants from the National Institutes of Health, the Retina Research Foundation, Fight for Sight, Inc., the Retinitis Pigmentosa Foundation Fighting Blindness, and Research to Prevent Blindness, Inc. Dr. Hollyfield has been active in the Association for Research in Vision and Ophthalmology as a past member of the Program Committee and is currently a member of the Board of Trustees and President-elect for 1993-94. He is the Immediate-past President and former Secretary of the International Society for Eye Research.

**Robert E. Anderson, Ph.D., M.D.,** is a Professor of Ophthalmology, Biochemistry and Neuroscience at Baylor College of Medicine, Houston, Texas. He received his Ph.D. degree in Biochemistry (1968) from Texas Agricultural and Mechanical University and was a postdoctoral fellow at Oak Ridge Associated Universities (1968). At Baylor, he was appointed Assistant Professor in 1969, Associate Professor in 1976, and Professor in 1981. While a faculty member at Baylor, he attended medical school and received his M.D. in 1975.

Dr. Anderson has published extensively in the areas of lipid metabolism in the retina and biochemistry of retinal degenerations. He has edited six books, five on retinal degenerations and one on biochemistry of the eye.

Dr. Anderson has received the Sam and Bertha Brochstein Award for Outstanding Achievement in Retina Research from the Retina Research Foundation (1980), the Dolly Green Award (1982), a Senior Scientific Investigator Award (1990) from Research to Prevent Blindness, Inc., and an Award for Outstanding Contributions to Vision Research from the Alcon Research Institute (1985). He has served on the editorial boards of *Investigative Ophthalmology and Visual Science* and is currently on the editorial boards of *Current Eye Research* and *Experimental Eye Research*. Dr. Anderson has received grants from the National Institutes of Health, the Retina Research Foundation, the Retinitis Pigmentosa Foundation Fighting Blindness, and Research to Prevent Blindness, Inc. He has been an active participant in the program committees of the Association for Research in Vision and Ophthalmology and has served on the Vision Research Program Committee and the Board of Scientific Counselors of the National Eye Institute. Currently, he is on the Faculty of the Basic and Clinical Science Series of American Academy of Ophthalmology and is a past Councillor and current Treasurer for the International Society for Eye Research.

**Matthew M. LaVail, Ph.D.,** is a Professor of Anatomy and Ophthalmology and Vice-Chairman of Anatomy at the University of California at San Francisco School of Medicine. He received his Ph.D. degree in Anatomy (1969) from the University of Texas Medical Branch in Galveston and was subsequently a postdoctoral fellow at Harvard Medical School. Dr. LaVail was appointed Assistant Professor of Neurology-Neuropathology at Harvard Medical School in 1973. In 1976, he moved to UCSF, where he was appointed Associate Professor of Anatomy. He was appointed to his current position in 1982, and in 1988, he also became director of the Retinitis Pigmentosa Research Center at UCSF.

Dr. LaVail has published extensively in the research areas of photoreceptor-retinal pigment epithelial cell interactions, retinal development, circadian events in the retina, genetics of pigmentation and ocular abnormalities, inherited retinal degenerations, and light-induced retinal degeneration. He is the author of more than 70 research publications and has edited five books on inherited and environmentally induced retinal degenerations.

Dr. LaVail has received the Fight for Sight Citation (1976), the Sundial Award from the Retina Foundation (1976), the Friedenwald Award from the Association for Research in Vision and Ophthalmology (1981), a Senior Scientific Investigator Award from Research to Prevent Blindness, Inc. (1988), a MERIT Award from the National Eye Institute (1989), and an Award for Outstanding Contributions to Vision Research from the Alcon Research Institute (1990), and the Award of Merit from the Retina Research Foundation (1990). He has served on the editorial board of *Investigative Ophthalmology and Visual Science* and is currently on the editorial board of *Experimental Eye Research*. Dr. LaVail has been an active participant in the program committees of the Association for Research in Vision and Ophthalmology and the International Society for Eye Research, and he is currently a Vice-President of the International Society for Eye Research.

# CONTENTS

# III. STUDIES OF RETINAL DEGENERATION USING TRANSGENIC MICE AND OTHER ANIMAL MODELS

# IV. AGENTS WHICH CAUSE OR PREVENT RETINAL DEGENERATION

# I. THE MACULA, AGING AND MACULAR DEGENERATION

The most important region of the retina for human vision is the macula, a small pit-like depression at the back of the eye. Approximately 50,000 cone photoreceptors are concentrated in this region which are responsible for all detailed visual activities such as reading, recognizing faces, and driving an automobile. Age-related macular degeneration is a major vision problem for the elderly and is the leading cause of blindness in persons over 50 years of age in industrialized nations. The cause(s) of death of the macula photoreceptors is not known. The chapters in this section deal with structural, biochemical and molecular biological studies directed at understanding age-related changes in the retina that may lead to macular degeneration.

# LIPOFUSCIN IN AGED AND AMD EYES

C. Kathleen Dorey [2,3] Giovanni Staurenghi [2,3,4]
and Francois C. Delori [1,3]


[1] Biomedical Physics and [2] Biochemistry and Cell Biology
Macular Degeneration Research Center
Schepens Eye Research Institute, Boston, MA 02114

[3] Department of Ophthalmology, Harvard Medical School, Boston

[4] Clinica Oculistica Universitá degli Studi di Milano
Istituto di Scienze Biomediche, Osp S. Paolo, Milano, Italy

## INTRODUCTION

AMD is a progressive degeneration of the retina which preferentially and initially affects the macula but ultimately involves the fovea (1). The leading cause of visual impairment for individuals over 65, it will affect an estimated 8 million Americans in the next 20 years (2). The majority of these will experience a gradual erosion of vision due to progressive atrophy of photoreceptors—causing decreased visual acuity, loss of color vision and tunnel vision; some will be legally blind (3). About 90% of legal blindness in AMD is due to exudative AMD, characterized by serous detachment of the retina, retinal pigment epithelial detachment and tears, aggressive growth of new blood vessels from the choroid into the subretinal space, and disciform scaring. Fortunately, only 5-10% of patients will develop exudative AMD (4,5) While laser photocoagulation can delay and/or reduce the consequent visual loss, the prognosis remains grim—most will be blind (6). Presently, there is no medical treatment to stop or slow the course of this disease nor any known prevention.

The earliest clinical manifestations of AMD are soft drusen, (Fig. 1) characterized in fundus photos as discrete structures at least 63 μm in diameter and with density at the center fading to *often* indistinct edges (1,5). There are other types of drusen, most notably the small (<63 μm) distinct hard drusen that are by far the most abundant in the posterior pole of all ages, but have no correlation with age or AMD (1,5). Soft indistinct drusen exhibit age related increases in size, number and total area occupied, and are strongly correlated with increased risk for atrophy, RPE detachments and/or neovascular AMD (5,7-9). Retinal regions are not equally involved with drusen: the percentage area occupied by drusen is generally higher in the region more than 500 μm but less than 1500 μm from the foveola than in areas closer or further (5). Since soft drusen are deposits within Bruch's membrane (1,7,8), it has long been considered that the *essential defect in AMD resides in the aging retinal pigment epithelium (RPE)* (1,10-11).

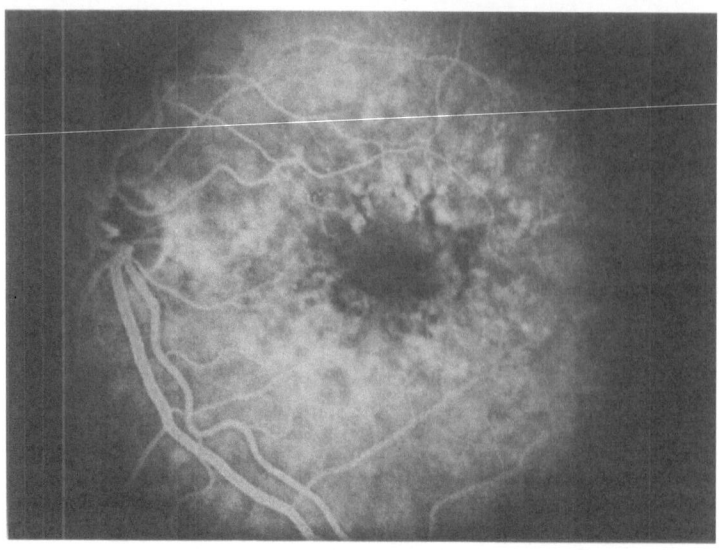

**Figure 1.** Fluorescein angiogram of an individual with AMD demonstrates numerous soft drusen (arrow) throughout the posterior pole. The macular pigment blocks the choroidal fluorescence in the fovea (white arrow). Mottling of the RPE melanin is seen at the lower edge of the macula equivalent in size to the optic disc (1500 um in diamerter).

The most striking feature of aging human RPE cells in the macula is the massive quantity of lipofuscin in the apical cytoplasm (11-14). The lipofuscin content of the RPE cells exhibits marked correlations with retinal region, aging, and race that have parallels in the risk for AMD (Summarized in Table I). Both are selective for the macula, increase with aging, and have their greatest expression in whites. The amount of lipofuscin in an aging RPE cell is significantly correlated with the quantity of debris (a component of drusen) deposited in "its" Bruch's membrane (12); however, these parameters may also be determined by.the retinal region. Lipofuscin accumulates throughout life and therefore precedes the presence of drusen and AMD.

**Table 1.** Greater lipofuscin levels are associated with greater risk for AMD.

| RPE Lipofuscin | Signs of AMD | References[1] |
|---|---|---|
| AGE: | | |
| Age-related increase. | Rare before age 65,10% of those over 65; 30% of those over 75. | 13, 14, 17, 18 |
| RACE | | |
| Higher in Caucasians than in Afro-Americans of comparable age | Unresolved: Prevalence *may or may not* be higher in Caucasians but AMD may be more severe in this group. | 13, 18,19, 20 Cf. 5 and 21 |
| TOPOGRAPHY | | |
| All aging RPE have increased lipofuscin. | Drusen are evidence of diffuse changes in macula | 1 |
| Maximum in area (5000 μm dia) around the fovea | Atrophy begins at the perimeter of the fovea. Drusen and atrophy are more frequent in the macula region less than 2000 μm from the fovea. | 14, 1, 16 |
| Low in fovea | Fovea is initially spared, involved only later. Most neovascularization starts close to fovea. | 13, 22, 16, 1 23, 24 |
| Maximum at 6° (~1800 μm) from fovea | No particular risk at 6-7° from fovea. | 13, 22 |

1. The first references on each line refer to lipofuscin, the last to AMD.

Therefore, we and others have hypothesized that lipofuscin contributes to the pathogenesis of AMD (11,13-15), and that lipofuscin topography explains the annular pattern of macular degeneration (16). In the support of the hypothesis that lipofuscin contributes to the pathogenesis of AMD, previous reports suggested the preferential attack on the macula and sparing of the fovea could be attributed to differences in lipofuscin content (13-16). However, the available data for lipofuscin distribution had large sampling intervals (13, 22) or large sampling areas (14) that did not permit comparison of lipofuscin topgography with specific areas such as the perifoveal ring of primary atrophy described by the Sarks (1). Higher resolution maps are needed to determine whether lipofuscin topography is related to perifoveal atrophy (1, 25), the higher concentration of large soft drusen in the macula, or the scattered irregular pattern of drusen.

## Is Lipofuscin Stored in Lysosomes?

It is generally accepted that lipofuscin is derived from photoreceptors (26-29), that premature loss of photoreceptors greatly reduces lipofuscin accumulation (28, 29), and that its formation is accelerated in antioxidant deficiency (27, 30, 31). Because the granule is a structural derivative of the phagolysosome, lipofuscin granules have been classically considered lysosomal containing indigestible residues of oxidative damage too large to diffuse out (27).

Newer evidence that lipofuscin resides in metabolically active lysosomes emphasizes the possibility that lipofuscin in the aging RPE could reach levels that compromise RPE functions essential for photoreceptor survival and thereby cause or accelerate AMD. Lysosomal characteristics of lipofuscin include: a) presence of hydrolytic enzymes such as acid phosphatase, aryl sulfatase (32) and cathepsin; b) fusion with primary lysosomes and phagosomes (17); c) co-localization with phagocytosed beads (33); and d) absence of heterogeneity in size or emission spectrum among lipofuscin granules in the same cell, an observation consistent with lysosomal fusion and intermixing of contents (34).

## Is AMD An Acquired Lysosomal Storage Disease?

Lysosomal accumulation of nontoxic, indigestible material causes tissue dysfunctions in numerous lysosomal storage diseases (35) such as Tay-Sach's disease (36) or ceroid lipofuscinosis (Batten's disease; 37). Lysosomal storage results from inability to digest the accumulated material (35-38). Unusually high levels of lipofuscin, lipofuscin-like, or ceroid accumulations in the RPE lysosomes are a major pathological change associated with macular degeneration due to Best's disease (39,40), Batten's disease (37), Stargardt's disease (fundus flavi) (41), Sjögren-Larsson syndrome (42), cone-rod dystrophy (43), and ovine ceroid lipofuscinosis (44).

Human RPE cells exhibit an age dependent increase in lipofuscin—whether measured by its fluorescence (13,22), the number of granules (14), or the percentage of total cytosolic free space occupied (14). In older individuals, lipofuscin can occupy up to 20% of the free cytosolic space (that not already occupied by nuclei, Golgi bodies, mitochondria, etc; 14). The lipofuscin model for the pathogenesis of AMD is intellectually satisfying because it integrates factors known to influence risk for AMD. For example, exposure to blue light, a risk factor for AMD (45), causes photooxidative damage to the photoreceptors (46) which can lead to increased lipofuscin (27). Elevated plasma antioxidants, that might prevent this damage, have been associated with lower risk for AMD (47). Increased risk for AMD has been correlated with smoking (47) which lowers plasma carotenoids (48); similarly, diets high in carotenoids have been associated with reduced risk for AMD (49). Macular pigment (derived from carotenoids in the diet) may regulate lipofuscin topography by absorbing damaging blue light (50).

## Research Approach.

*If lipofuscin is involved in the pathogenesis of AMD, greater complexity in the lipofuscin pattern is predicted both by the discrete nature of drusen, and by the high correlation between debris in Bruch's membrane and the amount of lipofuscin in overlying RPE cells* (12). Clearly more data was needed to clarify the relationship between lipofuscin accumulation and risk for AMD. Our ongoing research focuses on three specific questions. 1) Do accurate maps of lipofuscin suggest variability that would be consistent with the size and distribution of soft drusen in the macula? Is there a perifoveal ring of peak lipofuscin? 2) Do donor eyes with advanced AMD have higher levels of RPE lipofuscin than eyes with no evidence of AMD? This part of the study is being done in collaboration with Shirley and John Sarks (51). 3) Do high levels of RPE lipofuscin identify subjects at risk for AMD? In a subject with AMD, do lipofuscin levels correlate with risk for progression, or predict regions likely to atrophy? Are regions of atrophy bounded by marked elevations in lipofusin? Before this study could be initiated, it was first necessary to develop an accurate noninvasive measurement of RPE lipofuscin. The results described here represent a progress report in longterm projects; additional subjects are needed to arrive at statistically significant statements.

## METHODS

**Digital Imaging Fluorescence Microscopy (DIFM).** Lipofuscin fluorescence (Ex:488; 530HP) in paraffin sections of human eyes was studied at 400 X on a Zeiss Axiovert microscope equipped with infinity corrected planoapo objectives and a SIT (Silicon Intensified Target) video camera. After standardizing gain and voltage settings to minimize fluorescence from Bruch's membrane, digital images were acquired by averaging 20 sequential frames 120x128x8bits. (Each bit records the X and Y positions, and the light coming from that position in the microscope field.)

The fluorescence was mapped in contiguous fields from the optic nerve through the fovea and into the temporal macula, by manually tracing the area of interest for software determination of the mean and standard deviation of the fluorescence (ImageMeasure, Microsciences, Federal Way, WA). Care was taken to include only lipofuscin and not Bruch's membrane or RPE cell cytoplasm without fluorescence. For comparison of absolute lipofuscin content, data was expressed relative to the fluorescence of a uranium glass standard measured twice daily to correct for any changes in output of the exciting light, transmission of the filters, or sensitivity of the camera. The measurements and profiles obtained were quite reproducible; the mean coefficient of variation in repeated measurements was 2.4%.

**Lipofuscin profiles in vitro.** The lipofuscin measurement from each image sampled all of the lipofuscin in a 37 µm long strip of RPE. Measurements in 4 adjacent fields were averaged to obtain profiles with 150 µm (0.5°) resolution; data was then smoothed with a 5 point, center weighted smoothing process. The means of 16 fields were averaged to produce profiles with a 600 µm resolution, approximating the 2° sampling area of the in vivo measurement.

**Comparison of normal and AMD eyes.** In a collaboration with John and Shirley Sarks we have begun comparing lipofuscin levels in AMD and normal aging. Unstained paraffin sections that include the fovea and optic nerve have been selected from their extensive collection of eyes with documented clinical history. Based on histopathology (25), these eyes were classified these eyes by the Sarks as Group I (normal), Group II (normal with evidence of aging), Group IV (AMD with atrophy present) or Group V (AMD with neovascularization).

**Non-invasive Fundus Fluorescence Spectrometry.** Measurements of the intrinsic fluorescence of the fundus were made with the new Fundus Fluorophotometer (52,53). This technique allows measurements of the emission spectrum for discrete locations on the retina (sampling area: 600 µm in diameter). Each emission spectrum (spectral resolution: 7 nm) is acquired in 200 msec by a cooled image intensifier-multichannel spectral analyzer. Excitation is provided by a Xenon-arc lamp and interference filters centered at 430, 450, 470, 490, 510, and 530 nm (20 nm halfwidth). Retinal exposures are < 5% of the ANSI maximum permissible level. The technique incorporates corrections for contributions from crystalline lens fluorescence, and is fully calibrated for the spectral characteristics of excitation and detection.

**Lipofuscin measurements in vivo.** Fluorescence measurement in the fovea and at 7° temporal to the fovea were obtained iin 25 normal volunteers varying between 21 and 73 years of age. Fluorescence profiles along a horizontal line through the fovea were measured for 3 normal subjects (51).

# IN VITRO MEASUREMENTS OF RPE LIPOFUSCIN

## Lipofuscin Topography

High resolution profiles of lipofuscin topography confirmed the low level of lipofuscin in the fovea and broad areas of higher values at 6° or more from the fovea (Fig. 2A). In this particular example the lowest levels observed are in the fovea. The lipofuscin content of adjacent RPE cells was far more varied than previous studies had suggested. RPE lipofuscin levels in adjacent 37 μm fields occasionally varied by as much as 100% (Compare -3° and -4° in Fig. 2A ); this observation was consistent with the cell to cell variation observed in whole mounts and in sections. Simple smoothing operations emphasized the presence of local areas with high levels of lipofuscin (-2, 2 and 3° in Fig. 2B). Within the 1500 μm of the macula it was quite common to observe groups of cells with lipofuscin concentration as high or higher than that at 6-7°. These profiles were individually unique and quite reproducible, even when repeated at different emission wavelengths and with different camera settings.

In some, but not all eyes, the lipofuscin fluorescence increased sharply at the edge of the fovea to form a peak approximately 1.5° from the center of the fovea. Profiles prepared with 2° resolution (Fig. 2C) emphasized the macropattern of lipofuscin: fluorescence was minimal in the fovea, and the maximum mean macular RPE fluorescence occurred at 4 to 9° (Mean 7.2 ± 1.6° temporally, and 6.1 ± 1.1° nasally) consistent with Weiter et al., (13).

These data provide the first suggestion that the perifoveal drusen and atrophy might reflect earlier accumulation of lipofuscin. The evidence of variable lipofuscin content is quite consistent with the pattern of drusen formation and refutes one objection to the lipofuscin hypothesis for pathogenesis of AMD—that there was not sufficient complexity in the lipofuscin topography to account for the complex patterns of drusen, atrophy and neovascularization

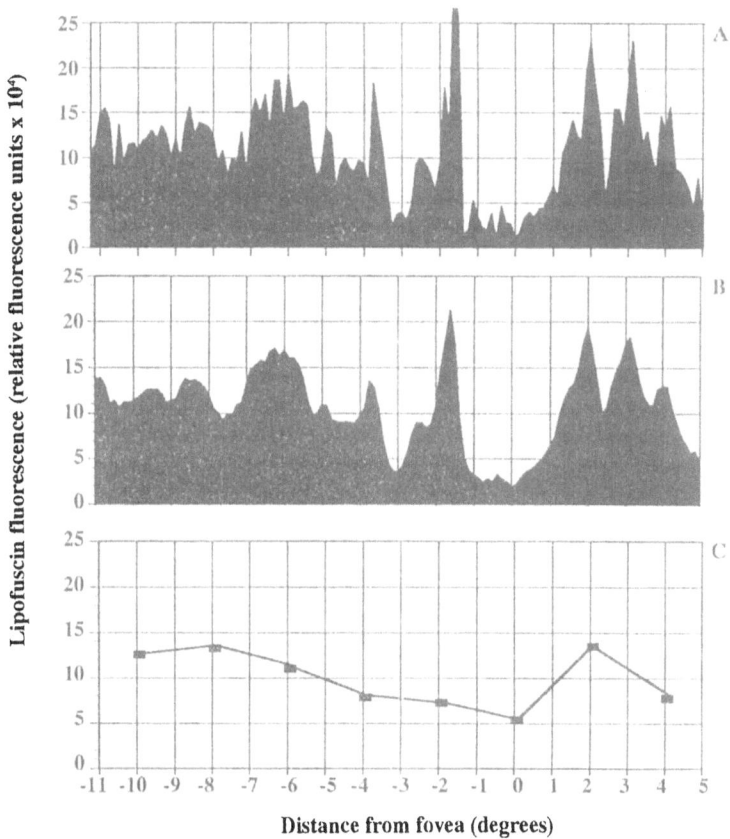

**Figure 2**. Profile of lipofuscin in 42 year old RPE from optic disc (left), through the fovea (0°) to 5° temporal. Resolution is A) 37 μm; B) 37 μm with curve smoothing; C) 600 μm to approximate 2°profiles obtained in vivo.

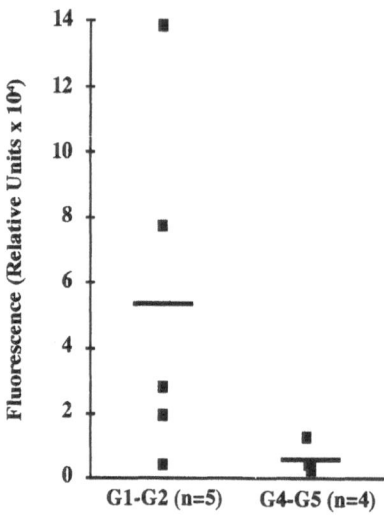

**Figure 3.** Lipofuscin concentration in the macula of normal and AMD eyes. Each point represents the mean of 32 measurements made across the diameter of a 1200 µm area centered on the fovea. Represented are Group 1/ Group2 (normal eyes that may or may not have hard drusen or other evidence of aging) and for Group 4/Group 5 (eyes with advanced AMD including atrophy and/or neovascularization).

### Comparison of Normal and AMD Eyes.

No obvious differences were noted in the shapes of the lipofuscin *profiles* of aging and AMD eyes, except for the minimal values associated with obvious areas of atrophy. Contrary to our hypothesis, the lipofuscin levels were not elevated in eyes with known AMD. In fact, the levels were so much lower that the fluorescence could not be visualized at the video camera settings used for normal eyes. (All variations in camera settings, strength of the excitation light, transmission of filters, etc were controlled by expressing all data relative to the fluorescence of a glass standard). Because of the local variability seen in both sets of eyes, the mean lipofuscin value for the macula (including the fovea) was determined for each eye.

As seen in Figure 3, macular RPE lipofuscin was markedly higher in normal eyes (53,000 ± 52,000) than in eyes with known AMD (6,000 ± 4,000). The mean age in the AMD eyes was older (80 vrs 57), but since lipofuscin increases with aging (13,14), it seemed unlikely that age differences could contribute to the decrease in lipofuscin. However, this possibility cannot at this time be completely discarded (see below).

One weakness of cross sectional studies like this is that they cannot reveal cause and effect, only correlations. We cannot know, for example, if an eye with exceptionally high lipofuscin levels would have developed AMD in subsequent years, or whether the RPE in eyes with AMD ever had lipofuscin levels comparable to—or greater than—normal RPE of comparable age.

### *IN VIVO* MEASUREMENT OF RPE LIPOFUSCIN:

To properly address the significance of RPE lipofuscin in the pathogenesis of AMD a longitudinal study is clearly necessary; any approach to that question will clearly require noninvasive measurement of intrinsic fundus fluorescence and the ability to discriminate lipofuscin from other fluorophores present in the light path. To be accurate the measurement must account for individual differences in transmission of the ocular media and absorption by other pigments (melanin, hemoglobin, macular pigment). Data presented below document the feasibility of a longitudinal study to determine whether RPE lipofuscin contributes to the pathogenesis of AMD.

## Spectral Characteristics

Fluorescence emission spectra from the fundus (Fig 4) vary little in shape among subjects and are consistent with lipofuscin being the predominant fluorophore detected when excitation in the 430-530 nm range are used (53,54). Maximum emission occurs between 610 and 635 nm. With excitation in blue and green light, the emission characteristics from the fundus suggest that it is Eldred's Fraction VIII that is being measured (Eldred, this volume).

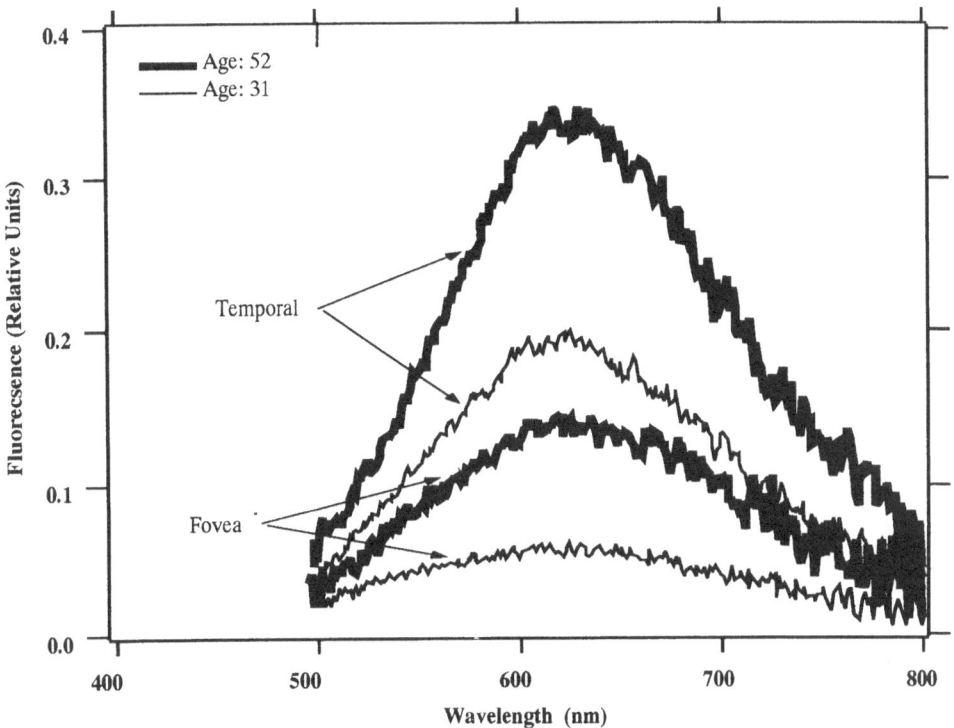

**Figure 4.** Fluorescence emission spectra from two subjects illustrates the regional differences and age related increases in lipofuscin content. Excitation at 470 nm (halfwidth: 20 nm)

## Topographic Distribution of Fundus Fluorescence:

The distribution of fundus fluorescence was recorded along a horizontal line through the fovea in 3 normal subjects less than 50 years of age (Fig 5). The fluorescence profiles show a well delineated minimum in the fovea, maximal fluorescence at 7 to 12° from the fovea, and a decrease in fluorescence towards the periphery. The fluorescence from the optic disc is low and has spectral emission characteristics different from those of the fundus. Since the sampling area was 2°, these in-vivo profiles must be compared to ex-vivo data at similar resolution (Fig. 2C). Reasonable correspondence is evident. Both profiles feature low foveal fluorescence and maximum fluorescence more than 1500 μm from the fovea. Note that absorption of the excitation light (470 nm) by the macular pigment (absorption 400-520 nm) reduces the measured amount of fluorescence and therefore exaggerates the depression in the fovea. Correction methods to account for this absorption are being developed, which will also yield a measure for the macular pigment density. Profiles of lipofuscin in vivo, being both more consistent and less likely to be incorporate aretefacts than those obtained from donor eyes, offer a sound basis for a prospective study.

Additional studies are necessary to determine to what extent variable extraction of fluorohores by organic solvents contributes to the local variation in sections, to explore any correlation of perifoveal lipofuscin peaks with changes in Bruch's membrane or other evidence of AMD, and if promising, to do the same in vivo.

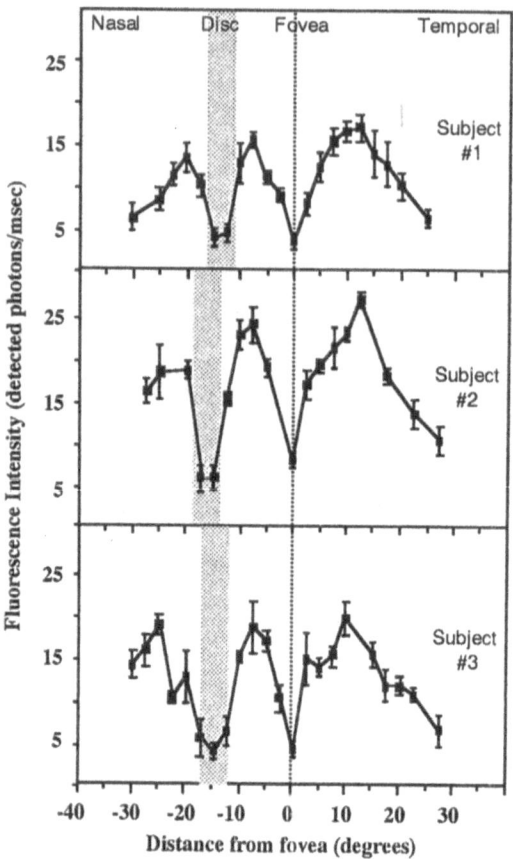

**Figure 5.** Topographical distribution of the fundus fluorescence in 3 subjects (resolution: 2°). The shaded area represents the optic disc. Reproduced from reference 52 with permission from the Optical Society of America.

### Age Related Increase in Fluorescence

The RPE lipofuscin content increases with age and may increase more rapidly in the first 2 decades, less rapidly in the next 4 or 5, and then again more rapidly (13,14). Noninvasive measurement of fundus fluorescence at 7° temporal and in the fovea also exhibits a strong age related increase, at least until age 60 or 65, when the measured fluorescence decreases (Fig. 6). The ratio of fluorescence at the two sites remains statistically constant with age. However, light scatter and absorption by the ocular media would cause a decrease in signal intensity that has not yet been corrected in these data. Correction will increase the fluorescence at old age, but it is unlikely to result in a continuous increase in fluorescence throughout the life span.

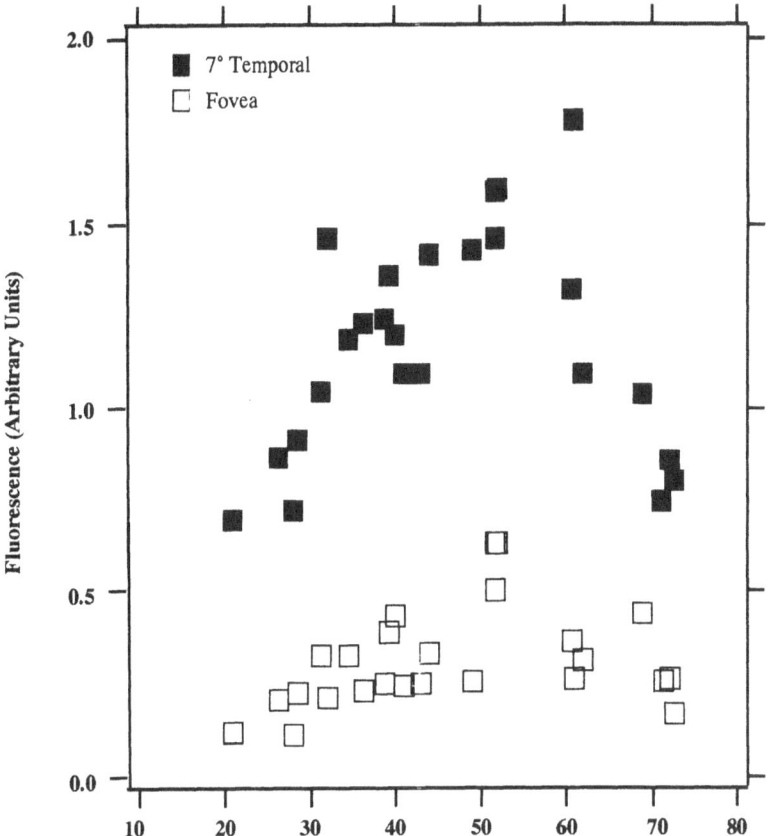

Figure 6. Changes in lipofuscin content with aging. The foveal RPE lipofuscin content is lower than the temporal in each of the 25 subjects shown.

## DISCUSSION

### Decreases in lipofuscin fluorescence

These in vivo and in vitro data suggest that lipofuscin fluorescence may decrease with advanced aging and/or advanced AMD. Because of age differences in our groups, we cannot discriminate the influence of advancing age and pathology. Among our older volunteers there is also a drop in lipofuscin fluorescence; some of this drop is due to ocular media, but some could also be evidence of early pathology. With a prevalence of 30%, it is probable that some of these eyes will develop AMD. Only longitudinal studies can ultimately determine the clinical significance of the decreased fluorescence.

Neither the lipofuscin model nor the previous literature suggested that lipofuscin levels might decrease. This unexpected reduction in fluorescence raises some interesting and testable questions. One possibility, that we view as unlikely, is that the reduced lipofuscin fluorescence is an artifact; further data acquisition will resolve this question. Another rationale is that only the *fluorescence* is decreased while the *mass* of lipofuscin is the same or larger in AMD eyes. Three conditions could result in lipofuscin with lower fluorescence but equivalent mass. 1) In vitamin A deficiency fluorescence is greatly reduced (30,55). One of the major fluorophores of lipofuscin is derived from vitamin A, and its emission would contribute to these measurements (55, Eldred, this volume). 2) With increasing concentration, the emission of some fluorophores is shifted to longer wavelengths. Emission spectra will resolve this issue for lipofuscin. 3) Although there is no supporting data, the possibility of a different metabolic pathway in subjects with AMD should not be overlooked.

Another feasible interpretation of low lipofuscin fluorescence in subjects with AMD is that lipofuscin is turned over and/or degraded. According to this view, photoreceptor atrophy would decrease the phagocytic load on the RPE cells, slow the rate of incoming material, and finally permit elimination of the lysosomal lipofuscin. The considerable volume of material passing through RPE cells on a daily basis was recently emphasized by the massive accumulation of fluorescent material (lipofuscin?) in rat RPE only 2 days after inhibition of protease activity (56).

Finally, a fascinating model for decreased fluorescence in AMD eyes predicts lysis of lysosomes when the lipofuscin reaches a critical concentration. This concept of Eldred is fully explained in another chapter. *If* RPE lipofuscin is lower in older eyes, it would provide strong support for the concept that lipofuscin is released by lysis. The provocative differences seen in our very preliminary study of lipofuscin in normal and AMD eyes cannot be interpreted until the influence of advanced age on lipofuscin fluorescence is also understood.


## CONCLUSIONS

1) Low or minimal lipofuscin levels were observed in the fovea of every eye examined, an unequivocal confirmation of previous reports (13, 22).

2) Elevated lipofuscin in perifoveal RPE suggest a more precise cellular relationship between peak lipofuscin content and the sites of primary atrophy or subretinal neovascularization in AMD.

3) Consistently high levels of lipofuscin are reliably measured at 7° from the fovea. In vivo measurements at this point should provide a reliable index of lipofuscin levels without a confounding influence of atrophy. It will be also be important to determine whether lipofuscin content at this site in tissue sections correlates strongly with the other markers for AMD such as drusen and the basal linear deposit in Bruch's membrane (1, 25).

4) Lipofuscin may be more variable than previous studies indicated. The concordance between profiles obtained in vivo and comparable maps of in vitro data confirm that regional patterns are retained in paraffin sections, and simultaneously provide additional verification of the heterogeneity in lipofuscin content of adjacent RPE cells. Large differences in lipofuscin content were frequently observed between adjacent RPE cells. While atrophy of *individual* RPE cells may have a low probability of clinical significance, loss of function in a *cluster* of RPE cells may lead to drusen or and/or atrophy of photoreceptors and RPE. Detection of clusters of RPE with peak lipofuscin levels could prove a useful clinical tool.

5) The Fundus Fluorophophotometer is a reliable tool for rapid in vivo quantification of lipofuscin. Data obtained to date confirm previous in vitro studies of lipofuscin spectra, topography and age-related increase. The limitations of cross sectional studies make it impossible to test the hypothesis that individuals with elevated lipofuscin concentration have increased *risk* for AMD. This tool provides a unique opportunity for longitudinal study of the significance of lipofuscin in the pathogenesis of AMD. If the lipofuscin model is correct, this noninvasive approach will also facilitate identification of medical intervention and prophylaxis in the early stages of the disease process, before significant loss of vision.


## ACKNOWLEDGMENTS

This work was supported in part by National Eye Institute Grants RO1-EY0-8121 and RO1-EY0-8511, and by institutional awards from Biomedical Research Support Grant RR-05527 from the Public Health Service. Parts were presented at the annual meeting of the Association for Research in Vision and Ophthalmology in May 1992, and at the Retinal Degeneration Meeting in Sardinia, September 1992.

# REFERENCES

1.  S.H Sarks and J.P. Sarks, Age-related macular degeneration: atrophic form, *in* "Volume Two Medical Retina," A.P. Schachat, R.P. Murphy, A. Patz, eds, *in* "Retina," S.J. Ryan, ed., C.V. Mosby Co., St. Louis,(1989).
2.  Pizarello, L.D. The dimensions of the problem of eye disease among the elderly., *Ophthalmol.*, 94:1191 (1987).
3.  Macular Photocoagulation Study Group. Argon laser photocoagulation for neovascular maculopathy, *Arch. Ophthalmol* 109:1109 (1991).
4.  L.G. Hyman, A.M. Lilienfeld, F.L. Ferris, III and S.L. Fine, Senile macular degeneration, a case-control study, *Am. J. Epidemiol.* 118:213 (1983).
5.  R. Klein, B.K. Klein, and K.L.P. Linton, Prevalence of age-related maculopathy. The Beaver Dam eye study, *Ophthalmol.* 99:933 (1992).
6.  Macular Photocoagulation Study Group, Subfoveal recurrent neovascular lesions in age-related macular degeneration. Guidelines for evaluation and treatment in the Macular Photocoagulation Study, *Arch. Ophthalmol.* 109:1242 (1991).
7.  S.H. Sarks, Drusen and their relationship to senile macular degeneration, *Aust J Ophthalmol.* 8:117 (1980).
8.  W.R. Green, P.H. McDonnell, and Y.H. Yeo, Pathologic features of senile macular degeneration. *Ophthalmol.* 92:615 (1985).
9.  D. Pauleikoff, M.J. Barondes, D. Minassian, I.H. Chisholm, A.C. Bird, Drusen as risk factors in age-related macular disease. *Am. J. Ophthalmol.* 109:38 (1990).
10. M.J. Hogan, Role of the retinal pigment epithelium in macular disease, *Trans. Am. Acad. Ophthalmol. Otolaryngol.* 76:64 (1972).
11. R.W. Young, Pathophysiology of age-related macular degeneration, *Surv. Ophthalmol.* 31:291 (1987).
12. L. Feeney-Burns and M.R. Ellersieck MR, Age-related changes in the ultrastructure of Bruch's membrane, *Am. J. Ophthalmol.* 100: 686 (1985).
13. J.J. Weiter, F.C. Delori, G.L. Wing and K.A. Fitch, Retinal pigment epithelial lipofuscin and melanin and choroidal melanin in human eyes. *Invest. Ophthalmol. Vis. Sci.* 27:145 (1986).
14. L. Feeney-Burns, E.S. Hilderbrand, and S. Eldridge, Aging human RPE: morphometric analysis of macular, equatorial and peripheral cells. *Invest Ophthalmol Vis. Sci.* 25:195, (1984).
15. C.K. Dorey, G. Wu, D. Ebenstein, A. Garsd, and J.J. Weiter, Cell loss in the aging retina. Relationship to lipofuscin accumulation and macular degeneration. *Invest. Ophthalmol. Vis. Sci.* 30:1691 (1989).
16. J.J. Weiter, F. Delori, and C.K. Dorey, Central sparing in annular macular degeneration, *Am. J. Ophthalmol.* 106:286 (1988).
17. B.E. Klein and R. Klein: Cataracts and macular degeneration in older Americans. *Arch. Ophthalmol.* 100:571 (1982).
18. F.L. Ferris, III. Senile macular degeneration: review of epidemiologic features, *Am. J. Epidemiol.* 118:132 (1983).
19. Z. Gregor and L. Joffe, Senile macular changes in the black African, *Br. J. Ophthalmol.* 62:547 (1978).
20. J.J. Weiter, F.C. Delori, G.L. Wing, K.A. Fitch, Relationship of senile macular degeneration to ocular pigmentation, *Am. J. Ophthalmol.* 99:185 (1985).
21. A.P. Schachat, M.C. Leske, A. Connell, V. Squicciarini, N. Oden, and the Barbados Eye Study Group, Prevalence of macular degeneration in a black population, *Invest. Ophthalmol. Vis. Sci.* Abstracts 33:801 (1992).
22. G.L. Wing, G.C. Blanchard, and J.J. Weiter, The topography and age relationship of lipofuscin concentration in the retinal pigment epithelium, *Invest. Ophthalmol. Vis. Sci.* 17:601 (1978).
23. N.M. Bressler, S.B. Bressler, and S.F. Fine, Age-related macular degeneration, *Surv. Ophthalmol.* 32:375 (1988).
24. M.J. Elman and S.L. Fine, Exudative age-related macular degeneration, *in* "Volume Two Medical Retina," A.P. Schachat, R.P. Murphy, A. Patz, eds, *in* "Retina," S.J. Ryan, ed., C.V. Mosby Co., St. Louis, (1989).
25. S.H. Sarks, Evolution of geographic atrophy, *Eye* 2:552 (1988).
26. G.E. Eldred and M.L. Katz, Lipofuscinogenesis in the RPE, in "Lipofuscin - 1987. State of the Art." Zs-I. Nagy, ed. Excerpta Medica, New York, pp 185 (1988).
27. L. Feeney-Burns and G.E. Eldred. The fate of the phagosome: conversion to "age pigment" and impact in human retinal pigment epithelium, *Trans. Ophthalmol. Soc. UK* 103:416 (1984).
28. M.L. Katz and G.E. Eldred, Retinal light damage reduces autofluorescent pigment deposition in the retinal pigment epithelium, *Invest. Ophthal. Vis. Sci.* 30:37 (1989).
29. M.L. Katz, C.M. Drea, G.E. Eldred, H.H. Hess, W.G. Robison, Jr, Influence of early photoreceptor degeneration on lipofuscin accumulation in the retinal pigment epithelium. *Exp. Eye Res.* 43:561 (1986).
30. W.G. Robison,Jr, T. Kuwubara, and J.B. Bieri, Deficiencies of Vitamins E and A in the rat: Retinal

damage and lipofuscin accumulation. *Invest. Ophthal. Vis. Sci.* 19: 1030(1980).

31. M.L. Katz, K.R. Parker, G.J. Handelman, T.L. Bramel, and E.A. Dratz, Effects of antioxidant deficiency on the retina and retinal pigment epithelium of albino rats: A light and electron microscopic study, *Exp Eye Res.* 34:339 (1982)

32. L. Feeney, Lipofuscin and melanin of human retinal pigment epithelium, *Invest. Ophthalmol. Vis. Sci.*, 17:583 (1978)

33. C.K. Dorey and S.A. Curran. Unpublished results.

34. C.K. Dorey and D.B. Ebenstein, Quantitative multispectral analysis of discrete subcellular particles by digital imaging fluorescence microscopy, in "Visual Communications and Image Processing," R. Tsing, ed.,Soc. Photoo-optical and Instrumentation Engineers, Bellingham, WA (1988).

35. E. Holtzman, "Lysosomes," Plenum Press, New York, (1989).

36. R.E.W. Watts and D.A. Gibbs, "Lysosomal Storage Diseases: Biochemical and Clinical Aspects," Taylor and Francis, London, pp 1-8, 201, (1986).

37. D. Armstrong, N. Koppang, and J. Rider, "Ceroid lipofuscinosis (Batten's disease)" Elsevier, Amsterdam, (1982).

38. R.O. Brady, Lysosomal storage diseases. *Pharmacol. Therapeutics* 19:327 (1982) (REVIEW).

39. S. O'Gorman, W.A. Flaherty, G.A. Fishman, and E.L. Berson, Histopathologic findings in Best's vitelliform macular dystrophy, *Arch. Ophthalmol.* 106:1261 (1988).

40. T.A. Weingeist, J.L. Kobrin, and K.E. Watz, Histopathology of Best's macular dystrophy, *Arch. Ophthalmol.* 100:108 (1982).

41. R.C. Eagle, Jr., A.C. Lucier, V.B. Bernadino, Jr., and M. Yanoff, Retinal pigment epithelial abnormalities in fundus flavimaculatus. *Ophthalmol.* 87:1189 (1980).

42. S.E. Nilsson and S. Jagell, Lipofuscin and melanin content of the retinal pigment epithelium in a case of Sjorgen-Larsson syndrome, *Br. J. Ophthalmol.* 71:224, (1987).

43. M.F. Rabb, M.O.M. Tso, and G.A. Fishman, Cone-rod dystrophy. A clinical and histopathologic report, *Ophthalmol.* 93:1443(1986).

44. D. Samuelson, W.W. Dawson, A.I. Webb, J. Dowson, R. Jolley, and D. Armstrong. Retinal pigment epithelial dysfunction in early ovine ceroid lipofuscinosis: electrophysiologic correlates, *Ophthalmologia* 190:150 (1985).

45. H.R. Taylor, S. West, B. Muñoz , F.S. Rosenthal, S.B. Bressler, N.M. Bressler, The long term effects of visible light on the eye. *Arch. Ophthalmol.* 110:99 (1992).

46. W.T. Ham, J.J. Ruffolo, Jr., H.A. Mueller, and D. Guerry, The nature of retinal radiation damage: dependence on wavelength, power level and exposure time. *Vis. Res.* 20:1163 (1980).

47. L. Hyman, O. He, R. Grimson, N. Oden, A.P. Schachat, M.C. Leske, and the Age-related Macular Degeneration Risk Factors Study Group, Risk factors for age-related maculopathy, *Invest. Ophthalmol. Vis. Sci (SUPPL)* 33:801 (1992).

48. R. Russell-Briefel, M.W. Bates, and L.H. Kuller, The relationship of plasma carotenoids to health and biochemical factors in middle-aged men. *Am. J. Epidemiol.* 122:741 (1985).

49. J. Goldberg, G. Flowerdew, E. Smith, J.A. Brody, and M.O. Tso, Factors associated with age-related macular degeneration. An analysis of data from the first National Health and Nutrition Examination Survey, *Am. J. Epidemiol.* 128:700, 1988.

50. C.K. Dorey, A. Elsner, G. Staurenghi, and F.C. Delori, What patterns the accumulation of RPE lipofuscin? *Exp. Eye Res (SUPPL).* 55:710 (1992).

51. C.K. Dorey, G. Staurenghi, F.C. Delori, J.R. Sarks, S.H. Sarks, Lipofuscin distribution in aging and AMD eyes. Invest. Ophthalmol. Vis. Sci. (SUPPL) 33:#2677 (1992).

52 F.C. Delori, K.A. Fitch, and J.M. Gorrand, In-Vivo characterization of intrinsic fundus fluorescence, *Noninvasive Assessment of the Visual System, Opt. Soc. Amer. Tech. Dig.* 3:72 (1990).

53. F.C. Delori, Fluorophotometer for noninvasive measurement of RPE lipofuscin. *Noninvasive Assessment of the Visual System, Opt. Soc. Amer. Tech. Dig.* 1:164 (1992).

54. G.E. Eldred and M.L. Katz, Fluorophores of the human retinal pigment epithelium: Separation and spectral characterization, *Exp. Eye Res.*, 47:71 (1988).

55. G.E. Eldred, Vitamins A and E in PRE lipofuscin formation and implications for age-related macular degeneration. Progr. Clin. Biol. Res. 314;113 (1989).

56. M.L. Katz and M. Norberg, Influence of dietary vitamin A on autofluorescence of leupeptin-induced inclusions in the retinal pigment epithelium *Exp. Eye Res.* 54:239 (1992).

# RETINOID REACTION PRODUCTS IN AGE RELATED RETINAL DEGENERATION

Graig E. Eldred

University of Missouri-Columbia
School of Medicine
Department of Ophthalmology
Columbia, MO 65212, U. S. A.

## INTRODUCTION

Great strides are being made toward defining the gene defects in retinitis pigmentosa[1-3], Best's vitelliform dystrophy[4], and animal models of retinal degeneration[5-7]. Yet, little progress has been made in identifying the biochemical bases of the leading cause of blindness in the elderly: age related macular degeneration (AMD)[8]. Truly age-related retinal pathologies may well be due to causes other than identifiable point mutations in well defined genes and/or gene products.

As with many retinopathies, the clinical entity losely termed age-related macular degeneration includes a variety of pathological conditions with the common characteristic that macular degeneration is preceded by subretinal neovascularization, which in turn is preceded by drusen deposition in Bruch's membrane[9]. Long before drusen appear, lipofuscin granules (age pigments) begin to accumulate in the RPE[10]. It has long been thought that this lipofuscin burden is somehow intimately associated with the root cause of AMD[11].

Yet age pigment accumulation is ubiquitous, while AMD, although prevalent, is not. Is there truly a causal connection between age pigment granules in the RPE and the deposition of drusen? What triggers the pathology? What causes age pigment accumulation in the first place? What is the chemical composition of the pigment granules? Recent findings suggest that the fluorescent components of RPE age pigments may be the key to many of these questions.

# RPE AGE PIGMENTS AND THEIR FLUOROPHORES

Accumulation of lipofuscin granules in the retinal pigment epithelium is one of the earliest age-related changes in the retina[12.] The granules have been identified as lysosomes in which ill-defined, nondegradable residues are trapped[13]. Most of the trapped material is derived from phagocytosed photoreceptor outer segments[14,15]. Lysosomal residual bodies start to appear before the twenty years of life and continue to accumulate throughout the lifespan of the individual[16]. Eventually, the granules come to pack the cells, and it is suspected that at some critical point, the RPE is stimulated to shed packets of its cytoplasm basally, and the cascade of events leading to age-related macular degeneration is triggered as a consequence[17,18].

While morphological studies have been performed to correlate the changes seen in photoreceptor cell numbers and the RPE lipofuscin burden with drusen deposition and neovascularization[19,20], very little is known of the biochemical mechanisms causing lipofuscin formation. Nor is the exact chemical composition of the granule contents known. Detailed biochemical knowledge in these matters could open new doors to understanding and treatment.

One characteristic of age pigments is their golden yellow autofluorescent emissions when excited by uv illumination[10]. Historically, this unusual fluorescence has been thought to be the key behind the cause of lipofuscin accumulation, and theories of RPE lipofuscinogenesis have centered on the suspected chemical sources of this fluorescent emission[21-23]. Yet, for many years no autofluorescent compound had been isolated, purified and identified to test these theories.

It is now known that the golden yellow emission is a composite of emissions from a variety of green, yellow, and orange-emitting fluorescent compounds[24]. While the green emissions are believed to originate from retinol and retinyl esters[25], the yellow to orange emissions come from compounds of previously unidentified structures.

The prominent orange-emitting fluorophores have now been chemically characterized and categorically identified. Results from mass spectral and NMR analyses suggested that the prominent orange-emitting fluorophore of human RPE age pigments is a direct reaction product of two molecules of retinaldehyde with one molecule of ethanolamine[26]. The general structure is that of an amphoteric quaternary nitrogen compound: N-retinyl-idene-N-retinylethanolamine (Fig. 1).

This retinoid derivative is generically similar to the structure proposed for the compound termed "indicator yellow" during the early days of research on visual pigment chromophores[27]. However, after rhodopsin was identified, structures related to that proposed for indicator yellow were abandoned. Until now, no further report of categorically similar structures has been made.

The only other natural products in which two molecules of retinoids have reacted are [2 + 4] cycloaddition products. These were first identified in whale liver oil[28], and subsequently noted as products of photosensitization reactions in synthetic retinoids[29]. The human RPE fluorophore cannot be such a cycloaddition product based upon the lack of proton signals from the cycloaddition ring structure (i.e., at 3.03 ppm) in the NMR spectrum.

The compound proposed for the RPE orange-emitting fluorophore has also now

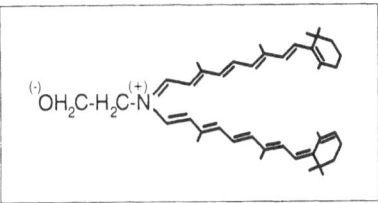

Figure 1. Proposed general structure of the predominant orange-emitting fluorophore of human retinal pigment epithelium lipofuscin granules (age pigments). Several isomeric and tautomeric forms probably exist, but have not yet been elucidated[26].

been synthesized *in vitro* confirming its structure[26]. An acid-catalyzed reaction between *trans*-retinaldehyde and ethanolamine in a molar ratio of 2:1 or greater, yields an orange-emitting autofluorescent product with identical spectral and chromatographic properties to those of the native fluorophore. In addition, other products of this reaction system are generated that appear similar to several more of the natural age pigment fluorophores. Two dimensional proton and $^{13}$C-NMR analyses of the orange-emitting product revealed that several isomers (i.e., *cis-/trans-* in the vitamin A moieties or *syn-/anti-* at the imine linkage) are present in the purified synthetic preparations. Such isomerism often makes exact conformational identification of this category of compound difficult or impossible[30]. Nonetheless, it may now be firmly concluded that N-retinyl-idene-N-retinylethanolamine is a major contributor to the fluorescence of RPE lipofuscin granules

## SUGGESTED MECHANISM OF FORMATION

The source of the ethanolamine moiety could either be from free ethanolamine formed via the decarboxylation of serine, or from phosphatidylethanolamine. The latter is the most likely source. This phospholipid makes up 30-40 mol% of the photoreceptor outer segment membrane lipids[31], and it has previously been shown to react with retinaldehyde in a 1:1 ratio[32-34].

To test this, dipalmitoylphosphatidylethanolamine was substituted for ethanolamine in the *in vitro* reaction system[26]. An orange-emitting fluorescent product was formed that chromatographed to a position much higher on the plate, but to the position of the vitamin A-related, orange-emitting fluorophore that develops in the photoreceptor outer segment (POS) debris that accumulates in degenerating RCS rat retinas[35]. A similar orange-emitting band is seen faintly in human age pigment fluorophore chromatograms[24]. This high migrating, orange-emitting fluorophore is ,therefore, chemically related to the low-migrating, orange-emitting fluorophore. It appears that retinaldehyde can react initially in the POS with disc membrane phosphatidylethanolamine (PE) to form N-retinylidene-N-retinylphosphatidylethanolamine (A$^2$-PE). After being shed and phagocytosed, the phosphoglycerolipid moiety is lost to yield the ethanolamine-linked vitamin A derivative (A$^2$-E).

Such a relationship had previously been suspected based upon the relative prevalence of these orange-emitting fluorophores in an age series of RCS rats[36]. During the time of peak outer segment accumulation and disorganization, the $A^2$-PE product predominates. Then as this substance disappears from the chromatograms in a time frame paralleling the disappearance of the POS debris, the $A^2$-E product starts to appear and is the only one remaining after the retina is fully degenerated.

This interpretation has one difficulty. Phospholipase D would cleave phosphatidylethanolamine at the proper location to yield the ethanolamine-retinaldehyde reaction product as opposed to the O-phosphoryl ethanolamine-retinaldehyde product. But, phospholipase D is not normally present in animal lysosomes. To test for the possible missed presence of a phosphate on the ethanolamine moiety, O-phosphorylethanolamine was reacted with *trans*-retinaldehyde. An orange-emitting fluorophore resulted, but it chromatographed to a different location from the native fluorophore. Thus, it appears that either the enanolamine-to-phosphate bond is broken by a different enzyme, or the thermodynamic stability of the products favor nonenzymatic cleavage at this location.

In sum, these observations lead to a new theory of lipofuscin fluorophore formation. When *trans*-retinaldehyde is released from rhodopsin upon capturing a photon, it is rapidly converted to *trans*-retinol by the photoreceptor outer segment, *trans*-specific retinol oxidoreductase[37]. Any retinaldehyde escaping reduction by this mechanism has two alternative pathways for reaction. First, free *trans*-retinaldehyde will most likely interact with the hydrophobic domain of the intact phospholipid membrane[38]. Under normal conditions, hydrophilic interactions at the ethanolamine end of the membrane phospholipid molecules would not be favored. In the case of the degenerating RCS rat outer segments, however, membrane disruption could allow this reaction to proceed with a higher likelihood.

If the aldehyde is allowed to approach the amine, a Schiff base reaction becomes possible. In a reversible reaction, water is eliminated through a hemiaminal intermediate to yield an imine bond. Chemically speaking, simple aliphatic Schiff bases (i.e., aliphatic imines) are inherently unstable, especially in aqueous solution, and have a tendency to either dissociate or further dimerize or polymerize to stable products[39]. Retinaldehyde, however, is known to readily react with amines to form relatively stable Schiff bases. This is the basis for the visual pigment linkage. Additionally, the 1:1, retinaldehyde:phosphatidylethanolamine reaction product (retinylidene phosphatidylethanolamine) has previously been isolated from retinas[40]. Thus, this pathway for *trans*-retinaldeyde reaction is a real possibility.

It is somewhat surprising to find that this retinylidene phosphatidylethanolamine would further react with another molecule of retinaldehyde to form the quaternary amine. Generally, secondary and tertiary amines will not react readily with aldehydes to form Schiff bases, unless they are conjugated to a system of double bonds[41], as is the case here. In the present instance, the initial Schiff base need not be first reduced before reaction with the second vitamin A as is evidenced by the presence of 12, rather than 11 double bonds in the molecule (5 from each retinaldehyde, and 1 each from the sequential Schiff base reactions).

There also seems to be nothing special about the ethanolamine or phosphatidylethanolamine that would favor this double Schiff base reaction. When methylamine,

ethylamine, n-propylamine, n-butylamine, n-amylamine and n-hexylamine were used instead of ethanolamine, very similar orange-emitting flurophores resulted. Thus, the electrophilic oxygen is not required for this reaction to proceed. Much more work needs to be done to elucidate the chemical reaction mechanisms involved here.

Once the N-retinylidene-N-retinylphosphatidylethanolamine molecule is formed in the outer segment membrane disk, it remains stable and is engulfed by the RPE. The phosphoglycerolipid is then lost by a mechanism yet to be determined, to leave behind the stable, nondegradable N-retinylidene-N-retinylethanolamine. This then appears as a predominant fluorophore of the RPE age pigments

## LYSOSOMOTROPIC THEORY OF LIPOFUSCINOGENESIS AND MACULAR DEGENERATION

The chemical structure of N-retinylidene-N-retinylethanolamine opens new insights into the potential role that it might play in the formation of lipofuscin and in the etiology of AMD.

Once formed, this amphoteric quaternary amine should assume a positive charge in the acidic environment of the lysosome. Upon being charged, it will not diffuse from the lysosome and will be trapped and accumulate there. This is precisely the type of compound that was predicted by deDuve[42] to accumulate within lysosomes. He termed such compounds lysosomotropic amines. As the concentration of these weak bases increases, the intralysosomal pH increases and the lysosomal acid lipases and protesases no longer function because they are outside their pH optima[43,44].

Many xenobiotic compounds have been found to behave as lysosomotropic amines[45-47]. The antimalarial drug, chloroquine, is one such compound, and one of its deleterious side effects is its induction of lysosomal inclusions within the RPE of the macula of the retina[48].

Pharmacologically, if long chain hydrocarbons (12 carbons or more) are bound to the lysosomotropic amines, they can display surfactant properties. Firestone and colleagues have described a variety of such lysosomotropic detergents, including retinoid derivatives, which are designed to accumulate within the lysosome and, upon reaching a critical micelle concentration, lyse the membrane and release the lysosomal enzymes into the cell causing cell death[49-51]. The dose response curve of such compounds is sigmoidal rather than linear because the detergent molecules must achieve their critical micelle concentration before they can lyse the membranes.

Prior to membranolysis, lysosomal proteolysis is disrupted interfering with the flow of amino acid subunits to the cellular anabolic machinery with a variety of deleterious consequences. Additionally, the lysosomes are reported to become leaky, allowing the release of certain metabolites into the cell which can further interfere with the nuclear and/or anabolic processes[52,53].

A key characteristic of cells under the influence of subacute doses of lysosomotropic detergents is plasma membrane blebbing and the shedding of cellular cytoplasm[54]. This behavior is very similar to that of the RPE cells overlying drusen. Thus, N-retinylidene-N-retinylethanolamine accumulation may ultimately be responsible for stimulating RPE apoptosis and drusen formation, which are the clinical hallmarks of the onset of age-related maculopathies.

Much work remains to be done to prove the theory that N-retinylidene-N-retinyl-ethanolamine acts as a lysosomotropic detergent and that it stimulates the RPE cell to bleb off parts of its cytoplasm to form drusen deposits in Bruch's membrane. Nonetheless, if proven correct, many potentially disparate observations could be explained by this theory:

1) Slow rate of accumulation: Under normal conditions, the rarity of the 2:1, retinaldehyde:phosphatidylethanolamine reaction could readily explain the slow rate of accumulation of lipofuscin granules in the RPE.

2) Lack of early effects of lipofuscin: A sigmoidal dose-response behavior would explain why the RPE is apparently able to function unimpeded with significant lipofuscin granule burdens for years. Only after a critical micelle concentration is reached would the deleterious effects become manifested.

3) Unusual homogeneity of the RPE lipofuscin granule contents as opposed to related granule types: The surfactant properties of this retinaldehyde-derived lyso-somotropic amine may explain the relative homogeneity of the RPE age pigment granule contents in electron microscopy. Other lysosomotropic amines that do not exhibit detergent properties stimulate the accumulation of materials in lysosomes that exhibit a large degree of heterogeneity and substructure[55]. Such is the case with the chloroquine-induced RPE granules[56,57]. Age pigments in other tissues in which retinaldehyde is not likely to play a role, also exhibit much more varied substructures[58].

4) Role of lysosomal protease inhibition: Application of lysosomal enzyme in-hibitors, such as the serine protease inhibitor, leupeptin, also causes heterogeneous granule accumulation in the RPE and other tissues[59,60]. This has served as the basis of a proteolytic decline theory of lipofuscinogenesis[61]. No specific enzyme inhibitors have ever been identified, however, so it has been speculated that either substrates or proteases are covalently modified (by free radical damage or by fluorescent adducts) causing their deactivation[61,62]. Formation of lysosomotropic amines and/or detergents via Schiff base reactions could explain general lysosomal protease inhibition without the need to invoke protease or substrate modification.

5) Macular susceptibility: Chloroquine stimulates RPE granule accumulation in the macular region preferentially[48]. This could reflect an greater lysosomal activity in this region, and may aid in explaining why age-related degeneration has a macular preference.

6) Influence of vitamin A: The proposed structure of the orange-emitting fluoro-phores is consistent with dietary studies implicating the involvement of vitamin A in lipofuscin fluorophore formation and granule accumulation in the retinal pigment epithelium[35,63]. These previous studies were unable to clarify whether vitamin A was serving as a direct precursor for the fluorophores or whether it served as an intermediary in the metabolic pathways that led to the fluorophore formation (eg., as a photosensitizer for lipid oxidation). The current results prove that vitamin A is a component of the fluorescent molecule.

7) Light as a predisposing factor: Excessive retinal light exposure has been sus-pected as a causative factor in age-related macular degeneration[64-66]. Exposure to intense light could come into play by releasing excessive retinaldehyde from rhodopsin and overwhelming the conversion pathway to retinol allowing more formation of the detergent precursor (i.e., the phosphatidylethanolamine reaction product).

8) Zinc as a predisposing factor: Zinc is a cofactor to the retinol oxidoreductases[37,67]. Disturbance of this enzyme activity could swing the balance of retinol/retinaldehyde toward retinaldehyde excess, which could then favor the reaction with phosphatidylethanolamine. Such a mechanism could readily be a factor in the reports of both the beneficial effects of dietary zinc in the etiology of age-related macular degeneration[68], and the formation of lipofuscin-like inclusion bodies in animal models of zinc deficiency[69,70].

9) Lack of an obvious genetic predisposition for AMD: While some forms of AMD have been suspected of demonstrating genetic inheritance[71,72], most seem not to do so. The self-assembling lysosomotropic detergent theory would explain a universal, non-genetically determined occurrence, while still allowing for a subset of specific genetically determined forms (eg., errors in zinc metabolism or point mutations in the POS, *trans*-specific retinaldehyde oxidoreductase enzyme).

10) Lipofuscin granules causing drusen deposition: Hogan first proposed that there was a link between age pigment accumulation in the RPE and drusen formation[11]. Yet no feasible explanation prior to this has been advanced to explain such a connection. Amphiphiles in general, including the lysosomotropic amines and detergents, can stimulate such shedding behavior[73-75].

In addition to all of the questions raised above, the age-related trapping of retinol, retinyl esters and other retinoids within lipofuscin granules raises very interesting questions with regard to consequences on the cellular processes involving retinoids: the retinal vitamin A cycle[76], glycolipid metabolism[77] and genetic control mechanisms responsible for maintaining differentiated function[78,79]. With a much clearer understanding of the chemistry of RPE age pigment formation, new insights into age related retinopathies are now possible.

## ACKNOWLEDGMENTS

Supported by USPHS grant EY-06458 and by a grant from Research to Prevent Blindness, Inc.

## REFERENCES

1. P. McWilliam, S. A. Jordan, P. Kenna, M. M. Humphries, R. Kumar-Singh, E. Sharp, and P. Humphries, Progress in the localisation of a late onset ADRP gene, in: "Retinal Degenerations," R. E. Anderson, J. G. Hollyfield, and M. M. LaVail, eds., CRC Press, Boca Raton, FL (1991).
2. S. S. Bhattacharya, R. Bashir, J. Keen, D. Lester, B. Lauffart, M. Jay, A. C. Bird, and C. F. Inglehearn, Linkage studies and rhodopsin mutation detection in autosomal dominant retinitis pigmentosa: An update, in: "Retinal Degenerations," R. E. Anderson, J. G. Hollyfield, and M. M. LaVail, eds., CRC Press, Boca Raton, FL (1991).
3. P. A. Hargrave and P. J. O'Brien, Speculations on the molecular basis of retinal degeneration in retinitis pigmentosa, in: "Retinal Degenerations," R. E. Anderson, J. G. Hollyfield, and M. M. LaVail, eds., CRC Press, Boca Raton, FL (1991).
4. E. M. Stone, B. E. Nichols, L. M. Streb, A. E. Kimura, and V. C. Sheffield, Genetic linkage of vitelliform macular degeneration (Best's disease) to chromosome 11q13, *Nature Genetics* 1: 246 (1992).

5.  D. B. Farber, M. Danciger, and C. Bowes, Studies on the gene defect of the *rd* mouse, *in*: "Retinal Degenerations," R. E. Anderson, J. G. Hollyfield, and M. M. LaVail, eds., CRC Press, Boca Raton, FL (1991).

6.  S. J. Pittler and W. Baehr, Identification of the precise molecular defect responsible for blindness in the mouse retinal degeneration mutant, *rd*, *in*: "Retinal Degenerations," R. E. Anderson, J. G. Hollyfield, and M. M. LaVail, eds., CRC Press, Boca Raton, FL (1991).

7.  G. Connell, L. L. Molday, D. Reid, and R. S. Molday, Molecular structure and properties of peripherin/rds the normal product of the gene responsible for retinal degeneration in the rds mouse, *in*: "Retinal Degenerations," R. E. Anderson, J. G. Hollyfield, and M. M. LaVail, eds., CRC Press, Boca Raton, FL (1991).

8.  F. L. Ferris, S. L. Fine, and L. Hyman, Age-related macular degeneration and blindness due to neovascular maculopathy, *Arch. Ophthalmol.* 102: 1640 (1984).

9.  S. H. Sarks, D. Van Driel, L. Maxwell, and M. Killingsworth, Softening of drusen and subretinal neovascularization, *Trans. Ophthalmol. Soc. U.K.* 100: 414 (1980).

10. L. Feeney, Lipofuscin and melanin of human retinal pigment epithelium. Fluorescence, enzyme cytochemical, and ultrastructural studies, *Invest. Ophthalmol. Vis. Sci.* 17: 583 (1978).

11. M. J. Hogan, Role of the retinal pigment epithelium in macular disease, *Trans. Am. Acad. Ophthalmol. Otolaryngol.* 76: 64 (1972).

12. G. L. Wing, G. C. Blanchard, and J. J. Weiter, The topography and age relationship of lipofuscin concentration in the retinal pigment epithelium, *Invest. Ophthalmol. Vis. Sci.* 17: 601 (1978).

13. L. Feeney-Burns and G. E. Eldred, The fate of the phagosome: Conversion to 'age pigment' and impact in human retinal pigment epithelium, *Trans. Ophthalmol. Soc. U.K.* 103: 416 (1983).

14. M. L. Katz, C. M. Drea, G. E. Eldred, H. H. Hess, and W. G. Robison, Jr., Influence of early photoreceptor degeneration on lipofuscin in the retinal pigment epithelium, *Exp. Eye Res.* 43: 561 (1986).

15. M. L. Katz and G. E. Eldred, Retinal light damage reduces autofluorescent pigment deposition in the retinal pigment epithelium, *Invest. Ophthalmol. Vis. Sci.* 30: 37 (1989).

16. L. Feeney-Burns, E. S. Hilderbrand, and S. Eldridge, Aging human RPE: Morphometric analysis of macular, equatorial, and peripheral cells, *Invest. Ophthalmol. Vis. Sci.* 25: 195 (1984).

17. R. P. Burns and L. Feeney-Burns, Clinico-morphologic correlations of drusen of Bruch's membrane, *Trans. Am. Ophthalmol. Soc.* 78: 206 (1980).

18. T. Ishibashi, R. Patterson, Y. Ohnishi, H. Inomata, and S. J. Ryan, Formation of drusen in the human eye, *Am. J. Ophthalmol.* 101: 342 (1986).

19. C. K. Dorey, G. Wu, D. Ebenstein, A. Garsd, and J. J. Weiter, Cell loss in the aging retina. Relationship to lipofuscin accumulation and macular degeneration, *Invest. Ophthalmol. Vis. Sci.* 30: 1691 (1989).

20. C. K. Dorey, G. Staurenghi, F. C. Delori, J. R. Sarks, and S. H. Sarks, Lipofuscin distribution in aging and AMD eyes, *Invest. Ophthalmol. Vis. Sci.* 33(Suppl.): 1229 (1992).

21. K. S. Chio, U. Reiss, B. Fletcher, and A. L. Tappel, Peroxidation of subcellular organelles: Formation of lipofuscinlike fluorescent pigments, *Science* 166: 1535 (1969).

22. L. Feeney-Burns, E. R. Berman, and M. S. Rothman, Lipofuscin of human retinal pigment epithelium, *Am. J. Ophthalmol.* 90: 783 (1980).

23. G. E. Eldred and M. L. Katz, The autofluorescent products of lipid peroxidation may not be lipofuscin-like, *Free Radical Biol. Med.* 7: 157 (1989).

24. G. E. Eldred and M. L. Katz, Fluorophores of the human retinal pigment epithelium: Separation and spectral characterization, *Exp. Eye Res.,* 47: 71 (1988).

25. G. E. Eldred, Vitamins A and E in RPE lipofuscin formation and implications for age-related macular degeneration, *in:* "Inherited and Environmentally Induced Retinal Degenerations," M. M. LaVail, R. E. Anderson, and J. G. Hollyfield, eds., Alan R. Liss, Inc., New York (1989).

26. G. E. Eldred and M. R. Lasky, Retinal age pigments caused by novel self-assembling lysosomotropic detergents, *submitted.*

27. F. D. Collins and R. A. Morton, Studies on rhodopsin. 2. Indicator yellow, *Biochem. J.* 47: 10 (1950).

28. K. Tsukida and M. Ito, The structure of kitol, *J. Nutr. Sci. Vitaminol.* 26: 319 (1980).

29. K-H. Pfoertner, G. Englert, and P. Schoenholzer, Photosensitized [4 + 2] cyclodimerizations of aromatic retinoids, *Tetrahedron* 44: 1039 (1988).

30. J. T. Cross, A critical review of techniques for the identification and determination of cationic surfactants, *in:* "Cationic Surfactants," E. Jungermann, ed., Marcel Dekker, Inc., New York (1970).

31. S. J. Fliesler and R. E. Anderson, Chemistry and metabolism of lipids in the vertebrate retina, *Prog. Lipid Res.* 22: 79 (1983).

32. H. Shichi and R. L. Somers, Possible involvement of retinylidene phospholipid in photoisomerization of all-*trans*-retinal to 11-*cis*-retinal, *J. Biol. Chem.* 249: 6570 (1974).

33. G. W. T. Groenendijk, C. W. M. Jacobs, S. L. Bonting, and F. J. M. Daemen, Dark isomerization of retinals in the presence of phosphatidylethanolamine, *Eur. J. Biochem.* 106: 119 (1980).

34. D. Lukton and R. R. Rando, Catalysis of vitamin A aldehyde isomerization by primary and secondary amines, *J. Am. Chem. Soc.* 106: 4525 (1984).

35. M. L. Katz, G. E. Eldred, and W. G. Robison, Jr., Lipofuscin autofluorescence: Evidence for vitamin A involvement in the retina, *Mech. Ageing Dev.* 39: 81 (1987).

36. G. E. Eldred, The fluorophores of the RCS rat retina and implications for retinal degeneration, *in:* "Retinal Degenerations," R. E. Anderson, J. G. Hollyfield, and M. M. LaVail, eds., CRC Press, Boca Raton, FL (1991).

37. W. F. Zimmerman, F. Lion, F. J. M. Daemen, and S. L. Bonting, Biochemical aspects of the visual process. XXX. Distribution of stereospecific retinol dehydrogenase activities in subcellular fractions of bovine retina and pigment epithelium, *Exp. Eye Res.* 21: 325 (1975).

38. S. Robert, P. Tancrede, C. Salesse, and R. M. LeBlanc, Interactions in mixed monolayers between distearoyl-L-phosphatidylethanolamine, rod outer segment phosphatidylethanolamine and all-*trans* retinal, *Biochim. Biophys. Acta* 730: 217 (1983).

39. R. W. Layer, The chemistry of imines, *Chem. Rev.* 63: 489 (1963).

40. R. E. Anderson and M. B. Maude, Phospholipids of bovine rod outer segments, *Biochemistry* 9: 3624 (1970).

41. S. Dayagi and Y. Degani, Methods of formation of the carbon-nitrogen double bond, *in:* "The Chemistry of the Carbon-Nitrogen Double Bond," S. Patai, ed., Interscience Publ., New York (1970).

42. C. deDuve, T. deBarsy, B. Poole, A. Trouet, P. Tulkens, and F. Van Hoof, Lysosomotropic agents, *Biochem. Pharmacol.* 23: 2495 (1974).

43. P. O. Seglen, Inhibitors of lysosomal function, *Meth. Enzymol.* 96: 737 (1983).

44. Y. Matsumoto, T. Watanabe, T. Suga, and H. Fujitani, Inhibitory effects of quaternary ammonium compounds on lysosomal degradation of endogenous proteins, *Chem. Pharm. Bull.* 37: 516 (1989).

45. D. Rideout, J. Jaworski and R. Dagnino, Jr., Environment-selective synergism using self-assembling cytotoxic and antimicrobial agents, *Biochem. Pharmacol.* 37: 4505 (1988).

46. C. J. Duncan and M. F. Rudge, Are lysosomal enzymes involved in rapid damage in vertebrate muscle cells? A study of the separate pathways leading to cellular damage, *Cell Tissue Res.* 253: 447 (1988).

47. S. P. F. Miller, S. A. French, and C. R. Kaneski, Synthesis and characterization of a novel lysosomotropic enzyme substrate that fluoresces at intracellular pH, *J. Org. Chem.* 56: 30 (1991).

48. M. F. Raines, S. K. Bhargava, and E. S. Rosen, The blood-retinal barrier in chloroquine retinopathy, *Invest. Ophthalmol. Vis. Sci.* 30: 1726 (1989).

49. R. A. Firestone, J. M. Pisano, and R. J. Bonney, Lysosomotropic agents. 1. Synthesis and cytotoxic action of lysosomotropic detergents, *J. Med. Chem.* 22: 1130 (1979).

50. J. M. Pisano and R. A. Firestone, Lysosomotropic agents. III. Synthesis of N-retinyl morpholine, *Synth. Comun.* 11: 375 (1981).

51. D. K. Miller, E. Griffiths, J. Lenard, and R. A. Firestone, Cell killing by lysosomotropic detergents, *J. Cell Biol.* 97: 1841 (1983).

52. M. O. Bradley, V. I. Taylor, M. J. Armstrong, and S. M. Galloway, Relationship among cytotoxicity, lysosomal breakdown, chromosome aberrations, and DNA double-strand breaks, *Mutation Res.* 189: 69 (1987).

53. S. Forster, L. Scarlett, and J. B. Lloyd, The effect of lysosomotropic detergents on the permeability properties of the lysosome membrane, *Biochim. Biophys. Acta* 924: 452 (1987).

54. C. M. Martinez, S. Ayala, A. Coquet, M. Lepinay, O. Michel, T. Robles, F. Chiang, K. Alexanderson, P. E. Sanchez, and J. Lever, Malignant cell autolysis caused by intracytoplasmic liberation of lysosomal enzymes, *Cell Biol. Internat. Rep.*, 14: 255 (1990).

55. A. L. Kovacs, A. Reith, and P. O. Seglen, Accumulation of autophagosomes after inhibition of hepatocytic protein degradation by vinblastine, leupeptin or a lysosomotropic amine, *Exp. Cell Res.*, 137: 191 (1982).

56. M. S. Ramsey and B. S. Fine, Chloroquine toxicity in the human eye. Histopathologic observations by electron microscopy, *Am. J. Ophthalmol.* 73: 229 (1972).

57. A. R. Rosenthal, H. Kolb, D. Bergsma, D. Huxsoll, and J. L. Hopkins, Chloroquine retinopathy in the rhesus monkey, *Invest. Ophthalmol. Vis. Sci.* 17: 1158 (1978).

58. K. R. Brizzee and J. M. Ordy, Cellular features, regional accumulation, and prospects of modification of age pigments in mammals, *in:* "Age Pigments," Elsevier/North-Holland Biomedical Press, New York (1981).

59. M. L. Katz and M. J. Shanker, Development of lipofuscin-like fluorescence in the retinal pigment epithelium in response to protease inhibitor treatment, *Mech. Ageing Devel.* 49: 23 (1989).

60. G. O. Ivy, F. Schottler, J. Wenzel, M. Baudry, and G. Lynch, Inhibitors of lysosomal enzymes: Accumulation of lipofuscin-like dense bodies in the brain, *Science* 226: 985 (1984).

61. G. O. Ivy, Y. Ihara, and K. Kitani, The protease inhibitor leupeptin induces several signs of aging in brain, retina and internal organs of young rats, *Arch. Gerontol. Geriatr.* 12: 119 (1991).

62. M. L. Katz, Incomplete proteolysis may contribute to lipofuscin accumulation in the retinal pigment epithelium, *in:* "Lipofuscin and Ceroid Pigments," E. A. Porta, ed., Plenum Press, New York (1990).

63. M. L. Katz, C. M. Drea, and W. G. Robison, Jr., Relationship between dietary retinol and lipofuscin in the retinal pigment epithelium, *Mech. Ageing Devel.* 35: 291 (1986).

64. R. W. Young, Solar radiation and age-related macular degeneration, *Surv. Ophthalmol.* 32: 252 (1988).

65. B. Munoz, S. West, N. Bressler, S. Bressler, F. S. Rosenthal, and H. R. Taylor, Blue light and risk of age-related macular degeneration, *Invest. Ophthalmol. Vis. Sci.* 31(Suppl.): 49 (1990).

66. J. J. Weiter, F. C. Delori, G. L. Wing, and K. A. Fitch, Relationship of senile macular degeneration to ocular pigmentation, *Am. J. Ophthalmol.* 99: 185 (1985).

67. J. C. Saari and L. Bredberg, Enzymatic reduction of 11-cis-retinal bound to cellular retinal binding protein, *Biochim. Biophys. Acta* 716: 266 (1982).

68. D. A. Newsome, M. Swartz, N. C. Leone, R. C. Elston, and E. Miller, Oral zinc in macular degeneration, *Arch. Ophthalmol.* 106: 192 (1988).

69. A. E. Leure-duPree, Electron-opaque inclusions in the rat retinal pigment epithelium after treatment with chelators of zinc, *Invest. Ophthalmol. Vis. Sci.* 21: 1 (1981).

70. A. E. Leure-duPree and C. J. McClain, The effects of severe zinc deficiency on the morphology of the rat retinal pigment epithelium, *Invest. Ophthalmol. Vis. Sci.* 23: 425 (1982).

71. K. W. Small, J. L. Weber, A. Roses, F. Lennon, J. M. Vance, and M. A. Pericakvance, North Carolina macular dystrophy is assigned to chromosome 6, *Genomics* 13: 681 (1992).

72. M. B. Gorin, T. O. Paul, J. Ngo, D. E. Weeks, C. Sarneso and C. Berkebile, Genetics of age-related maculopathy: Clinical phenotypes, intrafamilial concordance and theoretical considerations for molecular studies, in: "Retinal Degeneration: Clinical and Laboratory Applications," J. G. Hollyfield, M. M. LaVail, and R. E. Anderson, eds., Plenum Publ., New York (1993).

73. M. P. Sheetz and S. J. Singer, Biological membranes as bilayer couples. A molecular mechanism of drug-erythrocyte interactions, *Proc. Nat. Acad. Sci. USA* 71: 4457 (1974).

74. P. R. Cullis and B. de Kruijff, Lipid polymorphism and the functional roles of lipids in biological membranes, *Biochim. Biophys. Acta* 559: 399 (1979).

75. H. Hagerstrand and B. Isomaa, Morphological characterization of exovesicles and endovesicles released from human erythrocytes following treatment with amphiphiles, *Biochim. Biophys. Acta* 1109: 117 (1992).

76. D. Bok, Retinal photoreceptor-pigment epithelium interactions, *Invest. Ophthalmol. Vis. Sci.* 26: 1659 (1985).

77. L. M. DeLuca, The direct involvement of vitamin A in glycosyl transfer reactions of mammalian membranes, *Vitam. Hormones* 35: 1 (1977).

78. H. de The, A. Marchio, P. Tiollais, and A. Dejean, Differential expression and ligand regulation of retinoic acid receptor $\alpha$ and $\beta$ genes, *EMBO J.* 8: 429 (1989).

79. C. J. Tabin, Retinoids, homeoboxes, and growth factors: Toward molecular models for limb development, *Cell* 66: 199 (1991).

# HOW MANY CONES ARE REQUIRED TO "SEE?": LESSONS FROM STARGARDT'S MACULAR DYSTROPHY AND FROM MODELING WITH DEGENERATE PHOTORECEPTOR ARRAYS

Andrew M. Geller[*] and Paul A. Sieving

Retinitis Pigmentosa Center
University of Michigan
W.K. Kellogg Eye Center
Ann Arbor, MI 48105

## INTRODUCTION

In an era of attempts to restore vision by transplanting photoreceptors, it may be worthwhile to consider just how many photoreceptors are required to "see." "Seeing," of course, requires definition. From the patient's perspective, an essential goal is to achieve useful visual acuity. This in turn will require repopulating the fovea with an appropriate number of cones.

For the normal human fovea, best visual acuity is sustained over only the central 1 deg. Cones of the central fovea each subtend about 0.5 min arc, and the diameter of this 1 deg region is formed by fewer than 120 cones placed side by side, given the increase in cone spacing with eccentricity[1]. When cones are packed maximally in a regular hexagonal array, the central area contains about 11,000 photoreceptors. But is it necessary to repopulate the fovea with 11,000 functioning cones to recover reasonable acuity? Of course, restoring vision for only the central fovea would result in a severe visual field limitation. A 1 deg visual angle is equivalent to the width of a thumb held at arm's length, and such tunnel vision would be insufficient to orient oneself in a room or to navigate in a complex environment. But the issue of visual fields is peripheral to considering visual acuity.

---

*current address: Center for Environmental Medicine and Lung Biology, University of North Carolina at Chapel Hill, Chapel Hill, NC 27599-7310.

Acuity describes the ability to resolve two adjacent objects as being separate. Given the normal geometry of the fovea, three adjacent cones theoretically can resolve a tiny spot only 0.5 min arc with empty space on either side of it. This limits the highest, or Nyquist, spatial frequency of the visual system to 60 cycles/deg. In the Snellen acuity metric, this is equivalent to 20/10 vision. However, such super-acuity is rarely achieved because the refractive optics of the eye degrade the image and limit acuity to 20/15 or even to 20/20 (i.e. 30 cycle/deg), which is considered the clinical measure of excellent vision.

In a conventional Snellen acuity target of 20/20 size, two photoreceptors see each black arm and another two each empty space. The target could still be identified even if every second cone were removed, cutting the packing density of the regular foveal cone array in half. This maneuver would reduce the Nyquist frequency to 30 cycle/deg, still equivalent to 20/20 vision, approximating vision at about 1 deg eccentricity[1]. By extension, 20/160 acuity could still be supported if the fovea were repopulated with cones at a density of only one-eighth of normal. Such acuity would be better than the 20/200 or less that defines "legal blindness." Measurements of acuity vs. eccentricity in the normal fovea and near parafovea support this relationship between packing density and acuity[2]. We see in the following study, however, that this relationship gives only a conservative estimate of potential acuity with a given photoreceptor packing density and geometry.

## SIMULATING EFFECTS ON ACUITY BY A MODEL OF PHOTORECEPTOR LOSS

We explored the relationship between acuity and photoreceptor density by asking normal subjects to judge the orientation of gratings created by points on a CRT screen to simulate visual sampling by foveal cones[3]. A hexagonal array of points was displayed on the CRT and a grating pattern was superimposed. We reasoned that effects of photoreceptor loss might be explored by viewing gratings created with fewer points than in the fully packed hexagonal array. In this scheme, with a full array of points on the CRT, the bar pattern was fully sampled to simulate viewing with a normal fovea. Randomly removing points from the array would simulate viewing with photoreceptors missing from a diseased fovea.

Stimulus parameters were scaled to mimic acuity conditions of the human fovea while avoiding the optical limitations of the eye. First, spatial resolution was determined for normal subjects viewing the CRT gratings, and the narrowest bars that could be seen were considered equivalent to Nyquist spatial frequency for the simulation. Second, this model Nyquist frequency was used to scale the geometrical equivalent of anatomical cone packing and to limit the CRT screen area to the equivalent of the central 0.6 deg of the fovea for this simulation. Gratings were presented for brief periods to prevent volitional scanning of the pattern.

Frequency of seeing the correct orientation varied from 100% correct whenever the CRT gratings were fully populated to the mandatory 50% correct (chance-level performance in the two-alternative forced-choice experimental paradigm) when all points were eliminated from the grating (i.e. the screen was blank since no points were presented). The density of points required to reach threshold performance varied with grating frequency, as one would hope for an appropriate simulation. But surprisingly, observers reached threshold correct levels with only 12% of points presented, corresponding to a 3-fold increase in mean spacing, even for patterns at Nyquist model frequency (figure 1).

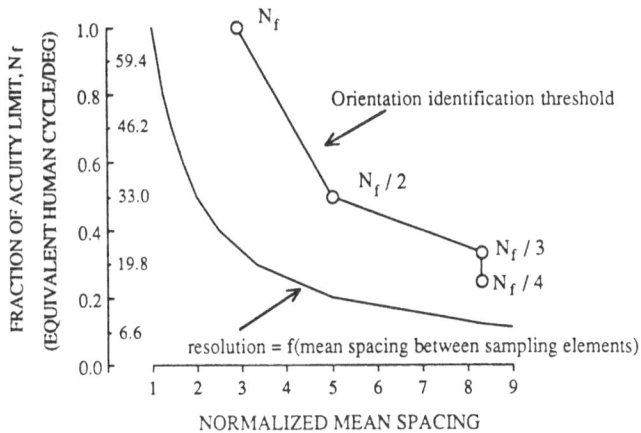

Figure 1. Acuity as a function of mean spacing. The lower curve shows expected acuity if it is determined by the mean spacing of the sampling elements. An intact array, i.e. one with no degeneration, has mean spacing of 1.0 and acuity of $N_f$. One-half acuity (0.5 $N_f$) results from doubling the mean spacing (i.e. 2.0). Open circles show mean spacing of array elements for 75%-threshold performance in orientation-identification task as a function of model spatial frequency. Equivalent spatial frequency in cycle/deg is provided to facilitate comparison to standard grating acuity tasks[1].

This should be encouraging for those interested in transplantation -- these results suggest that good visual performance can be achieved at cell counts far lower than those indicated by the acuity-eccentricity function. This can be understood in terms of the sampling statistics of the foveal cone array. The modal sampling frequency of a regularly-spaced array is determined by the "near neighbor" relationships of its sampling elements. These relationships remain fairly intact even in the face of random degeneration. In engineering terms, loss of sampling points resulted in the addition of broad band noise to the sampling spectrum but did not change the modal sampling frequency from that of the original array[3]. Additional corroborative evidence comes from the retinal histology of a young Stargardt's patient who had 20/30 acuity despite the post-mortem finding that the majority of foveal cones were missing[4].

There are limits to this model of sampling with degenerate arrays. This simulation assumed a disease model in which (1) receptors are each lost randomly and independently of neighboring receptors; (2) the loss of one receptor has no effect on the position of its neighbors; (3) each receptor is lost completely and emits no partial signaling. This simulation maintained strict hexagonal regularity of CRT points even for massive loss of photoreceptor-points.

For real patients with maculopathy, regularity of packing may not necessarily be maintained, since cone misalignment has been suggested for foveal vision of patients with retinitis pigmentosa tested by the Stiles-Crawford effect[5]. A reinterpretation of our empirical modeling might be required if the simulation incorporated random placement of photoreceptor-points to create the degenerate arrays. The message, however, remains that repopulating the fovea with very few cones may enable a patient to recover useful acuity.

## COUNTING PHOTORECEPTORS IN STARGARDT'S MACULOPATHY

Given the finding in our simulation that visual discriminations can be performed under conditions of massive degeneration, we sought empirical corroboration by attempting to determine the density of foveal cones remaining in Stargardt's maculopathy[6]. Studying the vision of Stargardt's patients was appealing, since, as already mentioned, good Snellen acuity could be maintained despite considerable photoreceptor loss shown in the post-mortem histology in a young patient tested just prior to accidental death[4].

Stargardt's disease is an autosomal recessive macular dystrophy, characterized by macular retinal pigment epithelium (RPE) atrophy, the deposition of lipofuscin in the RPE, and reduced central vision, typically by the first or second decade of life[7,8,9,10]. One might postulate that Stargardt's could cause discrete loss of individual photoreceptors at random or in small clumps, based on clinical observations that some patients complained of a fine meshwork character to their vision, as though they were looking through "cheese cloth."

The resistance of visual acuity to the effects of photoreceptor loss in the CRT simulation indicated that grating detection would not provide a particularly sensitive method to count photoreceptors remaining in disease conditions. In contrast to the redundancy inherent in a grating pattern, testing with tiny dot stimuli had the appeal of probing for tiny scotomata caused by small clumps of photoreceptors lost simultaneously, possibly from discrete loss of RPE cells, each of which underlie and support 10-15 cones in the fovea. As a clinical corollary, the fluorescein angiogram frequently shows tiny punctate areas of RPE loss early in Stargardt's.

The desirable stimulus condition would be equivalent to observing a minuscule point of starlight against the nighttime sky. In the laboratory, punctate monochromatic stimuli can appear as red or green, seemingly dependent upon whether an individual long or middle wavelength cone is stimulated[11]. Cicerone and Nerger[12,13] used this phenomenon as one basis of a model to deduce a density ratio for these two cone types in the fovea and to suggest an absolute count of the number of long and middle wavelength cones.

Their model was also based on an analysis of quantal detection, which has a venerable lineage extending back to Hecht, Schlaer and Pirenne's[14] use of Poisson statistics to deduce the threshold quantal sensitivities of rod photoreceptors. Cones operate differently from rods in two ways that enables using Poisson quantum counting to deduce the number of cones lying under a tiny spot. First, cones require multiple simultaneous quantal hits to be activated, normally believed to be about six[12,15,16,17]. Second, the foveal cones can act as independent sampling units, unlike the rods which pool their input. Consequently, by illuminating a limited foveal area with a small but known number of quanta and determining the threshold for "seeing," one can deduce the number of cones mediating detection using statistical arguments.

Elements of this analysis were adapted from Cicerone and Nerger[12] and Vimal, et. al.[16]. The strategy relies on determining conditions for which too few quanta are absorbed by an individual cone to enable detection of the tiny spot. Specifically, each single cone will fail to see the tiny spot if it absorbs fewer than six quanta per flash. This is convenient, since in the range of zero to five quanta associated with "not seeing" a flash, the statistics are described by a Poisson sum:

$$\Pi(x) = \sum_{k=0}^{m-1} \frac{e^{-x} x^k}{k!} \qquad (1)$$

where m = # quantal hits for "seeing," usually taken to be m=6, and x = the number of quanta incident on the retina.

Assuming that the cones act independently, for the small number (N) of cones illuminated by a tiny spot, the cumulative probability of "not seeing" a flash is simply the product of "not seeing" by each of (N) individual cones, i.e. $\Pi(x)^N$.

Finally, each time the tiny spot is flashed, either one sees it or not, and

$$P(yes|x) = 1 - \Pi(x)^N \qquad (2)$$

Normal subjects and Stargardt's patients were tested. The strategy was to show tiny spots to the fovea over a range of intensities and determine the frequency-of-seeing functions for foveal absolute (i.e. dark background) threshold (figure 2). Two spot sizes were used, one subtending about six foveal cones and a second subtending about 41 cones, according to the point spread function of the eye[6].

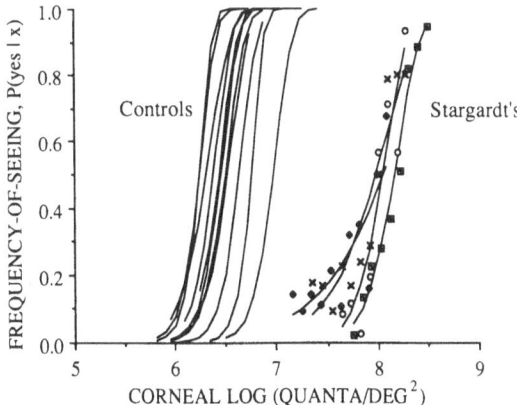

Figure 2. Frequency-of-seeing data from patients and control observers. Control observers' data are represented by curves fit to their data. Patients' data are fit with curves to differentiate individual observers. Data are from detection task with 1.125 min diameter test field. Comparable results were obtained with a 3.375 min diameter test.

Responses of normal subjects were readily modeled by the Poisson statistics described above. Maximum likelihood estimator fits gave the minimum number of simultaneous quantal hits required to activate a single cone as mean m = 5.4, which agreed well with the range of m = 4 - 7 described in the literature cited above. Further, to within a factor of two, the analysis identified the number of cones (N) under each spot in correct agreement to expectation from physiological optics and the anatomy.

Thresholds for Stargardt's patients were greatly elevated from normal. These subjects had corrected acuities ranging from 20/30 to 20/100, and they required intensities 1 - 1.5 log unit brighter than normals to detect the spot (figure 2). However, their transition zone from 0% detection to full detection was nearly as steep as the normals. To model these data, it was necessary to add a term to the Poisson model above which expressed the binomial probability that the incident light illuminated positions at which photoreceptors were present or absent[6]. This led to three disease models against which the data were tested (figure 3).

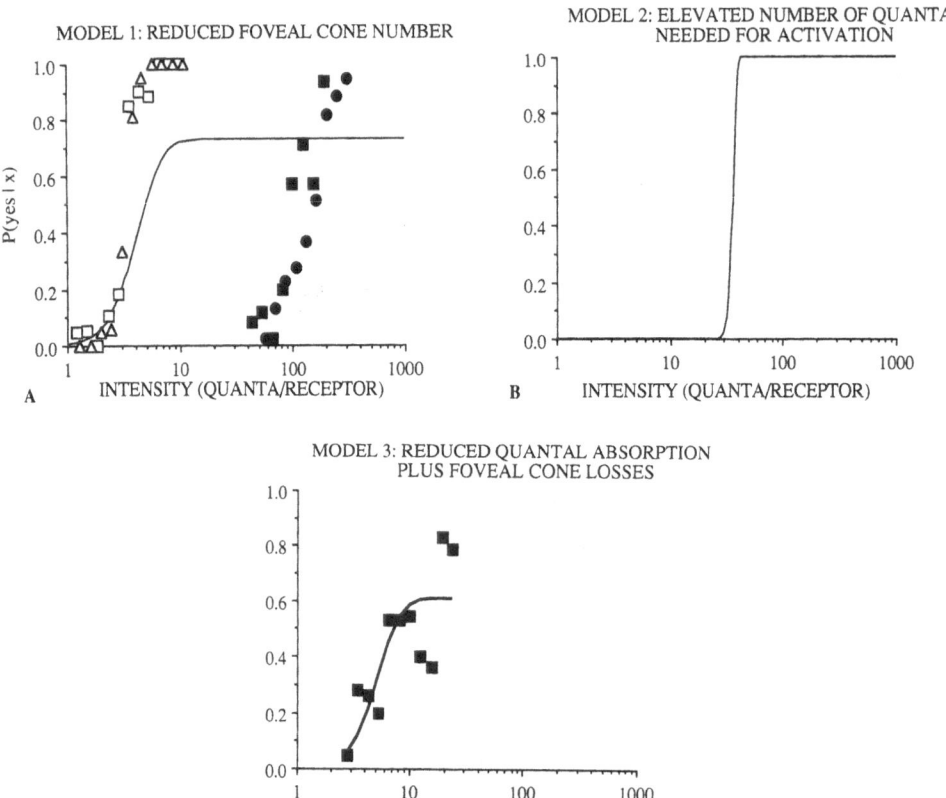

Figure 3. A. Solid curve shows theoretical detection function corresponding to a loss of 83% of the photoreceptors for a 3.375 min test. Open symbols show data of 2 control observers, solid symbols show data of 2 Stargardt's observers for comparison. Asymptote of theoretical function changes with field size[6]. B. Solid curve is calculated with m = 50 quanta necessary for activation. Postulating photoreceptor loss in addition to a change in the quantal requisite for activation produces an aymptote at less than p(yeslx) = 1.00, but does not significantly reduce the slope[6]. C. Data of one patient fit by estimating one parameter to account for decreased quantal absorption and another for the remaining number of photoreceptors[6].

Model 1 hypothesized that Stargardt's subjects had reduced numbers but otherwise normal photoreceptors. This model predicted response functions which were shifted to somewhat higher intensities but which would level off considerably below 100% correct. Both of these predictions are contradicted by the data from the Stargardt's subjects, which are shifted far more in intensity than is accounted for by cell loss, but do not necessarily asymptote (figure 3a).

Model 2 hypothesized that these photoreceptors required many more than 6 quantal hits for activation, thus increasing the value of m in the Poisson equation above. However, while this shifted the intensity range to that needed for detection by the patients, this forced the transition zone from approximately 0 to 1.00 probability of detection to be extraordinarily steep, far beyond that exhibited by the patients and even steeper than for normal controls (figure 3b).

Model 3 proposed that effective quantum catch was reduced and also that the number of cones subserving detection was reduced. Van Meel and van Norren[18] showed with *in vivo* densitometry that Stargardt's patients exhibited reduced two-way densities in the fovea, supporting tests of color-matching which had suggested that this was due to reduced pigment density[19]. This would account for much of the 1-1.5 log difference in effective quantum catch in our Stargardt's patients to account for their depressed detection thresholds. Maximum likelihood estimator analysis then indicated that our patients exhibited a loss of 85-92% of foveal cones (figure 3c). Overall, Model 3 provided the most satisfactory fit to the experimental data.

How well do these findings reflect anatomical reality? Stargardt's disease leads to profound macular photoreceptor loss and there is some anatomical evidence of cone outer segment shortening[4,20], yet visual acuity better than 20/40 can persist into the fifth decade of life[8]. While different from the human condition, psychophysical and histopathological studies in rats and mice with light-induced or inherited outer retinal degeneration conclude that these animals can perform light and pattern discrimination with very few or severely impaired photoreceptors[21,22,23].

## CONCLUSION

The convergence of results from modeling with degenerate arrays and from counting photoreceptors in Stargardt's patients was gratifying even if serendipity or coincidence was involved. Both methods suggested that a 1 log unit decrease in foveal cone density was still compatible with useful acuity, even to a level superior to legal blindness. The important lesson from this is that it is not merely the number or even the mean spacing of the receptors which are critical to discriminating spatial targets. The near neighbor relationships of the original foveal receptor mosaic set the sampling rate[2]. This in turn determines the spatial frequencies discriminable by the observer, even in the case of a degenerate array. For the reverse case of re-seeding a retina with healthy photoreceptors, the tightness of the receptor-packing in the transplanted patch is critical for the potential visual acuity even if all of the transplanted receptors do not ultimately survive.

In addition, the result with Stargardt's patients indicated that partially disabled cone photoreceptors might still function quite adequately for vision, whether by requiring greater quantal flux to overcome decreased quantum catch (e.g. because of shorter or disordered outer segments), as we suggest here, or by needing additional quantal absorptions for activation, as suggested by animal models of retinal degeneration which include reduced cGMP-phosphodiesterase activity[24].

There is still scarce evidence that transplanted photoreceptors are physiologically active[25], though recent reports demonstrate that a morphologically normal-looking photoreceptor level can be reconstructed with photoreceptor transplants[26]. Recovering usable vision is additionally contingent upon the re-linking of the receptors to retinal output channels and the ability of the visual system to relearn the locations of its new receptors on the retina[27,28]. Given these caveats, our results suggest that the "front end" of vision is potentially recoverable with far fewer than the usual complement of receptors.

## ACKNOWLEDGEMENTS

This research was supported by the National Retinitis Pigmentosa Foundation, Inc., Baltimore, MD and by National Institutes of Health T32-EY07022. We thank John Lee for his substantial contributions of time and energy.

## REFERENCES

1. J. Hirsch and C. Curcio, The spatial resolution capacity of human foveal retina, *Vision Research* 29:1095 (1989).

2. A.M. Geller, J. Lee, and P.A. Sieving, Effects on spatial sampling of simulated degeneration and retinal eccentricity, *Inv. Ophthal. Vis. Sci. (suppl.)* 33:1349 (1992).

3. A.M. Geller, P.A. Sieving, D.G. Green, Effect on grating identification of sampling with degenerate arrays, *J. Opt. Soc. Am. A* 9:472 (1992).

4. R.C. Eagle, A.C. Lucier, V.B. Bernardino, and M. Yanoff, Retinal pigment epithelial abnormalities in fundus flavimaculatis, *Ophthal.*, 87:1189 (1980).

5. D.G. Birch, M.A. Sandberg, and E.L. Berson, The Stiles-Crawford effect in retinitis pigmentosa, *Inv. Ophthal. Vis. Sci.*, 22:157 (1982).

6. A.M. Geller and P.A. Sieving, Assessment of foveal cone photoreceptors in Stargardt's macular dystrophy using a small dot detection task, In press, *Vis. Res.*

7. K.G. Noble and R.E. Carr, Stargardt's disease and fundus flavimaculatis, *Arch. Ophthal.*, 97:1281 (1979).

8. G.A. Fishman, M. Farber, B.S. Patel, and D.J. Derlacki, Visual acuity loss in patients with Stargardt's macular dystrophy, *Ophthal.* 94:809 (1987).

9. R.G. Weleber and A. Eisner, Cone degeneration ("bull's-eye dystrophies) and color vision defects, *in:* "Retinal Dystrophies and Degeneration," D.A. Newsome, ed., Raven Press, New York (1988).

10. A.C. Bird and J. Marshall, Retinal receptor disorders without known metabolic abnormalities, *in:* "Pathobiology of Ocular Disease, part B," Garner and Klintworth, eds., Marcel Dekker, New York (1982).

11. J. Krauskopf and R. Srebo, Spectral sensitivity of color mechanisms: derivation from fluctuations of color appearance near threshold, *Science* 150:1477 (1965).

12. C. Cicerone and J. Nerger, The relative numbers of long-wavelength sensitive and middle-wavelength-sensitive cones in the human fovea centralis, *Vis. Res.* 29:115 (1989).

13. C. Cicerone and J. Nerger, The density of cones in the fovea centralis of the human dichromat, *Vis. Res.* 29:1587 (1989).

14. S. Hecht, S. Schlaer, and M. Pirenne, Energy, quanta, and vision, *J. Gen. Physiol.* 25:819 (1942).

15. F.H.C. Marriott, The foveal absolute for short flashes and small fields, *J. Physiol.* 169:416 (1963).

16. R.L.P. Vimal, J. Pokorny, V.C. Smith, and S.K. Shevell, Foveal cone thresholds, *Vis. Res.* 29:61 (1989).

17. M.F. Wesner, J. Pokorny, S.K. Shevell, and V.C. Smith, Foveal cone detection statistics in color-normals and dichromats, *Vis. Res.* 31:1021 (1991).

18. G.J. van Meel and D. van Norren, Foveal densitometry as a diagnostic technique in Stargardt's disease, *Am. J. Ophthal.*, 102: 353 (1986).

19. J. Pokorny, V.C. Smith and J.T. Ernest, Macular color vision defects: specialized psychophysical testing in acquired and hereditary chorioretinal diseases, *in:* "Electrophysiology and psychophysics, their use in ophthalmic diagnosis, I.O.C. 20," S. Sokol, ed.,Little, Brown and Co., Boston (1980).

20. M. Järveläinen and A.H. Milam, unpublished data.

21. Z.M. Nagy and J.F. Misanin, Visual perception in the retinal degenerate C3H mouse, *J. Comp. Physiol. Psych.* 72:306 (1970).

22. K.V. Anderson W.K. O'Steen, Black-white and pattern discrimination in rats without photoreceptors, *Experimental Neurology* 34:446 (1972).

23. M.M. LaVail, M. Sidman, R. Rausin, and R.L. Sidman, Discrimination of light intensity by rats with inherited retinal degeneration, *Vis. Res.* 14:693 (1974).

24. D. Farber and T.A. Shuster, A proposed sequence of events leading to photoreceptor degeneration in the rd mouse retina, *in:* "Retinal Degeneration: Experimental and Clinical Studies," M.M. LaVail, J.G. Hollyfield, and R.E. Anderson, eds., New York, Alan R Liss, Inc. (1985).

25. H. Klassen and R.D. Lund, Retinal transplants can drive a pupillary reflex in host rat brains, *Proc. Nat. Acad. Sci USA* 84:6958 (1987).

26. P. Gouras, J. Du, H. Kjeldbye, S. Yamamoto, D.J. Zack, Reconstruction of degenerate rd mouse retina by transplantation of transgenic photoreceptors, *Inv. Ophthal. Vis. Sci.* 33:2579 (1992).

27. L.T. Maloney, Spatially irregular sampling in combination with rigid movements of the sampling array, *Inv. Ophthal. Vis. Sci. (suppl.)* 29:58 (1988).

28. A.J. Ahumada and J.I. Yellott, A connectionist model for learning receptor position, *Inv. Ophthal. Vis. Sci. (suppl.)* 29:58 (1988).

# THE GENETICS OF AGE-RELATED MACULOPATHY

Michael B.Gorin[1,2], Carmella Sarneso[2], T. Otis Paul[3], Julielani Ngo[3], and Daniel E. Weeks[2]

[1]The Eye & Ear Institute of Pittsburgh, Ophthalmology Department, University of Pittsburgh, Pittsburgh, PA 15213
[2]Department of Human Genetics, University of Pittsburgh Graduate School of Public Health, Pittsburgh, PA 15213
[3]Smith-Kettlewell Eye Research Institute, San Francisco, CA 94115

## INTRODUCTION

Age-related maculopathy[1](ARM) or age-related macular degeneration is the leading cause of irreversible vision loss in the elderly population in the United States and the Western world[2-6] and a major public health issue. While ARM has been observed to have a genetic component, it has not been the subject of genetic investigations due to difficulties in diagnosis, late-onset, and complexity of expression. With the development of new genetic and analytical methods, it is now feasible to study ARM. The localization of the gene or genes that contribute to ARM susceptibility will guide studies of the underlying cause(s) of ARM. Identification of individuals who have an increased genetic susceptibility for developing ARM will provide the basis for future therapeutic and preventative interventions.

ARM is a degenerative disorder involving the retinal pigment epithelium, choriocapillaris[7,8] and retina which primarily, but not exclusively, affects the macular region. Symptoms from ARM include metamorphopsia, impaired light adaptation and decreased central vision. ARM causes changes in the macula of the retina which include the presence of drusen[9-12] and/or changes in the retinal pigment epithelium (RPE), geographic atrophy, and subretinal, choroidal neovascularization[11,13-15]. Drusen are lipid and protein deposits that accumulate within and on the surface of Bruch's membrane adjacent to the retinal pigment epithelium. Decreased visual acuity has been included by several investigators as one of the criteria for the diagnosis of ARM in older individuals[9,16-20], though the visual acuity does not always correlate with the severity of the retinal/RPE pathology. ARM has been categorized in two forms, an exudative form (disciform or wet form) or an atrophic form (geographic or dry form). Either form may be present with hard drusen and pigment migration. Despite morphological differences, there is no evidence to suggest that these different clinical forms have distinct etiologies.

Despite a plethora of scientific studies, the etiology of ARM remains unclear. Deficiencies of serum zinc, copper and selenium levels have been postulated in the pathogenesis of ARM[21, 22]. Light exposure has been suggested as a causative factor[23-25] but the relationship appears to be circumstantial[23]. Our lack of understanding of the primary pathways and molecular mechanisms that underlie the pathogenesis of ARM prevents the use of a focused, candidate gene-based approach. Genetic linkage analysis methods allow us to search for susceptibility genes independently of the disease mechanisms.

## Genetic Epidemiology of ARM

The overall prevalence of ARM is estimated to be 9% in the population of individuals over 60, ranging from 2% in the 45-65 year-old age group to 28% in the 75-85 year-old age group with at least 12% of people between the ages of 65 and 75 affected[26-31] In all studies of incidence, prevalence and associated factors, the age of onset and increasing prevalence with advancing age are consistently correlated with ARM.

The Framingham Eye Study identified a number of risk factors associated with ARM[18]. The study reported higher age-specific prevalence rates for ARM in women compared with men[18]; however, this finding is controversial[4]. The association of ARM and hypertension and cardiovascular disease[32] has similarly been disputed by several investigators[5,33,34]. In addition, both refractive error and smoking have been further associated with ARM[35, 36] as well as chemical work exposures[5] and a family history of ARM[17,37-39].

Both clinical case studies and epidemiologic studies have supported a major contribution of genetics to the susceptibility of developing ARM. Francois[37] presented a number of case studies of families dating back to 1921 in which several family members were affected. He found several families with bilateral, central areolar atrophy of the choroid in the macula and onset between 20 and 50 years of age and suggested that they represent a different form distinct from the disciform diseases which are either dominant or recessive. Gass[38] performed a retrospective review of over 200 patients and their families. With careful questioning and investigation of available relatives he found a family history of loss of central vision in almost 20% of the patients. Hyman[17] in a study of 228 cases and 237 age-matched controls, reported a positive association between family history and expression of ARM, and examinations of first degree relatives in her study indicated these individuals displayed a 2.9-fold increased risk of ARM. In addition to these studies, two cases of ARM in monozygotic twins have been described[40,41]. The observed variability in drusen formation in the twin studies provides evidence that ARM is variably expressed, that exogenous factors may play a role, and that it is difficult to make clinical distinctions between types of ARM. The strong evidence that genetics plays an important role in ARM susceptibility does not contradict reports that effects of environmental and dietary factors contribute to ARM[21, 42].

The transmission of ARM in families in several studies (summarized in Francois[37]) suggest that dominant inheritance is not uncommon. It is difficult to determine whether families with isolated cases or affected siblings without affected parents are the result of recessive inheritance, nongenetic etiologies, or dominant inheritance with incomplete penetrance in the parental generation. Segregation analyses of ARM are difficult because diagnostic methods are inconsistent and unreliable; macular disease will be overlooked in individuals with relatively normal visual acuities. Confirmation of ARM in several generations is often unobtainable: individuals who are in their 50s and 60s generally do not have a parental generation available for evaluation, while the generation in their 20s and 30s are generally asymptomatic and the children of the latter often do not demonstrate any ophthalmological changes. In addition, identification of ARM by family history is often based on recollection. Given these difficulties, it is highly likely that ARM diagnosis is incomplete, that the familial nature of the disorder is markedly underestimated and that the actual prevalence in the general population over age 60 is higher than 9%.


## Clinical Diagnosis of ARM

Well-defined diagnostic criteria for ARM and exclusion of disorders with overlapping phenotypes are essential for genetic studies. There are three clinical diagnostic concerns that must be addressed:

1) Distinguishing clinical/anatomic subtypes of ARM that may represent different genetic and/or nongenetic etiologies.

2) Confirming that a patient with endstage macular degeneration has ARM and not another disorder, particularly when atrophic or exudative lesions have destroyed the drusen that are thought to be precursors of ARM.

3) Identifying mild or subclinical cases of ARM before the onset of visual loss, in order to genotype these individuals for the presence of the disease marker.

The impact of these specific diagnostic issues on the study of ARM is highly dependent on the type of genetic studies that are undertaken. If ARM is studied by linkage analysis in a few large families, then genetic heterogeneity is less critical unless spouses introduce ARM into the pedigree. For large-family studies, the identification of mild cases is essential to maximize the informativeness of the family.

In the Affected-Pedigree-Member method described below, it is critical to confirm a diagnosis of ARM for endstage macular degeneration. Ambiguous mild or subclinical cases of ARM can be excluded at the cost of reducing the power of the analysis, but without compromising the informativeness of the study. The presence of confounding clinical subtypes can limit the power of the study to detect a major genetic locus.

Clinically, there are many subtypes of ARM based upon the presence or absence and type of drusen, geographic atrophy, subretinal neovascular membranes and/or pigment epithelial changes and detachment. Whether these different features constitute genetically distinct subgroups of ARM can be tentatively inferred from concordance of the clinical phenotype among family members, including twin studies[37,40,41]. The concordance of distinctive features of ARM such as geographic atrophy or type of drusen among affected family members may reflect shared genetic and environmental backgrounds, while variations in the severity of ARM may reflect other modifying factors (gene interactions, diet, light, toxic exposure, etc.).

There are a variety of inherited and acquired disorders that result in exudative or atrophic macular patterns, e.g., toxoplasmosis, presumed ocular histoplasmosis, central areolar choroidal sclerosis, Stargardt's disease, Best's disease, juxtafoveolar telangiectasia, adult foveomacular dystrophy, or pattern dystrophies. These could seriously confound the genetic analysis of ARM. Fortunately, ARM can usually be distinguished from other causes of macular degeneration by the late onset of visual loss (after age 40), evaluation of the less-affected eye, and exclusion of specific types of retinal dysfunction such as lifelong nightblindness or severe photophobia. Many clinicians have observed geographic atrophy or subretinal neovascular membranes in individuals with minimal numbers of or no drusen. In at least some of these cases, it has been proposed that these patients have diffuse abnormalities of Bruch's membrane and the basement membranes of the choriocapillaris[43]. Under these circumstances, clinicians use the diagnosis of ARM for endstage macular atrophy or disciform scarring in the absence of significant macular drusen. It is unknown if these conditions reflect genetic or etiologic heterogeneity rather than phenotypic variation.

Extramacular manifestations of ARM such as multiple extramacular drusen (MED)[44] and/or reticular degeneration of the pigment epithelium (RDPE)[45], can serve as valuable features of ARM in the absence of macular drusen and the presence of macular exudation or atrophy causing significant disruption of the macular architecture. These peripheral changes have been found in association with both hard and soft drusen as well as with geographic atrophy and disciform degeneration. The severities of MED and the RDPE correlate closely with the extent of macular pathology. By inclusion of RDPE and MED as part of the clinical criteria for the diagnosis of ARM, there may be some legitimate cases of ARM that will be excluded; however, the family members that are included are more likely to have ARM as a primary process. If the presence of RDPE and/or MED constitute a distinctive subclass of ARM, then the use of these features as clinical criteria will tend to bias the study toward a more homogeneous group of disorders (thus increasing the power of the analysis to detect specific genetic loci). This hypothesis can be tested by evaluating the concordance of extramacular changes among family members. These clinical subgroups can be evaluated for pleiotropy of a major susceptibility gene or as distinct genetic entities.

The final dilemma of ARM diagnosis concerns subclinical individuals or individuals with some macular changes and normal vision. This problem is analogous to attempts to diagnose glaucoma in individuals with mildly-to-moderately elevated intraocular pressures in the absence of nerve fiber layer defects, optic nerve cupping or visual field loss. Diagnosis of ARM in asymptomatic individuals is essential for family studies and desirable for affected-pair analyses. ARM diagnosis of subclinical individuals is not yet possible with psychophysical approaches, which have tended to focus on sensitivity of detection rather than specificity. A few clinicians, including ourselves, have looked at drusen as an early indicator of ARM, but almost no data are available for the prevalence of drusen in the population under the age of 60. The age-dependent expression of drusen in normal eyes is unknown, and longitudinal studies assessing the progression of ARM in young adults with drusen have not been done. At this time, the reliability of these presymptomatic markers must be viewed as inconclusive for use in genetic studies.

# Genetic Analysis and Modeling of ARM

Classical linkage analysis methods have been used to localize and characterize single gene defects for diseases such as cystic fibrosis and Huntington disease. Unfortunately, for complex traits such as ARM, there are many factors that make classical linkage methods ineffective These variables include late onset, unknown gene frequency, reduced penetrance, possible multiple genes contributing to the development of the disorder (heterogeneity), and phenocopies (cases of ARM that arise from non-genetic factors).

Traditional linkage methods are sensitive to the above-mentioned variables; an error in estimation of a parameter can significantly affect the results of the analysis. Because these parameters for ARM are unknown, many assumptions would need to be made if classical linkage methods are used. The Affected-Pedigree-Member method developed by Weeks and Lange[46] avoids the use of these numerous, untested assumptions.

The Affected-Pedigree-Member linkage analysis uses only affected relatives of a family, thus avoiding the confusion of subclinical individuals or persons whose diagnosis is unclear. Like the sib-pair method[47-53], it assumes that affected relatives share portions of the genome that are responsible for the development of the disease. Identifying the shared portions is accomplished through the use of DNA markers, i.e., DNA segments whose location within the genome is known and which can detect individual polymorphisms.

The Affected-Pedigree-Member method attempts to identify intervals of the genome (flanked by two DNA markers) that are shared by affected individuals at a frequency greater than would be expected by random segregation of genetic material. Ideally one needs to establish that the intervals among these family members are identical based upon common inheritance. However, in reality one must approximate this "identity by descent" by determining the probability that family members have the same genetic interval when they have identical DNA marker alleles. Having established that the individuals have the same marker alleles "identical by state", one infers whether these shared alleles came from a common ancestor. This estimation is highly dependent on the population frequency of the marker alleles. If individuals within a family share marker alleles that are very rare in the population, then it is likely that they acquired these alleles through descent. However, if the alleles are very common in the population, the probability is increased that they were not acquired from a common ancestor. The Affected-Pedigree-Member method is based on identity by state. Genetic parameters defining the disease such as type of inheritance, disease gene frequency, penetrance or phenocopy rates need not be known to perform the analysis. The identity by state analysis is especially applicable to the study of late-onset diseases for which several generations cannot be studied and identity by descent determination is generally not feasible.

The Affected-Pedigree-Member method has recently been successfully applied to other complex genetic disorders such as early-onset familial breast cancer[54-56] and familial Alzheimer's disease (FAD)[57,58] Analysis of FAD, like that of ARM, is complicated by phenocopies, age-dependent penetrance, and late-onset which precludes multigenerational families. Similar to the different clinical forms of drusen in ARM, clinical heterogeneity of Alzheimer's disease has been documented by histopathology[59]. Linkage analyses similar to the Affected-Pedigree-Member method have been used to identify specific genes, angiotensin converting enzyme and angiotensinogen, as playing important roles in rat and human forms of hypertension, respectively [60,61].

The Affected-Pedigree-Member method relies solely on the genotyping of affected relatives and thus is very sensitive to marker allele frequencies. The power of Affected-Pedigree-Member method can be increased by genotyping unaffected family members and partially reconstructing parental haplotypes. Because the approach cannot detect recombination events, it lacks the ability to localize the disease-related loci by physical exclusion. However, one can confirm a significant result from the Affected Pedigree Member method by genotyping unaffected relatives to determine if the marker alleles really are inherited from a common ancestor. Classical linkage analysis can detect recombination events only in the context of a specific genetic model. By combining the Affected-Pedigree-Member approach with classical linkage analyses of extended families, one can increase the power of both methods.

# DNA Genotyping

The discovery of microsatellite repeats and the use of the polymerase chain reaction make the systematic search for disease genes a reasonable goal. The human genome contains numerous highly polymorphic simple repeat sequences (di-, tri-, and tetranucleotide repeats)[62,63,64] which are distributed randomly over the human genome and are easily typed by DNA amplification using the polymerase chain reaction (PCR) technique. It is estimated[65] that the human genome contains approximately 50,000 copies of an interspersed simple sequence repeat with the sequence (dT-dG)n, where n=10-60 and that these repeats are frequently highly polymorphic with average heterozygosities often exceeding 0.7. Use of the PCR allows for rapid typing using about 1/20 the amount of DNA typically required for standard Restriction Fragment Length Polymorphism typing, obviating the need to establish permanent cell lines on each individual as a source of DNA. These simple or microsatellite repeats have enabled the development of linkage maps of most human chromosomes with a resolution of 5-10 centiMorgans (cM).

## Characterization of ARM-affected families

Presently, we have enrolled 45 families with at least two members with confirmed diagnoses of ARM. The first 15 families recruited were used to create the data used in the computer simulations.

There are 17 families with 2 affected individuals, 10 families with 3 affected members and 10 families with 4 affecteds. We have three families with 5 affected individuals and single families with 6 and 8 affected members. Of the 135 affected individuals, 49 are male and 86 are female.

Comparison of the clinical features found in the affected individuals shows concordance within families. These are summarized for the first 22 families in Table 1. The differing types and number of drusen and subretinal neovascularization membranes (SRNVM) were the features most often concordant within families. Geographic atrophy and reticular disturbance of the pigment epithelium (RDPE) were also concordant in some families. In one family, no macular drusen were found, but multifocal pigment figures were identified in the 3 affected individuals. It is unclear whether or not the members of this family have a subtype of ARM or a distinct familial retinal/pigment epithelial disorder.

Of the 45 families, 20 have confirmed affected individuals in at least two generations while eight families have possibile affected individuals in two generations, but late onset precludes confirmation of the diagnosis in the parental generation. Eleven families could have three affected generations; however again the late-onset precludes confirmation in the grandparental generation and the younger generation is currently asymptomatic. The remaining 9 families have affected individuals in only one generation. Fifty-three percent (24/45) of our families are strongly suggestive of autosomal dominant inheritance while an additional 33% (15/45) provide historical support for dominant inheritance. These percentages are substantially higher than the 27% (4 of 15 familial cases) reported as dominant by Heinrich[47]. This difference may reflect our recruitment bias or be the result of the detailed family histories that we obtain. There are five families in which the penetrance of ARM appears to be reduced. We took this into account in our computer simulations by using a decreased penetrance that varies with age.

## 2. Computer simulations

The practical utility of the Affected-Pedigree-Member approach to ARM must first be evaluated by computer simulations to systematically assess the chances of false indication of linkage, and to predict the power to detect linkage. The power of a linkage study is determined by a number of disease parameters including mode of inheritance, gene frequency, penetrance rates, phenocopy rates, and degree of heterogeneity. Unfortunately, these parameters have not been clearly defined for ARM. We describe below the assumptions for these genetic parameters in our simulation study.

**Table 1.** Clinical features and concordance of the ARM-affected members from the first 22 ARM families evaluated at The Eye & Ear Institute of Pittsburgh and The Smith-Kettlewell Eye Research Institute.

| # | Relative pair | Sex | Age | Age onset | Visual Acuity | Ophthalmic Features | Concordance |
|---|---|---|---|---|---|---|---|
| 1 | proband | F | 61 | NA | 20/25+2;20/20 | RDPE, geographic atrophy OU | Geographic |
|   | brother | M | 63 | NA | 20/400;20/25 | RDPE, MED OU; geographic atrophy OD, mixed hard & soft drusen | atrophy RDPE |
| 2 | proband | F | 67 | 60 | CF@1';CF@3' | Mixed, nonconfluent, pigment figures, SRNVM, PED * | Drusen SRNVM |
|   | sister | F | 70 | 64 | HM; 20/20 | Drusen OU; RPE atrophy, disciform lesion OD | |
| 3 | proband | M | 75 | 74 | 20/100;20/40-2 | Hard drusen OU, SRNVM OD, * | Hard drusen |
|   | sister | F | 83 | 73 | 20/25; CF | Hard drusen OU, central atrophy OS * | |
| 4 | proband | F | 70 | 62 | 20/25;20/400 | Few hard drusen OU, disciform OS | SRNVM |
|   | mother | F | 90 | 86 | CF; 20/300 | Disciform lesions OU | |
|   | m. uncle | M | 84 | 81 | 20/30; CF | Peripapillary membrane OD | |
| 5 | proband | M | 61 | 48 | 20/300;20/300 | Few drusen, geographic atrophy, central choroidal sclerosis OU | No drusen Geographic |
|   | sister | F | 75 | 33 | 20/200;10/160 | Few drusen, central, peripapillary atrophy OU | atrophy |
| 6 | proband | F | 68 | 65 | 20/20; 20/20 | Pigment figures, multifocal RPE disturbances, no drusen OU | No drusen Pigment figures |
|   | m. cousin | F | 58 | 55 | 20/25; 20/20-1 | Pigment figures, granular RPE, no discrete drusen OU | |
| 7 | proband | F | 75 | | 20/40; 20/400 | Multiple drusen, RPE mottling, geographic atrophy OU | Geographic atrophy |
|   | m. aunt | F | 85 | | 20/500;20/500 | Geographic atrophy OU * | |
| 8 | proband | M | 70 | 69 | 20/200;20/80 | Hard drusen, pigment figures, SRNVM, RDPE OU * | SRNVM RDPE |
|   | brother | M | NA | 50 | 20/400;10/100 | Geographic atrophy OS; pigment figures OD; RDPE, probable SRNVM OU | Geographic atrophy |
|   | sister | F | 84 | 74 | CF@3' OU | Central choroidal sclerosis OU * | |
| 9 | proband | M | 74 | 73 | <20/400;20/60 | Multiple mixed drusen OS; SRNVM, PED OD | SRNVM |
|   | brother | M | 76 | 72 | 20/300;CF@4' | SRNVM OU * | |
| 10 | proband | F | 81 | 81 | 20/30,20/400 | Discrete hard drusen OU, SRNVM OS | SRNVM |
|   | brother | M | 76 | 75 | 20/40;CF@ 1' | Few pigment figures, no drusen, SRNVM OU * | |
| 11 | proband | F | 69 | | 20/30;20/30 | Almost confluent soft drusen, pigment figures OU * | |
|   | father | M | 96 | | 20/300;prsths. | Macular degeneration OD* | |

| # | | Sex | | | | | |
|---|---|---|---|---|---|---|---|
| 12 proband | F | 68 | 66 | 20/400;20/40 | Mixed drusen, geographic atrophy, SRNVM OU * | SRNVM Drusen |
| sister | F | 74 | 69 | 20/70;20/800 | Moderate hard drusen OU, SRNVM OU | |
| sister | F | 80 | 72 | CF@5';20/400 | Few large, hard drusen, atrophy, disciform lesions OU | |
| 13 proband | M | 61 | 61 | 20/20; 20/20 | Extensive hard drusen OU | Drusen |
| p. aunt | F | 79 | 74 | CF@5';20/60 | Numerous mixed drusen OU | |
| 14 proband | F | 75 | 73 | 10/80; 20/16 | Few hard drusen OU; rare pigment figures OS; SRNVM OD | SRNVM |
| sister | F | 68 | 67 | 20/25;20/30 | No drusen, SRNVM OD * | |
| aunt | M | 87 | 86 | HM;20/80 | PED, disciform OD, RPE changes OU, RDPE OS * | |
| 15 proband | F | 73 | 68 | 20/400;20/32 | Large drusen OS, SRNVM OD | SRNVM |
| sister | F | 85 | 78 | 10/160;10/160 | Geographic atrophy OU | |
| brother | M | 71 | 68 | HM;20/25-2 | Large disciform OD, RPE atrophy OS, fine moderate drusen OU * | |
| brother | M | 78 | 70 | 20/200;CF@2' | Atrophic scar OD, disciform scar OS * | |
| 16 proband | F | 63 | NA | 20/20; 20/20 | Numerous hard macular drusen OU | Hard drusen |
| sister | F | 65 | 63 | 20/25; 20/20 | Numerous hard macular drusen, RDPE OU | |
| 17 proband | M | 71 | 65 | 20/640;20/30 | SRNVM OD, no drusen OU | No drusen |
| mother | F | 95 | 80 | 20/200; HM | Geographic atrophy, no drusen OU | |
| 18 proband | M | 74 | 64 | 10/50;10/200 | Moderate hard drusen, SRNVM, extensive RDPE OU | Hard drusen SRNVM ,RDPE |
| brother | M | 73 | 65 | 20/80; 20/50 | Moderate hard drusen, geographic atrophy, RDPE,extensive MED OU | |
| 19 proband | F | 57 | 55 | 20/30; 20/16 | Numerous mixed drusen, SRNVM OD, extensive MED OU | Mixed drusen SRNVM RDPE/MED |
| mother | F | 84 | 72 | 20/200;20/40 | Numerous mixed drusen, atrophy, RDPE OU; probable SRNVM OD | |
| m. aunt | F | 82 | 70 | 20/60; 20/30 | Numerous mixed drusen, RDPE OU | |
| 20 proband | F | 78 | NA | 20/200;20/35 | Moderate mixed drusen, exudative, MED, RDPE OU* | |
| brother | M | 80 | NA | NA | Macular degeneration by history, not yet examined | |
| p. cousin | F | 90 | 80 | deceased | " " " " | |
| p. cousin | F | NA | NA | NA | " " " " | |
| 21 proband | M | 70 | NA | 20/50; 20/40 | Soft confluent drusen, MED OU; no SRNVM, RDPE | No RDPE MED, Drusen |
| mother | F | 91 | 75 | 20/200;20/25 | Hard drusen, geographic atrophy, prob MED, no RDPE OU | |
| 22 proband | F | 98 | 86 | 50/200 OU | Hard drusen, atrophy, pigment disruption OU* | Pigment disruption |
| daughter | F | 75 | 73 | 20/25; 20/20 | Hard drusen, MED, pigment disruption OU | Hard drusen |

Abbreviations

# = Family number    m. = maternal    p.= paternal    NA = not available    prsths = prosthesis    SRNVM = subretinal neovascular membrane    MED = multiple extramacular drusen    RDPE = reticular degeneration of the pigment epithelium    PED = pigment epithelial detachment    * = information from records

Mode of Inheritance: We assumed that there is a major susceptibility locus that is inherited in an autosomal dominant manner. ARM is a complex disease that may be the result of several genes interacting with each other and the environment. The observed distribution of affecteds in the families already recruited and previous family studies[37-39] supports the hypothesis that there is a major autosomal dominant susceptibility locus for ARM. The 15 families initially identified at The Eye & Ear Institute of Pittsburgh with at least two members with confirmed diagnosis of ARM were used to establish the family structures for our initial simulation. In 5 of these families, there are affecteds in two generations.

Disease Gene Frequency in ARM: We assumed a gene frequency of 0.01. It is difficult to calculate an accurate gene frequency directly from the prevalence data available, due to the confounding factors of phenocopies and reduced penetrance. It is therefore necessary to make assumptions concerning gene frequency and penetrance and use this as a means of calculating phenocopy rates. The choice of 0.01 is based upon the use of similar gene frequencies in previous studies of other complex dominant diseases such as human breast cancer and FAD. Williams and Anderson[56] used a gene frequency of 0.00756, which approximates 0.01, in their investigation of human breast cancer. Likewise, 0.01 has been used as the gene frequency of FAD by Pericak-Vance [57, 58].

Penetrance Rate: We assumed rates of 30% minimum at age 45 and 70% maximum at age 75. Analysis of our families suggests that the penetrance of ARM is age-dependent and relatively low. We have not performed formal segregation analysis because our recruitment introduces an ascertainment bias, the number of families is relatively low, and our evaluations of unaffected family members is incomplete. However, the proportion of affected offspring in a family with a normal and an affected parent is less than would be expected from full penetrance. An age-dependent penetrance was used to accommodate the established correlation between age and disease prevalence. Minimum penetrance was set at 30% at age 45 years. Maximum penetrance was set at 70% at age 75.

Phenocopy Rate: The phenocopy curve can be calculated based upon the observed age-specific prevalence rates and the assumed gene frequency, and assumed penetrance rate. At q=0.01 with a minimum penetrance of 30%, the phenocopy rate is 2% for those age 45. The phenocopy rate rises to a maximum of 27% as the age reaches 75, when the penetrance is at its maximum value. This is consistent with a model which assumes that the phenocopy rate increases with age due to increased exposure to environmental factors.

Heterogeneity: We assumed 70% of the families are linked and 30% are unlinked. Heterogeneity refers to the presence of more than one gene influencing the expression of ARM. We assume that 70% of the families in our study are segregating with the linked major susceptibility gene while 30% are unlinked (i.e., they have ARM due to causes other than the linked gene). Due to the complex nature of ARM we expect that more than one locus may contribute to ARM susceptibility and this is reflected in our assumption that heterogeneity will be observed. A lower amount of heterogeneity (<30%) would be expected to improve the ability of the linkage analyses to detect a major susceptibility locus.

Family Structure and Disease Status. We based our computer simulations on the structures of the recruited families. Each of the family structures was duplicated an appropriate number of times to produce the desired sample size. The simulations were carried out conditioned upon the observed disease status of each individual; thus, each simulated pedigree has not only the same structure but also the same disease status as the prototype pedigrees. We assumed that these families are representative of the families that will be recruited throughout the rest of the study.

Preliminary Simulations. The simulations were performed using the computer program SIMLINK[66,67] to generate 1000 sets of the family data. These data were then analyzed by the Affected-Pedigree-Member programs. The simulations were carried out on a SPARC station 2 Unix workstation. Initial computer simulations were run under the assumptions outlined above with markers having four codominant, equally frequent alleles. Single and flanking markers were evaluated for sample sizes of 75 and 150 families. The simulations shown in Table 2, used flanking markers with 4,6, or 8 alleles. The alleles either had the same frequency in the population (equifrequent) or there was a common allele with a population frequency of 0.5 and the rest of the alleles having equal frequencies.

## Simulation Results

The initial simulations evaluated the percent of family sets for which linkage would be correctly inferred by single marker analyses as compared to the flanking marker analyses for the same distance of the disease from the markers. For all conditions tested, the detection of linkage was better using two flanking markers rather than a single marker (data not shown). The simulations shown in Table 2 illustrate the percent of family sets for which linkage would be correctly inferred at a given significance value (p=0.05, 0.01, 0.001) and using flanking markers that vary with respect to their spacing and alleles (4,6,8 alleles, an equifrequent model versus a common allele model). When the markers are closely spaced the ability to detect linkage dramatically increases. Thus for the same total number of geno-typings (i.e., 25,000), it is better to genotype 75 families with markers 20 cM apart rather than genotype 150 families at 40 cM distances. However as one attempts to achieve greater resolution, it is better to genotype 75 more families at 20 cM marker spacing rather than further genotype the original 75 families. In this model, the ability to detect linkage is dramatically affected by the informativeness of the markers. This is most apparent when one

**Table 2.** Percent of family sets for which linkage was correctly inferred at a given significance level using flanking markers.*

| p value | centiMorgans | 5 | 10 | 20 | 40 |
|---|---|---|---|---|---|
| 0.05 | | 95/94 | 90/87 | 78/76 | 50/45 |
| | | 99/98 | 97/96 | 89/86 | 62/58 |
| | 75 | 100/100 | 99/99 | 94/91 | 68/64 |
| | Families | | | | |
| 0.01 | | 82/77 | 72/67 | 52/48 | 26/21 |
| | | 94/90 | 87/82 | 70/66 | 37/34 |
| | | 98/96 | 94/92 | 81/76 | 44/39 |
| 0.001 | | 53/51 | 42/38 | 25/23 | 8/7 |
| | | 75/70 | 60/55 | 39/38 | 13/13 |
| | | 87/80 | 78/72 | 52/46 | 18/15 |

| p value | 5 | 10 | 20 | 40 | centiMorgans |
|---|---|---|---|---|---|
| 0.05 | 100/100 | 100/100 | 95/96 | 73/70 | |
| | 100/100 | 100/100 | 100/99 | 88/83 | |
| | 100/100 | 100/100 | 100/100 | 93/89 | |
| | | | | | 150 |
| 0.01 | 100/98 | 98/95 | 85/83 | 46/44 | Families |
| | 100/100 | 100/99 | 97/94 | 65/62 | |
| | 100/100 | 100/100 | 99/98 | 77/71 | |
| 0.001 | 91/91 | 85/81 | 59/57 | 21/19 | |
| | 99/99 | 97/95 | 81/78 | 37/32 | |
| | 100/100 | 100/99 | 91/88 | 50/44 | |

| number of total | 100,000 | 50,000 | 25,000 | |
|---|---|---|---|---|
| genotypings | Assuming 2.2 persons/family and a 3000 cM genome | | | |

* The upper half of the table represent the simulations involving 75 families and the lower half are those for 150 families. The centiMorgan distances indicate the distances between the flanking markers with the ARM locus placed in the center of the interval. Each pair of numbers (a/b) represent the percentage of trials that linkage would be detected at a given p value using markers with equal allele frequencies (a) and those with an unequal distribution of allele frequencies (b). Each set of three pairs (a/b c/d e/f) represent markers having 4 alleles (a/b), 6 alleles (c/d), and 8 alleles (e/f). All of the simulations in a given column would require the same number of total genotypings. The boxed pairs represent the simulation trials that reflect a progressive strategy for genotyping (see text).

compares the 76% detection rate with the least informative markers used for 20 cM spacings and 75 families compared to the 94% detection rate that would be possible with markers that had eight, equifrequent alleles. The differences in detection rates due to the informativeness of the markers persists but becomes much less impressive as the number of genotypings are increased and the spacing between markers is narrowed. In all cases, equifrequent alleles were slightly better for detecting linkage than when a single allele was common in the population. However this difference was small compared to the effect of having different numbers of alleles for the markers.

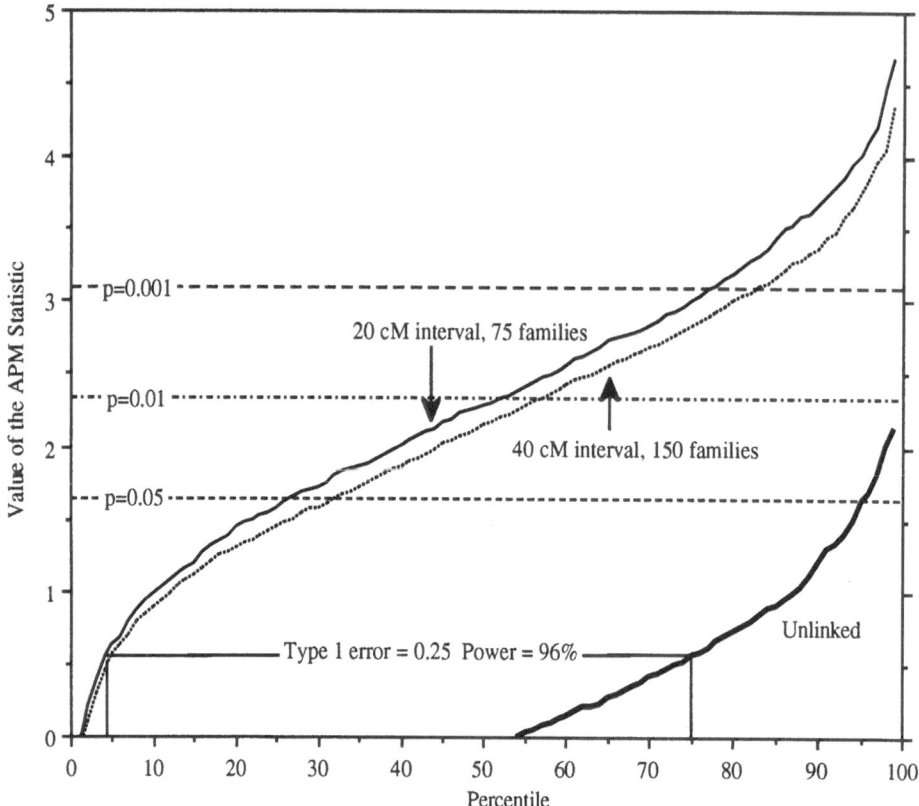

**Figure 1.** Distribution of Affected-Pedigree-Member, APM, statistic for linked and unlinked intervals. Simulations involved two flanking markers, each with four equifrequent alleles. For each set of conditions (markers linked to the disease locus, 75 families, 20 cM spacing; markers linked to the disease locus, 150 families, 40 cM spacing; and unlinked flanking markers), 1000 sets of family data were simulated. The APM statistic is calculated for each family set and the distribution of the 1000 APM values for each condition are shown. Using an APM value of 0.6 as a cutoff value, 96% of the time the linked markers would have an APM value greater than 0.6, while 25% of unlinked cases would also have a value greater than 0.6. At $p=0.05$, 5% of the unlinked marker pairs would have an APM statistic greater than the cutoff value of 1.7, while marker pairs flanking the disease locus would have APM values >1.7, 78% of the time.

The distribution of the Affected-Pedigree-Member statistic for flanking marker analysis with a 40 cM interval for 150 families and a 20 cM interval for 75 families is shown in Figure 1. These correspond to the simulations for the four, equifrequent allele conditions in table 2. The results of these simulations (Table 2 and Figure 1, above) suggest a strategy for focused genotyping after searching the genome using the first 75 families. This approach requires theoretical validation before implementation but illustrates how one can select regions for additional analysis with the expectation of identifying linkage at minimal cost and effort. These curves show that with a false positive rate of 0.25, one would select a subset of 20 cM intervals such that there is a 0.96 probability that a linked interval is within that set. If one then genotypes the selected 25% of the genome in an additional 75 families, the ability to identify the linked interval and discard the false positives will increase to a greater extent than if one simply types the original families with more markers. By genotyping the entire subset with an additional set of markers spaced 10 cM apart, the discrimination is further improved. Using this focused strategy one accepts an initial 4% chance of missing a linked interval but accomplishes the equivalent of 100,000 genotypings (the entire genome of 150 families with spacings at 10 cM) with only 43,750 genotypings (25% of the genome typed in 150 families at 10 cM + 75% of the genome typed in 75 families at 20 cM). The increase in the false-positive errors caused by selective genotyping approach is partially offset by the reduced number of tests that are performed. One can modify the selection process for any desired probability of detection and adjust the total number of genotypings accordingly.

## Conclusions

There is strong clinical evidence that many individuals with ARM demonstrate positive family histories and that clinical studies support a dominant inheritance in many, if not most, familial cases. The localization of ARM-susceptibility genes can be achieved, despite the late-onset and variable expressivity of this disease using the Affected-Pedigree-Method. One can assess the power of Affected-Pedigree-Method as well as design an appropriate research strategy, based upon computer simulations that incorporate realistic genetic models. These computer simulations which are based upon actual ARM pedigrees, incorporate known age-dependence and disease frequencies for ARM, and use allele frequencies comparable to those of available microsatellite-based markers. These simulations emphasize the importance of selecting maximally informative markers for Affected-Pedigree-Method linkage studies and provide a rational strategy of genotyping that will conserve resources during the search for a major susceptibility locus for ARM.

## REFERENCES

1. A. Jampolsky, Senile accommodative degeneration (SAD), *Am J Ophthalmol* 111:510 (1991).
2. W. Aclimandos and N. Galloway, Blindness in the city of Nottingham (1980-1985), *Eye* 2:431 (1988).
3. C. Banks and W. Hutton, Blindness in New South Wales, *Aust J Ophthalmol* 9:285 (1981).
4. I. Ghafour, D. Allan, and W. Foulds, Common causes of blindness and visual handicap in the west of Scotland, *Br J Ophthalmol* 67:209 (1983).
5. L. Hyman, Epidemiology of eye diseases in the elderly, *Eye* 1:330 (1987).
6. M. Yap and J. Weatherill, Causes of blindness and partial sight in the Bradford Metropolitan District from 1980 to 1985, *Ophthalmol Physiol Opt* 9:289 (1989).
7. S. Duke-Elder, ed. Diseases of the uveal tract. *in* "System of Ophthalmology", vol IX Klimpton, London p 613 (1966).
8. A. Kornzweig, Changes in the choriocapillaris associated with senile macular degeneration, *Ann Ophthalmol* 9:753 (1977).
9. N. Bressler, S. Bressler, and S. Fine, Age-related macular degeneration, *Surv Ophthalmol* 32(6):375 (1988).
10. S.H. Sarks, Aging and degeneration in the macular region: a clinicopathological study, *Br J Ophthalmol* 60:324 (1976).
11. S.H. Sarks, D. Van Driel, L. Maxwell,and M. Killingsworth, Softening of drusen and subretinal neovascularization, *Trans Ophthalmol Soc UK* 100:414 (1980).
12. S.H. Sarks, Drusen and their relationship to senile macular degeneration, *Aust J Ophthalmol* 8(2):117 (1980).
13. S.H. Sarks, Drusen in patients predisposing to geographic atrophy of the retinal pigment epithelium, *Aust J Ophthalmol* 10(2):91 (1982).
14. J. Sarks, S. Sarks, and M. Killingsworth, Evolution of geographic atrophy of the retinal pigment epithelium, *Eye* 2:552 (1988).

15. I. Chisholm, The recurrence of neovascularization and late visual failure in senile disciform lesions, *Trans Ophthalmol Soc UK* 103:354 (1983).

16. J. Goldberg, G. Flowerdew, E. Smith, J. Brody,and M. Tso, Factors associated with age-related macular degeneration, *Am J Epidemiol* 128(4):700 (1988).

17. L. Hyman, A. Lilienfeld, F. Ferris, and S. Fine, Senile macular degeneration: a case-control study, *Am J Epidemiol* 118(2):213 (1983).

18. H. Leibowitz, D. Krueger, L. Maunder, R. Milton, M. Kini, et al., The Framingham Eye Study monograph, *Surv Ophthalmol* 24(suppl):428 (1980).

19. T. Vinding, Visual impairment of age-related macular degeneration, *Acta Ophthalmol* 68:162 (1990).

20. L. Wu, Study of aging macular degeneration in China, *Jpn J Ophthalmol* 31:349 (1987).

21. D. Newsome, M. Swartz, N. Leon, R. Elston, and E. Miller, Oral zinc in macular degeneration, *Arch Ophthalmol* 106:192 (1988).

22. B. Silverstone, L. Landau, D. Berson, and J. Sternbauch, Zinc and copper metabolism in patients with senile macular degeneration, *Ann Ophthalmol* 17:419 (1985).

23. M.A. Mainster, Light and macular degeneration: a biophysical and clinical perspective, *Eye* 1:304 (1987).

24. J. Marshall, Radiation and the aging eye, *Ophthal Physiol Opt* 5(3):241 (1985).

25. R. Young, Pathophysiology of age-related macular degeneration, *Surv Ophthalmol* 31(5):291 (1987).

26. L. Hakkinen, Vision in the elderly and its use in the social environment, *Scan J Soc Med Suppl* 35:5 (1984).

27. H. Kahn, H. Leibowitz, J. Ganley, M. Kini, T. Colton, et al., The Framingham Eye Study. I. Outline and major prevalence findings, *Am J Epidemiol* 106:17 (1977).

28. B. Klein and R. Klein, Cataracts and macular degeneration in older Americans, *Arch Ophthalmol* 100:571 (1982).

29. G. Martinez and A. Campbell, Prevalence of ocular disease in a population study of subjects 65 years old and older, *Am J Ophthalmol* 94(2):181 (1982).

30. R. Sperduto and D. Seigel, Senile lens and senile macular changes in a population-based sample, *Am J Ophthalmol* 90:86 (1980).

31. T. Vinding, Age-related macular degeneration. Macular changes, prevalence, and sex ratio, *Acta Ophthalmol* 67:609 (1989).

32. H. Kahn, H. Leibowitz, J. Ganley, M. Kini, T. Colton, et al., The Framingham Eye Study. II. Association of ophthalmic pathology with single variables previously measured in the Framingham Heart Study, *Am J Epidemiol* 106:33 (1977).

33. W. Delaney and R. Oates, Senile macular degeneration: a preliminary study, *Ann Ophthalmol* 14:21(1982).

34. B. Maltzmann, M. Mulvihill, and A. Greenbaum, Senile macular degeneration and risk factors: a case control study, *Ann Ophthalmol* 11:1197 (1979).

35. Paetkau, T. Boyd, M. Grace, J. Bach-Mills, and B. Wenship, Senile disciform macular degeneration and smoking, *Canad J Ophthalmol* 13:67 (1978).

36. J. Weiter, C. Dorey, F. Delori, and K. Fitch, Relationship of senile macular degeneration to refractive error, *Invest Ophthalmol Vis Sci* 28((suppl)):119 (1987).

37. J. Francois, L'heredite des degenerescences maculaires seniles, *Ophthalmologica* 175:67 (1977).

38. J.D.M. Gass, Drusen and disciform macular detachment and degeneration, *Arch Ophthalmol* 90:206 (1973).

39. P. Heinrich, Senile degeneration of the macula, *Klin Monatsbl Augenheilkd* 162:3 (1973).

40. M.A. Melrose, L.E. Magargal, and A.C. Lucier, Identical twins with subretinal neovascularization complicating senile macular degeneration, *Ophth Surg* 16(10):648 (1985).

41. S.M. Meyers and A.A. Zachary, Monozygotic twins with age-related macular degeneration, *Arch Ophthalmol* 106:651 (1988).

42. N.M. Bressler, S.B. Bressler, S.K. West, S.L. Fine, and H.R. Taylor, The grading and prevalence of macular degeneration in Chesapeake Bay watermen, *Arch Ophthalmol* 107:847 (1989).

43. D. Pauleikhoff, M.J. Barondes, D. Minassian, I.H. Chisholm, and A.C. Bird, Drusen as risk factors in age-related macular disease, *Am J Ophthalmol* 109:38 (1990).

44. H. Lewis, B.R. Straatsman, and R.Y. Foos, Chorioretinal juncture. Multiple Extramacular drusen, *Ophthalmol* 93(8):1098 (1986).

45. H. Lewis, B.R. Straatsma, and R.Y. Foos, D. Lightfoot, Reticular degeneration of the pigment epithelium, *Ophthalmology* 92:1485 (1985).

46. D. Weeks and K. Lange, The affected-pedigree-member method of linkage analysis, *Am J Hum Genet* 42:315 (1988).

47. N. Day and M. Simons, Disease susceptibility genes - their identification by multiple case family studies, *Tissue Antigens* 8:109 (1976).

48. J. Green and J. Woodrow, Sibling method for detecting HLA-linked genes in a disease, *Tissue Antigens* 9:31 (1977).

49. J. Haseman and R. Elston, The investigation of linkage between a quantitative trait and a marker locus, *Behav Genet* 2(1):3 (1972).

50. L. Penrose, The detection of autosomal linkage in data consisting of pairs of brothers and sisters of unspecified parentage, *Ann Eugen* 6:133 (1935).

51. B. Suarez, J. Rice, and T. Reich, The generalized sib pair IBD distribution: its use in the detection of linkage, *Ann Hum Genet* 42:87 (1978).

52. B. Suarez and P. Van Eerdewegh, A comparison of three affected-sib-pair scoring methods to detect HLA-linked disease susceptibility genes, *Am J Med Genet* 18:135 (1984).

53. G. Thomson, Determining the mode of inheritance of RFLP-associated diseases using the affected sib-pair method, *Am J Hum Genet* 39:207 (1986).

54. E. Claus, N. Risch, and W. Thompson, Genetic analysis of breast cancer in the cancer and steroid hormone study, *Am J Hum Genet* 48:232 (1991).

55. J. Hall, M. Lee, B. Newman, J. Morrow, L. Anderson, et al., Linkage of early-onset familial breast cancer to chromosome 17q21, *Science* 250:1684 (1990).

56. W. Williams and D. Anderson, Genetic epidemiology of breast cancer: segregation analysis of 200 Danish pedigrees, *Genet Epidemiol* 1:7 (1984).

57. M. Pericak-Vance, L. Yamaoka, C. Haynes, M. Speer, J. Haines, et al., Genetic Linkage studies in Alzheimer's Disease families, *Exp Neurol* 102:271 (1988).

58. M. Pericak-Vance, J. Bebout, P. Gaskell, L. Yamaoka, W.-Y. Hung, et al., Linkage studies in familial Alzheimer Disease: evidence for chromosome 19 linkage, *Am J Hum Genet* 48:1034 (1991).

59. C. Joachim and J. Morris, Clinically diagnosed Alzheimer's Disease-autopsy results in 150 cases, *Ann Neurol* 24:50 (1988).

60. H.J. Jacob, K. Lindpaintner, S.E. Lincoln, K. Kusumi, R.K. Bunker, et al., Genetic mapping of a gene causing hypertension in the stroke-prone spontaneously hypertensive rat, *Cell* 67:213 (1991).

61. X. Jeunemaitre, F. Soubrier, Y.V. Kotelevtsev, R.P. Lifton, C.S. Williams, et al., Molecular basis of human hypertension: Role of angiotensinogen, *Cell* 71:169-180 (1992).

62. NIH/CEPH Collaborative Mapping Group, A comprehensive genetic linkage map of the human genome, *Science* 258:67 (1992).

63. J. Weissenbach, G. Gyapay, C. Dib, A. Vignal, J. Morissette, et al., A second-generation linkage map of the human genome, *Nature* 359:794 (1992).

64. A. Edwards, H. Civitello, H. Hammond, and C. Caskey, DNA typing and genetic mapping with trimeric and tetrameric tandem repeats, *Am J Hum Genet* 49:746 (1991).

65. R. Stallings, A. Ford, D. Nelson, D. Torney, C. Hildebrand, et al., Evolution and distribution of (GT)n repetitive sequences in mammalian genomes, *Genomics* 10:807 (1991).

66. M. Boehnke, Estimating the power of a proposed linkage study: A practical computer simulation approach, *Am J Hum Genet* 39:513 (1986).

67. L. Ploughman and M. Boehnke, Estimating the power of a proposed linkage study for a complex genetic trait, *Am J Hum Genet* 44:543 (1989).

# LINKAGE ANALYSIS IN MALATTIA LEVENTINESE, AN AUTOSOMAL DOMINANT FORM OF MACULAR DEGENERATION

Manfred Stuhrmann,[1] Astrid Spangenberg,[1]
Albert Schinzel,[2] and Jörg Schmidtke[1]

[1]Institut für Humangenetik
 Medizinische Hochschule Hannover
 3000 Hannover 61, Germany
[2]Institut für Medizinische Genetik
 Universität Zürich
 8001 Zürich, Switzerland

## INTRODUCTION

Malattia leventinese, an autosomal dominant form of macular degeneration, originates from the Leventine valley in the North of the Ticino Canton of Switzerland and was first published by Vogt and coworkers in 1925 (16). Normally between age 20 and age 30, a few round, brown-yellow, later whitish structures appear in the deeper retinal layers of the posterior pole in both eyes. Later, these spots show confluence and the retina in front of them becomes atrophic. The histological examination discloses round accumulations of eosinophilic hyaline bodies in the pigment epithelium. These structures are connected with the inner layer of Bruch's membrane and are called "drusen". Drusen can be divided into degenerative and hereditary drusen. Hereditary (dominant) drusen occur in malattia leventinese, Doyne's honeycomb dystrophy, Hutchinson-Tay chorioiditis and Holthouse-Batten chorioretinitis. Therefore, Deutman and Jansen (3) proposed the designation "dominant drusen" for all these diseases. The molecular abnormality which underlies the formation of dominant drusen is not known. Since elucidation of the molecular basis of malattia leventinese may also shed light on the pathogenesis of other forms of macular degeneration (e.g. age-related macular degeneration, which - in developed countries - is the most common cause of blindness in older

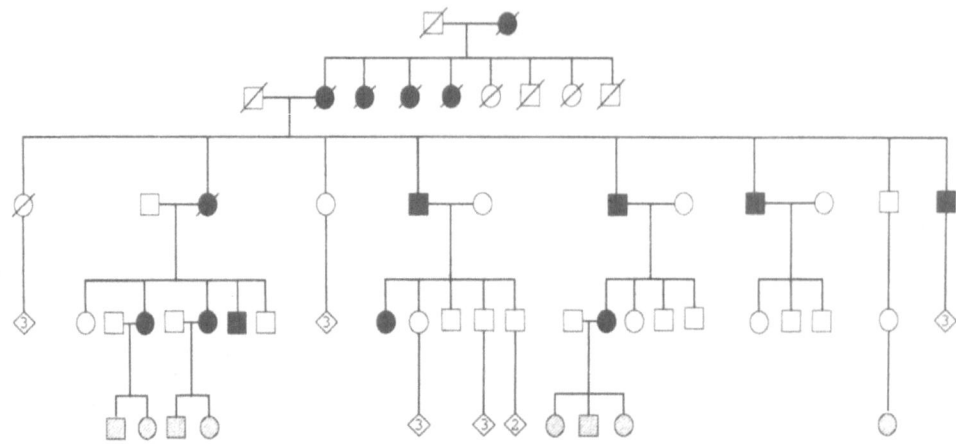

**Figure 1.** Malattia leventinese pedigree. Males are represented by squares, and females by circles. Affected individuals are represented by black symbols. Adult persons without symptoms and a normal retinal appearance are depicted as open symbols. Undiagnosed individuals are represented by hatched symbols.

individuals), we have recently initiated genetic linkage studies in a large affected family from the Ticino (Figure 1).

## LINKAGE ANALYSIS

Linkage analysis has the potential to identify the primary molecular defect of a Mendelian disorder without knowledge of its pathogenesis. This method has been applied successfully to the chromosomal mapping of several eye diseases, including three different forms of retinitis pigmentosa (1, 6, 7, 13), atypical vitelliform macular dystrophy (10) and very recently of vitelliform macular degeneration (Best's disease, ref.14). In some of the above disorders, chromosomal mapping was followed by the identification of disease causing gene mutations (4, 5, 8, 9, 11, 12).

Our current strategy is to study genetic loci that have been shown to be associated with retinal or macular diseases in previous studies. Special emphasis was put on the study of locus D11S527 because of its linkage to Best's disease, the only form of macular degeneration which has been mapped with highly polymorphic markers ("microsatellites") so far (14). Moreover, we are starting to use highly polymorphic markers (mainly dinucleotide repeats) spread over the entire autosomal genome. Table 1 gives an overview of the markers studied to date.

**Table 1.** No evidence for linkage of malattia leventinese to:

| Locus | chromosome | method | reference | remarks |
|---|---|---|---|---|
| RHO | 3q21-q24 | PCR,[1] dinuc.[2] | 17 | candidate gene |
| RDS | 6p21.2-cen | PCR, dinuc. | GDB[3] | candidate gene |
| TCRD | 14q11.2 | PCR, dinuc. | GDB | candidate region[5] |
| CA2 | 8q13-q22 | PCR, RFLP[4] | GDB | candidate gene |
| PENK | 8q23-q24 | PCR, dinuc. | 18 | candidate region[6] |
| D11S527 | 11q13 | PCR, dinuc. | 2 | candidate region[7] |

[1]polymerase chain reaction; [2]dinucleotide repeat; [3]genome data base; [4]restriction fragment length polymorphism; [5]close to the retinal specific gene NRS-1 (ref.15); [6]close to GPT (ref.10); [7]linkage to Best's disease (ref.14)

## RESULTS

At least two or more recombination events have been observed between malattia leventinese and each of the loci listed in table 1. It is remarkable that our two point analysis excludes linkage between D11S527 and malattia leventinese with a maximum lod (likelihood of the odds) score of -5.048 at a recombination fraction of 0.05. At a distance of 10 centimorgans (cM), the lod score is -3.105, whereas Stone et al. (14) confirmed linkage between D11S527 and Best's disease with a maximum lod score of 5.09 at a distance of 11.9 cM. We therefore assume that the clinical heterogeneity of macular degeneration is due to at least two genes at different chromosomal locations.

## Acknowledgments

We thank the family members for their participation in this research program, Johanna König for excellent technical assistance, and Michael Krawczak for expert advice on linkage analysis.

## REFERENCES

1. S.H.Blanton, J.R.Heckenlively, A.W.Cottingham, et al., Linkage mapping of autosomal dominant retinitis pigmentosa (RP1) to the pericentric region of human chromosome 8, *Genomics* 11;857-869 (1991)
2. D.L.Browne, J.Gault, M.B.Thompson, et al., Dinucleotide repeat polymorphism at the D11S527 locus, *NAR* 19;4790 (1991)
3. A.F.Deutman and L.M.A.A.Jansen, Dominantly inherited drusen of Bruch's membrane, *Brit.J.Ophthal.* 54;373-382 (1970)

4. T.P.Dryja, T.L.McGee, E.Reichel, et al., A point mutation of the rhodopsin gene in one form of retinitis pigmentosa, *Nature* 343;364-366 (1990)

5. T.P.Dryja, T.L.McGee, L.B.Hahn, et al. Mutations within the rhodopsin gene in patients with autosomal dominant retinitis pigmentosa, *New Engl. J. Med.* 323;1302-1307 (1991)

6. G.J.Farrar, P.McWilliam, D.G.Bradley, et al., Autosomal dominant retinitis pigmentosa: Linkage to rhodopsin and evidence for genetic heterogeneity, *Genomics* 8;35-40 (1990)

7. G.J.Farrar, S.A.Jordan, P.Kenna, et al., Autosomal dominant retinitis pigmentosa: Localisation of a disease gene (RP6) to the short arm of chromosome 6, *Genomics* 11;870-874 (1991a)

8. G.J.Farrar, P.Kenna, R.Redmond, et al., Autosomal dominant retinitis pigmentosa: A mutation in codon 178 of the rhodopsin gene in 2 ADRP families of Celtic origin, *Genomics* 11;1170-1171 (1991b)

9. G.J.Farrar, P.Kenna, S.A.Jordan, et al., A three-base-pair deletion in the peripherin-RDS gene in one form of retinitis pigmentosa, *Nature* 354;478-480 (1991c)

10. R.E.Ferrel, H.M.Hittner and J.H.Antoszyk, Linkage of atypical vitelliform macular dystrophy (VMD-1) to the soluble glutamate pyruvate transaminase (GPT1) locus, *Am. J. Hum. Genet.* 35;78-84 (1983)

11. C.F.Inglehearn, R.Bashir, D.H.Leister, et al., A 3bp deletion in the rhodopsin gene in a family with autosomal dominant retinitis pigmentosa, *Am. J. Hum. Genet.* 48;26-30 (1991)

12. K.Kajiwara, L.B.Hahn, S.Mukai, et al., Mutations in the human retinal degeneration slow gene in autosomal dominant retinitis pigmentosa, *Nature* 354;480-483 (1991)

13. P.McWilliam, G.J.Farrar, P.Kenna, et al., Autosomal dominant retinitis pigmentosa (ADRP): Localisation of an ADRP gene to the long arm of chromosome 3, *Genomics* 5;619-622 (1989)

14. E.M.Stone, B.F.Nichols, L.M.Streb, et al., Genetic linkage of vitelliform macular degeneration (Best's disease) to chromosome 11q13, *Nature genetics* 1;246-250 (1992)

15. A.Swaroop, J.Xu, N.Agarwal, et al., A novel human neural retina specific gene encodes a basic/leucine zipper motif homologous to the avian v-maf oncogene product, *Am.J.Hum.Genet.* suppl.49:433 (1991)

16. A.Vogt, Die Ophthalmoskopie im rotfreien Licht, in: Graefe-Saemisch Handbuch d. ges. Augenheilkunde. 3.ed. J.Springer, Berlin, vol 3;1-118 (1925)

17. J.L.Weber and P.E.May, Abundant class of human DNA polymorphisms which can be typed using the polymerase chain reaction, *Am.J.Hum. Genet.* 44;388-396 (1989)

18. J.L.Weber and P.E.May, Dinucleotide repeat polymorphism at the PENK locus. *NAR* 18;2200 (1990)

# MUTATIONS IN THE HUMAN RETINAL DEGENERATION SLOW (RDS) GENE CAN CAUSE EITHER RETINITIS PIGMENTOSA OR MACULAR DYSTROPHY

John Wells,[1] John Wroblewski,[1] Jeffrey Keen,[2] Christopher Inglehearn,[2] Christopher Jubb,[3] Anja Eckstein,[4] Marcelle Jay,[1] Geoffrey Arden,[4] Shomi Bhattacharya,[2] Fred Fitzke,[3] Alan Bird[1]

[1]:Department of Clinical Ophthalmology,[2]:Department of Molecular Biology; [3]:and Department of Visual Science, Institute of Ophthalmology, London; [4]: The Electrodiagnostic Department, Moorfields Eye Hospital

## SUMMARY

Mutations in the *RDS* gene have been sought in families with autosomal dominant retinal dystrophies. A cysteine deletion at codon 118/119 is associated with retinitis pigmentosa in one. Three families with similar macular dystrophy have mutations at codon 172, arginine being substituted by tryptophan in two and by glutamine in one. A stop sequence at codon 258 exists in a family with adult vitelliform macular dystrophy. These findings demonstrate that both retinitis pigmentosa and macular dystrophies are caused by mutations in the *RDS* gene, and that the functional significance of certain amino-acids in peripherin-rds may be different in cones and rods.

## INTRODUCTION

Retinitis pigmentosa(RP) denotes a group of inherited disorders which causes progressive loss of night vision and peripheral visual field with preservation of central visual function until late disease, and is characterized by migration of pigment into the neuroretina, attenuation of the retinal blood vessels and optic disc pallor. The diagnosis is confirmed by an abnormal or extinguished electroretinogram (ERG). By contrast, macular dystrophies result in loss of central vision with preservation of peripheral visual field, and changes in the ocular fundus are limited to the central fundus. The pattern of disease in macular dystrophies could be explained by selective affection of cones, possibly because a protein specific to them was abnormal. The peripheral loss of vision in RP, and night blindness might be due to primary rod disease, although it is known that in many cases cones are affected fairly early in the course of disorder, and in late disease both classes photoreceptors cease to function.

In 1989 a locus for autosomal dominant RP (ADRP) was identified on the long arm of chromosome 3[1]. Within a short time a proline to histidine mutation was detected at codon-23 of the rhodopsin gene, which maps to chromosome 3q[2]. Since then more than 30 mutations have been detected in the rhodopsin gene, and mutations in this gene account for about 25% of ADRP[3-6].

In several families with adRP, there is no linkage with markers in the long arm of chromosome 3[7], confirming heterogeneity of disease, and it has now been shown that mutations in genes other than that for rhodopsin may be responsible[8-12]. One is the gene on chromosome 6p for a photoreceptor specific glycoprotein known as peripherin-rds[11,12]. This is of particular interest since this gene which codes for the glyco-protein peripherin-rds, is implicated in the pathogenesis of retinal degeneration in the rds (retinal degeneration slow) mouse[13,14]. Peripherin-rds is localized to the outer segment disc membranes of both rods and cones, and is thought to be essential for the assembly, orientation and physical stability of the retinal photoreceptor outer segment discs[15-17].

In order to test for *RDS* mutations in patients with retinal dystrophies we examined exon sequences from 58 unrelated subjects with ADRP, 17 with recessive RP, 25 with sporadic RP and 13 with various macular dystrophies. Four mutations on the *RDS* gene have been identified. In one family it was associated with classical adRP, whereas in four the dystrophy affected predominantly the central retina.(18)


## PATIENTS AND METHODS

### DNA Analysis

Using the heteroduplex detection method we have described previously[10] the coding regions of the *RDS* gene were screened for point mutations and small deletion/insertion events in patients with retinal degenerations. The region around the mutations in exon 1 was amplified by PCR using the following primers: forward (derived from the middle of exon 1), 5'-GCCAAGTATGCCAGATGGAAG-3'; reverse (derived from the 5' end of intron 1), 5'-ATAGCTCTGACCCCAG-GACTG-3'. Exon 2 was amplified using the following primers: forward (derived from the 3' end of intron 1), 5'-AAGCCCATCTCCAGCTGTCTG-3'; reverse (derived from the 5' end of intron 2), 5'-CTTACCCTCTACCCCCAGCTG-3'. Following amplification the PCR products were purified using a Clontech chromapsin-30 column. The same primers end-labelled with [ -$^{32}$P] dATP were used for direct sequencing, carried out using the Sequenase II kit (US Biochemicals) by a method described previously[5].

**Clinical Evaluation:** Ophthalmoscopic examination, colour fundus photography and fluorescein angiography were performed to document the retinal findings. Retinal sensitivities were measured with a modified Humphrey field analyzer[19] using the 30-2 and 30/60-2 programs under photopic and scotopic conditions. Scotopic visual field testing was performed with a dilated pupil after 40 minutes of dark adaptation using 450 nm and 650 nm Goldman size V test spots to isolate rod and cone responses respectively. Recovery of retinal sensitivity after light exposure was measured at two locations at different eccentricities from fixation until prebleach thresholds were reached. Colour contrast sensitivity was measured by a system fully described elsewhere[20]. The minimum colour contrast between letter/annulus and background at which the identification is possible is between 6% centrally and approximately 25% peripherally for the protan,deutan and tritan axis in normal subjects. The ERG was recorded using standard testing protocols [21,22] with gold foil electrodes, using a Grass PS22 stroboscope as stimulator. The pattern evoked ERG was recorded as described in other reports[23]. The screen displayed checkerboards of squares which subtended $0.5^o$ or $8.5^o$ at the eye, reversing at 3 Hz. About 750 responses were averaged.

## RESULTS

In the five families, mutations were found in all affected members tested but not in unaffected relatives. The mutations gave rise to a Cysteine deletion at codon 118 or 119, Arginine to Glutamine and Arginine to Tryptophan mutations at codon 172, and a Tyrosine to stop sequence at codon 258. In all families but one there was evidence of visual loss in three or more generations without consanguinity and male to male transmission of disease implying autosomal dominant inheritance.

A further sixty unrelated control individuals with normal vision were tested by heteroduplex analysis for the presence of these mutations in the *RDS* gene and none were found.

**Cysteine 118/119 deletion:**  All five available members of this pedigree had symptoms of RP and each had the mutation in the *RDS* gene. They described night vision loss early in the second decade of life and peripheral visual field loss a short time thereafter. Central visual acuity was not affected until late in the third decade of life. By the early twenties, the fundus showed diffuse peripheral changes but the macula appeared normal. There was no pigment accumulation in the retina until the fourth decade of life. Attenuation of the retinal vessels and pallor of the optic disc was not present until later in life.

Each affected person showed a similar pattern of functional deficit. Rod and cone sensitivities were markedly and equally elevated peripherally but were near normal within the central $10^o$. Colour contrast thresholds were normal centrally, but were elevated between $2^o$ and $6^o$ of eccentricity, and beyond $6^o$ the coloured images were not seen even at the highest possible contrast.

Apart from a minimal cone response in 3 of the 4 patients, the flash ERG's were almost extinguished for every testing condition. The pattern evoked ERG (PERG) was present but abnormal, the P50 amplitude being reduced to 25%-50% of the normal value.

**Arginine 172 Glutamine:**  Three affected members and one unaffected member of the family were tested for the presence of the mutation which segregated with the phenotype. However, only one affected subject has had detailed functional testing to date. He had experienced visual difficulty when going from light to dark in early life, and central visual acuity loss with the left eye at 37 years. He denied photophobia or peripheral visual loss. At 46 years his best corrected visual acuity was 6/6 with his right eye and 6/18 with his left. The ocular fundi showed retinal atrophy paracentrally which was more severe in the left eye than the right.

Functional testing revealed dense loss in the central $10^0$ of visual field. On the edge of loss rod and cone threshold were elevated by approximately 2 log units, and the kinetics of dark adaptation were slowed. Peripheral rod and cone function were normal. The colour contrast threshold for colour perception was elevated by ten times centrally, from $2^0$ to $10^0$ eccentricity coloured images were not seen, but beyond $22^0$ the thresholds were almost normal.

The ERG responses were normal, with the exception of the PERG which was almost extinguished.

**Arginine 172 Tryptophan:** This mutation was identified in two apparently unrelated families. One is a seven member pedigree with four affected and three unaffected individuals, while the other has six members, 4 affected and 2 normal. The mutation segregated with the disease in each family. Typically, affected patients became symptomatic in the third decade of life with blurred central vision and photophobia. By 40 years the visual acuity was less than 6/60. None complained of night blindness or restricted peripheral visual fields. The ocular fundi showed irregular pigmentation at the level of the retinal pigment epithelium by the second decade of life and sharply demarcated atrophy affecting the central retina, pigment epithelium and choriocapillaris by 40 years.

After the age of 40 years, photopic perimetry revealed an absolute central scotoma up to $20^0$ in size. Under scotopic conditions, rod and cone thresholds were elevated approximately 2 log units at the edge of the affected area. Beyond $20^0$ visual thresholds under photopic and scotopic conditions were within the normal range and recovery from bleach had normal kinetics

Central colour contrast sensitivities could be measured in only one subject within the central $12^0$, and beyond $22^0$ thresholds were universally elevated.

The cone ERG responses were depressed in all but one young subject, and the rod responses depressed to a lesser degree in about half of them.

**258 stop:** Only one affected member of the family with the 258 mutation has been examined and tested, her father was dead but the disorder was verified by fundus photography. Her two children did not have the mutation but were too young for the disease to be excluded in them. The patient became aware of distorted left eye central vision in her mid thirties, but had no other symptoms. At 44 years her visual acuity was 6/6 with the right eye and 6/12 with the left. Fundus examination was normal except a for small discrete yellow deposit at the level of the pigment epithelium centred in the fovea of each eye.

The cone thresholds in the central $5^0$ were slightly elevated. The responses in the light adapted single flash ERG was at the upper limit of normal in voltage and prolonged. This was interpreted as a failure of rod desensitisation with light such that rods contributing to this response. This was supported by the normal implicit time of 30Hz. flicker response. All other functional attributes were normal.

## DISCUSSION

That the *RDS* mutations described are the cause of the eye disease in these patients is supported by several line of evidence. Firstly the disease segregates with the mutation in each of the five pedigrees. Secondly it would be a remarkable coincidence that the same rare amino acid polymorphisms occurred in two unrelated families with the same eye disease, and another polymorphism exist at the same codon in a third pedigree with almost identical phenotype. Thirdly, the codon 258 stop mutation would predictably have an effect on transcription similar to that in seen in the mouse heterozygous or the *rds* gene. Finally the fact that these mutations were not found in any of the sixty control subjects makes it unlikely that these represent silent polymorphisms segregating in the normal population.

These results show that different mutations in the *RDS* gene give rise to retinal dystrophies which differ markedly from one to another. The most striking variation is the selective losses of central vision in some families and of peripheral vision in others. The latter pattern of disease has been described previously[11,12], but limitation of involvement to the central retina is a new finding.

Because peripherin-rds is expressed in both rods and cones, and is believed to be important to structural stability in each class of photoreceptor, it is not surprising that both cone and rod photoreceptors may be affected by mutations in the RDS gene. There has been some controversy as to the location of this protein in the outer segment. With one antibody it appeared to be evenly distributed along the discs of the rod photoreceptor outer segment, and it was concluded that this protein was important in maintaining the close parallel arrangement of the lipid bilayers[16]. With another antibody the protein appeared to be limited to the disc rim in both rods and cone implying that its function was limited to maintaining the tight curve of the lipid bilayer at this site[17]. The observations in our families imply that the structural associations of this protein may be different in rods and cones. A potential explanation may be derived from the observations that peripherin-rds may be non-covalently linked to ROM1, a protein with a structure similar to peripherin-rds, and that this link may be functionally important[24]. ROM1 is found only on rods implying that this association is peculiar to rods and that an alternative arrangement exists in cones. It follows that a mutation which affects the ROM1/peripherin-rds binding site would give rise to functional defects might appear preferentially in rods and in peripheral retina. Another local region of peripherin-rds might bind to a protein found only in cones, (although such a protein has not yet been identified) and selectively disrupt the metabolism or structure of cones. This does not obviously explain the normality of peripheral cone function unless there are differences in the functional attributes of peripherin-rds between central and peripheral cones.

The disorder associated with the stop codon is different from the others in that there is little evidence of functional loss, and the changes in the ocular fundus are apparently at the level of the retinal pigment epithelium. It is possible that this mutation produces metabolic changes similar to those seen in the mouse heterozygous for the mutant *rds* gene in which there is a 10 kB insert in the gene at codon 238[16]. In the homozygous rds-mouse a relatively high molecular weight mRNA is produced demonstrating that the whole insert is transcribed[13]. On the other hand it is likely that no protein is expressed from the abnormal gene in the photoreceptor outer segment. If this were the case only half the normal amount of protein would be available in the heterozygous state as a result of expression of the normal gene. In the heterozygous mouse, the photoreceptor outer segments develop but contain long lengths of disc membrane[25] which is compatible with there being less than the normal quantity of peripherin-rds. However the ERG is well preserved, and 50% of the photoreceptors survive after 18 months of life[25,26] which is close to the life expectancy of the mouse. The outer segments appear to be unstable, and the retinal pigment epithelium contains large and abnormal phagosomes. Such a situation may exist in our family with adult vitelliform macular degeneration if there is no expression of the mutant gene in the outer segment. As in the heterozygous rds mouse the photoreceptors would receive only half the normal quantity of peripherin-rds. If the homology is close, it would be understandable that excessive shedding of the photoreceptor outer segments over many years would cause change in the retinal pigment epithelium, but little photoreceptor dysfunction.

### Acknowledgements

This study was generously supported by the Wellcome Trust, London (Grants 18468/1.5/G), the National Retinitis Foundation Fighting Blindness, USA, and the British Retinitis Pigmentosa Society.

### REFERENCES

1. McWilliam, P. et al. Autosomal dominant retinitis pigmentosa (ADRP): Localization of an ADRP gene to the long arm of chromosome 3. *Genomics* **5**, 619-622 (1989).
2. Dryja, T.P. et al. A point mutation of the rhodopsin gene in one form of retinitis pigmentosa. *Nature* **343**, 364-366 (1990).
3. Dryja, T.P. et al. Mutations within the rhodopsin gene in patients with autosomal dominant retinitis pigmentosa. *New Eng. J. Med.* **323**, 1302-1307 (1990).
4. Sung, C.H. et al. Rhodopsin mutations in autosomal dominant retinitis pigmentosa. *Proc. Nat. Acad. Sci.* **88**, 6481-6485 (1991).
5. Keen, T.J. et al. Autosomal dominant retinitis pigmentosa: four new mutations in rhodopsin, one of them at the retinal attachment site. *Genomics*, **11**, 199-205 (1991).
6. Inglehearn, C.F. et al. A completed screen for mutations of the rhodopsin gene in a panel of patients with autosomal dominant retinitis pigmentosa. *Hum. Mol. Genet.* **1**, 41-45 (1992).
7. Lester, D.H. et al. Linkage to D3S47 (C17) in one large dominant retinitis family and exclusion in another: confirmation of genetic heterogeneity. *Amer. J. Hum. Genet.* **47**, 536-541 (1990).
8. Blanton, S.H. et al. Linkage mapping of autosomal dominant retinitis pigmentosa (RP1) to the pericentric region of human chromosome 8. *Genomics* **11**, 857-869 (1991).
9. Farrar G.J. et al. Autosomal dominant retinitis pigmentosa; localization of a disease gene (RP6) to the short arm of chromosome 6. *Genomics* **11**, 870-874 (1991).
10. Keen, T.J., Lester, D.H., Inglehearn, C.F., Curtis, A. & Bhattacharya, S.S. Rapid detection of single base mismatches as heteroduplexes on hydrolink gels. *Trends Genet.* **7**, 5-10 (1991).
11. Farrar, G.J. et al. A three-base-pair deletion in the peripherin-RDS gene in one form of retinitis pigmentosa. *Nature* **354**, 478-480 (1991).
12. Kajiwara, K. et al. Mutations in the human retinal degeneration slow gene in

autosomal dominant retinitis pigmentosa. *Nature* **354**, 480-483 (1991).

13. Travis, G.H., Brennan, M.B., Danielson, P.E., Kozak, C.A. & Sutcliffe, J.G. Identification of a photoreceptor-specific mRNA encoded be the gene responsible for retinal degeneration slow (rds). *Nature* **338**, 70-73 (1989).

14. Connell, G. et al. Photoreceptor cell peripherin is the normal product of the gene responsible for the retinal degeneration in the rds mouse. *Proc. Natl. Acad. Sci. US.* **88**, 723-726 (1991).

15. Connell, G. & Molday, R.S. Molecular cloning, primary structure and orientation of the vertebrate photoreceptor cell protein peripherin in the rod disc membrane. *Biochemistry* **29**, 4691-4698 (1990).

16. Travis, G., Sutcliffe, J.G. & Bok, D. The retinal degeneration slow (rds) gene product is a photoreceptor disc membrane associated glycoprotein. *Neuron* **6**, 61-70 (1991).

17. Arokawa, K., Molday, L.L., Molday, R.S. & Williams, D.S. Localization of peripherin/rds in the disk membranes of cone and rod photoreceptors; relationship to disc membrane morphogenesis and retinal degeneration. *J. Cell. Biol.* **116**, 659-667 (1992).

18. Wells J, Wroblewski J, Keen J, Inglehearn C, Jubb C, Eckstein A, Jay M, Arden G, Bhattacharya S, Fitzke F, Bird A. Mutations in the human retinal degeneration slow (*RDS*) gene can cause either retinitis pigmentosa or macular dystrophy. Nature Genetics - in press.

19. Jacobson, S.G. et al. Automated light- and dark-adapted perimetry for evaluating retinitis pigmentosa. *Ophthalmology* **93**, 1604-1611 (1968).

20. Arden, G.B., Gunduz, K.& Perry, S. Colour vision testing with a computer system. *Clin. Vision Sci.* **2**, 303-320 (1988).

21. Marmor, M.F., Arden, G.B., Nilson, S.E. & Zrenner, E. Standard for clinical electroretinography. *Arch. Ophthal.* **107**, 816-819 (1989).

22. Arden, G.B. et al. A modified ERG technique and the results obtained in X-Linked retinitis pigmentosa. *Br. J. Ophthalmol.* **67**, 419-430 (1983).

23. Arden G,B. & Vaegan. Electroretinogramms evoked in man by local uniform and pattern stimulation. *J. Physiol.* **341**, 85-104 (1983).

24. Bascom, R.A. et al. Cloning of the cDNA for a novel photoreceptor membrane (rom-1) identifies a disk rim protein family implicated in human retinopathies. *Neuron.* **8**, 1171-1184 (1992).

25. Hawkins, R.K., Jansen, H.G. & Sanyal, S. Development and degeneration of retina in rds mutant mice: photoreceptor abnormalities in the heterozygotes. *Exp. Eye Res.* **41**, 701-720 (1985).

26. Sanyal, S. & Hawkins, R.K. Development and degeneration of retina in rds mutant mice. Altered disc shedding pattern in the albino heterozygotes and its relation to light exposure. *Vis. Res.* **28**, 1171-1178 (1988).

## II. RETINITIS PIGMENTOSA AND ALLIED RETINAL DEGENERATIONS: MOLECULAR, CELLULAR AND CLINICAL STUDIES

Since 1989 an explosion of information on the molecular basis of a variety of inherited retinal diseases has been forthcoming.  The first five chapters deal exclusively with autosomal dominant RP and describe defects in genes which are expressed uniquely in photoreceptors.  Additional chapters describe molecular genetic studies of Usher's syndrome, the molecular biology of Norrie disease and Leber's congenital amaurosis.  Two chapters describe technical advances in identification of candidate genes for eye disorders.  The final chapter describes the histopathology of retinitis pigmentosa.

# EXTENSIVE GENETIC HETEROGENEITY IN AUTOSOMAL DOMINANT RETINITIS PIGMENTOSA

G. Jane Farrar,[1] Siobhán A. Jordan,[1] Rajendra Kumar-Singh,[1]
Chris F. Inglehearn,[2] Andreas Gal,[3] Cheryl Greggory,[2]
May Al-Maghtheh,[2] Paul F. Kenna,[1] Marian M. Humphries,[1]
Elizabeth M. Sharp,[1] Denise M. Sheils,[1] Susanna Bunge,[3]
Paul A. Hargrave,[4] Michael J. Denton,[5] Eberhard Schwinger,[3]
Shomi S. Bhattacharya,[2] and Peter Humphries[1]

[1]Dept. of Genetics, Trinity College Dublin, Dublin 2, Ireland

[2]Dept. of Molecular Genetics, Institute of Ophthalmology,
University of London, Bath Street, London EC1V 9EL, England

[3]Institut fur Humangenetik, Medizinische Universitat zu Lubeck,
Ratzeburger Allee 160, D-W2400 Lubeck, Federal Republic of
Germany

[4]Dept. of Ophthalmology, University of Florida, College of
Medicine, Box 100284, J Hillis Miller Health Center, Gainesville,
Florida 32610, USA

[5]Dept. of Biochemistry, University of Otago, Dunaeden, New
Zealand

## SUMMARY

The most prevalent group of genetically determined progressive retinopathies, currently affecting approximately 1.5 million people, is collectively termed retinitis pigmentosa (RP). RP describes a heterogeneous group of disorders primarily involving photoreceptor degeneration. The rapid development of highly informative DNA polymorphisms as genetic markers throughout the human genome has facilitated the localisation of genes responsible for many human disorders such as RP. Furthermore, techniques for rapid identification of sequence variation have provided an effective means of investigating genes that are considered to be 'candidates' for a particular disease in a given patient population. Such techniques have been successfully applied to the study of RP. In this paper, we will deal mainly with the recent developments in the autosomal dominantly inherited forms of RP, in particular highlighting the extensive genetic heterogeneity which we now know to be inherent in this group of diseases.

## I. RHODOPSIN LINKED ADRP (CHROMOSOME 3q)

Systematic genetic linkage studies have been carried out over the last few years using large autosomal dominant RP (adRP) families. The first dominant gene was located in 1989 as a result of extensive work using an Irish family suffering from an early onset form of adRP and traditional genetic markers (RFLPs) in combination with Southern blotting techniques. Close linkage was observed between the locus for the DNA marker C17 [D3S47] on 3q and the adRP gene in this family (McWilliam et al., 1989; Farrar et al., 1990a). Confirmation of this linkage was reported by other laboratories (Lester et al., 1990; Olsson et al., 1990). The mapping of an adRP gene on 3q provided a clue as to the cause of the disease, the gene encoding rhodopsin having been mapped to 3q and hence prompted an intensive search for mutations within the rhodopsin gene in patients with adRP. The first mutation, Pro-23-His, was identified by Dryja and colleagues (Dryja et al., 1990). This mutation was common in the United States, but absent in Europe, probably due to a 'Founder Effect' in the US population (Farrar et al., 1990b). The first mutation identified in the European population was a 3 base pair (bp) deletion involving codon 255 reported by Inglehearn and colleagues (Inglehearn et al., 1991). Approximately 60 rhodopsin mutations (see Table 1, Figure 1) have now been identified in adRP and related conditions (reviewed in Humphries et al., 1992; Linsay et al., 1992). Most of these are point substitutions found in conserved regions throughout the molecule, that is, in the amino terminus, the transmembrane domains, the interdiscal and cytoplasmic loops and the carboxy terminus. However, a number of small deletions and a couple of frame-shift mutations which result in a radical alteration of the carboxy-terminus of the protein have also been encountered (Horn et al., 1992; Keen et al., 1991; Rastagno et al., 1993). The clinical expression of these mutations varies significantly in the age of onset and severity of the disease. This variability in phenotypic expression is highlighted by the recent report indicating that mutations in the rhodopsin gene have been implicated in autosomal recessive RP (arRP) (Rosenfeld et al., 1992). However, the genetic complexity in adRP may possibly be simplified at the cellular level, as some rhodopsin mutations have been shown to fall into two main classes on the basis of transport across the plasma membrane and regeneration with 11-cis retinal (Sung et al., 1991a).

Weak evidence suggesting the existence of a second adRP locus on 3q has previously been reported (Olsson et al., 1990; Inglehearn et al., 1992).

**Table 1.** Rhodopsin mutations identified in adRP, arRP and dominant congenital complete nyctalopia (DCCN).

| CODON | RHODOPSIN MUTATIONS | | | DISORDER | REFERENCE |
|---|---|---|---|---|---|
| 4 | Thr | --> | Lys | adRP | (Gal et al., 1992) |
| | | | | | (Bunge et al., 1993) |
| 15 | Asn | --> | Ser | adRP | (Gal, 1993) |
| 17 | Thr | --> | Met | adRP | (Dryja et al., 1991) |
| | | | | | (Sung et al., 1991b) |
| | | | | | (Sheffield et al., 1991) |
| | | | | | (Fishman et al., 1992a) |
| | | | | | (Fujiki et al., 1992) |
| 23 | Pro | --> | His | adRP | (Sung et al., 1991b) |
| | | | | | (Dryja et al., 1991) |
| | | | | | (Sheffield et al., 1991) |
| | | | | | (Stone et al., 1991) |
| | | | | | (Kemp et al., 1992) |
| | | | | | (Olsson et al., 1992) |
| 23 | Pro | --> | Leu | adRP | (Dryja et al., 1991) |
| 28 | Gln | --> | His | adRP | (Gal et al., 1992) |
| | | | | | (Bunge et al., 1993) |
| 45 | Phe | --> | Leu | adRP | (Sung et al., 1991b) |
| 46 | Leu | --> | Arg | adRP | (Rodriguez et al., 1993) |
| 51 | Gly | --> | Val | adRP | (Dryja et al,. 1991) |
| 51 | Gly | --> | Arg | adRP | (Dryja, 1992) |
| 53 | Pro | --> | Arg | adRP | (Bhattacharya et al., 1991) |
| 58 | Thr | --> | Arg | adRP | (Sung et al., 1991b) |
| | | | | | (Dryja et al., 1991) |
| | | | | | (Sheffield et al., 1991) |
| | | | | | (Bhattacharya et al., 1991) |
| | | | | | (Richards et al., 1991) |
| | | | | | (Moore et al., 1992) |
| 68-71 | 12bp | --> | Del | adRP | (Bhattacharya et al., 1991) |
| | | | | | (Keen et al., 1991) |
| 87 | Val | --> | Asp | adRP | (Sung et al., 1991b) |
| 89 | Gly | --> | Asp | adRP | (Sung et al., 1991b) |
| | | | | | (Dryja et al., 1991) |
| 90 | Gly | --> | Asp | DCCN | (Sieving et al., 1992) |
| 106 | Gly | --> | Trp | adRP | (Sung et al., 1991b) |
| 106 | Gly | --> | Arg | adRP | (Fishman et al., 1992b) |
| | | | | | (Moore et al., 1992) |
| 110 | Cys | --> | Tyr | adRP | (Dryja, 1992) |
| 125 | Leu | --> | Arg | adRP | (Dryja et al., 1991) |
| 135 | Arg | --> | Leu | adRP | (Sung et al., 1991b) |
| 135 | Arg | --> | Trp | adRP | (Sung et al., 1991b) |
| 135 | Arg | --> | Gly | adRP | (Gal et al., 1992) |
| | | | | | (Bunge et al., 1993) |
| 167 | Cys | --> | Arg | adRP | (Dryja et al., 1991) |
| 171 | Pro | --> | Leu | adRP | (Dryja et al., 1991) |
| 171 | Pro | --> | Ser | adRP | (Stone et al., 1993) |
| 178 | Tyr | --> | Cys | adRP | (Sung et al., 1991b) |
| | | | | | (Farrar et al., 1991c) |
| | | | | | (Bell et al., 1992) |
| 181 | Glu | --> | Lys | adRP | (Dryja et al., 1991) |
| 182 | Gly | --> | Ser | adRP | (Sheffield et al., 1991) |
| | | | | | (Fishman et al., 1992a) |
| 186 | Ser | --> | Pro | adRP | (Dryja et al., 1991) |
| 188 | Gly | --> | Arg | adRP | (Dryja et al., 1991) |
| 190 | Asp | --> | Gly | adRP | (Sung et al., 1991b) |
| | | | | | (Dryja et al., 1991) |

(continued)

Table 1. (Continued)

| 190 | Asp | --> | Asn | adRP | (Dryja et al., 1991) |
| | | | | | (Bhattacharya et al., 1991) |
| | | | | | (Keen et al., 1991) |
| 190 | Asp | --> | Tyr | adRP | (Fishman et al., 1992c) |
| 207 | Met | --> | Arg | adRP | (Farrar et al., 1992a) |
| 211 | His | --> | Pro | adRP | (Bhattacharya et al., 1991) |
| | | | | | (Keen et al., 1991) |
| 220 | Phe | --> | Cys | adRP | (Gal et al., 1992) |
| | | | | | (Bunge et al., 1993) |
| 222 | Cys | --> | Arg | adRP | (Gal et al., 1992) |
| | | | | | (Bunge et al., 1993) |
| 249 | Glu | --> | Stop | arRP | (Rosenfeld et al., 1992) |
| 255 / 256 | Ile | --> | Del | adRP | (Bhattacharya et al., 1991) |
| | | | | | (Artlich et al., 1992) |
| 267 | Pro | --> | Leu | adRP | (Sheffield et al., 1991) |
| | | | | | (Fishman et al., 1992c) |
| 296 | Lys | --> | Glu | adRP | (Bhattacharya et al., 1991) |
| | | | | | (Keen et al., 1991) |
| 296 | Lys | --> | Met | adRP | (Sullivan et al., 1993) |
| 340 | 1bp | --> | Del | adRP | (Horn et al., 1992) |
| 340-348 | 42bp | --> | Del | adRP | (Rastagno et al., 1992) |
| 341-343 | 8bp | --> | Del | adRP | (Horn et al., 1992) |
| | | | | | (Apfelstedt-Sylla et al., 1992a) |
| 341 | Glu | --> | Lys | adRP | (Richards & Sieving, 1993) |
| 342 | Thr | --> | Met | adRP | (Stone et al., 1993) |
| 344 | Gln | --> | Stop | adRP | (Sung et al., 1991b) |
| 345 | Val | --> | Met | adRP | (Dryja et al., 1991) |
| 347 | Pro | --> | Leu | adRP | (Sung et al., 1991b) |
| | | | | | (Dryja et al., 1991) |
| | | | | | (Bhattacharya et al., 1991) |
| | | | | | (Apfelstedt-Sylla et al., 1992b) |
| | | | | | (Fujiki et al., 1992) |
| 347 | Pro | --> | Ser | adRP | (Dryja et al., 1991) |
| 347 | Pro | --> | Arg | adRP | (Gal et al., 1991) |
| | | | | | (Niemeyer et al., 1992) |
| 347 | Pro | --> | Ala | adRP | (Stone et al., 1993) |

Inglehearn and colleagues observed 12% recombination between the rhodopsin gene and D3S47. Bunge and colleagues (Bunge et al., 1993) and Kumar-Singh and colleagues (Kumar-Singh et al., 1993a) have refined the genetic mapping of the rhodopsin gene, proximal to D3S47 and close to D3S20. These data together with the observation of families in which no rhodopsin mutations had been found but which showed tight linkage between the adRP and D3S47 loci was consistent with the hypothesis that there may be a second adRP gene on 3q. However, recent identification of a rhodopsin mutation in the original Irish family (TCDM1) used in the first linkage study (Farrar et al., 1992a), and in other adRP families initially believed to be without rhodopsin mutations (Bell et al., 1992; Bunge et al., 1993) has indicated that there is no longer evidence of a second adRP locus on 3q (Inglehearn et al., 1993a). Moreover, formal statistical analysis of available linkage data from 3q linked families reveals that from these data alone there is no significant evidence for non allelic genetic heterogeneity (Kumar-Singh et al., 1993a).

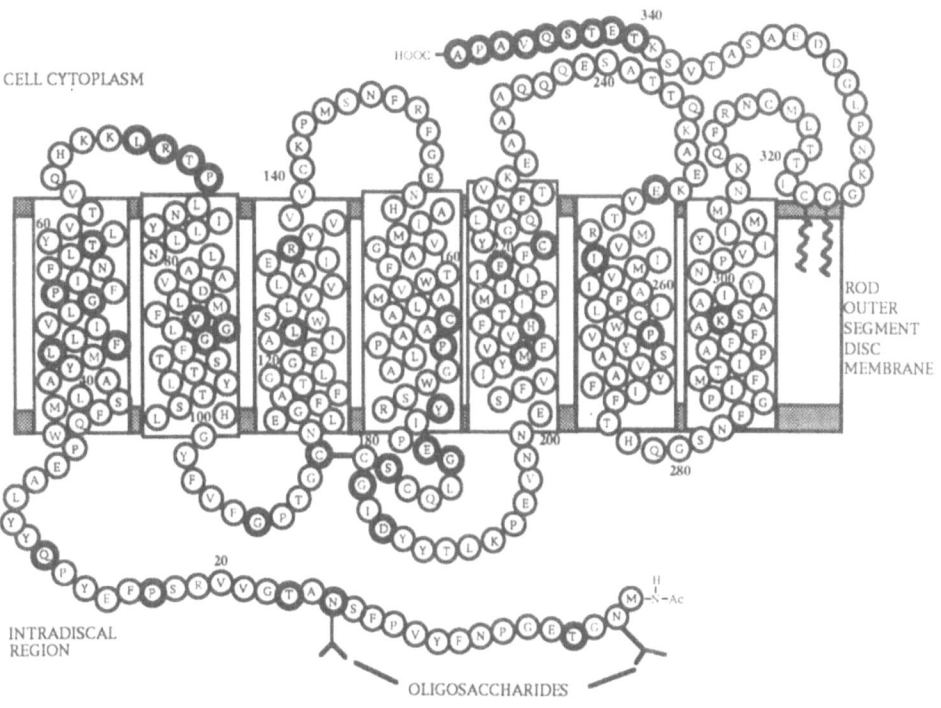

**Figure 1.** A diagramatic representation of the rhodopsin molecule. The amino acids which are mutated in patients suffering from autosomal dominant retinitis pigmentosa, autosomal recessive retinitis pigmentosa and dominant congenital complete nyctalopia are indicated by thicker black circles.

## II. PERIPHERIN / RDS LINKED ADRP (CHROMOSOME 6p)

In contrast to the speculations above, definitive evidence of non allelic genetic heterogeneity has been observed in adRP in that many adRP families have shown no linkage on 3q (Farrar et al., 1990a; Inglehearn et al., 1990; Jimenez et al., 1991). Continued genetic linkage studies using simple sequence repeat polymorphic markers detectable by the Polymerase Chain Reaction in an Irish family with a late onset form of adRP resulted in the localisation of another adRP gene in 1991 (Farrar et al., 1991a; Jordan et al., 1992a). This time the causative gene was located on 6p, close to the gene encoding the protein peripherin / *RDS*, a structural component of the rod cells which also resides in the membranes of the outer segment discs. A defective form of peripherin / *RDS* is also involved in the cause of the naturally occurring retinopathy of the mouse called *retinal degeneration slow* (Travis et al., 1991). A number of RDS mutations have now been identified in adRP patients (Farrar et al., 1991b; Kajiwara et al., 1991), including the original family used to localise an adRP gene to 6p (Farrar et al., 1992b) (Table 2, Figure 2). Interestingly, RDS mutations have now been implicated in autosomal dominantly inherited macular degenerations (Wells et al., 1993). Most of the RDS mutations identified to date are found in the large intradiscal loop of the protein or in transmembrane domains. Furthermore, it seems that the carboxy terminus of the protein is less well conserved. For example, we have found three amino acid changes in this region of the protein which are not associated with a retinopathy (Jordan et al., 1992b; Gruning et al., 1992).

Table 2. RDS mutations identified in adRP, autosomal dominant macular dystrophy (adMD) and autosomal dominant adult vitelliform macular dystrophy (adAVMD).

| CODON | RDS MUTATIONS | | | DISORDER | |
|---|---|---|---|---|---|
| 118/119 | Cys | --> | Del | adRP | (Farrar et al., 1991b) |
| 126 | Leu | --> | Arg | adRP | (Kajiwara et al., 1992) |
| 172 | Arg | --> | Glu | adMD | (Wells et al., 1993) |
| 172 | Arg | --> | Try | adMD | (Wells et al., 1993) |
| 173 | Asp | --> | Val | adRP | (Gal, 1993) |
| 185 | Leu | --> | Pro | adRP | (Kajiwara et al., 1991) |
| 212 | Ser | --> | Gly | adRP | (Farrar et al., 1992a) |
| 216 | Pro | --> | Leu | adRP | (Kajiwara et al., 1991) |
| 219 | Pro | --> | Del | adRP | (Kajiwara et al., 1991) |
| 258 | Tyr | --> | Stop | adAVMD | (Wells et al., 1993) |
| 266 | Gly | --> | Asp | adRP | (Kajiwara et al., 1992) |
| 307 | 1bp | --> | Del | adRP | (Gal, 1993) |

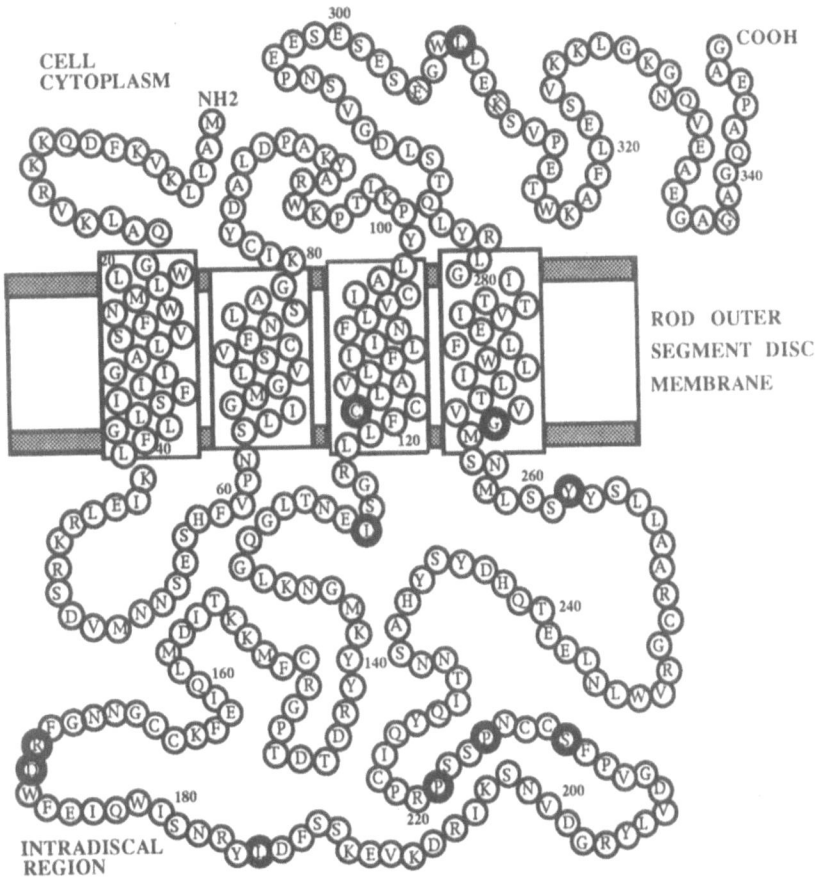

Figure 2. A diagramatic representation of the peripherin / RDS molecule. The amino acids which are mutated in patients suffering from autosomal dominant retinitis pigmentosa and macular dystrophy are indicated by thicker black circles.

## III. ADRP LOCI ON CHROMOSOMES 7 AND 8

Studies with a large American adRP family led to the localisation of a gene in the pericentric region of chromosome 8, with maximum Lod scores of 3.6 and 9.9 obtained with the markers D8S108 and D8S165 at zero recombination respectively (Blanton et al., 1991). A larger Australian family with adRP also shows linkage to the same markers and no linkage to any other regions to which adRP loci have been mapped (A. Gal, unpublished data). As yet no candidate genes have been localised to this region of chromosome 8.

Most recently, two adRP loci have been localised to chromosome 7, one on the short arm in an English adRP family (Inglehearn et al., 1993b) and the other on the long arm in a Spanish adRP family (Jordan et al., 1993). Shomi Bhattacharya and colleagues in the Institute of Ophthalmology, University of London obtained a maximum Lod score of 5.6 at zero recombination between the adRP gene segregating in the English pedigree and the 7p marker D7S460. Likewise, the RP research team in Trinity College Dublin led by Peter Humphries found a Lod score of 7.2 also maximising at zero recombination between the adRP gene and the 7q marker D7S480 using the Spanish family. In each case, no obvious candidate genes have been mapped close to the DNA markers showing tight linkage with adRP on 7p or 7q. The absence of obvious candidates for the adRP genes on chromosomes 7 and 8, will lead to the implementation of techniques such as positional cloning which have previously been elegantly used in the elucidation of the genes for disorders such as Cystic Fibrosis (Collins, 1992) and Choroideremia (Cremers et al., 1990).

## IV. FURTHER LOCUS HETEROGENEITY IN ADRP

The story of non allelic genetic heterogeneity in adRP does not end there. For example, the adRP gene segregating in a third large Irish adRP family has been excluded from each of the regions of the genome known to harbour adRP loci, that is, 3q, 6p, and the pericentric region of chromosome 8 (Kumar-Singh et al., 1993b) and more recently from 7p and 7q (R. Kumar-Singh, unpublished data). Furthermore most of these regions have been excluded in a number of other adRP families (Bashir et al., 1992). Interestingly, a number of families have shown small positive Lod scores with markers in the 7p region (A. Gal, unpublished data). Further work will be required to confirm / negate these initial linkage data. However, without doubt, it has now been established that there are at least six human genes which when mutated result in autosomal dominantly inherited forms of RP. The high levels of genetic heterogeneity inherent in adRP present enormous difficulties for efficient diagnosis, prognosis and future treatment of these disorders. However, one eventual benefit of such heterogeneity will be the characterisation of biologically important retinal specific genes and hence a greater understanding of the functioning of the healthy retina.

## V. FUTURE PERSPECTIVES

In summary, high levels of both non allelic and allelic genetic heterogeneity have been observed in adRP. Many different mutations have been identified in both the rhodopsin and peripherin / RDS genes in adRP patients. There is some evidence for the involvement of mutations in the gene encoding a third retinal specific protein (ROM1) in adRP, although this has yet to be confirmed (R. McInnis, personal communication). Moreover, as stated above, a minimum of six different adRP loci are involved. An objective inspection of the progress to date reveals that most of the developments in the field have been at the level of observation rather than providing an understanding of the underlying biological mechanisms which result in photoreceptor degeneration. The areas of research that are actively being pursued in our own laboratories and in those of others will aid us in making this transition in the future.

The harnessing of expression systems such as the insect infected baculo-virus system in which large quantities of biologically active protein can be produced will be useful in the investigation of the altered structural / functional relationships of mutant rhodopsin and RDS proteins. Similarly, previously developed models for rhodopsin from protein chemistry and site directed mutagenesis can be amalgamated to provide a crude representation of the 3-D structure of rhodopsin and used to speculate on the altered structure and function of mutant rhodopsin molecules (Farrar et al., 1992a). Such techniques have been used to investigate both the codons Tyr - 178 - Cys (Farrar et al., 1991c) and Met - 207 - Arg (Farrar et al., 1992a) rhodopsin mutations.

The development of animal models of RP using the powerful technique of gene targeting by homologous recombination will provide exact models for specific diseases. In this way the progression of the disease may be monitored in greater detail. Furthermore, these models provide good systems to test future possible gene and drug therapies. We can indeed apply gene targeting technology to address many different questions. Without doubt, 'knock out' targeting experiments may be used to establish more clearly the role of a particular protein in the functioning of the healthy retina. Similarly, this technology may also be used to identify further proteins which when mutated lead to retinopathies.

In conclusion, the past few years have seen the successful application of new techniques in molecular biology to the investigation of adRP. Two retinal proteins found in the rod outer segment disc membrane which when mutated result in photoreceptor degeneration have been identified. It has also been established that this group of diseases shows high levels of both non allelic and allelic genetic heterogeneity. The areas of research that are currently in development should lead to a substantial elucidation of the disease mechanisms in this important and debilitating group of hereditary conditions in the future.

# VI. ACKNOWLEDGEMENTS

The authors wish to thank The Wellcome Trust, The Retinitis Pigmentosa Foundation Fighting Blindness (United States), The Gund Foundation, RP Ireland-Fighting Blindness, The British and The German Retinitis Pigmentosa Societies, The Deutsche Forschungsgemeinschaft, Research to Prevent Blindness, The Ulverscroft Foundation, The Health Research Board of Ireland, and The Concerted Action, Science and Human Genome Analysis Programmes of the Commission of the European Communities for support of the programmes.

# VII. REFERENCES

Apfelstedt-Sylla, E., Kunisch, M., Horn, M., Ruther K., Gerdling, H., Gal, A., and Zrenner, E., 1992a, Clinical characteristics of a family with autosomal dominant retinitis pigmentosa and a frameshift gene deletion altering the carboxyl-terminus sequence of rhodopsin (submitted).

Apfelstedt-Sylla, E., Kunisch, M., Horn, M., Ruther K., Gal, A., and Zrenner, E., 1992b, Diffuse loss of rod function in autosomal dominant retinitis pigmentosa with Pro-347-Leu mutation of rhodopsin. *German J. Ophthalmol.* (in press).

Artlich, A., Horn, M., Lorenz, B., Bhattacharya, S.S., and Gal, A., 1992, Recurrent 3bp deletion at codon 255 / 256 of the rhodopsin gene in a German pedigree with autosomal dominant retinitis pigmentosa. *Am. J. Hum. Genet.* 50: 876.

Bashir, R., Inglehearn, C.F., Keen, T.J., Lindsey, J., Atif, U., Carter, S.A., Stephenson, A.M., Jackson, A., Jay, M., Bird, A.C., Papiha, S.S., and Bhattacharya S.S., 1992, Exclusion of chromosomes 6 and 8 in non-rhodopsin linked adRP families: further locus heterogeneity in adRP. *Genomics* 14: 191.

Bell, C., Converse, C.A., Collins, M.F., Esakowitz, L., Kelly, K.F., and Haites, N.E., 1992, Autosomal dominant retinitis pigmentosa (adRP) - A rhodopsin mutation in a Scottish family. *J. Med. Genet.* 29: 667.

Bhattacharya, S.S., Inglehearn, C.F., Keen, J., Lester, D., Bashir, R., Jay, M., and Bird, A.C., 1991, Identification of novel rhodopsin mutations in patients with autosomal dominant retinitis pigmentosa. *Invest. Ophthamol. Vis. Sci.* 32 (suppl): 890.

Blanton, S.H., Cottingham, A.W., Giesenschlag, N., Heckenlively, J.R., Humphries, P. and Daiger, S.P., 1991, Linkage mapping of autosomal dominant retinitis pigmentosa (RP1) to the pericentric region of human chromosome 8. *Genomics* 11: 857.

Bunge, S., Wedemann, H., David D., Terwilliger, D.J., Aulehla-Scholz, C.,

Samanns, C., Horn, M., Ott, J., Schwinger, E., Bleeker-Wagemakers, E-M., Schinzel, A., Denton, M.J., and Gal, A., 1993, Molecular analysis and genetic mapping of the rhodopsin gene in families with autosomal dominant retinitis pigmentosa. *Genomics* (in press).

Collins, F.S., 1992, Positional Cloning. *Nature Genetics* 1: 3.

Cremers, F.P.M., van de Pol, D.J.R., van Kerkhoff, L.P.M., Wieringa, B., and Ropers, H-H., 1990, Cloning of a gene that is rearranged in patients with choroideremia. *Nature* 347: 674.

Dryja, T.D., McGee, T.L., Reichel, E., Hohn, L.B., Cowley, G.S., Yandell, D.N., Sandberg, M.A., and Berson, E.L., 1990, A point mutation in the rhodopsin gene in one form of retinitis pigmentosa. *Nature* 343: 364.

Dryja, , T.P., Hahn, L.B., Cowley, G.S., McGee, T.L., and Berson, E.L., 1991, Mutation spectrum of the rhodopsin gene among patients with autosomal dominant retinitis pigmentosa. *Proc. Natl. Acad. Sci. USA* 88: 9370.

Dryja, T., P., 1992, Rhodopsin and autosomal dominant retinitis pigmentosa. *Eye* 6: 1.

Farrar, G.J., McWilliam, P., Bradley, D.G., Kenna, P., Lawler, M., Sharp, E.M., Humphries, M.M., Eiberg, H., Conneally, P.M., Trofatter, J.A. and Humphries, P., 1990a, Autosomal dominant retinitis pigmentosa: Linkage to Rhodopsin and evidence for genetic heterogeneity. *Genomics* 8: 35.

Farrar, G.J., Kenna, P., Redmond, R., McWilliam, P., Bradley, D.G., Humphries, M.M., Sharp, E.M., Inglehearn, C.F., Bashir, R., Jay, M., Watty, A., Ludwig, M., Schinzel, A., Samanns, C., Gal, A., Bhattacharya, S.S., and Humphries, P., 1990b, Autosomal dominant retinitis pigmentosa : Absence of the rhodopsin Proline-->Histidine substitution (codon 23) in pedigrees from Europe. *Am. J. Hum. Genet.*. 47: 941.

Farrar, G.J., Jordan, S.A., Kenna, P., Humphries, M.M., Kumar-Singh, R., McWilliam, P., Allamand, V., Sharp, E.M., and Humphries, P., 1991a, Autosomal dominant retinitis pigmentosa: Localisation of a disease gene (RP6) to the short arm of chromosome 6. *Genomics* 11: 870.

Farrar, G.J., Kenna, P., Jordan, S.A., Kumar-Singh, R., Humphries, M.M., Sharp, E.M., Sheils, D.M., and Humphries, P., 1991b, A three base pair deletion in the peripherin/RDS gene in one form of adRP. *Nature* 354: 478.

Farrar, G.J., Kenna, P., Redmond, R., Sheils, D.M., McWilliam, P., Humphries, M.M., Sharp, E.M., Jordan, S.A., Kumar-Singh, R., and Humphries P., 1991c, Autosomal dominant retinitis pigmentosa: a mutation in codon 178 of the rhodopsin gene in two families of celtic origin. *Genomics* 11: 1170.

Farrar, G.J., Findlay, J.B.C., Kenna, P., Kumar-Singh, R., Humphries, M.M., Sharp, E.M., and Humphries, P., 1992a, A novel rhodopsin mutation in the original 3q linked adRP family. *Hum. Mol. Genet.* 1: 769.

Farrar, G.J., Kenna, P., Jordan, S.A., Kumar-Singh, R., Humphries, M.M., Sharp, E., Sheils, D., and Humphries, P., 1992b, Autosomal dominant retinitis pigmentosa: A novel mutation at the peripherin/RDS locus in the original 6p-linked pedigree. *Genomics* 14: 805.

Fishman, G.A., Stone, E.M., Sheffield, V.C., Gilbert, L.D., and Kimura, A.E., 1992a, Ocular findings associated with a rhodopsin gene codon-17 and codon-182 transition mutations in dominant retinitis pigmentosa. *Arch. Ophthalmol.* 110: 54.

Fishman, G.A., Stone, E.M., Gilbert, L.D., and Sheffield, V.C., 1992b, Ocular findings associated with a rhodopsin gene codon 106 mutation glycine-to-Arginine change in autosomal dominant retinitis pigmentosa. *Arch. Ophthalmol.* 110: 646.

Fishman, G.A., Vandenburgh, K., Stone, E.M., Gilbert, L.D., Alexander, K.R., and Sheffield, V.C., 1992c, Ocular findings associated with a rhodopsin gene codon-267 and codon-190 mutations in dominant retinitis pigmentosa. *Arch. Ophthalmol.* 110: 1582.

Fujiki, K., Hotta, Y., Hayakawa, M., Sakuma, H., Shiono, T., Noro, M., Sakuma, T., Tamai, M., Hikiji, K., Kawaguchi, R., Hoshi, A., Nakajima, A., and Kanai, A., 1992, Point mutations of rhodopsin gene found in Japanese families with autosomal dominant retinitis pigmentosa. *Jpn. J. Hum. Genet.* 37: 125.

Gal, A., Artlich, A., Ludwig, M., Niemeyer, G., Olek, K., Schwinger, E., and Schinzel, A., 1991, Pro-347 Arg mutation of the rhodopsin gene in autosomal dominant retinitis pigmentosa. *Genomics* 11: 468.

Gal, A., Bunge, S., Wedemann, H., Aulenla-Scholz, D., Terwilliger, J.D., and Horn, M., 1992, Molecular analysis and mapping of the rhodopsin gene in patients with autosomal dominant retinitis pigmentosa. *Am. J. Hum. Genet.* 51 (suppl): A6.

Gal, A., 1993, unpublished.

Gruning, G., Meins, M., Bunge, S., Wedemann, H., Denton, M., Farrar, G.J., Schwinger, E., Li, Y., Humphries, P., and Gal, A., 1992, Screening for mutations in the peripherin gene of patients with autosomal dominant retinitis pigmentosa. *Am. J. Hum. Genet.* 51 (suppl): A96.

Horn, M., Humphries, P., Kunisch, M., Marchese, C., Apfelstedt-Sylla, E., Fusi, L., Zrenner, E., Kenna, P., Gal, A., and Farrar, G.J., 1992, Deletions in exon 5 of the human rhodopsin gene causing shift in the reading frame and autosomal dominant retinitis pigmentosa. *Human Genetics* 90: 255.

Humphries, P., Kenna, P., and Farrar, G.J., 1992, On the molecular genetics of retinitis pigmentosa. *Science* 256: 804.

Inglehearn, C.F., Jay, M., Lester, D.H., Bashir, R., Jay B., Bird, A.C., Wright, A.F., Evans, H.J., Papiha, S.S., and Bhattacharya, S.S., 1990, No evidence for linkage between late onset autosomal dominant retinitis pigmentosa and chromosome 3 locus D3S47 (C17): evidence for genetic heterogeneity. *Genomics* 6: 168.

Inglehearn, C.F., Bashir, R., Lester, D.H., Jay, M., Bird, A.C., and Bhattacharya, S.S., 1991, A 3bp deletion in the rhodopsin gene in a family with autosomal dominant retinitis pigmentosa. *Am. J. Hum. Genet.* 48: 26.

Inglehearn, C.F., Lester, D.H., Bashir, R., Atif, U., Keen, J., Sertedaki, A., Lindsey, J., Jay, M., Bird, A.C., Farrar, G.J., Humphries, P., and Bhattacharya, S.S., 1992, Recombination between rhodopsin and locus D3S47 (C17) in rhodopsin retinitis pigmentosa families. *Am. J. Hum. Genet.* 50: 590.

Inglehearn, C.F., Farrar, G.J., Denton, M., Gal, A., Humphries, P., and Bhattacharya, S.S., 1993a, Evidence against a second autosomal dominant Retinitis Pigmentosa locus close to rhodopsin on chromosome 3q. *Am. J. Hum. Genet.* (submitted).

Inglehearn, C.F., Carter, S.A., Keen, T.J., Lindsey, J., Stephenson, A., Bashir, R., Al-Maghtheh, M., Moore, A.T., Jay, M., Bird, A.C., and Bhattacharya, S.S., 1993b, A new locus for adRP on 7p. *Nature Genetics* (in press).

Jimenez, J.B., Samanns, C., Watty, A,. Pongratz, G., Olsson, J.E., Dickenson, P., Buttery, R., Gal, A., and Denton, M.J., 1991, No evidence of linkage between the locus for autosomal dominant retinitis pigmentosa and D3S47 (C17) in three Auustralian families. *Hum. Genet.* 86: 265.

Jordan, S.A., Farrar, G.J., Kenna, P., Kumar-Singh, R., Humphries, M.M., Allamand, V., Sharp, E.M. and Humphries, P. Autosomal dominant retinitis pigmenotsa (adRP, RP6): 1992a, Cosegregation of RP6 and the peripherin/RDS locus in a late-onset family of Irish origin. *Am. J. Hum. Genet.* 50: 634.

Jordan, S.A., Farrar, G.J., Kenna, P., and Humphries, P., 1992b, Polymorphic variation within 'conserved' sequences at the 3' end of the human RDS gene which results in amino acid substitutions. *Human Mutation* 1: 240.

Jordan, S.A., Farrar, G.J., Kenna, P., Humphries, M.M., Sheils D., Kumar-Singh R., Sharp E.M., Benitez J., Carmen A., and Humphries P., 1993, Localisation of an adRP gene to 7q. *Nature Genetics* (submitted).

Kajiwara, K., Hahn, L.B., Mukai, S., Travis, G.H., Berson, E.L. and Dryja, T.P., 1991, Mutations in the human retinal degeneration slow gene in autosomal dominant retinitis pigmentosa. *Nature* 354: 480.

Kajiwara, K., Hahn, L.B., Mukai, S., Berson, E.L., and Dryja, T.P., 1992, Mutations in the human RDS gene in patients with autosomal dominant retinitis pigmentosa. *Invest. Ophthalmol. Vis. Sci.* 33: 1396.

Keen, T.J., Inglehearn, C.F., Lester, D.H., Bashir, R., Jay, M., Bird, A.C., Jay, B., and Bhattacharya, S.S., 1991, Autosomal dominant retinitis pigmentosa: Four new mutations in rhodopsin, one of them in the retinal attachment site. *Genomics* 11: 199.

Kemp, C.M., Jacobson, S.G., Roman, A.J., Sung, C.H., and Nathans, J., 1992, Abnormal dark adaptation in autosomal dominant retinitis pigmentosa with Proline-23-Histidine rhodopsin mutation. *Am. J. Ophthalmol.* 113: 165.

Kumar-Singh, R., Wang, K., Humphries, P., and Farrar, G.J., 1993a, No evidence for non allelic genetic heterogeneity in 3q linked adRP. *Am. J.*

*Hum. Genet.* (in press).

Kumar-Singh, R., Kenna, P., Farrar, G.J., and Humphries, P., 1993b, Further evidence for non-allelic genetic heterogeneity in adRP. *Genomics* (in press).

Lester, D.H., Inglehearn, C.F., Bashir, R., Ackford, H., Esakowitz, L., Jay, M., Bird, A.C., Wright, A.F., Papiha, S.S., and Bhattacharya, S.S., 1990, Linkage to D3S47 (C17) in one large autosomal dominant retinitis pigmentosa family and exclusion in another: confirmation of genetic heterogeneity. *Am. J. Hum. Genet.* 47: 536.

Lindsay, S., Inglehearn, C.F., Curtis, A., and Bhattacharya, S.S., 1992, Molecular genetics of inherited retinal degenerations in '*Current Opinion in Genetics and Development*'' 12: 459.

McWiliam, P., Farrar, G.J., Kenna, P., Bradley, D.G., Humphries, M.M., Sharp, E.M., McConnell, D.J., Lawler, M., Sheils, D.M, Ryan, C., Stevens, K., Daiger, S.P., and Humphries, P., 1989, Autosomal Dominant Retinitis Pigmentosa: localisation of an ADRP gene to the long arm of chromosome 3. *Genomics* 5: 619.

Moore, A.T., Fitzke, F.W., Kemp, C.M., Arden, G.B., Keen, T.J., Inglehearn, C.F., Bhattacharya, S.S., and Bird, A.C., 1992, Abnormal dark adaptation kinetics in autosomal dominant sector retinitis pigmentosa due to rod opsin mutation. *Br. J. Ophthalmol.* 76: 465.

Niemeyer, G., Trub, P., Schinzel, A., and Gal, A., 1992, Clinical and ERG data in a family with autosomal dominant RP and Pro-347-Arg mutation in the rhodopsin gene. Documenta. *Ophthalmologica* 79: 303.

Olsson, J.E., Samanns, C.H., Jimenez, J., Pongratz, J., Chand, A., Watty, A., Seuchter, S.A. Denton, M.J., and Gal, A., 1990, Gene of type II autosomal dominant retinitis pigmentosa maps on the long arm of chromosome 3. *Am. J. Med. Genet.* 35: 595.

Olsson, J.E., Gordon, J.W., Pawlyk, B.S., Roof, D., Hayes, A., Molday, R.S., Mukai, S., Cowley, G.S., Berson, E.L., and Dryja, T.P., 1992, Transgenic mice with a rhodopsin mutation (Pro23His) - A mouse model of autosomal dominant retinitis pigmentosa. *Neuron* 9: 815.

Rastagno, G., Al-Maghtheh, M., Bhattacharya, S.S., Fernone, M., Garnerone, S., Samuelly, R., and Carbonara, A., 1993, A large deletion in the 3' end of the rhodopsin gene in an Italian family with a diffuse form of adRP. *Hum. Mol. Genet.* (in press).

Richards, J.E., Kuo, C.Y., Boehnke, M., and Sieving, P.A., 1991,
Richards, J.E., and Sieving, P., A. 1993, (personal communication)
Rhodopsin Thr58Arg mutation in a family with autosomal dominant retinitis pigmentosa. *Ophthalmology* 98: 1797.

Rodriguez, J.A., Herrera, C.A., Birch, D.G., and Daiger, S.P., 1993, A leucine to arginine amino acid substitution at codon 46 of rhodopsin is responsible for a severe form of autosomal dominant retinitis pigmentosa. *Human Mutation* (submitted)

Rosenfeld, P.J., Cowey, G.S., McGee, T.L., Sandberg, M.A., Berson, E.L., and Dryja, T.P., 1992, A null mutation in the rhodopsin gene causes rod photoreceptor dysfunction and autosomal dominant retinitis pigmentosa. *Nature Genetics* 1: 209.

Sheffield, V.C., Fishman, G.A., Beck, J.S., Kimura, A.E., and Stone, E.M., 1991, Identification of novel rhodopsin mutations associated with retinitis pigmentosa using GC-clamped denaturing gradient gel electrophoresis. *Am. J. Hum Genet.* 49: 699.

Stone, E.M., Kimura, A.E., Nichols, B.E., Khadivi, P., Fishman, G.A., and Sheffield, V.C., 1991, Regional distribution of retinal degeneration in patients with the proline to histidine mutation in codon 23 of the rhodopsin gene. *Ophthalmology* 98: 1806.

Stone, E.M., Vandenburgh, K., Kimura, A.E., Lam, B.L., Fishman, G.A., Heckenlively, J.R., Castillo, T.A., and Sheffield, V.C., 1993, Novel mutations in the peripherin (RDS) and rhodopsin genes associated with autosomal dominant retinitis pigmentosa (adRP). *Invest. Ophthalmol. Vis. Sci.* (submitted)

Sullivan, J.M., Scott, K.M., Falls, H.F., Richards, J.E., and Sieving, P.A., 1993, A novel rhodopsin mutation at the retinal binding site (Lys 296 Met) in adRP. *Invest. Ophthalmol. Vis. Sci.* 34: (suppl) (submitted)

Sung, C-H., Schneider, B.G., Agarwal, N., Papermaster, D.S., and Nathans, J., 1991a, Rhodopsin mutations in autosomal dominant retinitis pigmentosa. *Proc. Natl. Acad. Sci. USA* 88: 8840.

Sung, C-H., Davenport, C.M., Hennessey, J.C., Maumenee, I.H., Jacobson, S.G., Heckenlively, J.R., Nowakowski, R., Fishman, G., Gouras, P., and Nathans J., 1991b, Functional heterogeneity of mutant rhodopsins responsible for autosomal dominant retinitis pigmentosa. *Proc. Natl. Acad. Sci. USA* 88: 6481.

Travis, G.H., Christerson, L., Danielson, P.E., Klisak, I., Sparkes, R.S., Hahn, L.B., Dryja, T.P. and Sutcliffe, J.G., 1991, The human retinal degeneration slow (RDS) gene: Chromosome assignment and structure of the mRNA. *Genomics* 10: 733.

Wells, V., Wroblewski, V., Keen, T.J., Inglehearn, C.F., Jubb, C., Eckstein, A., Jay, M., Arden, G., Bhattacharya, S.S., Fitzke, F., and Bird, A.C., 1993, Mutations in the human RDS gene can cause either RP or macular dystrophy. *Nature Genetics* (in press).

# GENETIC AND EPIDEMIOLOGICAL STUDY OF AUTOSOMAL DOMINANT (ADRP) AND AUTOSOMAL RECESSIVE (ARRP) RETINITIS PIGMENTOSA IN SARDINIA

Maurizio Fossarello,[1] Antonina Serra,[1] David Mansfield,[2]
AlanWright,[2] Jorgos Loudianos,[3] Mario Pirastu,[3] and
Nicola Orzalesi[4]

[1]University of Cagliari, Department of Ophthalmology
I-09100 Cagliari, Italy
[2]MRC Human Genetics Unit, Western General Hospital, Edinburgh EH4
2XU, United Kingdom
[3]CNR Institute of Research on Thalassemias and Mediterranean
Anemias,
I-09100 Cagliari, Italy
[4]University of Milan, "San Paolo" Biomedical Institute,
Department of Ophthalmology,
I-20100 Milan, Italy

## INTRODUCTION

The recent advances and results in the application of recombinant DNA techniques to family linkage studies has opened new perspectives for the diagnosis and prevention of Retinitis Pigmentosa (RP)(see[1] for a review), and has renewed the interest in categorisation of families by recognized Mendelian pattern of inheritance with the aim to localize the genes of the disease on particular chromosomes and eventually to identify the genes and their products.

However, genetic linkage analysis may be hampered by genetic heterogeneity of RP population. Genetic heterogeneity in RP has been well established by family studies, by the recognition of several genetic syndromes of which RP is a feature, and by animal experiments. Moreover, althought it is generally accepted that RP can be classified into homogeneous groups according to to the mode of inheritance, it is possible to note many exceptions to this assumption, suggesting once more the existence of genetic heterogeneity within a RP subgroup on the basis of the electrophysiological pattern, varying severity, age of onset and rate of progression of the disease. Both dominant and recessive RP may encompass several different disease forms. Indeed, within each genetic category, the high degree of RP variability as regards onset age and natural history, may reflect different quantitative deficits at molecular levels or heterogeneity of disease mechanisms, possibly determined by a different gene penetrance (i.e. full and reduced) or different alleles.

For genetic linkage studies it is fundamental to select large RP families in which the condition is segregating, possibly with a relatively homogeneous genetic background, and to ascertain clinically and electrophysiologically the condition of all family members.

In Sardinia, socio-economic and geographical grounds have favoured in the past, and to some extent also nowadays, the segregation of the island from the continent, as well as of certain districts from the remainder of the territory, resulting in a forced endogamy for the majority of the population. Moreover, economic, social and religious factors have supported the establishment of very large families in Sardinia. Therefore is not surprising that in this island large pedigrees of several genetically determined disorders, including RP, are available, with a presumed very high homogeneity of the genetic material.

In the present study we report the characteristics of prevalence and distribution of the different forms of RP in Sardinia, and the preliminar results of the molecular genetics analysis performed in autosomal dominant (AD) and autosomal recessive (AR) RP.

## PATIENTS AND METHODS

The files of patients affected by RP examined at the University Eye Clinic of Cagliari during the last 13 years were utilized to determine the prevalence and geographical distribution of RP families and to ascertain the composition of the AD and AR family pedigrees. All families considered in this study have ancestors of Sardinian origin. Most of these patients were examined using standard procedure described elsewhere.[2] The files report the medical and family history, the complete ophthalmological examination, including biomicroscopy, applanation tonometry, fundus examination, electrophysiologic and visual field testing, and fluorescein angiography. The following RP phenotypes were included in the present study: typical RP, atypical RP, sector RP, paravenous pigmentary retinopathy, Usher's syndromes, progressive retinitis punctata albescens, Bardet-Biedl syndrome, Leber's amaurosis congenita.

Blood samples for molecular genetic studies were selectively obtained on 7 ADRP, 9 presumed ARRP, and 1 genetically undetermined family pedigrees. ADRP blood samples were jointly analyzed in the laboratories of the MRC Human Genetic Unit of Edinburgh, UK, and of the CNR Institute of Research on Thalassemias and Mediterranean Anemias of Cagliari, Italy, while ARRP molecular genetic analysis was carried out in the laboratory of the MRC Human Genetic Unit of Edinburgh, UK. Mutations were detected using the previously described method of DNA heteroduplex detection on ethidium bromide-stained hydrolink gels.[3,4] Pairs of oligonucleotide primers were synthesized surrounding exonic regions to give PCR products in the range of 200-300 bp. Genetic linkage analysis of ARRP families was carried out using the LINKAGE program package[5] to generate 2-point lod scores, which were used to exclude chromosomal regions using the EXCLUDE program.[6]

## RESULTS

We evaluated 298 patients affected by RP coming from 158 families. Of these families, 5 % appeared to have dominantly inherited disease based on careful family searches. We believe 21.5 % of our families have the recessive disease, either because they are a product of consanguineous marriage, or because they originate from areas with high inbreeding, or because only one generation over three or four appears to be affected, or because they fall into one of the well-known recessive syndromes such as Usher's syndrome or Leber's amaurosis. Two per cent of cases are X-linked, a report that is closest to Switzerland (1%)[7] and Finland (3.4%).[8] In 51.3% of cases we have no firm genetic diagnosis. These are either simplex cases or multiplex cases (Table I).

The vast majority of these families originate in central-southern Sardinia. The

**Table 1.** Prevalence of RP in central-southern Sardinia.

| RP type | n. families | | n. cases | | gender | |
| --- | --- | --- | --- | --- | --- | --- |
| | | | | | m | f |
| AD | 8 | (5%) | 51 | (17.1%) | 21 | 30 |
| AR | 34 | (21.5%) | 88 | (29.5%) | 41 | 47 |
| X-linked | 2 | (1.3%) | 6 | (2.1%) | 6 | |
| simplex | 85 | (53.8%) | 85 | (28.5%) | 46 | 39 |
| undeterm. | 29 | (18.4%) | 68 | (22.8%) | 40 | 28 |
| | 158 | | 298 | | 154 | 144 |

prevalence of RP in central- southern Sardinia is around 1:3355 (average number of inhabitants in the districts of Cagliari, Nuoro and Oristano = 1.000.000).

In dominant families, we identified 4 families with sectoral RP, 1 family with regional RP, and 3 families with diffuse RP (see Table 2), located in central-southern Sardinia (Figure 1). Three of these ADRP families have their place of origin in the same village, San Nicolò Gerrei, in the district of Cagliari, and all the affected members have sectoral RP. Anaware of being genetically related, they showed the same mutation in exon 3 of the rhodopsin gene, changing the codon 190 from GAC to AAC, resulting in the substitution of an asparagine residue for the normal aspartate (Table 2).

Fundus examination of affected members revealed degenerative pigmentary changes limited to the inferior retina; they were represented by hypopigmentation and sparse pigment clumping in a bone-spiculelike fashion, and by diffuse retinal atrophy. The foveas in general appeared normal, while the inferior retinal vessels were attenuated. The optic disc did not show a waxy-type of pallor, and interestingly, in two patients from family ADRP2 we observed a tilted disc with an atrophic crescent on the temporal edge (Figure 2). Moreover, older patients and women from family ADRP1 and ADRP3 which experienced a pregnancy, showed an extension of retinal degeneration to the upper retinal quadrants. The retinal changes were even clearer on fluorescein angiography (Figure 3).

**Table 2.** Summary of autosomal dominant RP families originating in central-southern Sardinia.

| family | place of origin | n affected members | Pheno type | Genotype mutation | aminoacid change |
| --- | --- | --- | --- | --- | --- |
| ADRP1 | 1 | 16 | sector | Rho/190 | Asp→Asn |
| ADRP2 | 1 | 7 | sector | Rho/190 | Asp→Asn |
| ADRP3 | 1 | 6 | sector | Rho/190 | Asp→Asn |
| ADRP4 | 2 | 7 | diffuse | ? | |
| ADRP5 | 3 | 6 | diffuse | Rho/347 | Pro→Leu |
| ADRP6 | 4 | 3 | diffuse | ? | |
| ADRP7 | 5 | 3 | regional | ? | |
| ADRP8 | 2 | 3 | diffuse | ? | |

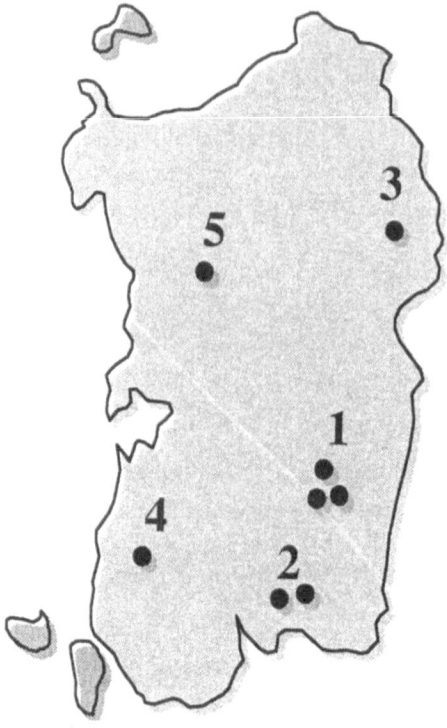

**Figure 1.** Places of origin of ADRP families in central-southern Sardinia. (1) San Nicolò Gerrei, (2) Quartu Sant'Elena, (3) Dorgali, (4) Iglesias, (5) Bortigali.

The ERG recording of younger patients showed a singleflash cone b-wave amplitude that was reduced approximately 40% below the lower limit of normal level, with a slight prolongation of the implicit time. The older patients had higher percentages of reduction of ERG amplitude and wider loss of visual field associated with the extension of retinal degeneration than younger individuals (Figure 4).

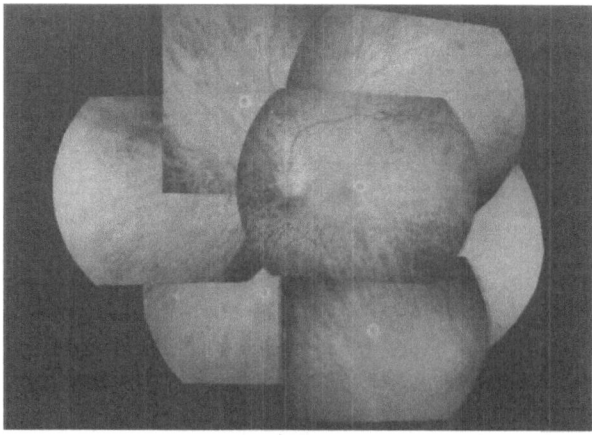

**Figure 2.** Fundus photograph of the left eye of a patient from family ADRP2 shows diffuse retinal atrophy, sparse pigment deposition in the inferior retina, and a pink, tilted disc.

One additional ADRP family originating from the town of Dorgali (District of Nuoro) showed a severe form of diffuse RP, with a point mutation in codon 347 in the rhodopsin gene changing proline to leucine (Table 1). In three additional ADRP families the mutation has not yet been found.

For 32 families presumed to be recessive, we analyzed the geographical origins of the parents, their degree of consanguinity, and the clinical phenotype, which was separated in two broad categories as "mild" and "severe" for sake of genetic comparison. In most cases the phenotype of these families was severe, being characterized by early age of onset of nightblindness and of difficulty with side vision, extinguished electroretinogram and intraretinal bone spicule pigment in the four quadrants of the retina (Table 3).

Due to the fact that in most of the ARRP families the place of residence has been remarkably stable throughout the last four or five generations, we could trace 10 geographical areas in central-southern Sardinia on the basis of their place of origin (Figure 5). In such a way, we may identify patients with a common pattern of inbreeding, thus permitting to select more homogeneous DNA samples for molecular genetics analysis.

The molecular genetic analysis was carried out on 52 members of 9 presumed autosomal recessive RP families consisting of 22 affected and 30 unaffected individuals.

The aim was to exclude or identify markers with measurable linkage of the ARRP locus in these families. A total of 33 markers were analysed, focusing on regions containing "candidate" genes. The markers used, together with their chromosomal localisations, are shown in Table 4.

The results of an exclusion map of chromosome 3 are shown in Figure 6. This chromosome contains the rhodopsin locus which has been shown to be a cause of both autosomal dominant and autosomal recessive RP in some families.[9] A microsatellite CA repeat polymorphism is present in the 1st intron of the rhodopsin gene[10] which was used to exclude this locus in the Sardinian families. Definite recombinants were observed in at least three of the nine families. Rhodopsin is therefore excluded as the site of mutation in families showing recombination and in the group of families as a whole, assuming a genetic homogeneity. The maximum likelihood value of the recombination fraction was 0.200 at a lod score of 0.41. Similarly, regions surrounding the D3S11, GLUT2 (glucose transporter) and D3S196 markers on chromosome 3 are excluded (Figure 6).

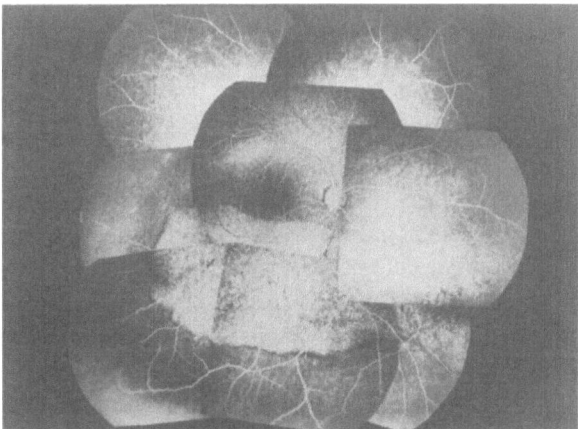

**Figure 3.** Fluorescein angiogram of the right eye of a patient from family ADRP1: an oval patch of RPE atrophy can be seen in the inferior retina, with sharp border between normal and abnormal RPE: the macula is grossly normal.

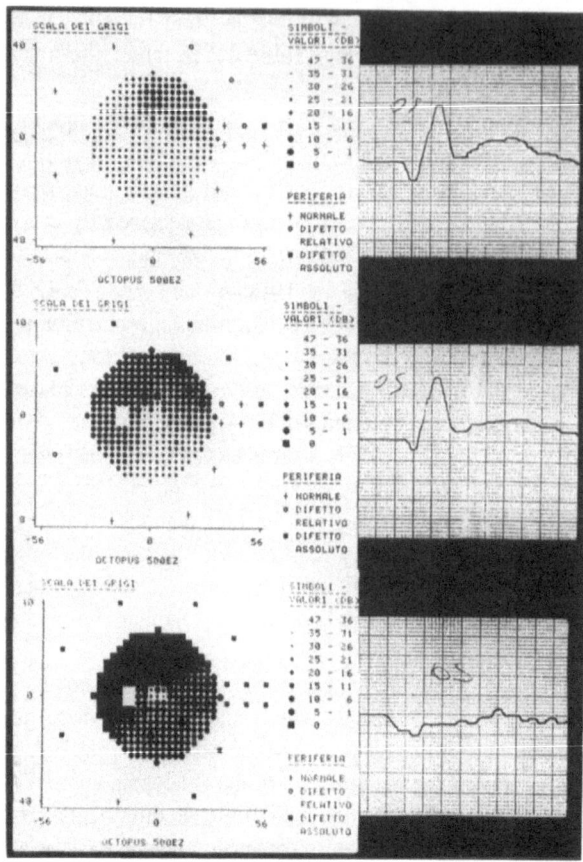

**Figure 4.** Automatic perimetry (Octopus 500EZ) and ERG recording for 3 patients from family ADRP1, aged 18 (top),34 (middle), and 62 years (bottom).

The human homologue of the mouse *rd* locus, the β subunit of rod cyclic GMP phosphodiesterase (β-PDE), on chromosome 4p16, is a candidate gene for ARRP. The homozygous *rd* mouse develops an early onset retinal degeneration similar to retinitis pigmentosa or Leber's congenital amaurosis. Ten markers on chromosome 4 were examined in the nine ARRP families, including the D4s179 and *hox7* homeobox (HOX7) loci which are located in the same chromosomal region as β-PDE. The results showed significant exclusions with all loci except HOX7. The maximum likelihood lod score with HOX7 was 1.15 at a recombination fraction of zero, which is not statistically significant. A distance of 18.5 centiMorgans on either side of the more distal locus D4S179 was excluded (lod <-2). Further work is therefore required to exclude the β-PDE locus.

The gene for peripherin/RDS on human chromosome 6p21.2-cen is another candidate gene for ARRP. It is a known cause of ADRP[11] and therefore, by analogy with the rhodopsin mutation responsible for ARRP and the homozygous *rds* mouse, two copies of certain mutant alleles could potentially lead to severe disease while a single copy of the same allele causes no or very mild disease. A mononucleotide (polyA/T) microsatellite repeat locus has been identified in the 3' untranslated end of the peripherin/RDS locus,[12] which again showed recombination with ARRP in at least five of the nine pedigrees. This candidate gene was therefore also excluded in the ARRP families.

**Table 3.** Summary of autosomal recessive RP families grouped in 10 geographical areas according to their place of origin.

| area | family | place of origin | n affected members | Pheno type | consanguineous marriage |
|------|--------|----------------|--------------------|-----------|-----------------------|
| 1 | ARRP1 | Ilbono | 1 | severe | |
|   | ARRP10 | Loceri | 4 | Leber's | |
|   | ARRP17 | Ulassai | 2 | severe | |
|   | ARRP23 | Urzulei | 2 | severe | yes |
| 2 | ARRP3 | Burcei | 1 | severe | |
|   | ARRP33 | Burcei | 2 | severe | |
|   | ARRP34 | Burcei | 3 | severe | yes |
| 3 | ARRP9 | Ballao | 3 | severe | yes |
|   | ARRP28 | Escalaplano | 2 | severe | |
| 4 | ARRP21 | Terralba | 3 | severe | |
|   | ARRP22 | Marrubbiu | 4 | variable | |
|   | ARRP27 | Marrubbiu | 2 | intermed. | |
|   | ARRP30 | Marrubbiu | 2 | severe | |
| 5 | ARRP11 | Gadoni | 2 | severe | |
|   | ARRP20 | Aritzo | 3 | severe | |
| 6 | ARRP15 | Gergei | 3 | variable | |
|   | ARRP18 | Senis | 3 | severe | |
|   | ARRP26 | Gesturi | 3 | severe | |
| 7 | ARRP2 | Donori | 2 | severe | |
|   | ARRP24 | Samatzai | 3 | Usher's | |
| 8 | ARRP6 | Collinas | 3 | variable | |
|   | ARRP25 | Sanluri | 2 | severe | |
| 9 | ARRP19 | Bacu Abis | 4 | severe | |
|   | ARRP32 | Domusnovas | 3 | severe | |
| 10 | ARRP5 | Cagliari | 5 | Usher's | |
|   | ARRP29 | Cagliari | 3 | Bardet-Biedl | |
|   | ARRP12 | Selargius | 2 | intermed. | |
|   | ARRP13 | Decimomannu | 2 | severe | |
|   | ARRP31 | Uta | 2 | severe | yes |

ARRP4, 7, and 14 are scattered

| ARRP16 origin was undetermined | | | | | yes |
|---|---|---|---|---|---|
| 34 | | | 88 | | 5 (15%) |

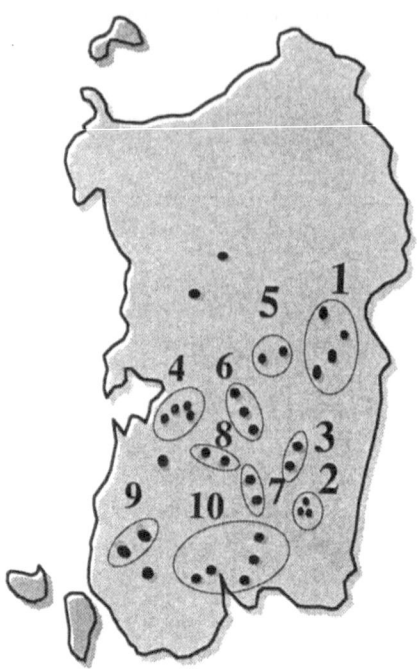

**Figure 5.**Identification of ten geographical areas in central-southern Sardinia grouping autosomal recessive RP families.

**Table 4.** List of microsatellite marker loci used in the linkage study of ARRP. The locus symbols are shown together with their chromosomal locations, given as a fraction of the length of each chromosome.[6]

| chromosome | locus | chromosome | locus |
|---|---|---|---|
| 2.509 | IL1A | 6.200 | TNF |
| 3.258 | D3S11 | 6.255 | FTHP1 |
| 3.800 | RHO | 6.300 | RDS |
| 3.860 | GLUT2 | 6.320 | D6S105 |
| 3.939 | D3S196 | 6.794 | ARG1 |
| 4.007 | D4S179 | 7.600 | W30 (CFTR) |
| 4.059 | HOX7 | 11.022 | TH |
| 4.243 | D4S230 | 11.579 | D11S527 |
| 4.303 | D4S190 | 11.868 | D11S490 |
| 4.303 | D4S174 | 16.719 | D16S261 |
| 4.336 | GABRB1 | 18.400 | D18S35 |
| 4.428 | D4S189 | 18.600 | D18S34 |
| 4.447 | ALB | 20.198 | D20S27 |
| 4.625 | D4S191 | 21.025 | D21S120 |
| 4.957 | D4S171 | 21.400 | D21S210 |
| 6.033 | F13A1 | 21.600 | D21S168 |
| 6.100 | | | |

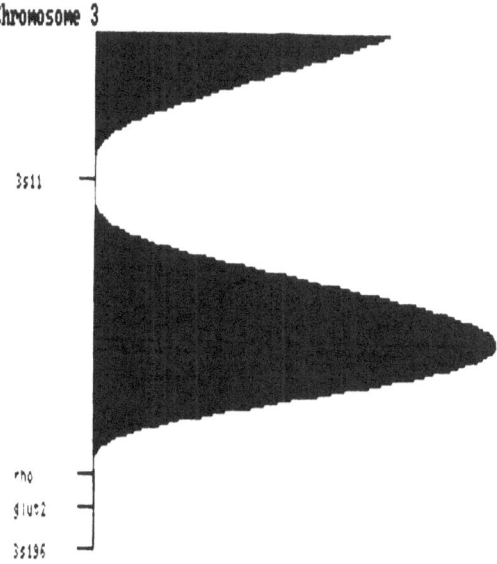

**Figure 6.** Exclusion map of chromosome 3, including the rhodopsin locus, as obtained with the EXCLUDE programme.[6]
The shaded area gives the likelihood distribution for the disease locus at each location on chromosome 3.

Other chromosome 6 markers excluded were microsatellites associated with the factor XIIIA (F13A1), D6S109, tumour necrosis factor (TNF), H-ferritin pseudogene (FTHP1), D6S105 and liver arginase (ARG1) loci.

The remaining 12 markers were located on chromosome 2, 7, 11, 16, 18 and 21. All these loci gave strongly negative lod scores with the exception of D11S490 which gave a maximum likelihood lod score of 0.81 at a recombination fraction of 0.20. Further work is

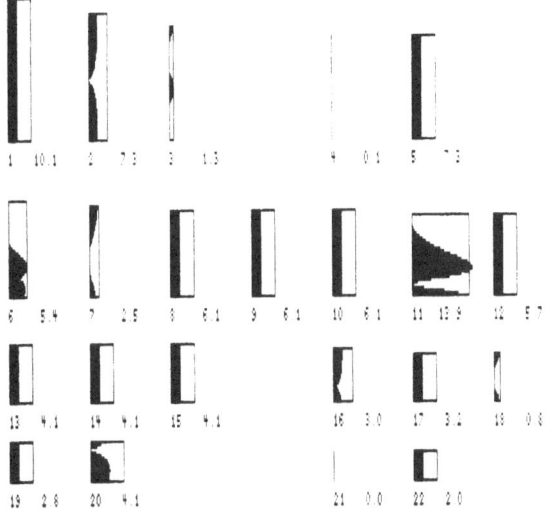

**Figure 7.** Exclusion map of the 22 autosomes using linkage data from 33 marker loci run against 9 ARRP families. The output from the EXCLUDE programme is shown, which demonstrates the final likelihood distribution for the disease locus at each chromosomal location.[6]

List of microsatellite marker loci used in the linkage study of ARRP. The locus symbols are shown together with their chromosomal locations, given as a fraction of the length of each chromosome (Edwards, 1987).

| chromosome | locus | chromosome | locus |
|---|---|---|---|
| 2.509 | IL1A | 6.200 | TNF |
| 3.258 | D3S11 | 6.255 | FTHP1 |
| 3.800 | RHO | 6.300 | RDS |
| 3.860 | GLUT2 | 6.320 | D6S105 |
| 3.939 | D3S196 | 6.794 | ARG1 |
| 4.007 | D4S179 | 7.600 | W30 (CFTR) |
| 4.059 | HOX7 | 11.022 | TH |
| 4.243 | D4S230 | 11.579 | D11S527 |
| 4.303 | D4S190 | 11.868 | D11S490 |
| 4.303 | D4S174 | 16.719 | D16S261 |
| 4.336 | GABRB1 | 18.400 | D18S35 |
| 4.428 | D4S189 | 18.600 | D18S34 |
| 4.447 | ALB | 20.198 | D20S27 |
| 4.625 | D4S191 | 21.025 | D21S120 |
| 4.957 | D4S171 | 21.400 | D21S210 |
| 6.033 | F13A1 | 21.600 | D21S168 |
| 6.100 | D6S109 | | |

Summary of exclusion data for candidate genes in 9 ARRP families, showing the two-point lod scores corresponding to each value of the recombination fraction ($\theta$). The results exclude the rhodopsin and peripherin/RDS loci but not the $\beta$-PDE locus in chromosomal region 4p16.

| LOCI | RECOMBINATION FRACTION ($\theta$) | | | | | | |
|---|---|---|---|---|---|---|---|
| | 0.00 | 0.01 | 0.050 | 0.10 | 0.20 | 0.30 | 0.40 |
| RHO | $-\infty$ | -3.42 | -0.89 | -0.04 | 0.41 | 0.33 | 0.11 |
| D4S179 | $-\infty$ | -6.36 | -3.02 | -1.72 | -0.65 | -0.22 | -0.05 |
| HOX7 | 1.15 | 1.13 | 1.04 | 0.90 | 0.59 | 0.30 | 0.08 |
| RDS | $-\infty$ | -10.06 | -4.70 | -2.60 | -0.89 | -0.25 | -0.04 |

**Figure 8**

**Table 5.** Summary of exclusion data for candidate genes in 9 ARRP families, showing the two-point lod scores corresponding to each value of the recombination fraction ($\theta$). The results exclude the rhodopsin and peripherin/RDS loci but not the $\beta$-PDE locus in chromosomal region 4p16.

| LOCI | RECOMBINATION FRACTION ($\theta$) | | | | | | |
|---|---|---|---|---|---|---|---|
| | 0.00 | 0.01 | 0.050 | 0.10 | 0.20 | 0.30 | 0.40 |
| RHO | $-\infty$ | -3.42 | -0.89 | -0.04 | 0.41 | 0.33 | 0.11 |
| D4S179 | $-\infty$ | -6.36 | -3.02 | -1.72 | -0.65 | 0.22 | 0.05 |
| HOX7 | 1.15 | 1.13 | 1.04 | 0.90 | 0.59 | 0.30 | 0.08 |
| RDS | $-\infty$ | -10.06 | -4.70 | -2.60 | -0.89 | -0.25 | -0.04 |

underway to exclude the rom-1 locus, another potential candidate gene, on this chromosome. This locus codes for a protein expressed in rod photoreceptors which is homologous to the peripherin/RDS gene product.[13] A summary of the candidate gene exclusions is shown in Table 5.

## DISCUSSION

The results of an epidemiological and genetic analysis of autosomal retinitis pigmentosa in central and southern Sardinia are presented. The overall prevalence of RP in Sardinia, 1 in 3355, is similar to published prevalence figures in other countries.[14] However, the proportions of identified autosomal dominant and X-linked families are less than in most other series, with the exception of Swiss and Finnish studies.[7,8] The ADRP group is interesting since it may contain a higher proportion of rhodopsin mutations than in more ethnically diverse populations. The proportion of ADRP families with rhodopsin mutations was at least 50% (4/8) which is higher than observed in other populations. In addition, only two mutant alleles were found to account for all four rhodopsin ADRP families, at codons 190 (n=3) and 347 (n=1).

The observed proportion of ARRP families (32/158); 20.25%) is probably an underestimate of the true figure since a significant proportion of ARRP families with only a single affected child is expected on segregation analysis, which would expand the simplex (51.3%) at the expense of the ARRP group. The exact proportion of observed ARRP families depends on the criteria used; the strictest criteria might might only include the consanguineous group (5/32 putative ARRP families), but the inclusion of multiplex families showing only a single affected generation out of three or four or coming from geographically isolated communities provides a more realistic estimate.

The results of exclusion mapping the ARRP locus in a subset of nine ARRP families are shown in Figure 8. A total of 33 microsatellite markers were used and no significant evidence of linkage was identified. However, two important candidate loci were excluded, rhodopsin and peripherin/RDS. The B-PDE and rom-1 loci remain to be excluded using further markers. The total chromosomal region excluded by analysis of 33 markers, using the criterion of 100:1 odds (z<-2), was about 15% of the total genetic length (ca. 4000 centiMorgans).

## ACKNOWLEDGEMENTS

This work was supported by grants from the Italian "Ministero della Pubblica Istruzione", the Concerted Action of the European Communities, and U.K. Human Genome Mapping Project.

## REFERENCES

1. M.A. Musarella, Gene mapping of ocular diseases, Surv.Ophthalmol. 36:285 (1992).
2. A.E. Krill. "Hereditary Retinal and Choroidal Diseases,Vol. I, Evaluation," Harper & Row, Hagerstown (1972).
3. T.J.Keen, C.F.Inglehearn, D.H. Lester, R. Bashir,M. Jay, A.C. Bird, B. Jay, and S.S. Bhattacharya, Autosomal dominant retinitis pigmentosa: four new mutations in Rhodopsin, one of them in the retinal attachment site, Genomics 11:199 (1991).
4. T.J. Keen, D.H. Lester, C.F. Inglehearn, A. Curtis, and S.S. Bhattacharya, Rapid detection of single base mismatches as heteroduplexes on hydrolink gels, Trends Genet. 7:5 (1991).

5. G.M. Lathrop, J.M. Lalouel, C. Julier, and J. Ott, Strategies for multilocus linkage analysis in humans, Proc. Natl. Acad. Sci. USA 81:3443 (1984).

6. J.H. Edwards, Exclusion mapping, J. Med. Genet. 24:539 (1987).

7. F. Ammann, D. Klein, and A. Franceschetti, Genetic and epidemiological investigation on pigmentary degeneration of the retina and allied disorders in Switzerland, J. Neurol. Sci. 2:183 (1965).

8. H. Voipio, V. Gripenberg, C. Raitta, and A. Horsmanheims, Retinitis pigmentosa; a preliminary report, Hereditas 52:247 (1964).

9. P.J. Rosenfeld, G.S. Cowley, T.L. McGee, M.A. Sandberg, E.L. Berson, and T.P. Dryja, A null mutation in the rhodopsin gene causes rod photoreceptor dysfunction and autosomal recessive retinitis pigmentosa, Nature Genetics 1:209 (1992).

10. J.L. Weber and P.E. May, Abundant class of human DNA polymorphisms which can be typed using the polymerase chain reaction, Am. J. Hum. Genet. 44:388 (1989).

11. G.J. Farrar, P. Kenna, S.A. Jordan, R. Kurnar-Singh, M.M. Humphries, E.M. Sharp, D. Sheils, and P. Humphries, A three base deletion in the peripherin/RDS gene in one form of retinitis pigmentosa, Nature 354:478 (1991).

12. R. Kumar-Singh, S.A. Jordan, G.J. Farrar, and P. Humphries, Poly (T/A) polymorphism at the human retinal degeneration slow (RDS) locus, Nucleic Acids Res. 19:5800 (1991).

13. R.A. Bascom, J. Garciaheras, C.L. Hsieh, D.S. Gerhard, C. Jones, U. Francke, H.F. Willard, D.H. Ledbetter, and R.R. McInnes, Localization of the photoreceptor gene *roml* to human chromosome-11 and mouse chromosome-19 - sublocalization to human 11q13 between PGA and PYGM, Am. J. Hum. Genet. 51:1028

14. J.R. Heckenlively. "Retinitis Pigmentosa," J.B. Lippincott Co., Philadelphia (1988).

# CLINICAL FEATURES OF AUTOSOMAL DOMINANT
# RETINITIS PIGMENTOSA ASSOCIATED WITH THE
# GLY-188-ARG MUTATION OF THE RHODOPSIN GENE

Giuseppe Del Porto[1], Enzo M. Vingolo[2], Dezsö David[3*], Katharina Steindl[1], Heike Wedemann[3], Renato Forte[2], Alessandro Iannaccone[2], Andreas Gal[3], Mario R. Pannarale[2]

[1] Medical Genetic Section of the Experimental Medicine Department, University of Rome "La Sapienza", c/o Ospedale L. Spallanzani, Via Portuense 292, 00149 Rome, Italy
[2] Institute of Ophthalmology, University of Rome "La Sapienza", Viale del Policlinico 1, 00161 Rome, Italy
[3] Institute of Human Genetics, Medical University of Lübeck, Ratzeburger Allee 160, 2400 Lübeck 1, Germany
* Permanent address: Laboratorio de Genetica Humana, INSA, Lisbon, Portugal

## INTRODUCTION

Several different rhodopsin gene mutations have been identified in the last years in pedigrees with autosomal dominant retinitis pigmentosa (adRP). In view of the differences in the molecular nature and location of these mutations, defining the phenotype has become increasingly important in order to identify the clinical counterpart to the different functional abnormalities of the photopigment molecule.[1-9]

Here we report the clinical features of the affected members of a large pedigree with adRP, in which a point mutation predicting the Gly-188-Arg change of rhodopsin, first described by Dryja et al.,[10] was identified. The clinical picture associated with this mutation was characterized by a well preserved visual acuity, a typical bone-spicule like sparse pigmentation affecting every retinal quadrants, a homogeneous concentric narrowing of the visual field of variable extent, and a typical ERG rod-cone pattern. The results of this clinical study might provide a deeper insight into pathomechanisms of retinal degenerations due to rhodopsin mutations.

## MATERIALS AND METHODS

The pedigree shown in Fig. 1 was elaborated from a propositus affected with adRP.

Six further family members were investigated as follows:

**A.** Detailed medical history to define the age of onset of symptoms of the disease, with particular interest for night vision impairment, difficulty with side vision, and progressive loss of Visual Acuity (VA);

**B.** Ophthalmological evaluation:

**B. 1** Best corrected distant and near VA with a standard Snellen chart and astigmometric charts modified according to Pannarale.[11]

**B. 2** Anterior and posterior segment were investigated (before and after pupil dilatation) with particular reference to the parameters listed in Tab. 1:

- Lens opacities (Posterior Subcapsular Cataracts, PSCs) were biomicroscopically graded according to the criteria proposed by Fishman[12] and modified by Pannarale.[13] Six progressive grades were adopted evaluating the concentric extension of the opacities from the axial area. Aphakic (AK) and pseudophakic (PFK) eyes, which had already undergone cataract surgery, were listed separately.

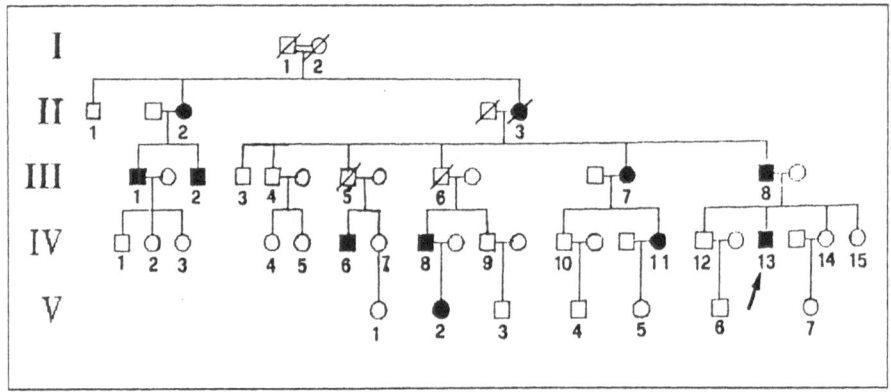

**Figure 1.** adRP pedigree in which a point mutation predicting the Gly-188-Arg change of rhodopsin was identified. Filled symbols indicate affected subjects in the pedigree. The propositus is highlighted by a bold diagonal arrow. Slashed symbols indicate dead persons. No informations were available about $I_1$ and $I_2$.

- Vitreal aspects were biomicroscopically graded according to a previously reported classification.[14]

- Fundus examination by indirect binocular ophthalmoscopy and slit lamp biomicroscopy, to assess possible central and/or mid-peripheral retinal changes. The aspects of the optic disc and the retinal vessels, the presence of pathological pigmentation, its topographical aspects and grading were particularly evaluated. The occurrence of macular alterations, such as epiretinal membranes, cystoid macular edema, subretinal neovascular membranes, macular atrophy or scarring, and pigmentary changes were also recorded.

**B. 3** Visual Field (VF) testing by Goldmann kinetic perimetry with I4e and III4e targets. Areas of the above targets were measured on conventional perimetry charts by

**Table 1.** Methods of classification of the major ophthalmological parameters.

| PARAMETERS | GRADING | | | | | | | |
|---|---|---|---|---|---|---|---|---|
| | **0** | **1** | **2** | **3** | **4** | **5** | **6** | **Other** |
| *Posterior Subcapsular Cataract (PSC)* | no PSC | Ø: 0.25 mm | Ø: <1mm<br>without sutural extensions | Ø: >1mm | extension not to equator | extension to equator | placoid | AK: aphakic PFK: IOL |
| *VITREOUS* | no alter. | fibrillary degeneration | cottonball floaters | non-pigmented particles | pigmentary particles | other aspects | | detachment |
| *OPTIC DISC* | normal | mild pallor | waxy pallor | sub atrophy | atrophy | other aspects | | (specify) |
| *RETINAL VESSELS* | normal | mild attenuation | moderate attenuation | severe attenuation | thread-like appearance | occlusive stage | | exudation, sheatings |
| *PIGMENTATION MORPHOLOGY* | absent (sine pigmento) | typical bone-spicule | atypical clumping | | | | | patchy RPE flecks |
| *PIGMENTATION EXTENSION* | absent (sine pigmento) | expressed in angular degrees depending on the number of quadrants involved (e.g., 90°, 180°, 270°, 360°) | | | | | | |
| *PIGMENTATION GRADING* | absent (sine pigmento) | sparse & thin | sparse & thick | moderate & thick | diffuse & thick | confluent | | |
| *MACULAR APPEARANCE* | normal | altered (specify) | | | | | | (specify) |

means of a light pen and expressed in square millimeters (mm$^2$), according to a previously described technique.[15] Scotomata within each isopter were measured and subtracted from the total area.

**B. 4** Electroretinographic analyses:

- Maximal ERG response was recorded with dilated pupils and after 30 minutes of dark-adaptation. After topical corneal anesthesia, Henkes-type corneal contact lens electrodes were placed on both eyes under dim red light and connected to a mechanical membrane suction pump to keep the electrodes stable on the corneal surface.[16,17] Ground and reference electrodes were fixed on the ear lobes. "Differential derivation" (DD) recording was performed in cases with amplitude <10 µV as described in detail elsewhere.[16,17] In brief, one corneal electrode was used as recording electrode, the other one, covered with a light-proof black bandage, as reference electrode. The signal was elicited with a 20 lux/sec-intensity white flash Ganzfeld stimulation at a 0.5 Hz repetition rate. Averaging of 50 to 100 iterations was performed, depending on the signal amplitude. Sophisticated post-processing analyses allowed separate averaging of groups of samples to verify the reproducibility of the signal. Amplitude and latency measurements were performed according to the international standards for clinical electroretinography.[18]

- Photopic ERG was recorded after the above testing and without modifying the patients' status in the faradic cage, elicited with a 20 lux/sec-intensity 30 Hz white flash Ganzfeld stimulation, and averaging 50 to 100 samples, depending on the signal amplitude. Amplitudes and latencies were measured as above.

Statistical analyses were performed on an Apple Macintosh PC with the Stat View SE+Graphics program.

Molecular genetic analysis was performed according to standard methods. DNA was extracted from peripheral blood samples obtained from affected and unaffected members of the pedigree. Methods used for detection and characterization of sequence variations of the rhodopsin gene have been extensively described elsewhere.[19] In brief, the five rhodopsin gene exons were amplified by polymerase chain reaction (PCR). PCR fragments were analysed with at least one of the following three screening methods: heteroduplex formation, single strand conformation polymorphism (SSCP), and restriction enzyme digestion. Finally, PCR fragments showing an aberrant pattern in any of the screening tests perfomed were sequenced according to standard protocols.


## RESULTS

The pedigree shows a typical autosomal dominant inheritance of the disease (Fig. 1). Six affected subjects between 8 and 64 yrs of age have been clinically examined. The data for each patient are summarized in Tab. 2.

**Patient III7.** This 64 yr-old female referred the onset of the first symptoms of RP at the age of 12, when she first experienced both night vision and side vision difficulties. Clinically, she presented with light perception in OD due to complications after cataract extraction (severe corneal opacities and edema, bullous keratopathy), and with a 20/20 (1.0) vision in OS. Color vision in the OS was normal. The patient had undergone cataract surgery in OU several years ago by an intracapsular procedure without secondary IOL implantation (i.e., aphakic). Surgical modifications in the vitreous did not allow its evaluation. Ophthalmoscopy disclosed typical RP with a thick and sparse bone-spicule like pigmentation involving all four quadrants. Optic discs were subatrophic (grade 3), retinal vessels severely attenuated with a thread-like appearance (grade 4). The macula appeared normal in OU. Visual field measurement was performable only on the left eye and showed extreme narrowing (only 163 mm$^2$ with the III4e target). The maximal ERG response (0.5 Hz) was severely reduced with a residual signal of 8 and 5 µV, respectively, for the two eyes. Cone 30 Hz ERG was not available.

**Table 2.** Summary of the clinical and functional findings in patients with the Gly-188-Arg aminoacid change.

| Pedigree position | Age | Sex | Age of Onset | Eye | Visual acuity | PSC | Vitreous | Optic Disc | Retinal Vessels | Morphol. | Pigmentation Extension (°) | Grading | Macula | I4 | III4 | 0.5 Hz | 30 Hz |
|---|---|---|---|---|---|---|---|---|---|---|---|---|---|---|---|---|---|
| | | | | | | | | **OPHTHALMOSCOPIC FINDINGS** | | | | | | **VISUAL FIELD** (area, mm²) | | **ERG** (µV) | |
| III7 | 64 | F | 12 | R | LP | AK | UN | 3 | 4 | 1 | 360 | 2 | 0 | UN | UN | 8 | UN |
| | | | | L | 1.0 | AK | UN | 3 | 4 | 1 | 360 | 2 | 0 | UN | 163 | 5 | UN |
| IV6 | 31 | M | 6 | R | 1.0 | 1 | 3 | 2 | 2 | 1 | 360 | 1 | 0 | 8737 | 17898 | 29.5 | UN |
| | | | | L | 1.0 | 0 | 3 | 2 | 2 | 1 | 360 | 1 | 0 | 8357 | 17917 | 27.7 | UN |
| IV8 | 42 | M | 30 | R | 1.0 | 4 | 2 | 1 | 1 | 1 | 360 | 2 | 0 | 3826 | 5036 | 17 | 4.7 |
| | | | | L | 0.6 | 4 | 2 | 2 | 1 | 1 | 360 | 2 | 0 | 1532 | 4517 | 19.7 | 7.5 |
| IV11 | 36 | F | 27 | R | 0.5 | 3 | 3 | 2 | 2 | 1 | 360 | 1 | 0 | 351 | 4833 | 24 | 1.8 |
| | | | | L | 0.4 | 3 | 3 | 2 | 2 | 1 | 360 | 1 | 0 | 180 | 4389 | 16 | 1 |
| IV13 | 38 | M | 20 | R | 1.0 | 4 | 4 | 2 | 2 | 1 | 360 | 2 | 0 | 107 | 683 | 6 | NR |
| | | | | L | 1.0 | 3 | 4 | 2 | 2 | 1 | 360 | 2 | 0 | 73 | 507 | 7 | NR |
| V2 | 8 | F | Unaware | R | 1.0 | 0 | 0 | 0 | — | 0 | 360 | 0 | 0 | UN | UN | 164.5 | 55 |
| | | | | L | 1.0 | 0 | 0 | 0 | — | 0 | 360 | 0 | 0 | UN | UN | 194 | 65 |

LEGEND (in alphabetical order)

AK: Aphakia
LP: light perception
UN: Undefinable
NR: Not Recordable
PSC: Posterior Subcapsular Cataract

95

**Patient IV$_6$.** This 31 yr-old male patient referred the onset of his first symptoms to the age of 6. He experienced night and side vision disturbances almost simultaneously, both difficulties having been primarily noted by his parents. Best corrected visual acuity was 1.0 in OU. No abnormality of color vision was found. The anterior segment examination was unremarkable, except an initial cataract on OD (grade 1 PSC). Vitreal non pigmented cells (grade 3) were seen in OU. Fundus examination revealed a typical RP condition with waxy pallor of the optic disc (grade 2), moderately attenuated retinal vessels (grade 2), and sparse and thin (grade 1) bone-spicule like pigmentation in all quadrants. Maculas were normal. Visual field testing revealed mild narrowing (8737 and 8357 mm$^2$ with the I4e target, and 17898 and 17917 mm$^2$ with the III4e target). Maximal ERG response was reduced (29.5 and 27.7 µV, respectively, for the two eyes). Cone 30 Hz ERG was not available.

**Patient IV$_8$.** This 42 yr-old male did not complain of any visual disturbance until the age of 30, when he first experienced both increasing night vision difficulties and side vision reduction. His best corrected visual acuity was 1.0 in OD and 0.6 in OS, despite a grade 4 PSC in OU. No color vision abnormality was detected in either eye. Vitreous cottonball-like condensations were seen in both eyes, while no particles were detectable. A waxy aspect of the optic discs and a mild attenuation of the retinal vessels were evident at ophthalmoscopic examination. Pigmentation occurred also in this case in a typical bone-spicule like, thick and sparse fashion in OU involving all 4 quadrants. Mild atrophic macular changes were observed in OS, while no such abnormality was found in OD. A moderate constriction of the visual field was seen, more prominently for the left eye with the I4e target (OD: 3826 mm$^2$; OS: 1532 mm$^2$), virtually superimposable with the larger target (5036 and 4517 mm$^2$, respectively). ERG results paralleled visual field findings showing a moderate impairment of both mixed and pure cone responses, the first one being proportionally more reduced than the latter one.

**Patient IV$_{11}$.** This 36 yr-old female referred the onset of night and side vision disturbances at the age of 27. Her best corrected visual acuity was 0.7 in OD and 0.8 in OS, and her color vision was normal. Slit lamp examination revealed grade 3 PSCs and non-pigmented vitreous particles in both eyes. Fundus appearance showed waxy pallor of the optic disc, moderate attenuation (grade 2) of the retinal vessels, typical bone-spicule like sparse and thin pigmentation in all 4 quadrants. No macular involvement was ophthalmoscopically evident. The I4e target visual field was severely constricted when compared to normal (351 and 180 mm$^2$, respectively), while the peripheral limits, as determined with the III4e target, were only moderately narrowed (4833 and 4389 mm$^2$). Similarly, the maximal ERG response and the 30 Hz cone ERG were equally impaired, implying a proportionally greater cone affection than in patient IV$_8$.

**Patient IV$_{13}$.** This 38 yr-old male patient did not notice any symptoms until the age of 20, when he first noted night vision difficulties and narrowing of the peripheral visual field. At the time of the present examination, his color vision was normal and his best corrected visual acuity was 1.0 in OU, despite grade 4 and 3 PSCs in OD and OS, respectively. Diffuse pigmented vitreous particles were detected in OU. The fundus demonstrated normal maculas, grade 2 optic discs and vessels, and typical, sparse and thick bone-spicules in the mid-periphery, affecting each quadrant. Visual field was severely constricted with both targets, and the maximal ERG response was markedly reduced as well. Cone ERG was not available.

**Patient V$_2$.** This 8 yr-old girl, daughter of patient IV$_8$, reported no subjective symptoms related to the disease. Her ophthalmological examination was unremarkable, with the exception of mildly attenuated retinal vessels and whitish RPE cloudiness in a patchy and reticular fashion in the mid-periphery. These aspects prompted further ERG analyses, which demonstrated a clearly subnormal maximal ERG response (50% of the lower normal limit) and a mildly attenuated and minimally delayed cone response.

A mutation in the rhodopsin gene at codon 188 that results in an aminoacid change (Gly-188-Arg) in the intradiscal domain was detected (Fig. 2). The Gly-188-Arg change in rhodopsin may cause profound functional disturbances of the photopigment. In fact, rhodopsin has two conserved cysteine residues, at positions 110 and 187, that are necessary to form and stabilize its correct structure by establishing a disulfide bridge. It is very likely that the aminoacid change next to Cys-187 prevents the formation of the disulfide bond. This disturbs the protein's structure and the assembly of a correct three-dimensional structure, which is essential for the formation of functionally active rhodopsin and binding of 11-cis retinal. Rhodopsin molecules mutated at (one of) the cysteine residues are expressed at reduced levels and are unable to generate the chromophore when incubated with retinal.

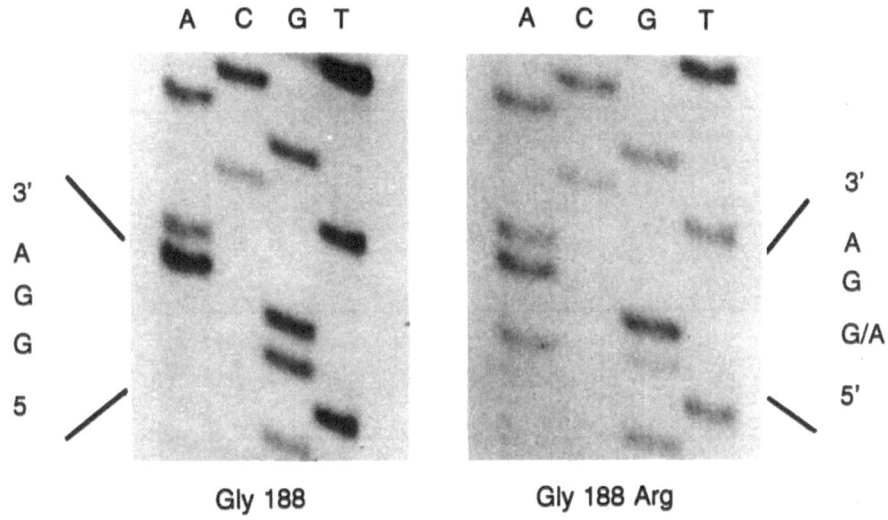

**Figure 2.** DNA sequences of PCR products from human rhodopsin gene. The wild type sequence is shown on the left, while the mutant sequence found both in patients III7 and IV13. on the right. The G-to-A transition predicts the replacement of Glycine-188 by Arginine.

## DISCUSSION

The examination of six patients from a family with adRP, caused by a point mutation of the rhodopsin gene (Gly-188-Arg), has allowed the accompanying clinical pattern to be defined. In Tab. 3, some phenotypic characteristics associated with Gly-188-Arg are summarized and compared to those observed in other mutations of the rhodopsin gene. In fact, this adRP pedigree shows several distinctive features.

Night blindness and visual field narrowing were referred as the first symptoms by all affected members of the pedigree but patient V2. Age of onset of night blindness and side vision disturbances ranged between 6 and 30 yrs (Tab. 2).

**Table 3.** Comparison of the phenotype associated with 7 different rhodopsin gene mutations (related references are given in brackets)

CLINICAL PARAMETERS

| RHODOPSIN MUTATIONS | AGE OF ONSET | VISUAL ACUITY | PSC | PIGMENTARY CHANGES | VISUAL FIELD | ERG AMPLITUDE | PROGNOSIS |
|---|---|---|---|---|---|---|---|
| Gly-188-Arg | Variable (6-30 yrs.) | Well preserved | Frequent | Sparse but diffuse (360°) | Concentric narrowing | Reduced | Benign |
| Thr-17-Met (ref. 8) | Late (> II decade) | Well preserved | Absent | Sparse and sectoral (inferior) | Regional defect (Superior) | Mildly Reduced | Benign |
| Pro-23-His (refs. 1,3,5) | Widely variable | Well preserved | Rare or absent | Sectoral (inferior and nasal) | Regional defect (Superior) | Moderately reduced | Fairly benign |
| Gly-106-Arg (ref. 9) | Late (IV decade) | Well preserved | Absent | Sectoral (inferior) | Regional defect (Superior) | Moderately reduced | Benign |
| Gly-182-Ser (ref. 8) | Early (I-II decade) | Moderately impaired | Absent | Sectoral (inferior) | Regional defect (Superior) | Variable Reduced | Benign |
| Pro-347-Leu (ref. 2) | Early (< 18 yrs.) | Quite impaired | Frequent | Diffused to all quadrants | Severely constricted | Severely reduced | Severe |
| Pro-347-Arg (ref. 6) | Early (< 11 yrs.) | Quite impaired | Frequent | Diffused to all quadrants | Concentric narrowing | Severely reduced | Severe |

BCVA ranged from 0.6 to 1.1 (Tab. 2). This is well in line with the values reported for some other rhodopsin defects (Pro-23-His, Thr-17-Met, Gly-182-Ser, Gly-106-Arg),[1,3,8,9] which were also characterized by a preserved visual acuity until late in the course of the disease (see Tab. 3).

PSCs at various stages were found in 4 ($IV_6$, $IV_8$, $IV_{11}$, $IV_{13}$) of the 6 patients (Tab. 2). A similar proportion of PSCs was reported for Pro-347-Leu [2] and Pro-23-His [1,5] mutations, while PSCs were rare or absent in patients with Thr-17-Met, Gly-182-Ser and Gly-106-Arg mutations (Tab. 3).[3,8,9]

Variable degrees of vitreal alterations were detected in 4 patients ($IV_6$, $IV_8$, $IV_{11}$, $IV_{13}$,Tab. 2). Vitreal aspects were investigated only in few studies. Patients with the Pro-347-Arg mutation presented moderate cellular particulation and fibrillar degeneration of the vitreous,[6] while no such alterations were found in patients with the Gly-106-Arg rhodopsin defect.[9]

Ophthalmoscopic examination showed typical intraretinal pigment clumping in a bone spicule-like fashion (type 1 of our classification) in all instances but one ($V_2$), who showed no pigmentary changes (Tab. 2). In all cases, chorioretinal pigmentary degenerative changes were extended from the vascular arcades to the mid-periphery in all four quadrants (360°). Berson et al.[2] and Niemeyer et al.[6] described a similar ophthalmoscopic pattern. However, these findings differ from other reports, [3,5,6,8] which described a preferential localization of typical degenerative alterations in the inferior retinal hemisphere for the Thr-17-Met, Pro-23-His, Gly-106-Arg and Gly-182-Ser mutations. Retinal vessels were mildly to moderately attenuated (grade 1-2) in all but one patient ($III_7$); this feature is very common in RP. Also $V_2$ showed mild attenuation of retinal vessels (grade 1) associated to whitish RPE cloudiness in a patchy and reticular fashion in the mid-periphery. Optic discs showed a waxy pallor (grade 2) in 4 ($IV_6$, $IV_8$, $IV_{11}$, $IV_{13}$) of the 6 patients. A sub-atrophic status (grade 3) was seen in case $III_7$, while subject $V_2$ had normal optic discs. Interestingly, Fishman and co-workers [8] reported waxy pallor only in patients with the rhodopsin Gly-182-Ser mutation, while members of the Thr-17-Met [8] and Gly-106-Arg [9] families showed normal discs.

Kinetic Goldmann visual field testing showed a homogeneous constriction in all patients with both the I4e and III4e targets (Fig. 4). This finding differs from those of other authors,[3,5,8,9] who described sectoral retinal degeneration and, accordingly, regionalized visual field impairments in families with mutation at codon 17, 23, 106, and 182. Visual field areas, expressed in $mm^2$, showed a substantial variability.

ERG amplitude was markedly subnormal to 0.5 Hz white flash stimulation in all instances. The most pronounced reduction in ERG amplitude was observed in cases $III_7$ and $IV_{13}$, who also had severe visual field loss. These findings are similar to those reported by Berson et al.[2] and Niemeyer et al.[6] in patients with Pro-347-Leu or Pro-347-Arg mutation, who also reported a comparably wider extension of retinal degenerative changes. Rod function seems to be proportionally more affected than cone function (Fig. 5), confirming a typical rod-cone pattern. Interestingly, the maximal ERG response was subnormal (50% of the lower normal limit) in the youngest patient ($V_2$), a sign strongly suggestive of a diffuse rod degeneration.

The diffuse photoreceptor impairement detected in this pedigree is similar to that seen in families with rhodopsin codon 347 mutation, [2,6] that is, in general, more severe than the one observed in regionalized degeneration ("sector" RP).[1,3,5,8,9] The clinical pattern of the Gly-188-Arg mutation appears to be comparable to type-1 adRP described by Massof and Finkelstein.[20]

In summary, the clinical picture related to the Gly-188-Arg mutation is characterized, from a functional point of view, by a well preserved visual acuity, homogeneous concentric narrowing of the visual field of variable extent, and a typical rod-cone ERG pattern. Age of onset of the first symptoms was quite variable, ranging from 6 to 30 yrs, but in every case patients reported that night blindness and visual field defects occurred and progressed simultaneously.

The results of this report demonstrate that a thorough evaluation of fundus parameters together with a standardized classification allow a better definition of the phenotypes related to each genotype, and a precise evaluation of the clinical variability within genealogies.

## ACKNOWLEDGEMENTS

Molecular genetics studies were financially supported by the Deutsche Forschungsgemeinschaft (Ga 210/5-2) and the WTZ-Program (BMFT, Bonn, Germany).

## REFERENCES

1. E.L. Berson, B. Rosner, M.A. Sandberg, and T.P. Dryja. Ocular findings in patients with Autosomal Dominant Retinitis Pigmentosa and a Rhodopsin gene defect (Pro-23-His). *Arch. Ophthalmol.* 109: 92 (1991).

2. E.L. Berson, B. Rosner, M.A. Sandberg, C. Weigel-Di Franco, and T.P. Dryja. Ocular findings in patients with Autosomal Dominant Retinitis Pigmentosa and Rhodopsin, Pro-347-Leu. *Am. J. Ophthalmol.* 111: 614 (1991).

3. J.R. Heckenlively, J.A. Rodriguez, and S.P. Daiger. Autosomal Dominant Sectoral Retinitis Pigmentosa. Two families with transversion mutation in codon 23 of rhodopsin. *Arch. Ophthalmol.* 109: 84 (1991).

4. J.E. Richards, C.Y. Kuo, M. Boehnke, and P.A. Sieving. Rhodopsin Thr58Arg mutation in a family with autosomal dominant Retinitis Pigmentosa. *Ophthalmology*, 98: 1797 (1991).

5. E.M. Stone, A.E. Kimura, B.E. Nichols, P. Khadivi, G.A. Fishman, and V.C. Sheffield. Regional distribution of retinal degeneration in patients with the Proline to Histidine mutation in codon 23 of the Rhodopsin gene. *Ophthalmology*, 98: 1806 (1991).

6. G. Niemeyer, P. Trüb, A. Schinzel, and A. Gal. Clinical and ERG data in a family with autosomal dominant RP and Pro-347-Arg mutation in the rhodopsin gene. *Doc. Ophthalmol.* 79: 303 (1992).

7. G.A. Fishman, E.M. Stone, L.D. Gilbert, P. Kenna, and V.C. Sheffield. Ocular findings associated with a rhodopsin gene codon 58 transversion mutation in autosomal dominant retinitis pigmentosa. *Arch. Ophthalmol.* 109: 1387 (1991).

8 G.A. Fishman, E.M. Stone, V.C. Sheffield, L.D. Gilbert, and A.E. Kimura. Ocular findings associated with Rhodopsin gene codon 17 and codon 182 transition mutations in Dominant Retinitis Pigmentosa. *Arch. Ophthalmol.* 110: 54 (1992).

9 G.A. Fishman, E.M. Stone, L.D. Gilbert and V.C. Sheffield. Ocular findings associated with a rhodopsin gene codon 106 mutation. Glycine-to-Arginine change in autosomal dominant retinitis pigmentosa. *Arch. Ophthalmol.* 110: 646 (1992).

10. T.P. Dryja, L.B. Hahn, G.S. Cowley, T.L. Mc Gee, and E.L. Berson. Mutation spectrum of the rhodopsin gene among patients with autosomal dominant Retinitis Pigmentosa. *Proc. Natl. Acad. Sci. USA* 88: 9370 (1991).

11. M.R. Pannarale. La determinazione della correzione astigmatica. Indicazioni e tecniche d'uso di quadranti astigmometrici personali. *Boll. Ocul.* 56: 71 (1977).

12. G.A. Fishman, R.J. Anderson, and P. Lourenco. Prevalence of posterior subcapsular lens opacities in patients with retinitis pigmentosa. *Br. J. Ophthalmol.* 69: 263 (1985).

13. L. Pannarale, E. Rispoli, R. Forte, E.M. Vingolo, G. De Ruggieri, C. Giusti, K. Steindl, and M.R. Pannarale. Posterior Subcapsular Cataract in retinitis pigmentosa: incidence in our casistic and proposal of a photographic system for a quantitative grading. Proceedings of the Workshop on Retintis Pigmentosa, Rome, March 2nd, 1991. Boll. Ocul. 72 (Suppl. 1): 111 (1993).

14. P. Onori, L. Pannarale, G. De Propris, C. Carlevale, and E.M. Vingolo. Vitreal alterations in Retinitis Pigmentosa. Proceedings of the International Symposium on Retinitis Pigmentosa, Napoli, April 13-15, 1992 (in press).

15. E.M. Vingolo, A. Iannaccone, P. Onori, K. Steindl, and E. Rispoli. Correlations between perimetric and electroretinographic measurements in Retinitis Pigmentosa. *Invest. Ophthalmol. Vis. Sci.* 33: 840 (1992).

16. E. Rispoli, A. Iannaccone, and S. Moretti. Problematiche di registrazione dell'ERG nella retinite pigmentosa: innovazioni tecnologiche, *in:* "Retinite Pigmentosa, movimenti oculari e ambliopia, glaucoma", proceedings of the VI Course on Electrophysiological Techniques in Ophthalmology (Parma, March 6-8, 1991), M. Cordella, G. Baratta, and C. Macaluso, eds., Casa Editrice Mattioli, Fidenza (1991).

17. A. Iannaccone, P. Tanzilli, and M. Rispoli. ERG recording: low-noise techniques in Retinitis Pigmentosa. *Invest. Ophthalmol. Vis. Sci.* 33: 840 (1992).

18. M.F. Marmor, G.B. Arden, S.E. Nilsson, and E. Zrenner. Standard for Clinical Electroretinography. *Arch. Ophthalmol.* 107: 816 (1989).

19. S. Bunge, H. Wedemann, D. David, D.J. Terwillinger, C. Aulchla-Scholz, C. Sammans, M. Horn, J. Ott, E. Schwinger, E.M. Bleeker-Wagemakers, A. Schinzel, M.J. Denton and A. Gal. Molecular analysis and genetic mapping of the rhodopsin gene in families with autosomal dominant retinitis pigmentosa. *Genomics* (in press).

20 R.W. Massof and D. Finkelstein. Two forms of autosomal dominant primary retinitis pigmentosa. *Doc. Ophthalmol. Proc. Ser.* 51: 289 (1981).

# OCULAR FINDINGS IN PATIENTS WITH AUTOSOMAL DOMINANT RETINITIS PIGMENTOSA DUE TO A 3-BASE PAIR DELETION AT CODON 255/256 OF THE HUMAN RHODOPSIN GENE

Birgit Lorenz,[1,2] Christine Hörmann,[2] Monika Horn,[3] Thomas Berninger,[2] Monika Döring,[2] and Andreas Gal[3]

[1]Dept. of Ophthalmology, University of Regensburg,
  Franz-Josef-Strauß-Allee 11, D-W8400 Regensburg, F.R.Germany
[2]Dept. of Ophthalmology, University of München,
  Mathildenstr. 8, D-W8000 München 2, F.R.Germany
[3]Dept. of Human Genetics, Medical University of Lübeck,
  Ratzeburger Allee 160, D-W2400 Lübeck 1, F.R.Germany

## INTRODUCTION

Retinitis pigmentosa (RP) is a genetically heterogeneous group of progressive hereditary retinal dystrophies. All three modes of Mendelian inheritance, autosomal dominant, autosomal recessive, and X-linked have been described. Even within the same mode of inheritance further genetic heterogeneity has been proven during the last three years as different genes/loci on different chromosomes have been identified for autosomal dominant retinitis pigmentosa (adRP) (for review see ref.1, and Farrar et al., this volume). The gene for rhodopsin which is localized on the long arm of chromosome 3 was the first adRP gene identified[2]. Since then an ever increasing number of allelic rhodopsin mutations have been described[3]. Different types and methods of subgrouping of adRP had been suggested on the basis of clinical observations[4-8]. The most widely accepted classification is that of Massof and Finkelstein who divided adRP in two types[5]. Type I is characterized with a diffuse loss of rod function and with an onset of night blindness usually during the first decade of life. The cone function is relatively well preserved throughout life. In contrast, type II ("sector") of adRP shows a regional loss of both rod and cone function, and a much later onset of night blindness. Type I is identical to type "D", and type II to type "R" according to Lyness and coworkers[8]. As this classification is based solely on phenotypic features, identification of the primary genetic defect may allow a more appropriate subtyping.

This paper describes the RP phenotype associated with a 3-base pair (bp) deletion at codon 255/256 of the human rhodopsin gene in a German family that has been reported recently[9]. The same deletion has also been identified in a British kindred[10].

*Retinal Degeneration*, Edited by J.G. Hollyfield *et al.*
Plenum Press, New York, 1993

The phenotypes of the two families will be compared to each other and to the phenotypes of other allelic mutations in the human rhodopsin gene.

## FAMILY DATA AND METHODS

In the family described, retinitis pigmentosa is known for at least five generations (Figure 1). The pattern of inheritance clearly shows autosomal dominant transmission of the trait. Penetrance appears to be high. Out of 23 documented members, 16 are reported to be affected. A total of 4 affected subjects were tested for the presence of the 3-bp deletion at codon 255/256 of the human rhodopsin gene that eliminates an isoleucine residue without changing the reading frame. A detailed description of the molecular genetic method has been published[9].

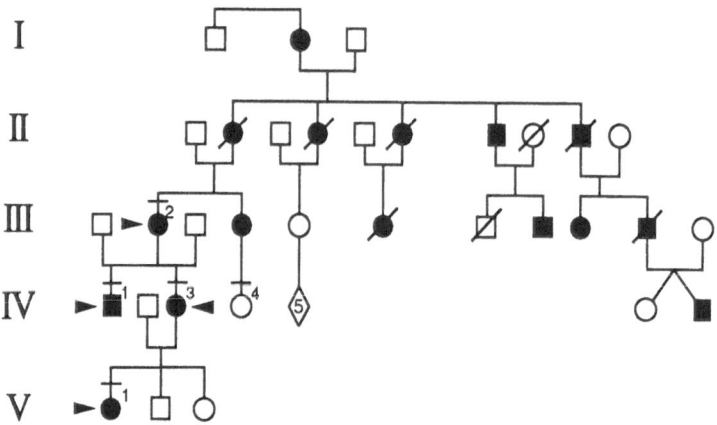

**Figure 1.** Pedigree of the German family with adRP due to a 3-bp deletion at codon 255/256 of the human rhodopsin gene. Solid symbols indicate affected individuals as known from history. Arrowheads indicate individuals examined clinically. Horizontal bars indicate individuals in whom DNA analysis was performed.

Four affected members were examined clinically. The family and patients´ history was recorded with respect of both, the age of onset of visual dysfunction and of nonocular medical problems such as hearing impairment. Ophthalmologic examination included best visual acuity at distance, slit lamp examination, fundoscopy, fundus photography, computerized static and kinetic perimetry, color vision testing, dark adaptometry, and electroretinography.

### Perimetry

Computerized perimetry was performed using an Octopus 2000 perimeter. Two programs were used, program 24 (grid of 15°) and program 32 (grid of 6°). The stimulus size was 4 mm$^2$ with a maximum intensity of 318 cd /m$^2$ on a 1.72 cd/m$^2$ white background. The stimulus duration was 100 ms. The range of attenuation was 0 to 51 dB. In one case, visual fields were too restricted. Therefore, kinetic perimetry was performed using a calibrated Goldmann perimeter. Targets V (64 mm$^2$ area) and I (0.25 mm$^2$ area) were used at 318 cd/m$^2$ intensity on a 10 cd/m$^2$ white background. Stimuli were moved from unseeing to seeing areas.

## Color vision

Color vision was tested with the Arden color contrast sensitivity test. This test is able to measure retinal color sensitivity alone by eliminating age-related influences from opacities in the optical media of the eye[11,12]. The technique has been described in detail elsewhere[13] and was modified in both, hardware and software, as follows: Hardware: The colors were displayed on monitors with a 90 Hz refresh rate. The monitors contain inbuilt computers which constantly monitor the characteristics of the video amplifiers, to ensure a constant luminous output. The software was rewritten in "C" for use in a PC, using an 80286 CPU with a mathematical co-processor. Software: Threshold was determined by a modified Binary Search scheme. Instead of gratings, the software presented an isoluminant alphabetic letter, and the subject's task was to read it corrrectly. The software ensured that the luminance remained constant while the color contrast was changed in all test situations. The relative luminance of the red/green and blue/green guns was first determined by heterochromatic flicker photometry, and then color contrast thresholds were measured in protan and tritan color confusion axes, which were orthogonal to each other in CIE color space. In the youngest patient, color vision was tested with the desaturated Lanthony Panel D 15 test.

## Dark adaptation

Dark adaptation was tested using a calibrated Goldmann Weekers adaptometer. Patients were dark adapted for a period of ten minutes of dark adaptation, followed by a 5 minute bleach of 1400 to 2100 asb. Dark adaptation was measured every minute during 20 minutes using an 11° black and white stripe target at maximum contrast, and with a luminance of 6 lux. The target was presented at the center of the visual field.

## Electroretinography

Full-field ERGs were recorded with a Nicolet Visual Compact, with Ganzfeld Controller (Nicolet Biomedical Instruments, Madison, Wisconsin, USA), and Henkes electrodes according to the international ERG standard[14]. In one case (V/1, 6 y), ERG was recorded using lid electrodes. ERG responses were amplified and filtered through a band pass. After the pupils were dilated with 0.5% tropicamide eye drops and 30 minutes of dark adaptation, rod responses were elicited by 0.5 Hz white stimuli (duration of 100 μs). The standard flash intensity of 2.5 cds/m$^2$ was reduced to 0.4, 4 and 40 mcds/m$^2$ in order to record isolated rod responses. At each intensity, 6 flashes were recorded and averaged. Combined rod cone responses were recorded by averaging six 0.2 Hz white stimuli of 2.5 cds/m$^2$ intensity. Photopic responses of L-cones (long wavelength cones) were obtained after 10 minutes of light adaptation to a ganzfeld dome background illumination of 30 cd/m$^2$. Single white (standard) and red flashes (Kodak Wratten gelatine filter 29, 629 nm) with 0.063, 0.11, 0.63, and 2.5 cds/m$^2$ intensities were presented on a blue background illumination of 22 cd/m$^2$. At each intensity, 20 responses were averaged. For flicker ERG, 50 responses to a 30-Hz stimulus at standard intensity were averaged. Photopic responses of S-cones (short wavelength cones) were obtained using blue flashes produced by filtering the standard white stimulus (2.5 cds/m$^2$) with a Kodak Wratten gelatine filter 98 (450 nm) on a yellow background illumination of 54 cd/m$^2$. A total of 100 responses were averaged.

## RESULTS

The patients´ ophthalmologic histories are summarized in Table 1. In all patients, night blindness was evident from childhood on. In the youngest proband, V/1, night blindness was suspected as early as in the first year of life. However, this early onset might not represent very early manifestation and variable expressivity of thease but could rather be due to increased awareness of this phenomenon by the parents. Unrelated to adRP, there was a history of squint in III/2 and V/1 with amblyopia in the squinting eye (III/2 L.E., V/1 R.E.). There was no history of hearing impairment or other medical problem in any of the 4 probands.

**Table 1.** Ophthalmological history of 4 subjects of the family with adRP (see Figure 1) due to a 3-bp deletion at codon 255/256 of the human rhodopsin gene.

| Patient Age | Onset of night blindness | Onset of day vision problems (visual field constriction) |
|---|---|---|
| V/1, 6 years | first year of life | no complaints at 6 years |
| IV/3, 32 years | childhood | 20 years |
| IV/1, 41 years | childhood | 20 years, cataracts at 39 years |
| III/2, 71 years | childhood | 3rd decade |

### Ocular findings

Ocular findings are summarized in Table 2. Representative fundus photographs are shown in Figure 2. Definite peripheral changes of the fundus were present in all quadrants as early as by the age of 6. The macular area appeared to be well preserved even at the age of 71, but peripheral changes had significantly progressed. Disease severity and progression

**Table 2.** Clinical data of the 4 probands.

| Patient Age | Eye | V.A | Refraction | Optical media | Central fundus | Peripheral fundus (all quadrants) |
|---|---|---|---|---|---|---|
| V/1 6 y | R.E. | 20/40 | not available | clear | macula near normal | RPE mottling/ thinning, prominent choroidal vessels |
| | L.E. | 20/30 | not available | clear | R.E. = L.E. | R.E. = L.E. |
| IV/3 32 y | R.E. | 20/30 | +1.25/-1.25x165 | clear | macula well preserved | mild choroidal sclerosis, sparse bone spicules |
| | L.E. | 20/30 | +1.25/-1.0x35 | clear | R.E. = L.E. | R.E. = L.E. |
| IV/1 41 y | R.E. | 20/50 | +1.5 | post.subcaps. cataract | macular area preserved | choroidal sclerosis, sparse bone spicules |
| | L.E. | 20/40 | +1.75 (over IOL) | posterior chamber IOL | R.E. = L.E. | R.E. = L.E. |
| III/2 71 y | R.E. | 20/60 | +8.5/-0.75x170 | nuclear/post. subcaps.cataract | macular area preserved | choroidal sclerosis, bone spicules |
| | L.E. | 20/150 | +11.25/-0.5x140 | synchisis scintillans | obscured by synchisis | ⇒ |

as to the retinal changes appeared rather uniform in this pedigree although the number of affected individuals examined was limited. The severe vitreous changes in patient III/2 are most likely secondary to the retinal changes, and therefore, not indicative of a more severe disease. The age of onset of cataract formation appeared to be variable in the small sample tested. Cataract formation requiring surgery was present at the age of 39 in one patient (IV/1), whereas his mother (III/2), at the age of 71, showed only mild lens changes.

**Figure 2a.** Fundus photographs of the 2 younger probands (always right eye shown). At the age of 6 (top), choroidal vessels are prominent outside the big arcades due to RPE thinning and mottling. At the age of 32 (bottom), sparse bone spicules are present in all quadrants; choroidal vessels are visible surrounding the macula. There is also thinning of the retinal arteries and optic disc pallor.

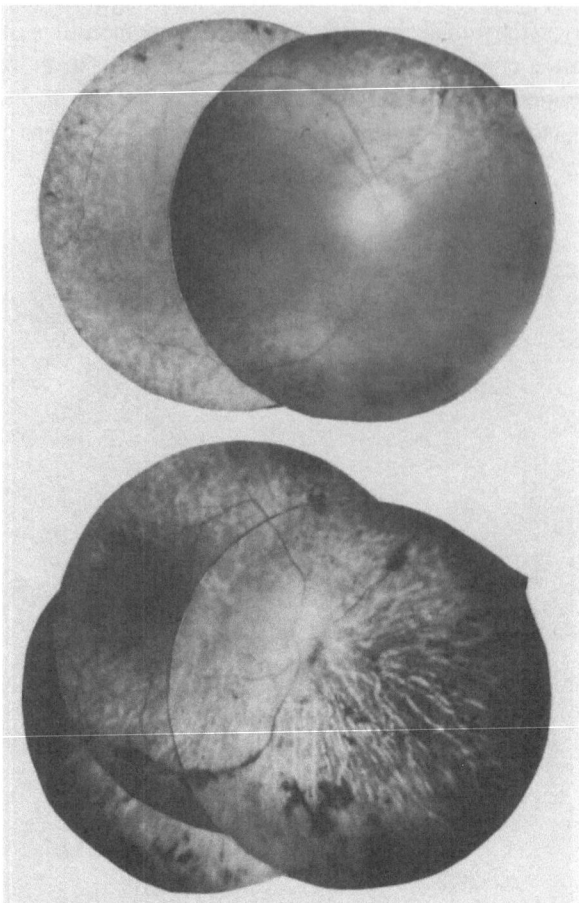

**Figure 2b**. Fundus photographs of the 2 elder probands (always right eye shown). At the age of 41 (top), choroidal sclerosis is even more prominent but pigment migration into the retina is still sparse, and the macula appears to be preserved although the view is obscured by a posterior capsular cataract. At the age of 71 (bottom), choroidal sclerosis is most prominent, and pigment migration into the neuroretina is more pronounced; the macular area is still well preserved. Retinal artery thinning and optic disc pallor in both probands.

## Psychophysical findings

Static perimetry was performed in subjects III/2, IV/1, and IV/3 (Figure 3). The visual fields were best preserved in subject IV/3, aged 32. In subject IV/1, aged 41, visual fields were constricted to about 10 degrees with preserved temporal islands. In subject III/2, aged 71, a central visual field of about 10 degrees was still measurable in the right eye but no peripheral islands were left. Visual field in the left eye was extremely constricted, and could not be tested by static perimetry. Kinetic perimetry with a standard Goldmann perimeter revealed a concentric constriction to 10 degrees for the stimulus V/4 (result not shown). Progression between the 4th and the beginning of the 8th decade appeared to be slow.

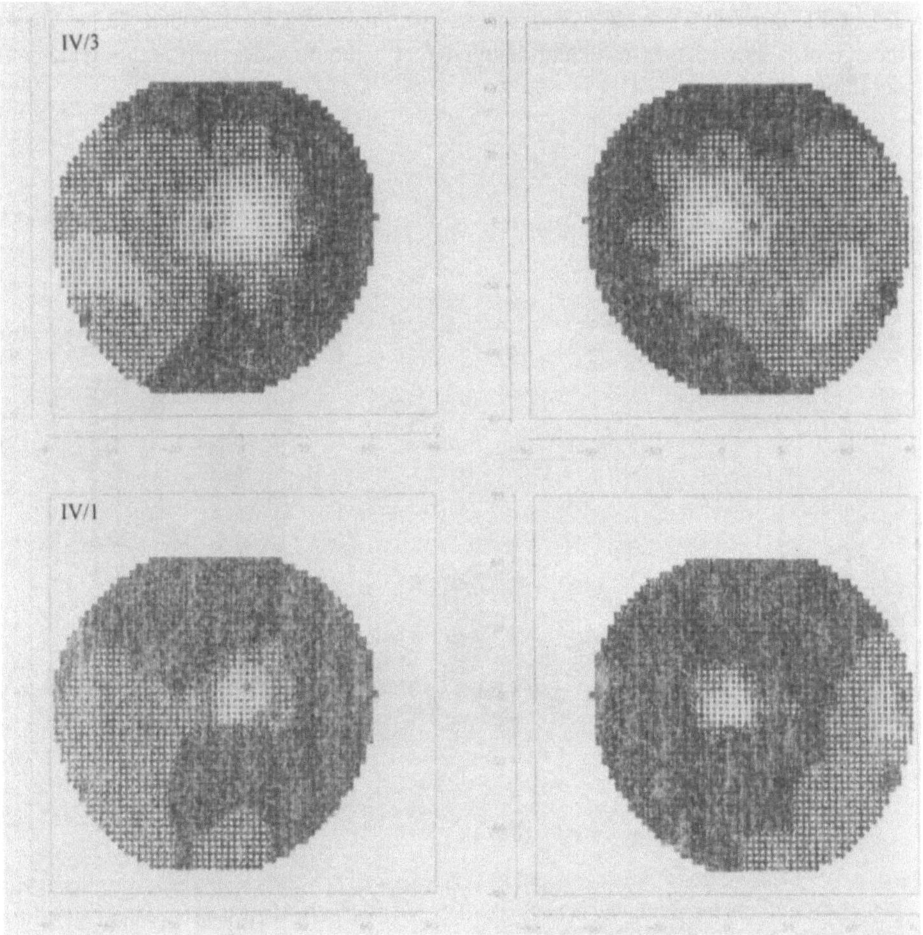

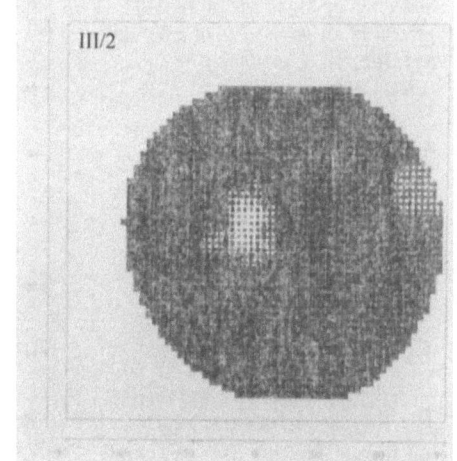

**Figure 3.** Visual fields in probands IV/3, IV/1, and III/2 with a computerized static perimeter Octopus 2000 (programs 24 and 32 combined). In proband IV/3, aged 32, a general depression in all quadrants is observed, with some residual sensitivity at nearly all eccentricities. In proband IV/1, a concentric constriction to about 30° is present, with some residual sensitivity temporally and inferiorly. In proband III/2, a severe concentric constriction to about 10° (R.E.) is observed. In the L.E., the visual field could be measured only with a kinetic Goldmann perimeter (not shown).

Color vision as tested with the Arden color contrast sensitivity test showed almost normal values for green and red cones up to the 4th decade (Figure 4). However, blue cones were already compromised at that age showing an increasing tritanomaly. At the age of 71, thresholds for all 3 cone systems were considerably elevated. Color vision at the age of 6 showed a diffuse color vision defect without clear axis (Lanthony Panel D15 desaturated test).

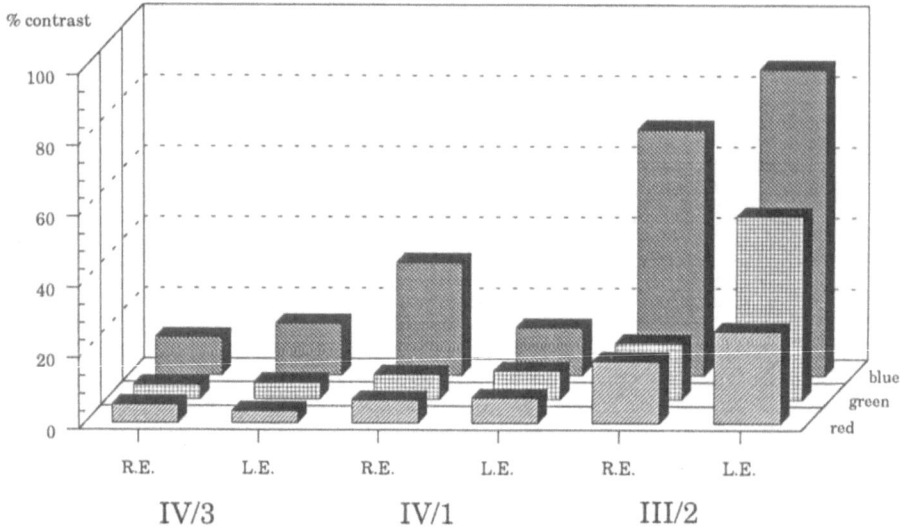

**Figure 4.** Color contrast sensitivity in probands III/2, IV/1, and IV/3. Normal color contrast sensitivity requires less than 7.5% contrast for all three cone systems. Almost normal color contrast sensitivity for red and green up to the age of 41 (IV/3, IV/1), but significantly reduced at the age of 71 (III/2). Tritan defect present in all probands, but most severe at the age of 71 (III/2). Interocular differences in probands IV/1 and III/2 can be explained by significant posterior capsular cataract (IV/1, R.E.) and synchisis scintillans (III/2, L.E.).

Dark adaptation for rods was severely compromised in all probands (Figure 5). At the age of 6 (V/1), the threshold was elevated by 2 to 2.5 log units (after 20 minutes of dark adaptation). At the age of 41 (IV/1), the threshold was elevated by about 3 log units. The threshold for cones (after 10 minutes of dark adaptation) was elevated by about 0.5 log unit at the age of 6 (V/1)and 32 (IV/3), by about 1.0 log unit at the age of 41 (IV/1), and by about 1.5 log units at the age of 71 (III/2) indicating progressive cone involvement.

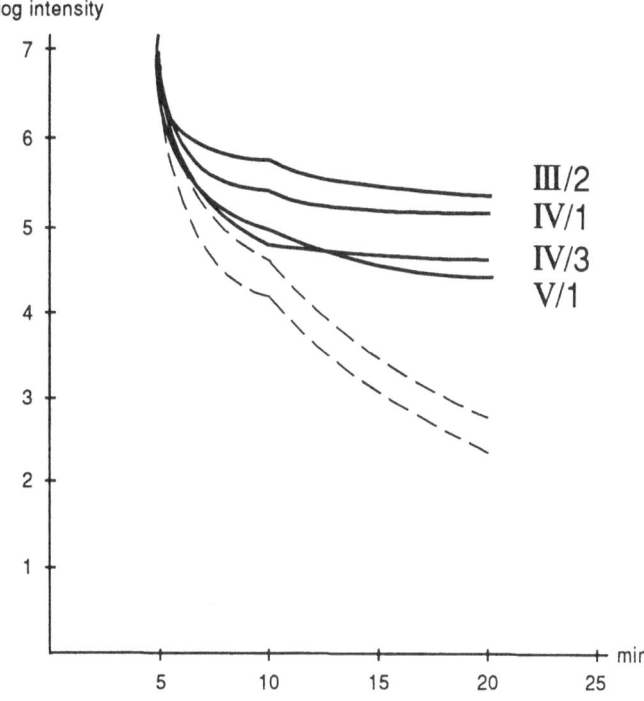

**Figure 5.** Dark adaptation curves in the 4 probands. The normal range for the age of 20 to 40 is indicated by hatched lines. V/1 6 y, IV/3 32 y, IV/1 41 y, III/2 71 y. For details see text.

### Electrophysiological findings.

Rod ERGs were absent or highly reduced in all probands even in the youngest aged 6. Cone ERGs were reduced in amplitude and delayed in implicit times, but could still be recorded at the age of 71 (Figure 6). Cone ERGs were preserved best in the youngest proband aged 6. In probands IV/3, IV/1, and III/2 cone flicker ERG amplitudes were reduced to about 10 to 15 $\mu$V (normal range 112-266 $\mu$V); and latencies ranged from 61 to 70 ms (normal range 60-64 ms), but were not age related. It is remarkable that the S-cone (shorth wavelength cone) ERG was better preserved than the L-cone (long wavelength cone) ERG although color contrast sensitivity was more reduced for blue than for green and red (compare Figure 4).

### DISCUSSION

We evaluated the phenotype of 4 affected individuals in a German family with adRP due to a 3-bp deletion at codon 255/256 of the rhodopsin gene. This small deletion has been previously described only in a British kindred[9]. As both inter- and intrafamilial phenotypic variability has been reported for mutations in the human rhodopsin gene, e.g. for the Pro-347-Leu mutation[15,16] a comparison of the phenotypes in the British and German kindreds segregating for the same mutation is important. Although a detailed clinical description of the British kindred has not been published yet, the limited amount of information available to us[10] (also personal communication F. Fitzke) together with our data

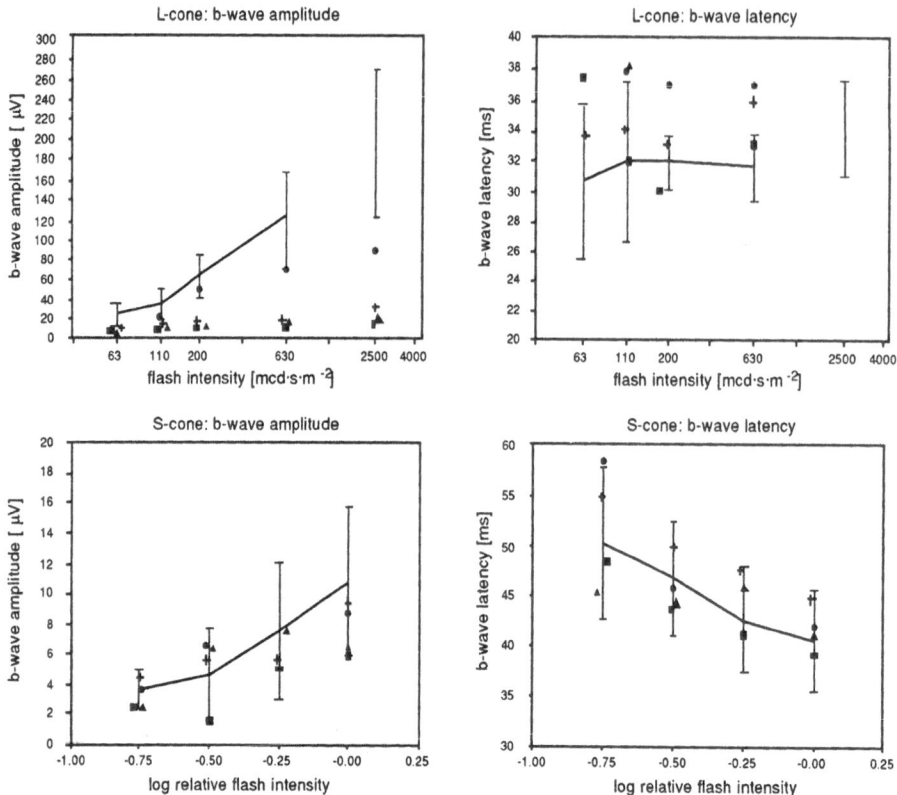

**Figure 6.** L- and S-Cone ERG responses in the 4 probands (●6y; +32y; ▲41y; ■71y). Bars indicate normal range (±2SD). Amplitudes and implicit times are within the normal range for S-cones. Amplitudes are reduced to about 5% for L-cones at standard flash intensity, latencies are within normal range or prolonged. In some instances, implicit times were prolonged over 40 ms (to 45 ms at maximum, not shown).

on the German family suggest that the phenotypes are rather similar in the two families. In all cases, the onset of night blindness was during the first decade in all cases, and might actually be present from birth on. Significant field restriction was evident by the age of 30. However, central cone function was relatively well preserved into the beginning of the 8th decade of life with still measurable (photopic) visual fields. By that time, visual fields were constricted to about 10 degrees. It is remarkable that cone ERG was still recordable even then. At that stage, pigment migration into the neuroretina was still relatively mild. In contrast, some cone dysfunction was already apparent as early as by the age of 6 as evidenced by reduced visual acuity and color vision abnormalities. Furthermore, distinct peripheral fundus changes were present in all quadrants. Choroidal vessels were prominent suggesting RPE thinning or depigmentation, possibly related to rod photoreceptor dystrophy. All these findings are consistent with type I adRP according to Massof and Finkelstein[5], or D-type according to Lyness and coworkers[8].

Little is known about the functional consequences of the 3-bp deletion eliminating one of the sequential isoleucine residues at position 255 or 256. It has been speculated that omission of this single amino acid will have the effect of replacing each of the following 20 amino acids with a different amino acid in helix 6 of rhodopsin. The effect of this shift can be equivalent to 20 sequential mutations in this part of the molecule[17]. Recently, the biochemical defects associated with different rhodopsin mutations have been analysed by in vitro expression assays in tissue culture cells[18]. The 13 mutations studied fall functionally into two major classes. Class I mutants behave similar to wild type rhodopsin in all assays tested. In contrast, class II mutants have a significantly lower level of yield, regenerate with 11-cis-retinal variably or not at all, and remain in the endoplasmatic reticulum instead of being transported to the plasma membrane. It seems that mutations affecting residues in the transmembrane or intradiscal portions of the rhodopsin molecule represent often class II mutants while those in the intracellular domain belong mainly to class I mutants. Del 255/256 has not been classified yet. However, as this mutation may considerably alter the structure of the 6th transmembrane helix of rhodopsin, it could resemble class II mutants. Although, a reliable correlation between the clinical course of the RP and the underlying mutation has not been documented yet, it is worth mentioning that the overall severity of the disease in del 255/256 can be best compared to that associated with the Thr-17-Met,[18] Pro-23-His,[20,21] Thr-58-Arg,[22] and Gly-106-Trp[23] mutations, all class II mutants. Conversely, the del 255/256 phenotype appeared to be less severe, even at a later stage, than that related to mutations in codon 347, class I mutants, for which a more severe course of the disease has been consistently reported[15,16,24].

## Acknowledgements

Molecular genetic studies were financially supported by the Deutsche Forschungsgemeinschaft (Ga 210/5-2). Thanks are due to the Concerted Action of the CEC "Prevention of Blindness: Molecular Research and Medical Care in Retinitis Pigmentosa (RP)". Computer graphics and photographical work were carried out by K. Heinfling and M. Rosner. Perimetry and dark adaptometry were performed by M. Marx and A. Rottenwöhrer.

## REFERENCES

1. P. Humphries, P. Kenna, and G.J. Farrar, On the molecular genetics of retinitis pigmentosa, *Science* 256:804 (1992)

2. R.R. McInnes and R.A. Bascom, Retinal genetics: a nullifying effect for rhodopsin, *Nature Genet.* 1:155 (1992)

3. T.P. Dryja, T. McGee, E. Reichel, L.B. Hahn, G.S. Cowley, D.W. Yandell, M.A. Sandberg, and E.L. Berson, A point mutation of the rhodopsin gene in one form of retinitis pigmentosa, *Nature* 343:364 (1990)

4. E.L. Berson, P. Gouras, and R.D. Gunkel, N.C. Myrianthopoulos, Dominant retinitis pigmentosa with reduced penentrance, *Arch. Ophthalmol.* 81:226 (1969)

5. R.W. Massof and D. Finkelstein, Two forms of autosomal dominant primary retinitis pigmentosa, *Doc. Ophthalmol.* 51:289 (1981)

6. G.B. Arden, R.M. Carter, C.R. Hogg, D.J. Powell, W. Ernst, G.M. Clover, A.L. Lyness, and M.P. Quinlan, Rod and cone activity in patients with dominantly inherited retinitis pigmentosa: comparison between psychophysical and electroretinographic measurements, *Br. J. Ophthalmol.* 67:405 (1983)

7. G.A. Fishman, K.R. Alexander, and R.J. Anderson, Autosomal dominant retinitis pigmentosa: a method of classification, *Arch. Ophthalmol.* 103:366 (1985)

8. A.L. Lyness, W. Ernst, M.P. Quinlan, G.M. Clover, G.B. Arden, R.M. Carter, A.C. Bird, and J.A. Parker, A clinical, psychophysical and electroretinographic survey of patients with autosomal domiant retinitis pigmentosa, *Br. J. Ophthalmol.* 69:326 (1985)

9. A. Artlich, M. Horn, B. Lorenz, S. Bhattacharya, and A. Gal, Recurrent 3-bp deletion at codon 255/256 of the rhodopsin gene in a German pedigree with autosomal dominant retinitis pigmentosa, *Am. J. Hum. Genet.* 50: 876 (1992)

10. C.F. Inglehearn, R. Bashir, D.H. Lester, M. Jay, A.C. Bird, and S.S. Bhattacharya, A 3-bp deletion in the human rhodopsin gene in a family with autosomal dominant retinitis pigmentosa, *Am. J. Hum. Genet.* 48: 26 (1991)

11. G.B. Arden, K. Gündüz, and S. Perry; Colour vision testing with computer graphics system, *Clin. Vis. Sci.* 12:303 (1988)

12. K. Gündüz and G.B. Arden; Changes in contrast sensitivity associated with operating argon lasers, *Br. J. Ophthalmol.* 73:241 (1989)

13. G.B. Arden, T.A. Berninger, C.R. Hogg, S. Perry, and O.E. Lund, A survey of colour discrimination in German ophthalmologists: change associated with the use of lasers and operating microscopes, AGARD, *Conf. Proc.* 492 (1991)

14. M.F. Marmor, G.B. Arden, S.E.G. Nilsson,and E. Zrenner, Standard for clinical electroretinography, *Arch. Ophthalmol.* 107:816 (1989)

15. E.L. Berson, B. Rosner, M.A. Sandberg, M.A. Weigel-DiFranco, and T.P. Dryja, Ocular findings in patients with autosomal dominant retinitis pigmentosa and rhodopsin, proline-347-leucine, *Am. J. Ophthalmol.* 111:614 (1991)

16. E. Apfelstedt-Sylla, M. Kunisch, M. Horn, K. Rüther, A. Gal, and E. Zrenner, Diffuse loss of rod function in autosomal dominant retinitis pigmentosa with pro-347-leu mutation in rhodopsin, *German J. Ophthalmol.* 1:319 (1992)

17. P.A. Hargrave and P.J. O´Brien, Speculations on the molecular basis of retinal degeneration in retinitis pigmentosa, in: "Retinal Degenerations," R.E. Anderson, J. Hollyfield, M.M. LaVail, eds., CRC Press, Boca Raton (1992)

18. C.H. Sung, B.G. Schneider, N. Agarwal, D.S. Papermaster, and J. Nathans Functional heterogeneity of mutant rhodopsins responsible for autosomal dominant retinitis pigmentosa, *Proc. Natl. Acad. Sci. USA* 88:8840 (1991)

19. G.A. Fishman, E.M. Stone, V.C. Sheffield, L.D. Gilbert, and A.E. Kimura, Ocular findings associated with rhodopsin gene codon 17 and codon 182 transition mutations in dominant retinitis pigmentosa, *Arch. Ophthalmol.* 110:54 (1992)

20. J.R. Heckenlively, J.A. Rodriguez, and S.P. Daiger, Autosomal dominant sectoral retinitis pigmentosa: two families with transversion mutation in codon 23 of rhodopsin, *Arch. Ophthalmol.* 109:84 (1991)

21. E.L. Berson, B. Rosner, M.A. Sandberg, and T.P. Dryja, Ocular findings in patients with autosomal dominant retinits pigmentosa and a rhodopsin gene defect (Pro-23-His), *Arch. Ophthalmol.* 109:92 (1991)

22. E.M. Stone, A.E. Kimura, B.E. Nichols, P. Khadivi, G.A. Fishman, and V.C. Sheffield, Regional distribution of retinal degeneration in patients with the proline to histidine mutation in codon 23 of the rhodopsin gene, *Ophthalmology* 98:1806 (1991)

23. G.A. Fishman, E.M. Stone, L.D. Gilbert, and V.C. Sheffield, Ocular findings associated with a rhodopsin gene codon 106 mutation, *Arch. Ophthalmol.* 110:646 (1992)

24. G. Niemeyer, P. Trüb, A. Schinzel, and A. Gal, Clinical and ERG data in a family with autosomal dominant RP and Pro-347-Arg mutation in the rhodopsin gene, *Doc. Ophthalmol.* 79:303 (1992)

# PHENOTYPES OF CARBOXYL-TERMINAL RHODOPSIN MUTATIONS IN AUTOSOMAL DOMINANT RETINITIS PIGMENTOSA

Eckart Apfelstedt-Sylla[1], Susanna Bunge[2], Dezsö David[2*],
Klaus Rüther[1], Andreas Gal[2], and Eberhart Zrenner[1]

[1] Department of Pathophysiology of Vision and Neuro-Ophthalmology,
University Eye Hospital, Tuebingen, Germany; [2] Department of Human
Genetics, Medical University, Luebeck, Germany; * Permanent address:
Laboratorio de Genetica Humana, INSA, Lissabon, Portugal

## ABSTRACT

We compared the ocular findings in 13 patients from 4 families with autosomal
dominant retinitis pigmentosa and 4 different mutations in the carboxyl-terminal sequence of
rhodopsin. Phenotypic similarities were found among patients with point mutations
predicting the amino acid changes valine-345-methionine, proline-347-serine or proline-347-
leucine in the rhodopin molecule. These patients had no measurable rod function and an early
impairment of cone function, which was most profound in the proline-347-serine genotype.
One patient with a valine-345-methionine mutation showed a regional predilection of fundus
abnormalities and cone sensitivity loss. A different phenotype with relatively mild disease
expression could be observed in a family with a deletion of 8 base pairs (codons 341-343).
One 34 year-old member showed regionally varying rod sensitivity loss, which was less
severe in the peripheral visual field, and well maintained cone function, as measured by
electroretinography and psychophysical tests.

## INTRODUCTION

In 1990, Dryja and coworkers first reported a point mutation of the rhodopsin gene in
families with autosomal dominantly inherited retinitis pigmentosa (ADRP)[1]. A cytosine-to-
adenine transversion was identified, which should result in a substitution of histidine for
proline in the 23rd amino acid of the rhodopsin molecule. Since then, a variety of mutations
in the rhodopsin gene have been shown to cosegregate with clinical signs of RP[2-6], and are
therefore looked upon as the primary cause of the disease.

As there are well-defined genotypes of ADRP, an interesting question is whether or not
there are characteristic phenotypic features associated with each of the rhodopsin mutations.
If so, this would considerably improve counselling of patients with respect to their visual
prognosis. In addition, clinical assessment of patients might help to better understand the

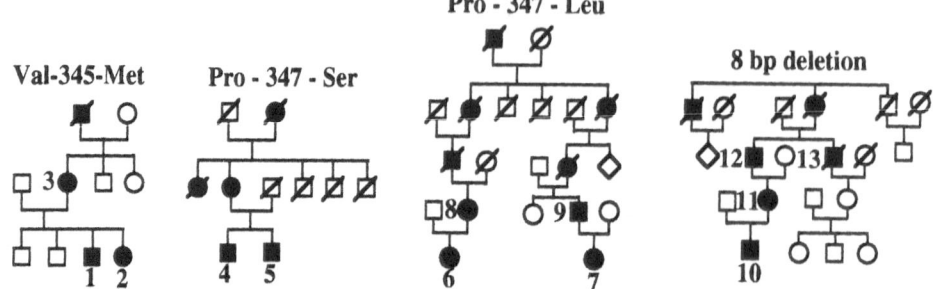

**Fig.1.** Pedigrees of the families with autosomal dominant retinitis pigmentosa. The rhodopsin amino acid change or the rhodopsin gene mutation, respectively, is indicated above each pedigree. Filled symbols are family members affected with retinitis pigmentosa; unfilled symbols are unaffected members; slashed symbols are deceased members. Examined persons are labeled with the numbers used throughout this article.

mechanisms underlying disease expression, as rhodopsin comprises several domains with different biological functions [7,8].

We were interested in the clinical phenotype of patients with mutations in the carboxyl(C)-terminal sequence of rhodopsin, as only few of the mutations found thus far are located in this domain of the molecule[2-4,6], and as these mutations cause a wide range of symptoms[9-13].

## PATIENTS AND METHODS

### Patients

By screening unrelated European ADRP patients for rhodopsin mutations, 17 different rhodopsin mutations were detected[4] . Out of this sample, 4 patients and 9 affected relatives representing 4 families with distinct C-terminal rhodopsin mutations were included in this study (see pedigrees in Fig.1). Three of the families carry a point mutation which leads to a single amino acid substitution in the rhodopsin molecule: valine-345-methionine (Val-345-Met), proline-347-serine (Pro-347-Ser), or proline-347-leucine (Pro-347-Leu), respectively. A detailed description of the latter family was the subject of a prior publication[11]. The 4th family segregates for a deletion of 8 base pairs (bp) spanning codons 341-343, which results in a shift of the reading frame and a substantially altered C-terminal sequence of the rhodopsin molecule. A detailed description of this mutation is given elsewhere[14].

### Molecular genetic analysis

Screening for mutations was carried out by single strand conformation and heteroduplex analyses, restriction digestion and genomic sequencing following PCR amplification of the five rhodopsin exons, as previously described[4,14].

### Clinical examination

The patients were clinically examined by the authors E. A-S., K.R. and E. Z. For each proband, an extended questionaire was completed with respect to age at onset of night blindness, side vision impairment, glare sensitivity and reading difficulties. Ophthalmologic examination included distance visual acuities, slit lamp examination with respect to lens opacities and vitreous changes, direct and indirect fundoscopy, fundus photography, color vision test, kinetic perimetry, electroretinography, two-color dark-adapted threshold perimetry, and, in some patients, dark adaptation.

On slit lamp examination, lens opacities were estimated following the classification of Fishman et al.[15]. Color vision was examined with the Lanthony Panel D 15 test. Kinetic visual fields were obtained with a Goldmann perimeter or a Tübinger Perimeter, using 30′ and 10′ white test spots of 200 cd · m$^{-2}$ on a background of 3.3 cd · m$^{-2}$. Stimuli were moved from non-seeing to seeing areas.

Dark adaptation was tested over a period of 40 minutes after dilating the pupils with tropicamide (0,5 %) and after a preadaptation period of 10 minutes to a ganzfeld of 890 cd · m$^{-2}$. The dark adaptation curve was obtained at 20° eccentricity on the horizontal meridian of one eye, determining the luminance necessary to detect a 106′ white stimulus (stimulus duration of 500 msec, maximum luminance: 272 cd · m$^{-2}$) in a Tübinger Perimeter, using 0.1 log unit increments. Dark-adapted two-color threshold perimetry was performed along the horizontal meridian with a 500 nm and a 656 nm 106′ stimulus, following the method of Massof and Finkelstein[16]. Thresholds were measured at the foveola, at 2°, 4°, 6°, 8°, 10° and then in 10° steps towards the nasal and temporal border.

After 30 minutes of dark adaptation, full-field ERGs were recorded from both eyes with a Nicolet Compact Four with Ganzfeld Controller (Nicolet Biomedical Instruments, Madison Wisconsin, USA), and DTL electrodes[17], according to the international ERG standard[18]. ERG responses were amplified, bandpass filtered and computer-avaraged. An artifact rejection was applied to eliminate disturbances by blinking reflexes. Rod responses were elicited with 0,5 Hz white stimuli (duration 0.1 msec.), by attenuating a standard flash of 2.5 cds/m$^2$ intensity by 2.5 log units with a Kodak Wratten neutral density filter. A mixed rod/cone response was obtained with a 0,2 Hz white flash at standard flash intensity. Oscillatory potentials were recorded by using a standard flash and cut-off filters at 100 and 250 Hz. Photopic responses were obtained after 10 minutes of light adaptation to a ganzfeld dome background illumination of 30 cd · m$^{-2}$ with white 30 Hz flicker stimuli and white single flashes at standard flash intensity. Responses to red single flashes were obtained with a standard flash and by interposing a Kodak Wratten W 25 filter.

## RESULTS

A summary of the clinical findings of the patients is given in Table 1. Except probands 4 and 7, all patients with Val-345-Met, Pro-347-Ser or Pro-347-Leu mutations reported early onset of night-blindness, i.e, within the first life decade. In patients with the 8 bp deletion, however, impaired night vision did not occur before the age of 16. In all patients, night-blindness preceeded the subjective onset of side vision problems.

Within each family, visual acuity in general appeared to be a function of age, but there were considerable differences between families, as individuals 11-13 with the 8 bp deletion retained equal or better visual acuities than most of the probands with codon 345 or codon 347 mutations, which were much younger of age. Visual acuity loss was most marked in patient 4 with the Pro-347-Ser mutation. Likewise, on slit lamp examination, only patients 4 and 5 with a Pro-347-Ser mutation displayed significant posterior subcapsular lens opacities. No characteristic macular alterations were seen. Younger individuals (1, 2, 5, 10, 11) showed near-normal maculae or some wrinkling of the inner limiting membrane, while older ones had diffuse atrophy or a bull′s eye appearance of the macular pigment epithelium. A cyst-like maculopathy could only be detected in patient 3 (Val-345-Met). Peripheral bone-spicule pigmentations were most marked in patients 4 and 5 (Pro-347-Ser), but absent or very sparse in younger probands (1, 2, 10, 11) carrying the Val-345-Met mutation or the 8 bp deletion. Degenerative pigment epithelium alterations were present in all cases except the youngest proband (10, age 4, 8 bp deletion). In patient 2 (Val-345-Met), there were greater pigment epithelium abnormalities in the inferior than in the superior retina. All other patients did not show a regional predilection of pigmentary retinopathy.

**Table 1.** Clinical characteristics of the patients

| Patient, age (yrs), sex | Rhodopsin Mutation | Age of onset of night blindness (yrs) | Visual acuity | Lens Opacities (Grade)* | Fundus findings Macula | Peripheral bone spicules |
|---|---|---|---|---|---|---|
| **1**, 15, M | Val-345-Met | 7 | R.E.: 20/20 | 0 | surface wrinkling | - |
| | | | L.E.: 20/20 | 0 | surface wrinkling | - |
| **2**, 14, F | Val-345-Met | 7 | R.E.: 20/25** | 0 | surface wrinkling | - |
| | | | L.E.: 20/20 | 0 | surface wrinkling | - |
| **3**, 41, F | Val-345-Met | 10 | R.E.: 20/60 | +1 | cystoid lesions | ++ |
| | | | L.E.: 20/200 | +1 | PE atrophy | ++ |
| **4**, 34, M | Pro-347-Ser | 18 | R.E.: 20/100 | +3 | PE atrophy | +++ |
| | | | L.E.: 9/200 | +4 | PE atrophy | +++ |
| **5**, 32, M | Pro-347-Ser | near birth | R.E.: 20/40 | +3 | surface wrinkling | +++ |
| | | | L.E.: 20/40 | +3 | surface wrinkling | +++ |
| **6**, 29, F | Pro-347-Leu | near birth | R.E.: 20/30 | +1 | PE atrophy | ++ |
| | | | L.E.: 20/50 | +1 | PE atrophy | ++ |
| **7**, 30, F | Pro-347-Leu | 12 | R.E.: 20/30 | +1 | Bull´s eye lesion | ++ |
| | | | L.E.: 20/30 | +1 | Bull´s eye lesion | ++ |
| **8**, 50, F | Pro-347-Leu | near birth | R.E.: 20/60 | +2 | PE atrophy | ++ |
| | | | L.E.: 20/60 | +2 | PE atrophy | ++ |
| **9**, 52, M | Pro-347-Leu | 2 | R.E.: 20/50 | +2 | Bull´s eye lesion | ++ |
| | | | L.E.: 9/200 | x | Bull´s eye lesion | ++ |
| **10**, 4, M | 8 bp deletion | no complaints | R.E.: 6/6 | 0 | broadened peri- | - |
| | | | L.E.: 6/6 | 0 | macular reflexes | - |
| **11**, 34, F | 8 bp deletion | 16 | R.E.: 20/25** | 0 | surface wrinkling | + |
| | | | L.E.: 20/20 | 0 | surface wrinkling | + |
| **12**, 76, M | 8 bp deletion | 22 | R.E.: 20/60 | +2 | PE atrophy | ++ |
| | | | L.E.: 20/60 | +2 | PE atrophy | ++ |
| **13**, 59, M | 8 bp deletion | 30 | R.E.: 20/60 | 0 | PE atrophy | ++ |
| | | | L.E.: 20/100 | 0 | PE atrophy | ++ |

*) following the classification of Fishman et al.[15]; x = pseudophakia; ** indicates strabism; PE = pigment epithelium; Peripheral bone spicules: - = absent; + = sparse; ++ = moderate; +++ = marked.

When examined with the desaturated Lanthony Panel D-15 test, only subjects 6 (Pro-347-Leu),10 and 11 (8 bp deletion) showed perfect color discrimination. All other patients had tritanomalous color vision defects.

On kinetic perimetry, essential differences could be observed between subjects with different rhodopsin mutations. Representative findings are illustrated in Figure 2.

In the family with the Val-345-Met mutation, the 15 year-old proband 1 had full fields to both the III-4e and the I-4e target (not shown), while his 14-year-old sister had full fields to the III-4e target, but showed a superior field depression to the I-4e stimulus. Patient 3 (age 41) had central islands of 10 - 20° extension.

In comparison to patient 3 (Val-345-Met), younger individuals 4 and 5 (age 34 and 32, respectively) with a Pro-347-Ser mutation showed even more reduced central fields. Patients 6-9 with a Pro-347-Leu mutation had concentric visual field constriction of variable extent, depending on their age. The functional loss in patients 6 and 7 (ages 29 and 30, respectively) roughly corresponded to that seen in 41-year-old subject 3 with the Val-345-Met mutation. Visual fields of patients 6 and 9 are shown in figure 2.

A relatively mild degree of visual field loss was detected in members affected with the rhodopsin gene 8 bp deletion. The 34-year-old proband 11 had normal outer field borders to both targets III 4e and I 4e, and small pericentral scotomata. Visual fields constricted to 20°

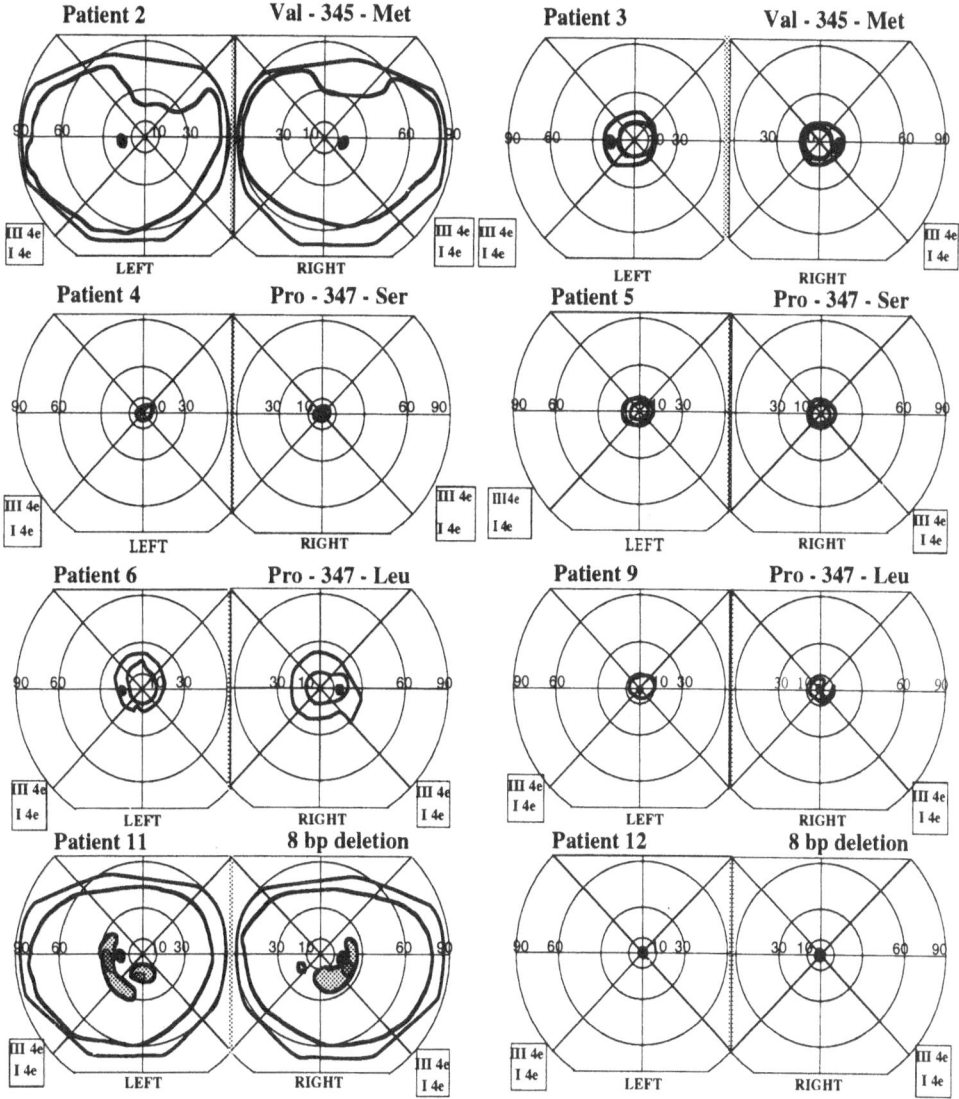

**Fig. 2.** Kinetic visual fields of patients with mutations in the carboxyl-terminal sequence of rhodopsin. For convenience, targets 30'-200 cd/m$^2$ and 10'-200 cd/m$^2$ of the Tuebinger perimeter are referred to as targets III/4e and I/4e, since they are equivalent to the test spots of our Goldmann Perimeter.

for the III-4e stimulus were observed in 59 year-old patient 13 (not shown), whereas individual 12 (age 76) had very small central islands of about 3° extension to the III 4e target, comparable to those detected in 34-year-old patient 4 with a Pro-347-Ser mutation.

Dark adaptation was tested in patients 1 (Val-345-Met), 7 (Pro-347-Leu) and 11 (8 bp deletion), respectively. All showed monophasic dark adaptation curves with no detectable rod function. Final threshold elevation was 3.9 log units in patient 1, and 4.5 log units in patients 7 and 11. There was a normal timing of the cone-mediated adaptation curve in all individuals tested. Patient 1 reached pre-bleach final sensitivity within 3 minutes.

Two-color dark-adapted threshold perimetry was performed along the horizontal meridian of one eye to determine the relation of rod to cone sensitivity loss (see representative examples in Figure 3). Patients with Val-345-Met, Pro-347-Ser or Pro-347-

**Table 2.** Electroretinographic findings in the patients

| Patient, age (yrs), sex | Rhodopsin mutation | | Rod ERG | Mixed cone-rod ERG | | Cone-flicker ERG | |
|---|---|---|---|---|---|---|---|
| | | | | amplitude (μV) | latency (msec) | amplitude (μV) | latency (msec) |
| **1**, 15, M | Val-345-Met | R.E.: | Not detectable | Not detectable | | 26 | 70 |
| | | L.E.: | Not detectable | Not detectable | | 21 | 65 |
| **2**, 14, F | Val-345-Met | R.E.: | Not detectable | 68 | 56 | 31 | 76 |
| | | L.E.: | Not detectable | 70 | 54 | 47 | 77 |
| **3**, 41, F | Val-345-Met | R.E.: | Not detectable | Not detectable | | Not detectable | |
| | | L.E.: | Not detectable | Not detectable | | Not detectable | |
| **4**, 34, M | Pro-347-Ser | R.E.: | Not detectable | Not detectable | | Not detectable | |
| | | L.E.: | Not detectable | Not detectable | | Not detectable | |
| **5**, 32, M | Pro-347-Ser | R.E.: | Not detectable | Not detectable | | Not detectable | |
| | | L.E.: | Not detectable | Not detectable | | Not detectable | |
| **6**, 29, F | Pro-347-Leu | R.E.: | Not detectable | Not detectable | | Not detectable | |
| | | L.E.: | Not detectable | Not detectable | | Not detectable | |
| **7**, 30, F | Pro-347-Leu | R.E.: | Not detectable | Not detectable | | Not detectable | |
| | | L.E.: | Not detectable | Not detectable | | Not detectable | |
| **8**, 50, F | Pro-347-Leu | R.E.: | Not detectable | Not detectable | | Not detectable | |
| | | L.E.: | Not detectable | Not detectable | | Not detectable | |
| **9**, 52, M | Pro-347-Leu | R.E.: | Not detectable | Not detectable | | Not detectable | |
| | | L.E.: | Not detectable | Not detectable | | Not detectable | |
| **10**, 4, M | 8 bp deletion | R.E.: | Not available | Not available | | Not available | |
| | | L.E.: | Not available | Not available | | Not available | |
| **11**, 34, F | 8 bp deletion | R.E.: | Not detectable | 83 | 46 | 90 | 67 |
| | | L.E.: | Not detectable | 93 | 46 | 76 | 67 |
| **12**, 76, M | 8 bp deletion | R.E.: | Not detectable | Not detectable | | Not detectable | |
| | | L.E.: | Not detectable | Not detectable | | Not detectable | |
| **13**, 59, M | 8 bp deletion | R.E.: | Not detectable | Not detectable | | Not detectable | |
| | | L.E.: | Not detectable | Not detectable | | Not detectable | |

ERG normal ranges (5th through 95th percentile) are 117 - 272 μV for rod amplitudies, 68 - 90 msec for rod amplitude implicit time, 292 - 622 μV for mixed cone-rod responses, 34 - 45 msec for mixed cone-rod response implicit time, 75 - 179 μV for 30 Hz cone-flicker amplitudes, and 58 - 63 msec. for 30 Hz flicker implicit time, respectively.

Leu mutations (1-9) showed similar patterns of function loss. In these persons, thresholds to 500 nm and 656 nm were mediated by cones at all loci examined. In individual 4 (Pro-347-Ser), however, no thresholds could be measured outside the foveal area due to the severe visual field constriction. Foveal sensitivities to 656 nm were low normal in most of the younger probands (1, 2, 6 and 7), but were reduced at a various extent to 500 nm. Probands 3, 4, 8 and 9 showed reduced foveal sensitivity to both 500 nm and 656 nm stimuli. Outside the macular area, all probands showed sensitivity loss ranging from about 3 to 4.5 log units for 500 nm and from about 0.5 to 2.5 log units for 656 nm. Some regional variation was found in individual 2 (Val-345-Met) in whom cone sensitivity loss was more profound in the nasal than in the temporal retina (not shown).

A degenerative pattern different from that described above could be detected in patient 11 carrying the 8 bp deletion (see Figure 3). Cones detected both 500 nm and 656 nm targets within the central 30 degrees. Sensitivity loss to 500 nm was about 3 log units in this area. In contrast, rod sensitivity to the 500 nm target could be clearly measured in the temporal and the nasal periphery, where sensitivity loss was only 1-1.5 log units. Cone function as tested by the 656 nm stimulus was considerably less reduced at all loci measured. Sensitivities to 656 nm were low normal in the fovea as well as in the far periphery, whereas a sensitivity loss ranging from 0.5 to 1.0 log unit was revealed in the perimacular and in the midperipheral area.

ERG findings are summarized in Table 2. Rod ERGs were nondectable in all patients, regardless of the mutation. Reduced scotopic mixed cone-rod responses with increased implicit times could be measured in patients 2 (age 14, Val-345-Met) and 11 (age 34, 8 bp deletion), but were nonrecordable in the other probands. Reduced cone amplitudes to white 30 Hz flicker stimuli and single flashes, as well as to red single flashes, could be obtained from patients 1 and 2 (Val-345-Met, ages 15 and 14). Implicit times were increased in both persons. Individual 11 (age 34, 8 bp deletion) however showed low normal cone amplitudes, with slightly increased latencies.

## DISCUSSION

We compared the phenotypic expression of different carboxyl-terminal rhodopsin mutations. Striking similarities were found among the phenotypes of the the Val-345-Met, the Pro-347-Ser and the Pro-347-Leu mutation: in each case there was early, diffuse loss of rod function, as evidenced by patients' history, electroretinography, dark-adaptation and dark-adapted two-color threshold perimetry. The findings correspond to type 1 RP according to the classification system of Massof and Finkelstein[19]. Cone function was less reduced, but showed some interfamilial variability. The most profound function loss, in relation to the patients' age, was found in the Pro-347-Ser phenotype. The data derived from this family, however, are too limited to conclude that early, severe cone function loss is a characteristic of this genotype, and, to our knowledge, the phenotype of this mutation has not been described in detail by others so far.

Our findings in the families with a Val-345-Met or a Pro-347-Leu mutation agree well with those reported by others on patients with Val-345-Met[10], Pro-347-Leu[9] or with Pro-347-Arg[12] mutation of rhodopsin. Berson et al.[10] found, that patients with the Val-345-Met mutation had, on average, larger ERG amplitudes than those with Pro-347-Leu or Pro-347-Ser mutation. In our study, the small number of patients prevents a comparison between these subgroups. It is, however, striking, that individual 3 (Val-345-Met) retained visual field areas of about the same size as did younger probands with the Pro-347-Leu mutation.

One patient with the Val-345-Met mutation showed some regional predilection of fundus alterations and cone sensitivity impairment, which was more profound in the inferior and nasal retina, while two other affected relatives did not show this regional pattern. Likewise, regional variability of retinal function appears to be an inconstant finding in the Pro-23-His phenotype[20-23], whereas it seems to be common in codon 17, codon 58, codon 106 and codon 182 mutations [13, 24-26]. These genotypes, however, also show regional patterns of **rod** sensitivity loss.

The family with the 8 bp deletion presented with a "mild" phenotype fairly different from those of the other families. Cone function was well maintained, causing severe impairment of vision only in a 76-year -old male. While there was no detectable rod function in older individuals, a 34 year-old patient showed a regional variation of rod sensitivity, beeing well preserved in the peripheral retina and virtually abolished within the central 30 degrees.This pattern of rod function impairment is different from that observed in patients with certain rhodopsin mutations located in the intradiscal and transmembrane domains of the molecule (e.g., codons 17, 23, 58 106, and 182 mutations, respectively), as in these genotypes, maximum rod function loss is located in the inferior or the nasal rather than in the pericentral retina[13,22,24-26].

Jacobson and coworkers reported on patients with a carboxyl-terminal stop-codon mutation of rhodopsin (Glu-344-stop)[13] and mild disease expression. In contrast to individual 11 (8 bp deletion) in our study, members of about the same age had well-detectable ERG rod b-waves, and only slightly reduced rod sensitivity throughout the retina, when tested by two-color dark-adapted threshold perimetry. These individuals were virtually non-affected, as they had normal visual acuity and full visual fields.

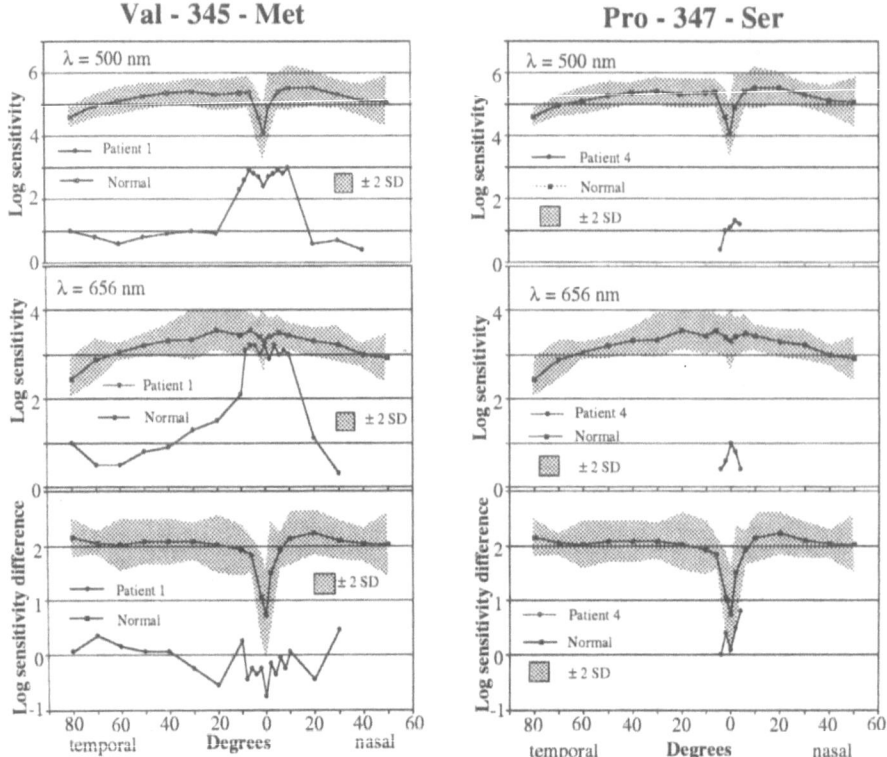

**Fig. 3.** Dark-adapted two-color threshold profiles along the horizontal meridian in normal observers (upper curves, mean ± 2 SD, n = 8) and patients representing families with different rhodopsin mutations, as labeled above each diagram. Top, sensitivities to 500 nm; middle, sensitivities to 656 nm; bottom, sensitivity differences between the two stimuli in patients 1, 4, 6, and 11, respectively. The photoreceptor that mediates the threshold for each stimulus can be determined from the sensitivity differences: rods detect both stimuli, if the difference is greater than 1.6 log units (lower limit of normal); rods detect 500 nm and cones detect 656 nm, if the sensitivity difference ranges from 1.6 to 0 log units, and cones detect both stimuli, if the difference is 0 or negative.

The reasons for the phenotypic differences among carboxyl-terminal rhodopsin mutations remain obscure. Nongenetic factors may be of importance. In order to understand the actual influence of the underlying genotype, the biological role of rhodopsin's carboxyl-terminal region has to be elucidated. Data from cell culture experiments suggest, that some rhodopsin mutants may lead to disturbed transport of synthesized molecules to the plasma membrane, but this apparently does not hold for C-terminal mutants Pro-347-Leu and Glu-344-stop[27]. The carboxyl-terminal region of rhodopsin contains phosphorylation sites (threonine and serine residues), which influence the binding of transducin[8]. These residues, however, are unaffected by codon 347, codon 345 as well as codon 344 mutations. Out of the 7 threonines and serines within codons 334-343, only two are removed in the mutant molecule resulting from the rhodopsin gene 8 base pair deletion. It therefore seems unlikely, that this mutation has a strong effect on the efficacy of rhodopsin phosphorylation.

In vertebrates, Val-345 and Pro-347 are highly conserved residues of the rhodopsin molecule, suggesting an important role of these amino acids. They are abolished in the codon 347 and codon 345, as well as the glu-344-stop mutation, but in the 8 bp deletion a proline is still present in the second last position of the extended mutant molecule.

The carboxyl-terminus of rhodopsin should be readily accessible to immunological reactions; a pathogenetic role of immunological mechanisms, however, has not been shown thus far.

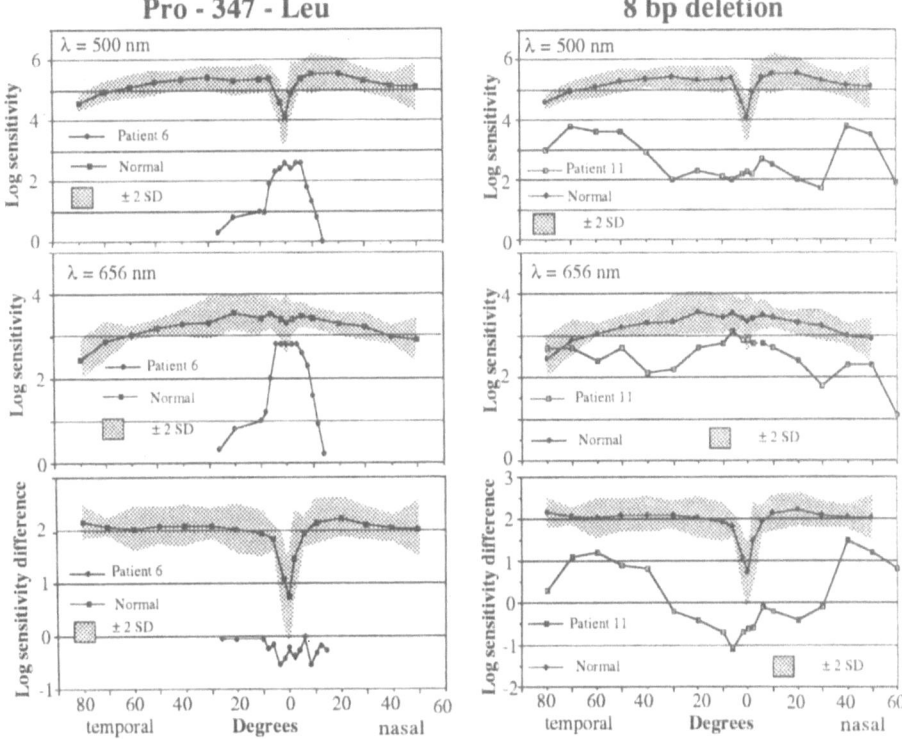

At the present time, it is impossible to draw definite conclusions concerning individual disease processes in the various genotypes. Our knowledge may expand, when larger numbers of individuals can be tested clinically, and when in vitro experiments and transgenic animal models lead to better understanding of the functional defects at the molecular and cellular level caused by alterations of the rhodopsin molecule.

## ACKNOWLEDGMENTS

This study was supported by grants Zr 1/7-1, Ru 457/1-1 and Ga 210/5-2 of the "Deutsche Forschungsgemeinschaft" (DFG).

## REFERENCES

1. T.P. Dryja, T.L. McGee, E. Reichel, L.B. Hahn, G.S. Cowley, D.W. Yandell, M.A. Sandberg, and E.L. Berson, A point mutation of the rhodopsin gene in one form of retinitis pigmentosa, Nature 343, 346-366 (1990).

2. C.F. Inglehearn, T.J. Keen, R. Bashir, M. Jay, F. Fitzke, A.C. Bird, A. Crombie, and S. Bhattacharya, A completed screen for mutations of the rhodopsin gene in a panel of patients with autosomal dominant retinitis pigmentosa, Hum. Mol. Genet. 1, 41-45 (1992).

3. T.P. Dryja, L.B. Hahn, G.S. Cowley, T.L. McGee, and E.L. Berson, Mutation spectrum of the rhodopsin gene among patients with autosomal dominant retinitis pigmentosa, Proc. Natl. Acad. Sci. USA 88, 9370-9374 (1991).

4. S. Bunge, H. Wedemann, D. David, D.J. Terwillinger, C. Aulehla-Scholz, C. Sammans, M. Horn, J. Ott, E. Schwinger, E.M. Bleeker-Wagemakers, A.Schinzel, and A. Gal, Molecular analysis and genetic mapping of the rhodopsin gene in families with autosomal dominant retinitis pigmentosa, submitted for publication.

5. V.C. Sheffield, G.A. Fishman, J.S. Beck, A.E. Limura, and E.M. Stone, Identification of novel rhodopsin mutations associated with retinitis pigmentosa using GC-clamped denaturing gel electrophoresis, Am. J. Hum. Genet. 49, 699-706 (1991).

6. C.-H. Sung, C.M. Davenport, J.C. Hennessey, I.H. Maumenee, S.G. Jacobson , J.R. Heckenlively, R. Nowakowski, G. Fishman, P. Gouras, and J. Nathans, Rhodopsin mutations in autosomal dominant retinitis pigmentosa, Proc. Natl. Acad. Sci. USA 88, 6481-6485 (1991).

7. P.A. Hargrave, and P.J. O'Brien, Speculations on the molecular basis of retinal degeneration in retinitis pigmentosa, in: "Retinal Degenerations", R.E. Anderson, J.G. Hollyfield, and M.M. LaVail, Eds., Boca Raton, Florida, CRC Press Inc., 517-528 (1991).

8. P.A. Hargrave, and J.H. McDowell, Rhodopsin and phototransduction: a model system for G protein-linked receptors, FASEB 6, 2323-2331 (1992).

9. E.L. Berson, B. Rosner, M.A. Sandberg, C. Weigel-DiFranco, and T.P. Dryja, Ocular findings in patients with autosomal dominant retinitis pigmentosa and rhodopsin, proline-347-leucine, Am. J. Ophthalmol. 111, 614-623 (1991).

10. E.L. Berson, M.A. Sandberg, and T.P. Dryja, Autosomal dominant retinitis pigmentosa with rhodopsin, valine-345-methionine, Tr. Am. Ophth. Soc. 89, 117-130 (1991).

11. E. Apfelstedt-Sylla , M. Kunisch, M. Horn, K. Ruether, A. Gal, and E. Zrenner, Diffuse loss of rod function in autosomal dominant retinitis pigmentosa with pro-347-leu mutation of rhodopsin. German.J. Ophthalmol. 1: 319-327 (1992).

12. G. Niemeyer, P. Trueb, A. Schinzel, and A. Gal, Clinical and ERG data in a family with autosomal dominant RP and pro-347-arg mutation in the rhodopsin gene, Doc. Ophthalmol. 79, 303-311 (1992).

13. S.G. Jacobson, C.M. Kemp, C.-H. Sung, and J. Nathans, Retinal function and rhodopsin levels in autosomal dominant retinitis pigmentosa with rhodopsin mutations. Am. J. Ophthalmol. 112, 256-271 (1991).

14. M. Horn, P. Humphries, M. Kunisch, C. Marchese, E. Apfelstedt-Sylla, L. Fugi, E. Zrenner, P. Kenna, A. Gal, and J. Farrar, Deletions in exon 5 of the human rhodopsin gene causing shift in the reading frame and autosomal dominant retinitis pigmentosa. Hum. Genet. 90, 255-257 (1992).

15. G.A. Fishman, K.R. Alexander, and R.J. Anderson, Autosomal dominant retinitis pigmentosa: a method of classification, Arch. Ophthalmol. 103, 366-374 (1985).

16. R.W. Massof, and D. Finkelstein, Rod sensitivity relative to cone sensitivity in retinitis pigmentosa, Invest. Ophthalmol. Vis. Sci. 18, 263-272 (1979).

17. W.W. Dawson, G.L. Trick, and C. Litzkow, Improved electrode for electroretinography, Invest. Ophthalmol. Vis. Sci. 18, 988-991 (1979).

18. M.F. Marmor, G.B. Arden, S.E.G. Nilsson, and E. Zrenner, Standard for clinical electroretinography, Arch. Ophthalmol. 107, 816-819 (1989).

19. R.W. Massof, and D. Finkelstein, Two forms of ausosomal dominant primary retinitis pigmentosa, Doc. Opthalmol. 51, 289-346 (1981).

20. E.L. Berson, B. Rosner, M.A. Sandberg, and T.P. Dryja, Ocular findings in patients with autosomal dominant retinitis pigmentosa and a rhodopsin gene defect (pro-23-his), Arch. Ophthalmol. 109, 92-101 (1991).

21. J.R. Heckenlively , J.A. Rodriguez , and P. Daiger, Autosomal dominant sectoral retinitis pigmentosa; two families with transversion mutation in codon 23 of rhodopsin, Arch. Ophthalmol. 109, 84-91 (1991).

22. C.M. Kemp, S.G. Jacobson, A.J. Roman, C.-H. Sung, and J. Nathans, Abnormal rod dark adaptation in autosomal dominant retinitis pigmentosa with proline-23-histidine rhodopsin mutation, Am. J. Ophthalmol. 113, 165-174 (1992).

23. E.M. Stone, A.E. Kimura, B.E. Nichols, P. Khadivi, G.A. Sheffield, and V.C. Sheffield, Regional distribution of retinal degeneration in patients with the proline to histidine mutation in codon 23 of the rhodopsin gene, Ophthalmology 98, 1806-1813 (1991).

24. J.E. Richards, K. Chen-Yu, M. Boehnke, and P.A. Sieving, Rhodopsin thr58arg mutation in a family with autosomal dominant retinitis pigmentosa, Ophthalmology 98, 1797-1805 (1991).

25. G.A. Fishman, E.M. Stone, V.C. Sheffield, L.D.Gilbert, and A.E. Kimura, Ocular findings associated with rhodopsin gene codon 17 and codon 182 transition mutations in dominant retinitis pigmentosa, Arch. Ophtalmol. 110, 54-62 (1992).

26. G.A. Fishman, E.M. Stone, L.D. Gilbert, and V.C. Sheffield, Ocular findings associated with a rhodopsin gene codon 106 mutation, Arch. Ophthalmol. 110, 646-653 (1992).

27. C.-H. Sung, B.G. Schneider, N. Agarwal, D.S. Papermaster, and J. Nathans, Functional heterogeneity of mutant rhodopsins responsible for autosomal dominant retinitis pigmentosa, Proc. Natl. Acad. Sci. USA 88, 8840-8844 (1991).

# HETEROGENEITY OF USHER SYNDROME TYPE I

Radha Ayyagari[1], Richard J.H. Smith[2], Elizabeth C.
Lee[2], William J. Kimberling[3], Marcelle Jay[4],
Alan Bird[4], and J. Fielding Hejtmancik[1]

[1]National Eye Institute, National Institutes of
Health, Bethesda, MD 20892
[2]Molecular Otolaryngology research laboratories,
Department of Otolaryngology–Head and Neck  Surgery,
The University of Iowa, Iowa City, IA 52242
[3]Boys Town National Research Hospital, Omaha,
NE 68131
[4]Institute of Ophthalmology  and Moorfields Eye
Hospital, London, UK

## INTRODUCTION

Usher syndrome consists of a group of autosomal recessive diseases characterized by congenital sensorineural hearing loss and the progressively blinding disorder, retinitis pigmentosa.  In 1914 Usher was the first to describe the hereditary nature of this syndrome (1).  Usher syndrome is the most common cause of deaf–blindness, causing at least 50% of all reported cases of combined deafness and blindness in developed countries.  Its prevalence in the general population ranges from 3 to 5 per 100,000 and estimates of gene frequency variy from 1/70 to 1/100 (2).  The Diagnosis of Usher syndrome is generally made following ophthalmoscopic examination of congenitally deaf patients.  In Usher syndrome patients  patients retinitis pigmentosa is the most important ocular finding while sensorineural hearing loss is the most frequent non–ocular abnormality.

Although in many earlier studies Usher syndrome was presumed to be due to a recessive mutation at a single locus (3), detailed clinical studies have shown distinct phenotypic differences suggesting the likelihood of genetic heterogeneity (4).  Thus, three major types of Usher have been described (Table 1). Usher syndrome type I(USH1) is the most severe form, consisting of profound congenital sensorineural deafness associated with vestibular dysfunction and retinitis pigmentosa with progressive blindness occurring in childhood.  The age at which the patient first walks and his other motor milestones are delayed. The patient's  gait is unsteady.  In Usher  syndrome type II (USH2) retinitis pigmentosa is associated with moderate to severe deafness and normal vestibular function.  Retinal changes are milder  and tend to occur later in USH2 than in USH1.  Usher syndrome type III, also known as Hallgren's syndrome, consists of retinitis pigmentosa with progressive hearingloss.(5,6).  Usher type I and type II are the most common forms, whereas type III is relatively rare. An X–linked form may also exist.

**Table 1.** Clinical types of Usher syndrome

| Clinical feature | Type I | Type II | Type III |
|---|---|---|---|
| Hearing loss/severity | Congenital/ Severe– Profound | Congenital/ Moderate– Severe | Congenital Variable/ Progressive |
| Vestibular function | Absent | Normal | Normal |
| Retinitis Pigmentosa | Prepubertal | Postpubertal | Variable |
| Mental retardation | Occasional | Occasional | No |
| Inheritance | Autosomal Recessive | Autosomal Recessive | Autosomal Recessive |

Localization of USH2 to the long arm of chromosome one and exclusion of this region in USH1 families established the presence of genetic heterogeneity in Usher syndrome (7,8,9,10). In addition, USH2 is genetically heterogenous, with a small percentage of families failing to show linkage to the 1q32–1q41 region (11). Usher syndrome type I has been mapped to the long arm of chromosome 14 in one group of families from a small region of western France (12). This linkage was not confirmed in the pooled data of the Usher syndrome consortium (13), suggesting genetic heterogeneity for USH1 as well. Here we review results of recent linkage analyses localizing two genes for USH1 to chromosome 11, thus demonstrating the genetic heterogeneity among USH1 with similar clinical symptoms.

## CLINICAL STUDIES

Families with members affected by USH1 were evaluated clinically by three groups: W.J.K. and his co-workers evaluated a mixed population with families from the United States, Sweden, Ireland, and South Africa (group I) (14); R.J.H.S. and his co-workers evaluated families from the Acadian population of Southwestern Louisiana (group II) (10); and A.B. and his co-workers evaluated a group of families from Great Britain (group III) (15). Clinical studies began with a medical history including a questionnaire and evaluation of all available medical records. Families were traced historically as far as possible to identify possible consanguinity. Diagnosic criteria were consistent with recomendations of the Usher consortium.

Physical examination consisted of a general physical examination, examination of the ear, nose and throat, and otoneurological examination. Ophthalmologic studies included visual acuity, visual field testing and electroretinography. Audiological examination included standard audiometry and speech discrimination. A battery of tests including the Bruininks-Oseretsky sub tests of balance function, an otologic examination, and ice-water calorics using Frenzel's glasses were performed to determine cerebellar and vestibular function. Diagnostic criteria included severe to profound congenital hearing loss at all frequencies, absent vestibular reflexes bilaterally, and a diagnosis of retinitis pigmentosa based on the presence of night blindness, restricted visual fields, the presence of pigmentary retinopathy, and a diminished or absent ERG.

Here we will review results of linkage analyses performed on three groups of patients. These include twenty-seven families of Irish, Swedish, South Afrian and American population (Group I), eight Acadian families who are mainly descendants of a small group of French-Acadians who settled in Louisiana in the nineteenth century (Group II), and eleven families ascertained in the United Kingdom (Group III).

## LINKAGE ANALYSIS OF USH1

Although initial results suggested that a gene for USH1 might be located on chromosome 4, this chromosome was excluded in group II families (6). General screening for USH1 by linkage analysis was then carried out co-operatively, co-ordinated through the Usher Syndrome Consortium sponsored by the Retinitis Pigmentosa Foundation (13). A total of 167 markers were analyzed, excluding 1251 cM or more than 30% of the human genome. As part of this study, linkage analysis of USH1 in families from Groups I, II and III was performed using microsatellite markers on chromosome 11 (Table 2), (14,15).

### Table 2A
LOD scores of chromosome 11 markers versus USH1 in Group I families

| Marker | Recombination | | | | | | Zmax | $\theta$est |
|--------|------|------|------|------|------|------|------|------|
| | 0.0 | 0.05 | 0.10 | 0.20 | 0.30 | 0.40 | | |
| D11S419 | $-\infty$ | -4.05 | -2.04 | -0.53 | -0.07 | 0.01 | .. | 0.500 |
| D11S860 | $-\infty$ | 5.80 | -3.08 | -0.97 | -0.25 | -0.03 | .. | 0.500 |
| FGF3 | $-\infty$ | 1.35 | 1.78 | 1.46 | 0.79 | 0.22 | 1.79 | 0.115 |
| D11S527 | – | 4.00 | 4.15 | 3.01 | 1.54 | 0.42 | 4.20 | 0.079 |
| D11S35 | $-\infty$ | -2.54 | -0.70 | 0.30 | 0.30 | 0.10 | .. | 0.500 |
| D11S490 | $-\infty$ | -5.51 | -2.77 | -0.82 | -0.21 | -0.03 | .. | 0.500 |
| CD3D | – | 0.59 | 1.63 | 1.62 | 0.93 | 0.27 | 1.79 | 0.140 |
| D11S528 /S420 | $-\infty$ | -1.12 | 0.97 | 1.64 | 1.05 | 0.32 | 1.66 | 0.180 |

### Table 2B
LOD scores of chromosome 11 markers versus USH1 in Group II families

| Marker | Recombination | | | | | | Zmax | $\theta$est |
|--------|------|------|------|------|------|------|------|------|
| | 0.0 | 0.05 | 0.10 | 0.20 | 0.30 | 0.40 | | |
| D11S419 | 4.20 | 3.74 | 3.25 | 2.24 | 1.23 | 0.37 | 4.20 | 0.00 |
| D11S875 | – | 0.92 | 1.13 | 0.90 | 0.46 | 0.12 | 1.14 | 0.11 |
| D11S527 | $-\infty$ | -5.73 | -3.19 | -1.15 | -0.37 | -0.07 | 0.00 | 0.50 |
| D11S35 | $-\infty$ | -2.49 | -1.29 | -0.39 | -0.10 | -0.02 | 0.00 | 0.50 |
| D11S490 | $-\infty$ | -6.45 | -3.76 | -1.46 | -0.51 | -0.11 | 0.00 | 0.50 |
| CD3D | $-\infty$ | -8.33 | -4.87 | -1.93 | -0.69 | -0.15 | 0.00 | 0.50 |
| D11S420 | $-\infty$ | -4.71 | -2.67 | -1.01 | -0.35 | -0.08 | 0.00 | 0.50 |
| D11S871 | $-\infty$ | -1.03 | -0.03 | 0.44 | 0.33 | 0.11 | 0.00 | 0.50 |
| D11S874 | $-\infty$ | -2.22 | -1.22 | -0.43 | -0.14 | -0.03 | 0.00 | 0.50 |
| INT2 | $-\infty$ | -3.13 | -1.68 | -0.57 | -0.17 | -0.03 | 0.00 | 0.50 |

LOD scores of chromosome 11 markers versus USH1 in Group III families

| Marker | Recombination | | | | | | Zmax | $\theta$est |
|--------|------|------|------|------|------|------|------|------|
| | 0.0 | 0.05 | 0.10 | 0.20 | 0.30 | 0.40 | | |
| D11S419 | $-\infty$ | -6.01 | -0.36 | 0.01 | 0.05 | 0.02 | 0.00 | 0.50 |
| D11S875 | $-$ | 0.90 | 1.02 | 0.74 | 0.37 | 0.09 | 1.03 | 0.09 |
| D11S527 | 6.03 | 5.22 | 4.39 | 2.79 | 1.39 | 0.38 | 6.04 | 0.00 |
| D11S871 | $-\infty$ | -1.86 | -0.91 | -0.23 | -0.04 | 0.01 | 0.00 | 0.50 |
| D11S874 | $-\infty$ | -3.58 | -2.05 | -0.71 | -0.28 | -0.06 | 0.00 | 0.50 |
| INT2 | $-\infty$ | -0.11 | 0.37 | 0.46 | 0.28 | 0.08 | 0.49 | 0.19 |
| CD3D | $-\infty$ | -3.55 | -2.00 | -0.74 | -0.25 | -0.05 | 0.00 | 0.50 |

The two point linkage results for USH1 in families from Group I showed positive lod scores of 4.20 with D11S527, 1.79 with FGF3, 1.79 with CD3D, and 1.66 with D11S528. Analysis with D11S419 showed a negative lod score which excluded at least the surrounding 20 cM with a lod score of less than $-2$. Similarly, the linkage results for USH1 in families from Group III showed positive lod scores of 6.03 with D11S527 ($\theta=$o), 1.03 with D11S875 ($\theta=0.09$), and 0.49 with INT2 ($\theta=0.19$). The results obtained with USH1 in group II families excluded 3.5 cM around D11S419 with a lod score of less than $-2$. In contrast, the two point linkage results for USH1 in families from group II showed a positive lod score of 4.2 with D11S419 ($\theta=0$) and 1.14 with D11S875 ($\theta=0.11$). Analysis with D11S527 excluded a region of 14 cM on either side of this marker in group II families. When the linkage results for D11S419 and D11S527 in the families in groups II and III were analyzed using the M-test (16), heterogeneity is supported with $\chi^2 = 24.255$, giving a likelihood ratio of $1.8\times10^5$. When the results of linkage analyses with USH1 and D11S419 and D11S527 in families from groups II and III were combined and analyzed with the HOMOG2 program (16), heterogeneity was supported with $\alpha= 0.65$ and a likelihood ratio of 220.76, corresponding to a p < .0023. Taken together, these results strongly support the existence of two loci capable of causing USH1 on chromosome 11.

The mapping of USH1 on chromosome 11 was refined and the evidence for genetic heterogeneity was strengthened using multipoint linkage analysis. Analysis of USH1 in families of group I used markers FGF3 and D11S527, which are known to be linked on 11q13.5 with $\theta = .067$. The maximum multipoint lod score was 4.5 occuring 10 cM telomeric to D11S527 (away from FGF3). No evidence for heterogeneity was detectable in this population using the HOMOG program with the multipoint linkage results (p = 0.49). Multipoint analysis with markers on 11p was not carried out in this population.

Analysis of USH1 in families of group II was carried out using the 11q markers D11S527 and INT2 which are known to be linked with $\theta = 0.07$. A maximum multipoint lod score of 5.51 was observed at D11S527. Similar analysis with the families of group III excluded a region of 30 cM stretching from 9 cM proximal to D11S527 to 14 cM cM distal to INT2. A similar analysis was carried out on families of groups II and III using 11p14 markers in the map: D11S875–0.10–D11S419–0.16–D11S871. Analysis using the families of group II gave a maximum multipoint lod score of 4.6 at D11S419. Families of group III, however, gave a negative lod score throughout this region, although it was not uniformly less than $-2$. When the linkage data from groups II and III were analyzed using the M-test, heterogeneity was supported with a $\chi^2 = 28$, giving a likelihood ratio R = $7.8 \times 10^5$. Similarly, analysis of the combined multipoint data from families in groups II and III with HOMOG2 supported heterogeneity with a likelihood ration R = 676 corresponding to a p < 0.007. Thus, while the results obtained in families from both groups I and III are consistent with a locus in 11q13 capable of causing USH1, the results obtained with families from group II exclude this region and indicate the existence of a separate locus in 11p14. Thus, these data confirm the existence of two separate genetic loci which can cause USH1, and refine the location of the USH1 genes to widely separated regions in or near 11p14 and 11q13.

## DISCUSSION

These observations clearly demonstrate genetic heterogeneity among patients with Usher syndrome type I, although the clinical symptoms in the different groups are not distinctly different. Taken together with the report by Kaplan et al., (12) of linkage of USH1 to the marker D14S13 on chromosome 14q in French families living in the Poitou-Charentes region of western France, these data would suggest the existence of at least three loci capable of causing USH1. In addition, genetic heterogeneity has been demonstrated in USH2 by excluding chromosome 1q region to where Usher type II was localized (11).

Several genes have been mapped to the short arm of chromosome 11, including genes for brain-derived neurotrophic factor (17) and aniridia (18). The 11q13 region is also interesting as genes causing autosomal dominant neovascular inflammatory vitreoretinopathy (19), autosomal dominant familial exudative vitreoretinopathy (20) and Best disease (21) also map to it. A retina specific gene, ROM 1, also maps to 11q13 region. This is a 37kDa membrane protein in the outer segment of the rod photoreceptors and which shows a 35% homology was observed to the human retinal degeneration slow (rds) protein (22). Although mutations in ROM 1 have been described in patients with autosomal dominant retinitis pigmentosa, none have been found in 21 patients with Usher syndrome (23). The presence of these genes on the long arm of chromosome 11 leads to speculation of the localization of a family of retina specific genes to this region.

Photoreceptors, auditory hair cells, and vestibular hair cells develop from ciliated progenitors, and axonemes are present in mature photoreceptors and vestibular hair cells. Abnormalities in connecting cilia, in sperm and photoreceptor axonemes have been reported in Usher patients ( 24,25). These abnormalities may play a significant role in the pathogenesis of the congenital deafness and progressive dysfunction and loss of rods and cones that are characteristic of this disease. A defective cilium has been speculated to cause outer segment atrophy, degeneration, and ultimately photoreceptor death. Evidence of abnormal nasal cilia in Usher syndrome has been presented by Arden and Fox (26). Thus genes encoding the group of proteins required for ciliary and axonemal functions represent logical candidates for the USH1 locus. To date none of these genes has been mapped to chromosome11.

In summary, Usher syndrome is group of autosomal recessive diseases affecting two distinct sensory functions indicating the occurrence of a defect in a system that is involved in the normal functioning of both auditory/vestibular and visual systems. So far Usher type I has been mapped to three different loci on 1114, 11q13 and 14q32 (12,14,15) whereas in type II, one group of families with USH2 map to 1q and another does not. The degree genetic heterogeneity observed in the two major types of Usher syndrome suggests wide diversity of the genes involved and is consistent with the complexity of the tissues affected.

## REFERENCES

1. C.H.Usher : On the inheritance of Retinitis Pigmentosa, with Notes of Cases, Roy. Lond. Ophthal. Hosp. Rep., 19:130-256 (1914).

2. J.A.Boughman, M.Vernon and K.A.Shaver, Usher Syndrome: Definition and Estimate of Prevalence from Two High Risk populations, J. Chronic Dis., 36:595-603 (1983).

3. H.Lindenov, The etiology of deaf-mutism with special reference to heredity. Op. Ex Domo Hered. Hum. Univ. Hafniensis 8:1-268.(1945).

4. G.A.Fishman, A.Kumar, M.E.Joseph, N.Torok, R.J.Anderson, Usher's syndrome-Ophthalmic and neuro-otologic findings suggesting genetic heterogeneity, Arch. Ophthalmol. 101:1367-1374 (1983).

5. C.J.Moller, W.J.Kimberling, S.L.H.Davenport, I.Priluck, V.White, K.Biscone-Halterman, L.N.Odkvist, P.K.Brookhouser, G.Lund, T.J.Grissom, Usher syndrome: An otoneurologic study, Laryngoscope 99:73-79 (1989).

6. R.J.H.Smith, J.D.Holcomb, S.P.Daiger, T.Caskey, M.Z.Pelias, B.R.Alford, D.D.Fontenot, J.F.H.and Hejtmancik, Exclusion of the Usher syndrome gene from much of chromosome 4. Cytogen. Cell Genet. 50:102-106 (1989).

7. J.Kaplan, G.Guasconi, D.Bonneau, J.Melki, M.Briard, A.Munnich, J.Dufier, and J.Frezal, Usher syndrome type I is not linked to D1S81 (pTHH33): evidence for genetic heterogeneity, Annales De Genetique. 33:105-108 (1990).

8. W.J.Kimberling, M.D.Weston, C.Moller, S.L.H.Davenport, Y.Y.Shugart, I.A.Priluck, A.Martini, M.Milani, R.J.H.Smith, Localization of Usher syndrome type II to chromosome 1q, Genomics 7:245-249 (1990).

9. R.A.Lewis, B.Otterud, D.Stauffer, J.M.Lalouel, M.Leppert, Mapping recessive ophthalmic diseases:Linkage of the locus for Usher syndrome type II to a DNA marker on chromosome 1q, Genomics 7:250-256(1990).

10. R.J.H.Smith, M.Z.Pelias,S.P.Daiger,B.Keats,W.J.Kimberling, and J.F.Hejtmancik, Clinical variability and genetic Heterogeneity within the Acadian Usher population, Amer. J. Med. Genet. 43: 964-969 (1992).

11. S.P.Dahl, M.D.Weston, W.J.Kimberling, M.B.Gorin, Y.Shugart, and J.B.Kenyon, Possible genetic heterogeneity of Usher syndrome type II: a family unlinked to chromosome 1q markers. 8th International Congress of Human Genetics, Washington D.C. Abstract 1077 (1991).

12. J.Kaplan, S.Gerber, D.Bonneau, J.Rozet, M.Briard, J.Dufier, A.Munnich, J.Frezal, Probable location of Usher type I gene on chromosome 14q by linkage with D14S13 (KLJ14 probe), Cytogen. Cell. Genet. 58:1988(1991).

13. B.J.B.Keats, A.A.Todorov, L.D.Atwood, M.Z.Pelias, J.F.Hejtmancik, W.J.Kimberling, M.Leppert, R.A.Lewis, and R.J.Smith, Linkage studies of Usher syndrome type I:Exclusion results from the Usher syndrome consortium, Submitted to Genomics.

14. W.J.Kimberling, C.G.Moller, S.Davenport, I.A.Priluck, P.H.Beighton, W.Reardon, J.B.Kenyon, J.A.Grunkmeyer, S.P.Dahl, L.D.Overbeck, D.J.Blackwood, A.M.Browe, D.M.Hoover, R.Rowland, and R.J.H.Smith, Linkage of Usher syndrome type I to the long arm of chromosome 11, Genomics. in press.

15. R.J.H.Smith, E.C.Lee, W.J.Kimberling, S.P.Daiger, M.Z.Pelias, B.J.B.Keats, M.Jay, A.Bird, W.Reardon, M.Guest, R.Ayyagari, and J.F.Hejtmancik, Localization of two genes for Usher syndrome type I to chromosome 11. Genomics. in press.

16. J.Ott, <u>Analysis of Human Genetic Linkage</u>. The Johns Hopkins University Press, Baltimore, 1991.

17. T.Ozcelik, A.Rosenthal, U.Franke, Chromosomal mapping of brain-derived neurotrophic factor and neurotrophin-3 genes in man and mouse, Genomics 10:569-571 (1991).

18. C.Junien and V.Van Heyningen,Report of the committee on the genetic constitution of chromosome 11. Cytogenet. Cell Genet.

19. Sheffield, V.C., Kimura, A.E., Folk,J.C., Bennett, S.R., Streb, L.M., Nichols, B.E., and Stone, E.M., The Gene for autosomal dominant neovascular inflammatory vitreoretinopathy maps to 11q13. Am.J.Hum.Genet. 51 (Suppl): A35 (1992).

20. U.Orth, B.Muller, C.Duviqneau, C.Julier, Y.Li, C. Fuhrmann, C.E.Van Nouhuys, U.Schonherr, H.Laqua, E.Schwinger, A.Gal, Autosomal dominant familial exudative vitreoretinopathy locus is closely linked to D11S388 on 11q, Am.J.Hum.Genet.51 (Suppl):Abstract 127 (1992).

21. E.M.Stone, B.E.Nichols, L.M.Streb, A.E.Kimura, and V.C.Sheffield, Genetic linkage of vitelliform macular degeneration (Best's disease) to chromosome 11q13. Nature Genet. 1:246-250 (1992).

22. R.R.McInnes, R.A.Bascom, R.S.Molday, V.I.Kalnins, ROM1: a retinopathy candidate gene implicated in human retinal structure and development, Amer.J. Human Genet. 49 (Suppl) :10a (1991).

23. R.A.Bascom, L.Liu, J.Chen, A.Duncan, W.J.Kimberling, C.G.Moller, P.Humphries, J.Nathans,and R.R.McInnes, ROM1: a candidate gene for autosomal dominant retinitis pigmentosa (ADRP), Usher syndrome type I, and Best vitelliform macular dystrophy. Am.J.Hum.Genet. 51 (Suppl):A6 (1992).

24. D.G.Hunter, G.A.Fishman, R.S.Mehta, F.L.Kretzer, Abnormal sperm and photoreceptor axonemes in Usher's syndrome, Arch. Ophthalmol. 104:385-389 (1986).

25. D.B.Shawn, M.H.Chaitin, S.J.Fliesler, D.E.Possin, S.G. Jacobson, A.H.Milam, Ultrastructure of connecting cilia in different forms of retinitis pigmentosa, Arch. Ophtholmol. 110:706-710 (1992).

26. G.B.Arden, and B.Fox, Increased incidence of abnormal nasal cilia in patients with retinitis pigmentosa, Nature 279:534-536 (1979).

# THE NORRIE DISEASE GENE: POSITIONAL CLONING, MUTATION ANALYSIS AND PROTEIN HOMOLOGIES

Thomas Meitinger,[1] Alfons Meindl,[1] Wolfgang Berger,[2] and Hans-Hilger Ropers[2]

[1]Abteilung für Pädiatrische Genetik, Kinderpoliklinik der Universität München, Goethestr.29, 8000 München 2, Germany
[2]Department of Human Genetics, University Hospital Nijmegen, P.O. Box 9101, 6500HB Nijmegen, The Netherlands

## SUMMARY

A candidate gene for Norrie disease, a rare X-linked eye disorder (McKusick no.310600) frequently accompanied by deafness and mental disturbances, has been isolated by positional cloning. The locus for Norrie disease has been assigned to Xp11.3 through both linkage analysis and deletion mapping. The DXS7 locus and both the monoamine oxidase genes were found to be linked to the disease locus. By subcloning a YAC clone encompassing these loci and overlapping the most proximal breakpoint in Norrie deletion patients, evolutionary conserved cDNA´s were isolated from retinal as well as brain cDNA libraries. One isolated cDNA has a length of 1.9 kb, contains an open reading frame of 399 bp which codes for a 133 amino acids long peptide and is disrupted in two Norrie patients with intragenic deletions.

The gene consists of three exons spanning over 27 kb with the first exon untranslated. Sequencing all three exons from different Norrie patients revealed missense and nonsense mutations as well as one splice-site mutation. Neither null alleles nor specific mutations can be correlated to a specific Norrie phenotype even in a single family suggesting a strong influence of genetic background. The major part of the Norrie protein contains a cysteine rich domain and by focussing the homology search on the number and the spacing of cysteine residues, we have detected similarities between the predicted product of the Norrie gene and a carboxyl-terminal domain which is common to a group of proteins including extracellular mucins. Missense mutations replacing evolutionary conserved cysteine residues emphasize the functional importance of these sites. Striking similarities to extracellular factors as well as histopathological findings point to a possible role of the Norrie gene product in retinal cell-cell interactions.

*Retinal Degeneration*, Edited by J.G. Hollyfield *et al.*
Plenum Press, New York, 1993

# NORRIE DISEASE - THE CLINICAL SPECTRUM

Norrie disease (ND) is a rare X-linked recessive disorder (McKusick no. 310600) which is characterized by congenital blindness due to degenerative and proliferative processes in the neuroretina (1, 2, 3). Retinal dysgenesis involves cells in the inner wall of the optic cup. At birth, a retrolental mass is the most prominent sign of this disease which ultimately leads to phthysis of the globe. Blindness is usually observed shortly after birth, with cataracts, retrolental membranes and retinal detachment. While the ocular symptoms are usually quite uniform, there is a report of a patient with some vision left until the age of twelve (3). The differential diagnosis of Norrie disease includes retrolental fibroplasia after oxygen therapy, intrauterine infections like toxoplasmosis, retinoblastoma and primary hyperplastic vitreous.

The disorder is not confined to the eye but also involves other tissues of neuroectodermal origin. A third of Norrie patients develop sensorineural deafness and more than half of the patients present mental retardation, often with psychotic features (3). Hearing loss and mental disturbances show marked intra- and interfamilial variation. Patients have been described with hearing loss and severe mental retardation who have sibs with ocular symptoms of the disease only (4). A characteristic physical appearance of ND patients has also been described (5). It includes a narrow nasal bridge, a flattened malar region, thin upper lips and large ears. Microcephaly, growth disturbances and cryptorchidism complement the list (6,7,8). Heterozygous female carriers do not show any physical, visual or auditory signs of the disease.

## CLONING THE GENE

Close genetic linkage between the ND gene and the DXS7 locus has been established by several groups (9,10). Positional cloning of the gene was facilitated by the description of deletions involving the DXS7 locus and adjacent Xp11.3 loci in Norrie patients (5,6,7,11,12). A 640 kb Yeast artifical chromosome (YAC: YL1.28) spanning the DXS7 locus, the monoamino-oxidase genes and a 250 kb proximal extension was used to obtain genomic sequences from this region (Fig.1).

The human cosmid clones were ordered into a contig (13) and the clones mapping proximal to the MAO genes were selected for a search of genomic rearrangements in DNA samples of ND patients. Intragenic deletions in three Norrie patients were detected with the cosmids G8 and A10. Furthermore, these two cosmids detected cross-species conserved sequences in a Zoo blot, containing DNAs from different species including non-mammalian DNAs. Consequently, these cosmids were used as hybridization probes for screening fetal brain, fetal retina and adult retina cDNA libraries. Several clones have been obtained, the largest of them with a length of 1,9 kb was used to test whether expressed sequences are deleted in the two Norrie patients. Figure 2 demonstrates two different intragenic deletions in two Norrie patients, confirming that we have isolated a transcribed gene which is disrupted in Norrie patients (13). By a very similar approach, Chen et al. (14) have also cloned the gene from a fetal retina cDNA library and have established its role as a candidate gene for Norrie disease by demonstrating intragenic deletions.

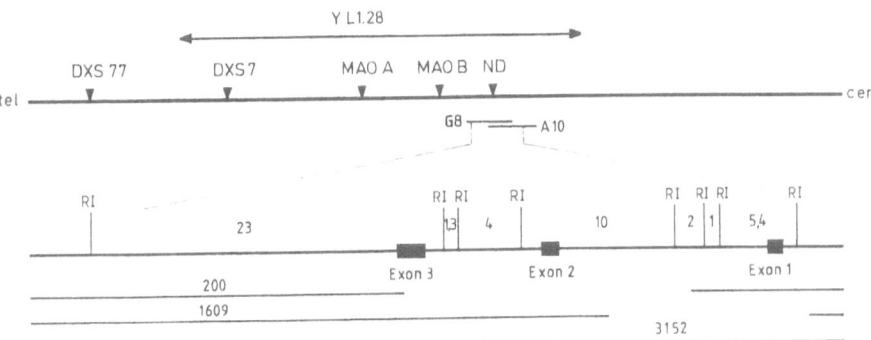

**Figure 1.** Physical map of the ND gene region. A schematic representation of the ND gene region shows MAO A and MOAB genes, DXS7 and DXS77 loci. The YAC clone YL1.28 (640kb) drawn to scale is depeicted above, the cosmid clones G8 and A10 below. A genomic EcoRI map (kb) of the ND gene shows exons 1-3 (blackboxes) The extent of deletions in three patients (200, 1609, 3152) is indicated by gaps within bold lines.

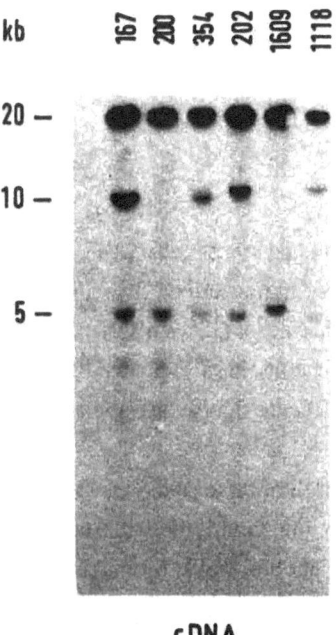

cDNA

**Figure 2**. Detection of microdeletions in Norrie patients. Genomic DNA samples of 6 Norrie patients digested with EcoRI were hybridizied with cDNA clone C2. The 10 kb fragment (polymorphism in lane 202) is deleted in patients 200 and 1609.

137

To establish the genomic structure of the Norrie gene, a restriction map of the two overlapping cosmids G8 and A10 has been obtained. Exon containing fragments were identified by hybridization of genomic DNA sequences with the full-length cDNA C2, subcloned and then sequenced with oligonucleotide primers corresponding to cDNA sequences. By alignment of genomic and cDNA sequences, exon-intron boundaries were defined.

The untranslated exon 1 is part of the 5.4 kb EcoRI fragment indicated in Fig.1. Exon 2 is in the 10 kb EcoRI fragment and contains the first 58 codons of the open reading frame. Exon 3 is the largest exon and is within a 23 kb EcoRI fragment. It contains amino acid residues 59 - 133 of the open reading frame and a 917 bp long untranslated 3´-region (Fig.3).

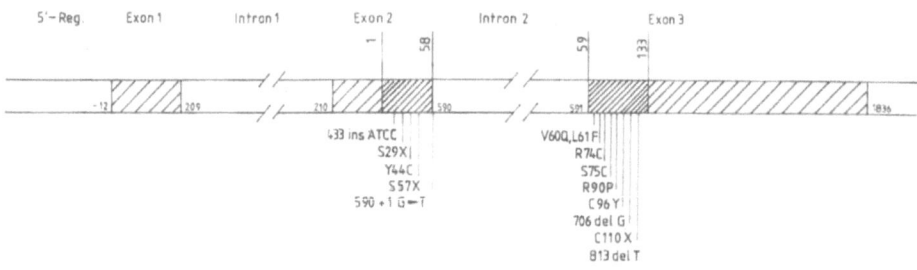

**Figure 3**. Gene structure of the Norrie gene and mutations identified. Amino acid positions and nucleotide positions at the exon-intron boundaries are indicated. Exons 1-3 are hatched, the coding regions are densely hatched. Observed mutations are given below.

The splice sites as well as the putative transcriptional start site ACAT correspond to general rules described (15). Sequencing of the adjacent untranscribed 5´-region of exon 1 revealed a putative TATA-box and upstream of this box lies a GC-rich DNA stretch, with restriction sites for the rare cutters EagI and NotI and a consensus sequence for eukaryotic DNA binding proteins (16). These sequence motifs indicate that we have also cloned the regulatory region of the Norrie gene.

Expression studies by means of Northern blot hybridisation revaeled an RNA transcript of 1.9kb in chorioid, retina and brain, but not in liver, lung and lymphoblasts (13, 14). In contrast, expression has been reported in lung and lymphoblasts (17) by making use of the more sensitive RT-PCR method.

# MUTATION ANALYSIS IN PATIENTS WITH NORRIE DISEASE

The full-length cDNA as well as four pairs of oligonucleotide primers were used for mutation screening in Norrie patients. Four out of 22 Norrie patients analysed showed deletions which should represent functional null alleles. Sequencing the 18 amplified DNA samples of non-deletion Norrie patients, we identified stop mutations in four families, missense mutations in seven families, frameshift mutations in three families, and a splice-site mutation in one family. In three families, we haven´t identified a mutation yet neither in the three exons nor in the regulatory region (18,19).

In summary, mutations occur in all three exons and involve point mutations as well as deletions. These mutations have failed to reveal any evidence for a genotype-phenotype correlation, not only when patients with different point mutations were compared, but also when phenotypes resulting from deletion patients were included in the study (Table 1). This makes the hypothesis attractive, that progression of the disease is influenced by other genes and environmental factors.

**Table 1.** Genotype-phenotype relationship in 19 ND patients

| patient | mutation | mut. type | hearing loss | mental disturb. |
|---------|----------|-----------|--------------|-----------------|
| 289 | del (ex1,2,3) | deletion | - | - |
| 2844[1] | del (ex2) | deletion | + | + |
| 200 | del (ex2u3) | deletion | - | - |
| 1609 | del (ex1) | deletion | - | - |
| P.Ru. | delG706 | frameshift | ? | + |
| 9500 | delT813 | frameshift | ? | + |
| 1443 | insATCC443 | frameshift | - | - |
| 204 | S29X (nt502C->G) | stop | ? | - |
| 1650 | Y44C (nt547A->G) | missense | - | + |
| 2248 | 590+1G->? | splice | - | +/- |
| 2843 | S57X (nt586C->G) | stop | + | +/- |
| 167[1] | V60Q (nt595T->A) | missense | - | - |
| 2942 | L61F (nt597C->T) | missense | + | - |
| 3710 | R74C (nt636C->T) | missense | + | - |
| 1245[1] | S75C (nt640C->G) | missense | - | - |
| 1127 | R90P (nt685G->C) | missense | - | - |
| Cyp | C96Y (nt703G->A) | missense | - | + |
| G | C110X (nt746C->A) | stop | ? | - |
| A.La.[1] | C110X (nt746C->A) | stop | - | - |

[1]sporadic case

# THE NORRIE PROTEIN: HOMOLOGIES AND POSSIBLE FUNCTION

The predicted Norrie gene product is a small, extracelluar protein with homologies to the globular C-terminal domains of mucins (19). Hydrophobicity analysis indicate a signal peptide for the amino-terminal part, suggesting a secretory status of the Norrie protein (20). To identify relationships to other proteins, homology searches were done with the FASTA program (21). The proteins identified include brain derived mucins, a family of extracellular proteins (22,23,24), the drosophila protein slit, an extracellular protein necessary for development of midline glia (25), van Willebrand factor, an extracellular coagulation factor (26, 27), and the growth factor binding proteins CEF-10 and cyr61 (28, 29).

A multiple alignment (30) of the protein sequences (Fig.5) displays the homologous domain ranging from the second cysteine residue at pos.55 of the Norrie protein to the cysteine-rich carboxyl-terminal end. Comparison of the last 95 amino acid residues of NDP with the carboxyl-terminal end of three mucins focused at the cysteine residues shows that number and relative positions of the last seven cysteine residues correspond in these proteins.

This strongly suggests an identical arrangement of disulfid bridges probably resulting in a similar tertiary structure of this domain. A remarkable feature is the increasing sequence identity towards the extreme carboxyterminus (see Fig.5 bottom line). Including conservative amino acid substitutions, the homology between mammalian mucins and the carboxyl-terminal part of the Norrie protein exceeds 70%. In all mutations analysed except for one (L61F), cysteine or other conserved residues have been replaced.

```
mucin-xen      CSAN--IIMAKCSGQCQHK------LTYDTIDNK-VVTKCRCCKAGRVEPRKAHLVCGNGKKKIYKYKHITSCKC--TSCTA-N(5)-COOH

mucin-bov      CKKK--VEMARCAGECKKT------IKYDYDIFQ-LKNSCLCCQEENYEYREIGLDCPDGGTIPYRYRHIITCSCL-DICQQ-N(4)-COOH

mucin-porc     CTIK--VEMARCVGECKKT------VTYDYDIFQ-LKNSCLCCQEEDYEFRDIVLDCPDGSTLPYRYRHITACSCL-DPCQQ-N(3)-COOH

norrie-hum     CSSK-MVLLARCEGHCSQASRSEPLVSFSTVLKQPFRSSCHCCRPQTSKLKALRLRCSGGMRLTATYRYILSCHCE-E-CNS-COOH

consensus      C*-*--*-*A*C-G*C-*------****----*--***C-CC*-*----*---L-C-*G------Y**I-*C-C--*-C---COOH
norrie C-pos.  55      65---69                 93-95/96        110          126-128 131--COOH
```

**Figure 5.** Comparison of the amino acid sequence of the Norrie protein with the C-terminal domains of brain derived mucin-like proteins from frog, cattle and pig. The deduced amino acid sequence of the Norrie protein (norrie hum) starts at the second cysteine residue (codon55). The consensus sequence depicts identical and conserved (*) amino acids.

Knowledge of the Norrie protein structure will help to unravel its role in normal neurodevelopment. In an eleven weeks old fetus with Norrie disease no major primary neuroectodermal maldevelopment could be observed on histopathological investigation of the eye (31). The disease status had been determined by indirect genotype analysis. The authors suggest that a fibrovascular proliferation could be induced by abnormal responses to angiogenic growth factors.

Analysis of the primary amino acid structure supports the model of a soluble extracellular factor with folding characteristics of a growth factor. Consequently, the Norrie protein should bind to neuroectodermal as well as neuronal receptors. Tissue in-situ hybridisation using antibodies directed against the Norrie protein will elucidate its pattern and timing of expression. The cloning and analysis of mouse or rat homologues will allow the construction of transgenic animal models for the disease.

## ACKNOWLEDGEMENTS

We are gratefull to patients and their clinicians in particular to E.M. Bleeker-Wagemakers, A. Gal, B. Lorenz, A. Mayerova, B. Peterlin, P. Schmitz-Valckenberg, M. Warburg and L. Zergollern. This work was supported by the Deutsche Forschungsgemeinschaft.

## REFERENCES

1. Norrie, G. Causes of blindness in children. *Acta Ophthalmol.* 5:357-386 (1927)

2. Warburg, M. Norrie´s disease: A new hereditary bilateral pseudotumor of the retina. *Acta Ophtalmol.* 39, 757-772 (1961)

3. Warburg, M. Norrie´s disease: A congenital progressive oculo-acoustico-cerebral degeneration. *Acta Ophtalmol.* (Suppl. 89) 1-147 (1966)

4. Zergollern, L. and Cupak, K. Norrie-Warburg syndrome. *Acta med.iug.* 40:263-273 (1986)

5. Donnai, D., Mountford, R.C. and Read, A.P. Norrie disease resulting from a gene deletion clinical features and DNA studies. *J. Med. Genet.* 25:73-78 (1988)

6. Moreira-Filho, C.A., Neustein, I. A presumtive new variant of Norrie disease. *J. Med. Genet.* 16:125-128 (1979)

7. Gal, A., Wieringa, B., Smeets, D.F.C.M., Bleeker-Wagemakers, L.M. and Ropers, H.-H. Submicroscopic interstitial deletion of the X-chromosome explains a complex genetic syndrome dominated by Norrie disease. *Cytogenet. Cell. Genet.* 42:219-224 (1986)

8. Townes, P.L. and Roca, P.D. Norrie´s disease (hereditary oculo-acoustico-cerebral degeneration). *Am.J. Ophthalmol.* 76:797-803 (1973)

9. Gal, A., Stolzenberger C., Wienker, T.F., et al. Norrie´s disease: close linkage with genetic markers from the proximal short arm of the X chromosome. Clin. Genet. 27:282-283 (1985)

10. Bleeker-Wagemakers, L.M., Friedrich, U., Wienker, T.F., Warburg, M. and Ropers, H.H. Close linkage between Norrie disease, a cloned DNA sequence from the proximal short arm and the centromere of the X chromosome. *Hum. Genet.* 71:211-214 (1985)

11. Sims, K.B, Lebo, R.V., Benson, G. et al. The Norrie disease maps to a 150 kb region on chromosome Xp11.3. *Hum. molec. Genet.* 1:83-89 (1992)

12. De la Chapelle, A., Sankila, E.-M., Lindlöf, M., Aula, P. and Norio, R. Norrie disease caused by a gene deletion allowing carrier detection and prenatal diagnosis. *Clin. Genet.* 28:317-320 (1985)

13. Berger, W., Meindl, A., van de Pol, T.J.R. et al. Isolation of a candidate gene for Norrie disease by positional cloning. *Nature Genetics* 1:199-203 (1992) [1]

14. Chen, Z.-Y., Hendriks R.W., Jobling, M.A. et al. Isolation and characterization of a candidate gene for Norrie disease *Nature Genetics* 1:204-208 (1992)

15. Breathnach, R. and Chambon, P. Organization and expression of eucaryotic split genes coding for proteins *Ann. Rev. Biochem.* 50:349-383 (1981)

16. Rupp, R.A.W. and Sippel, A.E. Chicken liver TGGCA protein purified by preparative mobility shift electrophoresis shows a 36.8 to 29.8 kd microheterogeneity. *Nucl. Acids Res.* 15:9707-9726 (1987)

17. Sims, K.B., Schuback, D., Solc, C.K. et al. The Norrie disease gene: predictions about the encoded protein norrin, and RNA expression studies. *Am. J. Hum. Genet.* 51:A43 abstract

18. Berger, W., van de Pol, D., Warburg, M., et al. Mutation in the candidate gene for Norrie disease. *Hum. molec. Genet.* 7:461-465 (1992)

19. Meindl, A., Berger, W., Meitinger, T. et al. Norrie disease is caused by mutations in an extracellular protein resembling C-terminal globular domains of mucins. *Nature Genetics* 2:139-143 (1992)

20. Kyte, J. and Doolittle, R.F. A simple method for displaying the hydropathic character of a protein *J. Mol. Biol.* 157:105-132 (1986)

21. Pearson, W.R. and Lipman, D.J. Improved tools for biological sequence comparison. *Proc. Natl. Acad. Sci. USA* 85:2444-2448 (1988)

22. Eckhardt, A.E., Timpte, C.S., Abernethy, J.L., Zhao, Y. and Hill, R.L. Porcine sub-maxillary mucin contains a cysteine-rich, carboxyl-terminal domain in addition to a highly repetitive, glycosylated domain. *J. Biochemistry* 266:9678-9686 (1991)

23. Probst, J.C., Gertzen, E.M. and Hoffmann, W. An integumentary mucin (FIM-B.1) from Xenopus laevis homologous with von Willebrand factor *Biochemistry* 29:6240-6244 (1990)

24. Bhargava, A.K., Woitach, J.T., Davidson and Bhavanadan, V.P. Cloning and cDNA sequence of a bovine submaxillary gland mucin-like protein containing two distinct domains. *Proc. Natl. Acad. Sci. USA* 87:6798-6802 (1990)

25. Slit: an extracellular protein necessary for development of midline glia and commisural axon pathways contains both EGF and LRR domains. *Genes and Development* 4:2169-2187 (1990)

26. Titani, K. et al. Amino acid sequence of human von Willebrand factor.*Biochemistry* 25:3171-3184 (1986)

27. Bonthron, D.T. et al. Structure of pre-pro-von Willebrand factor and its expression in heterologous cells.*Nature* 324:270-273 (1986)

28. Simmons, D.L., Levy, D.B. Yannoni, Y. and Eriksson, R.L. Identification of a phorbol ester-repressible v-src inducible gene. *Proc. Natl. Acad. Sci. USA* 86:1178-1182 (1989)

29. O´Brien, T.P., Yang, G.P., Sanders, L. and Lau, L.F. Expression of cyr61, a growth factor-inducible immediate-early gene. *Mol. and Cell. Biol.* 10:3569-3577 (1990)

30. Higgins, D.G. and Sharp, P.M. CLUSTAL a package for performing multiple sequence alignment on a microcomputer *Gene* 73:237-244 (1988)

31. Parsons, M.A., Curtis, D., Blank, C.E., Hughes H.N., McCartney A.C.E., The ocular pathology of Norrie disease in a fetus of 11 weeks. *Graefe´s Arch. Exp. Ophthalmol.* 230:248-251 (1992)

# CLINICAL AND GENETIC HETEROGENEITY

# OF LEBER'S CONGENITAL AMAUROSIS

Hélène Dollfus, Jean-Michal Rozet, Oliver Delrieu, Alain Vignal[1], Imhad Ghazi[2], Jean-Louis Dufier[2], Marie-Geneviève Mattei[3], Jean Weissenbach[1], Jean Frézal, Josseline Kaplan, Arnold Munnich

Unité de Recherches sur les Handicaps Génétiques de l'Enfant, INSERUM U12, Hôpital des Enfants Malades, 149 rue de Sèvres, 75743 Paris Cedex 15, France
[1]Unité de Génétique Moléculaire Humaine, CNRS URA 1445, Institut Pasteur, 75724 Paris Cedex, France and Généthon, 1 rue de l'Internationale, 91000 Evry, France
[2]Consultation d'Ophtalmologie, Hôpital Laënnec, 24 rue de Sèvres, 75007 Paris, France
[3]Unité INSERM U 242, Hôpital d'Enfants de la Timone, Bld Jean Moulin, 13385 Marseille Cedex 5, France

## INTRODUCTION

Leber's congenital amaurosis (LCA), originally described by Leber in 1869, is an autosomal recessively inherited congenital retinal blindness (OMIM 204000, 204001). LCA has been regarded as one of the most frequent cause of blindness in institutions for visually handicaped children, 10% to 20% of the children being affected by this disease (Alström, 1957). LCA is clinically considered as a separate entity from retinitis pigmentosa and is characterized by its precocity and severity (Foxman, 1985). The diagnosis is usually made at birth or during the first months of life when a child presenting signs of very severely impaired vision is found to have an extinguished electro-retinogram (ERG) (Franchescetti, 1954).

Little is known about the pathophysiology of LCA but this disease could be due to an abnormality of differentiation and development of the photoreceptor cells or to an extremely early degeneration of these normally developed cells.

Genetic heterogeneity has been suspected for a long time (Wardenburg 1963, Mc Kusick 1989). To date, no gene that could be responsible for this disease has been localized or identified.

Here we report genetic mapping of a gene for LCA to the long arm of chromosome 17 (17 q 24) by linkage to probe LEW 101 at the D17S40 locus and provide evidence for genetic heterogeneity of this condition.

## FAMILIES AND METHODS

### FAMILIES

Thirty-nine affected individuals (sex ratio 0.55) and thirty-eight relatives belonging to twenty LCA families were identified from three sources: i) Department of Clinical Genetics (Hôpital des Enfants-Malades, Paris), ii) Ophthalmological Consultation (Hôpital Laennec, Paris), iii) the National Register of the French Retinitis Pigmentosa Association. Among the twenty families, fourteen had more than one affected sib. First or second degree consanguinity was established in six families. Thirteen families originated from various regions of France, six families originated from North Africa (two of Sephardic Jewish origin and four of Arab origin) and one from Benin. Due to the severity of the disease, the distinction between affected and non-affected individuals was clearcut and unambiguous.

For each affected individual, ophthalmological data were available; a detailed history was taken and a pedigree established. The minimal criteria for diagnosis were: i) severe impairment of visual function detected at birth or during the first months of life, usually accompanied by pendular nystagmus or rowing eye movements, eye poking (Franceschetti's digito-ocular sign) and inability to follow lights or objects, ii) extinguished electroretinogram (ERG), iii) exclusion of ophthalmological or systemic diseases sharing ophthalmological features with LCA (Lambert, 1989). Mental retardation has been frequently described as associated with LCA (Nickel, 1982). All patients with LCA but three (in family 7 and family 13) had normal psychomotor development. However these two families were included in this study as the other sibs affected with LCA were not mentally retarded, suggesting independent segregation of the trait or variable expressivity of the same gene.

### METHODS

#### Southern blot analysis

Genomic DNA was extracted from circulating polymorphonuclears, digested using restriction enzymes according to the manufacturer instructions, electrophoresed on 0.7% agarose gel and transferred onto nylon filters (Hybond N+). The filters were hybridized with ($^{32}$P) dCTP-labeled probes (2. $10^6$ cpm/ml) and autoradiographed.

#### Microsatellite analysis

For genotyping using the polymerase chain reaction (PCR) based on hyper-variable microsatellites, 200 ng of genomic DNA was submitted to PCR amplification using 0.5 units of TAQ polymerase ( Amersham ) in the buffer provided by the supplier containing 1.5 mM $MgCl_2$, 20uM of each deoxynucleotide and 20uM surrounding primers in a final volume of 20 ul. Amplification conditions were: 95°C for 10 min., followed by 30 cycles of 94°C for 30 sec., 55°C for 30 sec., 72°C for 30 sec. and a final extension of 10 min. at 72°C. One ul of amplified DNA was mixed with 2ul formamide and 0.4 ul loading buffer (xylene cyanol FF/bromophenol blue/glycerol), electrophoresed at 1400 volts for two hours on 6% denaturing polyacrylamide gel and transferred onto nylon membranes (Hybond N+).

Membranes were hybridized with a dCTP-labeled (CA)$_{12}$ oligonucleotide by terminal deoxynucleotide transferase (Boehringer,Mannheim). Hybridization was carried out for 1h 30 at 42°C in 0.13 M sodium phosphate buffer (pH 7.2), 0.25 M NaCl, 7% SDS and 10% polyethyleneglycol. Blots were washed for 5 min. at room temperature in 2xSSC, 0.1 % SDS and autoradiographed.

## Physical mapping by *in situ* hybridization

For physical localization of the cGMP-PDEg cDNA (retinal cyclic GMP-phosphodiesterase gamma subunit) and LEW101 probe, *in situ* hybridization was carried out on metaphasic chromosomes. The probes were tritium labeled by nick translation and hybridized to the chromosome spreads to a final concentration of 100 ng/ml of hybridization solution (Mattei, 1985). After coating with nuclear track emulsion, the slides were exposed for ten days and read.

## Linkage study

For the linkage study, highly polymorphic markers including two microsatellites and two VNTRs were tested. The characteristics and genetic distances between markers are given in Table 1 and Figure 1. Although the average female-to-male ratio of genetic distance is thought to differ sightly in different regions of chromosome 17 (Nakamura 1989, Fain 1991), distances between markers were derived from the sex-averaged recombination fraction.

**Table 1 : DNA markers used for linkage analyses.**

| Nature of polymorph. | Probe Name | Locus Symbol | Physical mapping | Enz. | Alleles/Allele frequency | Contact |
|---|---|---|---|---|---|---|
| RFLP | LEW102 | D17S41 | 17q12-q24 | PstI | 12-8 kb/0.75-0.25 | Barker (1990) |
| Microsat. | AFM107yb8 | D17S789 | 17q | - | 162-154 bp/ PIC = 0.82 | Weissen-bach (1992) |
| RFLP | LEW101 | D17S40 | 17q12-q24 | MspI | 15-7 kb/0.7-0.3 | Barker (1990) |
| Microsat. | AFM049xc1 | D17S785 | 17q | - | 200-189 bp/ PIC = 0.80 | Weissen-bach (1992) |
| VNTR | THH59 | D17S4 | 17q23-q25.3 | PvuII | VNTR with 6 alleles from 0.8-1.8 kb | Nakamura (1989) |
| RFLP | PDEG | PDEG | 17q24-q25 | BsteII EcoRI | 14-8.7 kb/0.6-0.4 7.8-6.8 kb/0.83-0.17 | ATCC (1989) |
| VNTR | pRMU3 | D17S24 | 17q22-qter | PvuII | VNTR with > 10 alleles from 0.7-1.3 kb | Nakamura (1989) |

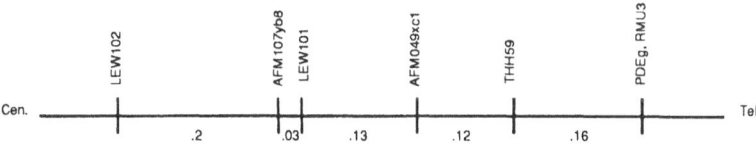

**Figure 1 : Map order and recombination estimates of the DNA markers used in our study**

The genetic localization of cGMP-PDEg was determined by linkage analysis with the data of our families using the telomeric markers.

Linkage analysis was performed using the MLINK and LINKMAP options of version 5.1 of the LINKAGE program (Lathrop, 1985). The HOMOG test version 2.4 was also performed on the total data derived from all twenty families (Ott, 1985).

## RESULTS

Combining all twenty families, pairwise linkage analysis was positive for probe LEW 101 at the D17S40 locus (Z= 2.34 , $\theta$ = 0.05 ) (Fig. 2). Nine families out of the twenty were informative, eight families were linked at this locus and one was recombining.

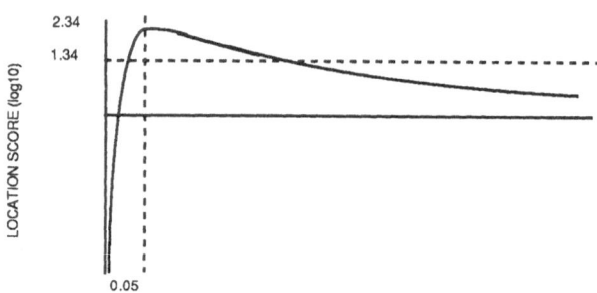

**Figure 2 : Lod score for all families with LEW101.**

To increase the informativity in this chromosomal region, two highly informative microsatellites AFM 107yb8 and AFM 049xc1 were used at loci D17S789 and D17S785 respectively. The number of recombining families increased among the previous non-informative families.

As the informativity of the DNA markers among the families was uneven, we decided to perform a HOMOG test on the data of the multipoint analysis of all families. The test determines conditional probabilities for each family of being linked or not. Two groups of families were determined, one group of ten linked families and another group of ten unlinked families. For each linked family, the conditional probabilities of being linked to the D17S40 locus were (for an alpha value of 0.37): 0.63(fam1), 0.44(fam 3), 0.88(fam 5), 0.45(fam 9), 0.63(fam 10), 0.65(fam 13), 0.38(fam 16), 0.70(fam17), 0.38(fam 19), 0.37(fam20). The remaining ten families had inconsistent conditional probabilities of being linked. The hypothesis of heterogeneity versus homogeneity was significant with a p value of 0.03 and the hypothesis of heterogeneity *versus* no linkage was significant with a p value of 0.05. When the twenty LCA families were split into the two groups according to the HOMOG test, the maximum location-score was 4.47 over the D17S40 locus ($\theta$=0) for the families of the first group (Fig. 3a).

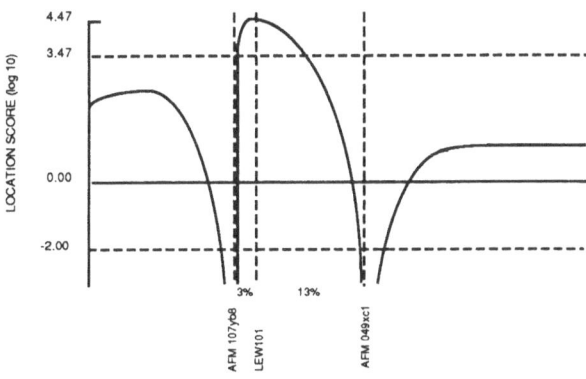

Figure 3a : Multipoint analysis of group one families (linked to LEW101).

Consistently, negative values excluded this region for linkage at this locus with the families of the second group (Fig. 3b).These data suppport the view that ten out of twenty families were linked at the D17S40 locus suggesting genetic heterogeneity for LCA.

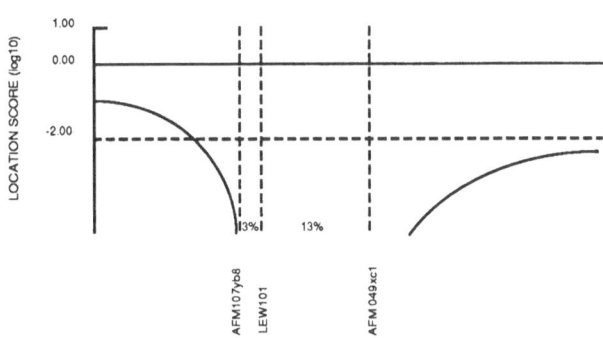

Figure 3b : Multipoint analysis of group 2 families (not linked to LEW101).

The haplotypes (Fig. 4a and 4b) built up for all the families with the three markers mentioned above and two more distal markers (one centromeric LEW102 at the D17S41 locus and one telomeric, cGMP-PDEg) showed that the affected sibs in the linked families had the same haplotypes. It is worth noting that all the North-African families of Arab origin and the Benin family were in the unlinked group.

The physical position of cGMP-PDEg was confirmed to be on chromosome 17 by *in situ* hybridization as reported by Tuteja (1990). The distribution of the grains showed a maximum on the 17q25 band. In-situ hybridization with the LEW101 probe gave the physical localization on the 17q24 band. Multipoint analysis of the data of our families with probes LEW101 (D17S40), THH59 (D17S4), RMU3 (D17S24) gave the genetic position of cGMP-PDEg at the D17S24 locus with a location score of 5.69 with no recombination event. Pairwise linkage analysis for the cGMP-PDEg locus with the LCA families showed no linkage for either of our patient group, excluding cGMP-PDEg as a candidate gene.

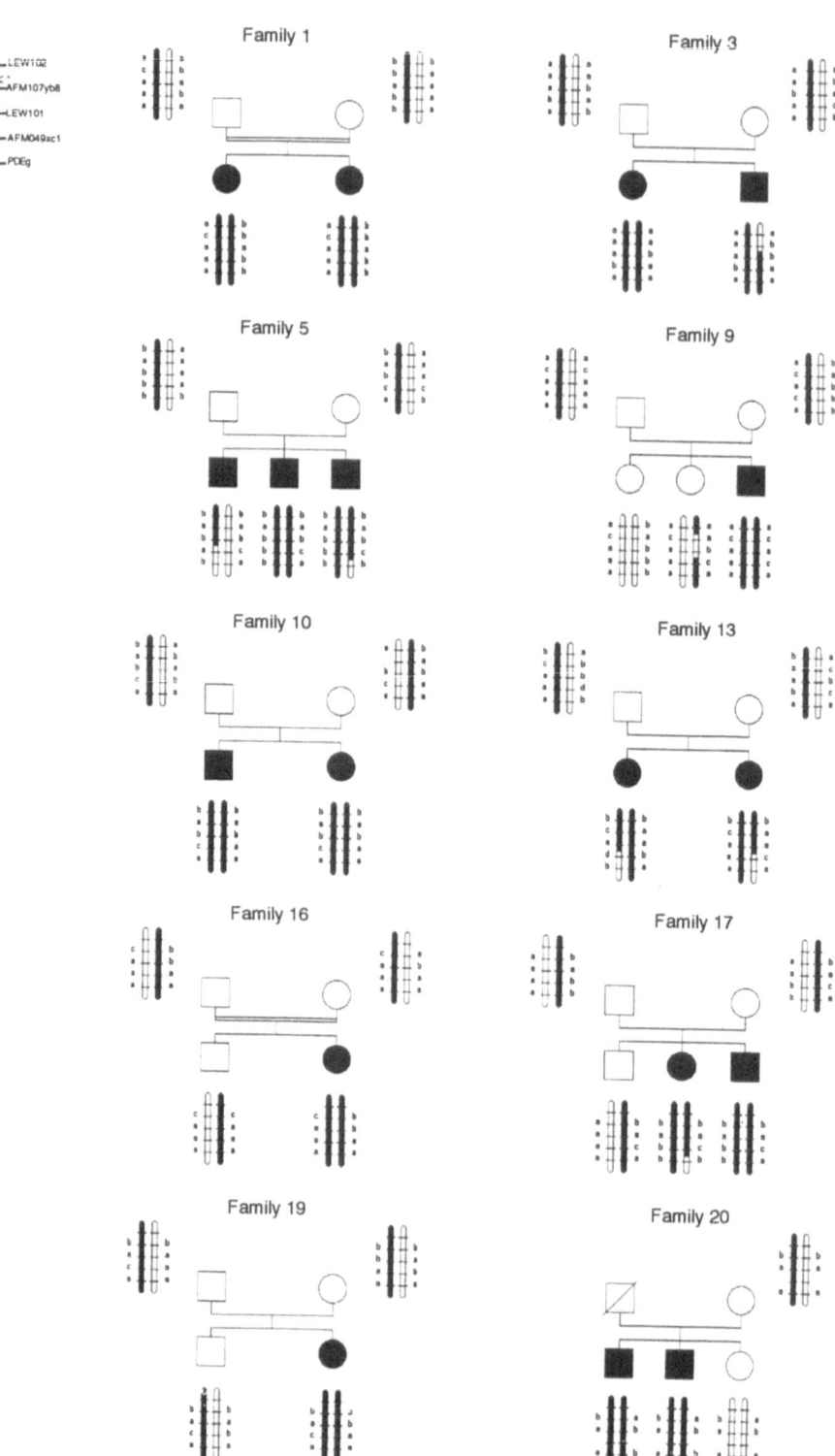

**Fig. 4a : Haplotypes of group 1 families (linked to D17S41)**

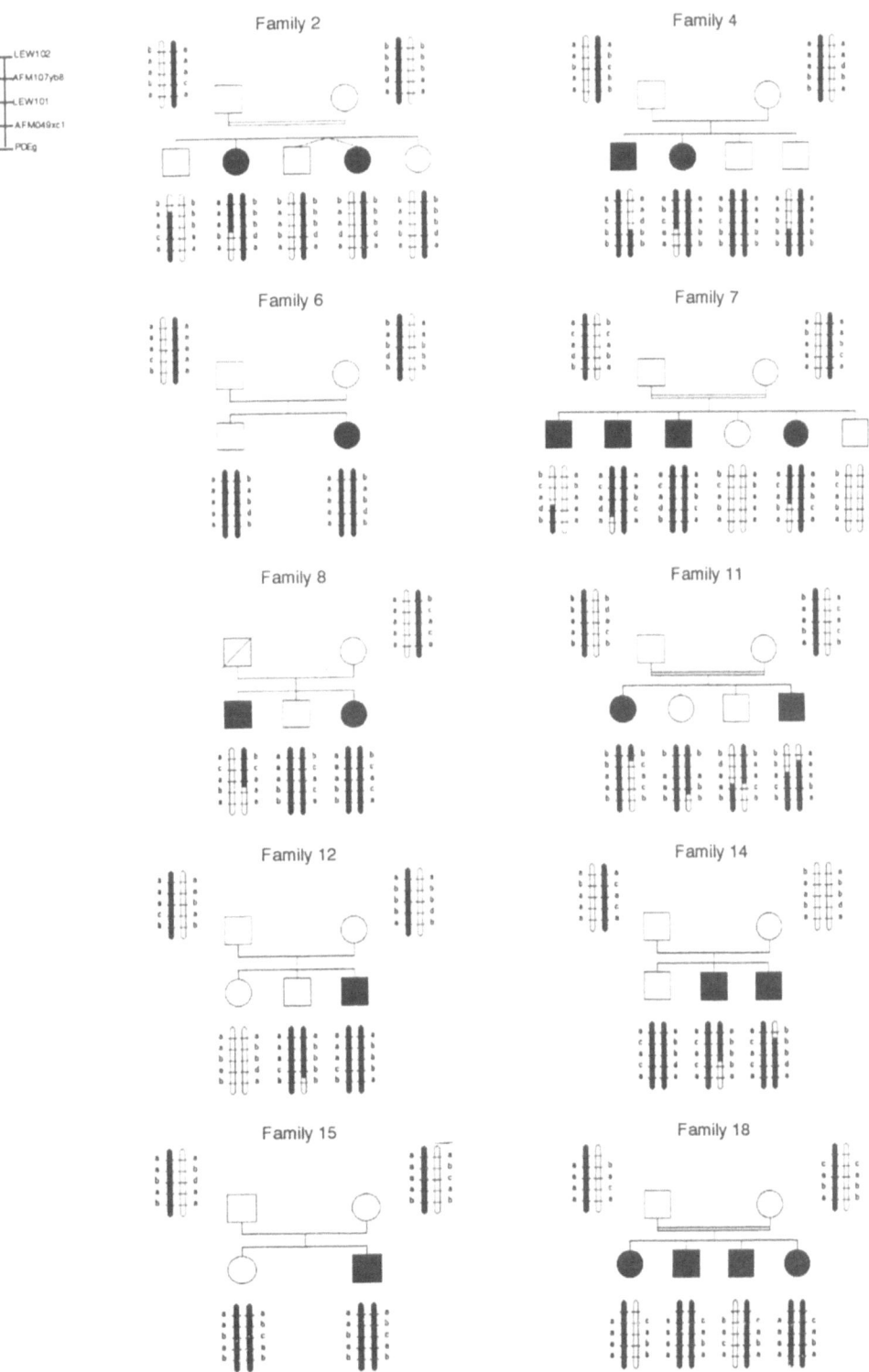

**Fig. 4b : Haplotypes of group 2 families (not linked to D17S40)**

## DISCUSSION

Up till now, to our knowledge, no information on the genetic localization of LCA was available. Here we report the mapping of a gene for LCA to the long arm of chromosome 17 at the D17S40 locus for ten out of twenty of our families selected by the HOMOG test. This test was done on the data of a multipoint analysis maximizing the information obtainable using two highly informative microsatellites flanking the D17S40 locus (AFM107yb8 at the D17S789 locus and AFM049xc1 at the D17S785 locus). The null hypothesis of homogeneity was rejected with a significant probability of heterogeneity (p=0.03) and defined at least two genetic sub-groups of families. This finding demonstrates genetic heterogeneity in LCA and the involvement of at least two loci. Genetic heterogeneity for LCA was first suggested by Waardenburg who reported a pedigree with normal children of two affected parents (Waardenburg,1963). Non-allelism has also been demonstrated for three animal models of early-onset retinal degeneration resembling human LCA (Acland, 1989).

This feature is not surprising if one considers the heterogeneity recently described for the inherited retinopathies especially autosomal dominant retinitis pigmentosa (for review see Humphries 1992). For the recessive forms of inherited retinal dystrophies, two mutations have recently been identified in the rhodopsin gene in patients with the recessive form of retinitis pigmentosa (Rosenfeld, 1992): similarly genetic heterogeneity has been reported in Usher's syndrome (Dahl, 1991; Kaplan, 1992). The small size of the nuclear families represents a limit to linkage analysis for the recessive forms and also limits the heterogeneity tests in which extensive data are required to achieve highly significant results.

Since the original report of Leber, LCA has been regarded as a congenital disorder defined by the severity of the visual impairment at birth or during the first months of life and the extinguished electroretinogram. Despite these strictly determined criteria the phenotype is somewhat variable and accounts for clinical heterogeneity (Schroeder , 1987; Lambert,1989; Smith, 1990). The eye fundus, which is classically normal initially (apart from narrowing of the retinal arterioles) can develop a wide spectrum of retinochoroid abnormalities during the course of the disease (Franceschetti, 1963). Additional associated ocular anomalies include hyperopia, keratoconus, macular colobomas and cataracts. These findings were present among our patients in about the same proportions already published by others and no significant clinical differences were found by comparing the linked with the unlinked group of families.

The cGMP-PDEg gene which codes for a major enzyme in the visual transduction cascade in the photoreceptor cells was ruled out as a candidate gene by linkage analysis in our panel of LCA families. This gene has been assigned to the long arm of human chromosome 17q21 (Tuteja, 1990). In situ hybridization allowed us to confirm this localization on 17q but located this gene in a more telomeric position (17q25) in comparison with Tuteja's localization.

The identification of the gene mapping to 17q24 as well as the other genes responsible for LCA should provide insight into the development and the differentiation of the  photoreceptor cells.

### Acknowledgements

We are greatful for the precious collaboration of the patients. We would like to thank A Rötig and J Melki for their very helpful advice, and

Dr Bonnemaison and Dr Leowski for cooperation. This work was supported by grants from Association Française contre les Myopathies (AFM), Association Française Retinitis Pigmentosa (AFRP), Société d'Etudes et de Soins pour les Enfants Polymalformés et Handicapés (SESEP).

## REFERENCES

Acland GM, Fletcher RT, Gentleman S, Chader GJ, Aguirre GD. Non-allelism of three genes (rcd1, rcd2 and erd) for early-onset hereditary retinal degeneration. Exp Eye Res 49, 993-998.(1989).

Alström C, Olson O. Heredo-retinopathia congenitalis monohybrida recessiva autosomalis. Hereditas 43, 1-77. (1957).

Dahl SP, Weston MD, Kimberling WJ, Gorin MB, Shugart YY, Kenyon JB. Possible genetic heterogeneity of Usher syndrome type 2: a family unlinked to chromosome 1q markers. 8th International Congress of Human Genetics. Washington 6,11 October:1077.

Fain PR, Solomon E, Ledbetter DF. Second International Workshop on human chromosome 17. Cytogenet Cell Genet, 57, 65-77. (1991).

Foxman SG, Heckenlively JR, Batemen BJ, Wirtschafter JD. Classification of congenital and early-onset retinitis pigmentosa. Arch Ophthalmol 103, 1502-1507. (1985).

Franceschetti A, Dieterle P. L'importance diagnostique de l'électrorétinogramme dans les dégénérescences tapedo-rétiniennes avec rétrécissement du champ visuel et héméralopie. Conf Neurol, 14, 184-186. (1954).

Franceschetti A, Babel J, François J. Les hérédo-dégénerescences chorio-rétiniennes. Rapport de la Société française d'Ophtalmologie. 1, 391-400. (1963).

Humphries P, Kenna P, Farrar J. On the molecular genetics of retinitis pigmentosa. Science, 256, 804-808.(1992).

Kaplan J, Gerber S, Bonneau D, Rozet JM, Delrieu O, Briard ML, Dollfus H, Ghazi I, Dufier JL, Frézal J, Munnich A. A gene for Usher syndrome type 1 maps to chromosome 14q. Genomics (in press).

Lambert SR, Taylor D, Kriss A. The infant with nystagmus, normal appearing fundi but an abnormal ERG. Surv Ophthal 34, 3-8 (1989).

Lambert Sr, Kriss A, Taylor D, Coffey R, Soc B, Pembrey M. Follow-up and diagnostic reappraisal of 75 patients with Leber's congenital amaurosis. Am J Ophth, 107, 624-631. (1989).

Leber T. Uber retinitis pigmentosa und angaborene amaurose. Graefes Arch Klin Exp Ophtalmol 15, 13-20. (1869).

Lathrop GM, Lalouel JM, Julier C, Ott J. Multilocus linkage

analysis in humans : Detection of linkage and estimation of recombination. Amer J Hum Genet, 37, 482-498. (1985).

McKusick VA. Mendelian inheritance in man. Eigth edition. The Johns Hopkins University Press, Baltimore. 804. (1989).

Mattei MG, Philip N, Passage E, Moisan JP, Mandel JL, Mattei JF. DNA probe localization at 18p11.3 band by in-situ hybridization and identification of a small supernumerary chromosome. Hum Genet 69, 268-271. (1985).

Nakamura Y, Lathrop M, O'Connel P, Leppert M, Barker D, Wright E, Skolnick M, Kondoleon S, Litt MK, Lalouel JM, White R. A mapped set of DNA markers for human chromosome 17. Genomics, 2, 302-309. (1988).

Nickel C, Hoyt CS. Leber's congenital amaurosis. Is mental retardation a frequent associated defect? Arch Opht 100, 1089-1091.(1982).

Ott J. Variability of the recombination fraction. Analysis of human genetic linkage. The John Hopkins University Press, Baltimore and London, 112-115. (1985).

Rosenfeld P J, Cowley G S, Mc Gee T L, Sandberg M A, Berson E L, Dryja T P. A *Null* mutation in the rhodopsin gene causes rod photoreceptor dysfunction and autosomal recessive retinitis pigmentosa. Nature genetics 1, 209-213 .(1992).

Schroeder R, Baird Mets M, Maumenee IH. Leber's congenital amaurosis, retrospective review of 43 cases and a new fundus finding in two cases. Arch Ophthalmol, 105, 356-359. (1987).

Smith D, Oestreicher J, Musarella M. Clinical spectrum of Leber's congenital amaurosis in the second to fourth decades of life. Ophthalmology, 97, 1156-1161. (1990).

Tuteja N, Dancinger M, Klisack I, Tuteja R, Inana G, Mohandas T, Sparkes R, Farber DB. Isolation and characterization of cDNA encoding the gamma-subunit of cGMP phosphodiesterase in human retina.Gene, 88, 227-232. (1990).

Waardenburg PJ, Schappert-Kimmijser J. On various recessive biotypes ofLeber's congenital amaurosis. Acta Ophtal (Kbh) 41. 317-320. (1963).

# STUDIES TOWARD THE ISOLATION OF THE RP3 GENE

Anne-Françoise Roux,[1] Johanna Rommens,[1] and Maria A. Musarella[1,2,3]

Departments of [1]Genetics, [2]Paediatrics and [3]Ophthalmology,
the Research Institute,
The Hospital for Sick Children
555 University Avenue
Toronto, Ontario, Canada   M5G 1X8

## INTRODUCTION

The understanding of the molecular basis of genetic diseases requires the identification of the affected genes or their altered functions. For each disease, one of several strategies can be used to identify the affected gene. If the biochemical defect is known, the defective protein may be immediately identified. A second strategy may involve examining known genes that have been characterized to be physiologically compatible with being the cause of a disease; these candidate genes are then tested for the presence of mutations in affected individuals. Finally, if the biochemical defect is unknown, a disease gene may be identified on the basis of its chromosomal location. In this approach, a candidate region on a chromosome must first be defined, either by linkage studies of the affected families with DNA markers or by association of chromosomal rearrangements with disease. Once a region has been identified, efforts can be directed toward isolating genes originating from this marked interval. Their candidacy can then be evaluated by examining the gene in affected individuals.

In X-linked retinitis pigmentosa (XLRP), the biochemical defect is unknown; the cloning of the genes so far depends on their physical location on the short arm of the X chromosome. This chapter describes our continuing efforts to isolate the XLRP gene located at Xp21 (RP3 gene).

### Retinitis Pigmentosa:  Genetic Heterogeneity

Retinitis pigmentosa comprises a heterogeneous group of disorders. When the inheritance can be shown (50% of cases), 20% of cases are classified as autosomal dominant (adRP) and another 20% as autosomal recessive (arRP), and a further 7-20% are due to an X-linked form of inheritance. XLRP affects about 1 in 20,000 of the population.[1-4]

Molecular studies based on the candidate gene approach and linkage studies have shown that at least three major genes and loci are involved in adRP. The rhodopsin gene at 3q24, has been found responsible in many cases, with 75 mutations described thus far[5] (Bhattacharya, personal communication). As well, 12 mutations have been found in the peripherin/rds gene at 6p12.[6] A third locus on chromosome 8, for which no candidate gene is available, has also been implicated for some cases.[7]

Genetic heterogeneity for the X-linked cases is supported by linkage studies of affected families. Two-point and multipoint linkage analyses based on a large number of polymorphic DNA markers have unambiguously assigned one XLRP locus to Xp21.[8-11] Refined genetic analyses with recombinations in our families clearly indicate that this locus is proximal to the Duchenne Muscular Dystrophy (DMD) gene and distal to the ornithine transcarbamylase (OTC) gene, consistent with an Xp21 locus. Further evidence for localization of the gene at Xp21 is provided by analysis of the Xp21 deletions found in two male patients with multiple disease phenotypes. One patient, BB, had four X-linked disorders, including chronic granulomatous disease (CGD), DMD, McLeod phenotype (XK), and RP.[12] The position of the deletion is shown on Figure 1 and is discussed in the following section. The other patient, SB, exhibited CGD, XK, and RP.[13] A locus for XLRP has been described at Xp11.3 by others.[14,15]

A large collaborative study of linkage mapping of XLRP revealed that there are at least three XLRP loci on the short arm of the X-chromosome.[16] In less than 30% of the affected families, the XLRP locus appears to map centromeric to Xp11 and in the majority of the affected families the XLRP locus mapped at Xp21, approximately 1cM distal to OTC. The disease locus of one family, however, mapped to neither of these locations but to a third between DXS28 and DMD.[11] In summary, evidence was found for three XLRP loci, designated as RP2 (Xp11), RP3 (Xp21), and RP6 (distal to DMD)[17] (Figure 1).

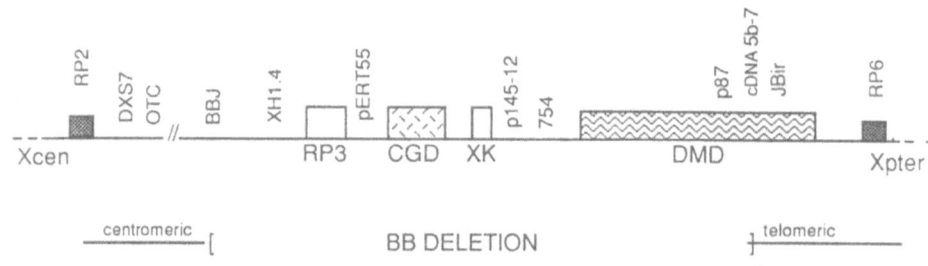

Figure 1. The approximate positions of known genes and genetic loci are indicated along the normal Xp chromosome as well as the localization of the three potential XLRP loci: RP2, RP3 and RP6. BB deletion is represented schematically.

## LOCALIZATION OF THE RP3 GENE BY PHYSICAL STUDIES

### Identification and Cloning of a Deletion Junction Fragment of the BB Deletion

The relative order of probes and genes loci, shown in Figure 1, has been derived by molecular analyses and long-range mapping of the deleted DNA of the patients BB and SB[18,19] and other Xp21 deletions present in non-RP patients. The deletions are consistent with an order as follows: Xcen-OTC-RP3-pERT55(DXS140)-CGD-XK-pERT145(DXS141)754(DXS84)-DMD-Xptel. Based on these results, the RP3 locus is positioned in the most proximal region of the DNA deleted in BB, flanked by the centromeric deletion breakpoint in BB and pERT55 (Figure 1). Cloned DNA fragments absent from the DNA of BB have been used to isolate the genes involved in CGD and DMD, but the loci responsible for XK and RP have not yet been identified.[19]

The telomeric endpoint of the deletion in the patient BB was mapped within the DMD locus,[20] as shown in Figure 1. In order to initiate cloning studies and identify the genomic DNA fragment spanning the proximal deletion junction, we used the DMD cDNA probe 5b-7[21, 22] to screen a series of restriction enzyme digests of BB genomic DNA for fragments of unique size that would have resulted from the joining of the deletion breakpoints. Digestion of HindIII of genomic DNA from a normal person resulted in detection of a restriction fragment of 18kb but in genomic DNA from BB a 10kb band was detected.

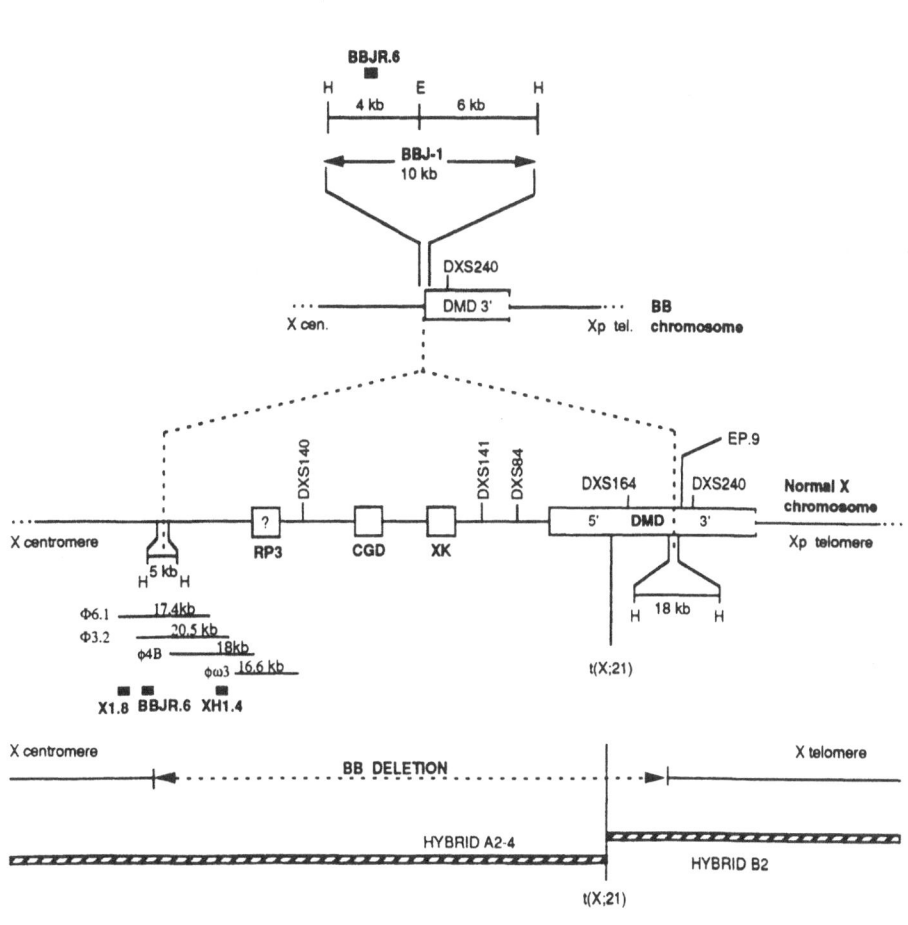

Figure 2. Normal X chromosome is depicted with relevant loci between the centromere and DMD region. Cloning of BBJ-1 and BBJR-6 from the centromeric part of BB deletion is shown as well as the 5 and 18kb fragments corresponding to the centromeric and telomeric junction fragments from normal X chromosome. Hybrids A2-4 and B2 were used to separate centromeric and telomeric fragments.

The novel HindIII fragment of 10kb was selected to be cloned from a size-selected BB genomic DNA library constructed in the phage vector, λ DASH. This library was screened with EP9 (part of the cDNA 5b-7) (Figure 2), and a phage clone containing a single 10kb HindIII insert was purified: BBJ-1. The identity of BBJ-1 was confirmed by hybridization of this sequence to the same Southern blot in which the BB junction fragment was originally detected. The same 10 kb HindIII band was detected, but in DNA of a normal subject a new 5kb Hind III fragment was detected in addition to the original 18kb fragment (Figure 2).

Two hybrid cell lines, B2 and A2-4,[23,24] containing the corresponding portions of an X-chromosome translocation, were used to orient sequences from the BB junction clone (Figure 2).The translocation breakpoint is located within the BB deletion. The hybridization patterns with these cell lines were consistent with the orientation shown in Figure 2. The 4kb HindIII/EcoRI double digestion product of the 10kb BBJ-1 hybridized to genomic DNA of the A2-4 cell line but not to DNA of B2 cell line. A unique sequence probe BBJR-6 was isolated from this fragment and used for subsequent work.

## Chromosome Walking

The probe BBJR.6 was used to screen a human genomic library to obtain sequences from the normal X chromosome. Several positively hybridizing clones were purified and ordered relative to one another by restriction enzyme mapping (φ6.1, φ3.2, φ4B , φω3, Figure 2). Each walk step isolated was tested for X chromosome specificity by hybridization to restricted DNA from patient BB, from the translocation mouse-human hybrids, A2-4 and B2, normal male, normal female, a mouse-human hybrid, AHA, containing a single intact X chromosome and a 49, XXXXY individual. The set of overlapping phage inserts covered about 25 kb of genomic DNA, including the BB deletion junction region. Two sub-clones were identified from the end portions of the cloned region. The end probe, XH1.4, was found to be 15 kb telomeric to the proximal deletion breakpoint (Figure 2). This probe was deleted in BB and showed specific hybridization to the X chromosome.

## Pulsed Field Gel Electrophoresis Mapping

A restriction mapping study using Pulse Field Gel Electrophoresis (PFGE) resulted in a detailed, long-range physical map of the RP3 region.[25] The map was generated with a cDNA from the CGD locus and with XH1.4, a unique sequence probe generated from the chromosome walking effort. This probe is positioned 15kb from the cloned junction and is in the BB deletion, as shown in Figure 2. Hybridizations were performed on partial, single, and double digestions of lymphoblastoid DNA of a normal female. The resultant composite map is shown in Figure 3. A single SfiI fragment of 205kb in size is the smallest band detected by both probes and so indicates the maximum size of the critical interval.

The CGD cDNA probe spans between 30 and 60 kb of genomic DNA (determined by adding individual EcoRI fragments detected by this probe in a standard Southern blot) within the SfiI band, as no additional SfiI products were seen in complete or double digestions. Although the size of the CGD gene is estimated, the RP3 locus therefore appears restricted to a region spanning 100-150 kb. A cluster of three GC-rich recognition sites, NotI, SacII and BssHII, is present within the SfiI fragment (see Figure 3), approximately 35kb from the proximal deletion endpoint of BB patient DNA. As a SalI site separates the CGD locus from the CpG island. This makes it unlikely that the discovered CpG island is part of the CGD locus and suggests that it marks a new gene.[25] CpG islands are known to be associated with the 5' ends of many housekeeping genes and a number of tissue-specific genes.[26]

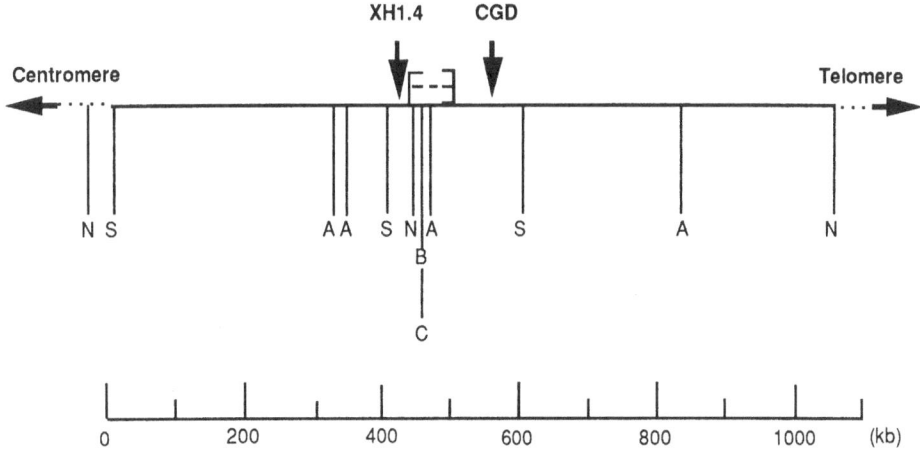

Figure 3: Restriction map around the RP3 locus, with the localization of the two probes XH1.4 and CGD used for the PFGE as well as the discovered CpG island. S, SfiI; N, NotI; B, BssHII; A, SalI [----] indicates where the cDNA maps.

## IDENTIFICATION OF A CANDIDATE GENE

Having identified the CpG island, we analyzed DNA fragments near it for cross-hybridization with other species. A DNA segment was found to have strong cross-hybridization signals in horse, dog, and pig DNA as well as CG-rich sequences. A unique portion of this conserved fragment was then used to screen cDNA libraries made from human retina and retinal pigment epithelium (RPE). Five RPE and three retinal cDNA clones were isolated. All cDNAs were sequenced and found to vary in length in either the 5' or 3' directions. The longest cDNA of 2.1kb was studied further; its map position is indicated in Figure 3.

The cDNA was used as a hybridization probe on a series of DNAs, including that from BB, the translocation mouse-human hybrids (A2-4 and B2), normal male, normal female, a mouse-human hybrid, AHA, containing a single intact X chromosome, and a 49,XXXXY individual. Both X-specific and autosomal hybridizing bands were detected.

The cDNA was also used to probe Northern blots of RNA from various tissues to determine approximate transcript size and tissue distribution. A 2.1kb transcript could be visualized in bovine retina, human brain, human fetal and adult retina, human fibroblast, and human lymphoblast RNAs. The transcript was more abundant in the RNA of adult human retina than in the fetal retina, arguing for an important function in the retina. This gene is therefore an excellent candidate, especially in view of its map position. The distance remaining in the interval first determined to contain the RP3 gene is at most 100kb (from this gene to CGD), restricting but not eliminating the possibility of additional candidate genes.

Genomic structure studies of this gene are in progress. A d(C-A) repeat polymorphism (pXRP3.1 at DXS1110) has been found within the candidate gene providing the closest marker to RP3. It will be valuable for family studies. Evaluation of the gene in the RP3 patients, together with the genetic studies being completed with the CA polymorphism in all our RP3 families, should provide a clue as to whether this candidate gene is the RP3 gene.

## Acknowledgments

The authors thank Cathy McDowell and Lynn Anson-Cartwright for their technical assistance. This research was supported in part from the RP Eye Research Foundation of Canada, Medical Research Council of Canada, and the Canadian Genetic Diseases Network. This paper was prepared with the assistance of Medical Publications, The Hospital for Sick Children, Toronto, Ontario, Canada.

## REFERENCES

1. B. Jay. Hereditary aspects of pigmentary retinopathy. *Trans. Ophthalmol. Soc. U.K.* 92:173-178 (1972).
2. A.C. Bird. X-linked retinitis pigmentosa. *Br. J. Ophthalmol.* 59:177-199 (1975).
3. J.A. Boughman, P.M. Conneally, and W.E. Nance. Population genetic studies of retinitis pigmentosa. *Am. J. Hum. Genet.* 32:223-235 (1980).
4. J.R. Heckenlively. Retinitis Pigmentosa. JB Lippincott, Philadelphia, (1988).
5. C.F. Inglehearn, T.J. Keen, R. Bashir, M. Jay, F. Fitzke, A.C. Bird, A. Crombie, and S. Bhattacharya. A completed screen for mutations of the rhodopsin gene in a panel of patients with autosomal dominant retinitis pigmentosa. *Hum. Mol. Genet.* 1:41-45 (1992).
6. S.A. Jordan, G.J. Farrar, R. Kumar-Singh, P, Kenna, M.M. Humphries, V. Allamand, E.M. Sharp, and P. Humphries. Autosomal dominant retinitis pigmentosa (adRP; RP6): Cosegregation of RP6 and the peripherin-RDS locus in a late-onset family of Irish origin. *Am. J. Hum. Genet.* 50:634-639 (1992).
7. S.H. Blanton, J.R. Heckenlively, A.W. Cottingham, J. Freidman, L.A. Sadler, M. Wagner, L.H. Freidman, and S.P. Daiger. Linkage mapping of autosomal dominant retinitis pigmentosa (RP1) to the pericentric region of human chromosome 8. *Genomics* 11:857-869, (1991).
8. M.J. Denton, J.D. Chen, S. Serravalle, P. Colley, F.B. Halliday, and J. Donald. Analysis of linkage relationships of X-linked retinitis pigmentosa with the following Xp loci: L1.28, OTC, 754, XJ-1.1, pERT87, and C7. *Hum. Genet.* 78:60-64 (1988).
9. M.A. Musarella, A. Burghes, L. Anson-Cartwright, M.M. Mahtani, R. Argonza, L.-C. Tsui, and R. Worton. Localization of the gene for X-linked recessive type of retinitis pigmentosa (XLRP) to Xp21 by linkage analysis. *Am. J. Hum. Genet.* 43:484-494 (1988).
10. M.A. Musarella, L. Anson-Cartwright, A. Burghes, R.G. Worton, J.G. Lesko, and R.L. Nussbaum. Linkage analysis of a large Latin-American family with X-linked retinitis pigmentosa and metallic sheen in the heterozygote carrier. *Genomics* 4:601-605 (1989).

11. M.A. Musarella, L. Anson-Cartwright, S.M. Leal, L.D. Gilbert, R.G. Worton, G.A. Fishman, and J. Ott. Multipoint linkage analysis and heterogeneity testing in 20 X-linked retinitis pigmentosa families. *Genomics* 8:286-296 (1990).

12. U. Francke, H.D. Ochs, B. De Martinville, J. Giacalone, V. Lindgren, C. Distèche, et al. Minor Xp21 chromosome deletion in a male associated with expression of Duchenne muscular dystrophy, chronic granulomatous disease, retinitis pigmentosa, and McLeod syndrome. *Am. J. Hum. Genet.* 37:250-267 (1985).

13. G. de Sainte-Basile, M.C. Bohler, A. Fischer, J. Cartron, J.L. Dufier, C. Griscelli, and S.H. Orkin. Xp21 DNA microdeletion in a patient with chronic granulomatous disease, retinitis pigmentosa, and McLeod phenotype. *Hum. Genet.* 80:85-89 (1988).

14. S.S. Bhattacharya, A.L. Wright, J.F. Clayton, W.H. Price, C.I. Phillips, C.M.E. McKeown, M. Jay, A.C. Bird, P.L. Pearson, E.M. Southern, and H.J. Evans. Close genetic linkage between X-linked retinitis pigmentosa and a restriction fragment length polymorphism indentified by recombinant DNA probe L1.28. *Nature* 309:253-255 (1984).

15. A.F. Wright, S.S. Bhattacharya, J.F. Clayton, M. Dempster, P. Tippett, C.M.E. McKeown, M. Jay, and A.C. Bird. Linkage relationships between X-linked retinitis pigmentosa and nine short-arm markers: Exclusion of the disease locus from Xp21 and localization to between DXS7 and DXS14. *Am. J. Hum. Genet.* 41:635-644 (1987).

16. J. Ott, S. Bhattacharya, J.D. Chen, M.J. Denton, J. Donald, C. Dubay, G.J. Farrar, et al. Localizing multiple X chromosome-linked retinitis pigmentosa loci using multilocus homogeneity tests. *Proc. Natl. Acad. Sci. USA* 87:701-704 (1990).

17. K.E. Davies, J.-L. Mandel, A.P. Monaco, R.L. Nussbaum, and H.F. Willard. Report of the committee on the genetic constitution of the X chromosome. *Cytogenet. Cell Genet.* 55:254-313 (1990).

18. G.-J.B. van Ommen, J.M.H. Verkerk, M.H. Hofker, A.P. Monaco, L.M. Kunkel, P. Ray, et al. A physical map of 4 million bp around the Duchenne muscular dystrophy gene on the Human X-chromosome. *Cell* 47:499-504 (1986).

19. C.J. Bertelson, A.O. Pogo, A. Chaudhuri, W.L. Marsh, C.M. Redman, D. Banerjee, et al. Localization of the McLeod locus (XK) within Xp21 by deletion analysis. *Am. J. Hum. Genet.* 42:703-711 (1988).

20. A.P. Monaco, C.J. Bertelson, C. Coletti-Freener, and L.M. Kunkel. Localization and cloning of Xp21 deletions breakpoints involved in muscular dystrophy. *Hum. Genet.* 75:221-227 (1987).

21. A.P. Monaco, R.L. Neve, C. Colletti-Feener, C.J. Bertelson, D.M. Kurnit, and L.M. Kunkel. Isolation of candidates cDNAs for portion of the Duchenne muscular dystrophy gene. *Nature* 323:646-650 (1986).

22. A.H.M. Burghes, C. Logan, X. Hu, B. Belfall, R.G. Worton, and P.N. Ray. A cDNA clone from the Duchenne/Becker muscular dystrophy gene. *Nature* 328:434-437 (1987).

23. C. Verellen-Dumoulin, M. Freund, R. De Meyer, C. Laterre, J. Frederic, M.W. Thompson, et al. Expression of an X-linked muscular dystrophy in a female due to translocation involving Xp21 and non-random inactivation of the normal X chromosome. *Hum. Genet.* 67: 115-119 (1984).

24. R.G. Worton, C. Duff, J.E. Sylvester, R.D. Schmickel, and H.F. Willard. Duchenne muscular dystrophy involving translocation of the *dmd* gene next to ribosomal RNA genes. *Science* 224: 1447-1449 (1984).

25. M.A. Musarella, C.L. Anson-Cartwright, C. McDowell, A.H. Burghes, S.E. Coulson, R.G. Worton, and J.M. Rommens. Physical mapping at a potential X-linked retinitis pigmentosa locus (RP3) by pulsed-field gel electrophoresis. *Genomics* 11:263-272 (1991).

26. A.P. Bird. CpG islands as gene markers in the vertebrate nucleus. *Trends Genet.* 3:342-347 (1987).

# COMMENTS ON GENE SYMBOLS AND TERMINOLOGY

Stephen P. Daiger

Graduate School of Biomedical Sciences
The University of Texas Health Science Center
Houston, TX 77030    USA

## INTRODUCTION

Two questions regarding nomenclature and terminology were raised at the International Symposium on Retinal Degeneration in Sardinia. First, what are the correct symbols for identifying the cloned and/or mapped forms of inherited retinal degeneration, and, second, what are the most appropriate terms for referring to mutations and genetic variation? This commentary addresses these questions.

## HOW GENE SYMBOLS ARE DETERMINED

For many years symbols for mapped human genes, that is, genes assigned to specific chromosomal sites, have been determined by the Human Gene Mapping Workshops. A second force for standardization has been the extensive genetic catalog maintained by Dr. Victor McKusick called "Mendelian Inheritance in Man" (Johns Hopkins Press). The catalog is a collection of autosomal dominant, autosomal recessive and X-linked phenotypes, known informally as "McKusick's Catalog". Together, the Workshops and the Catalog have largely established the nomenclature for human genes and genetic syndromes, including degenerative retinal diseases.

The first Human Gene Mapping Workshop, referred to as "HGM1", was held in 1973; the most recent, held in London in 1991, was HGM11. The Workshop proceedings are published as special editions of the journal *Cytogenetics and Cell Genetics*. The proceedings of HGM11 fill two volumes of over 2200 pages[1]. Similarly, the McKusick Catalog has gone through many editions: the first was published in 1968 and the most recent, the 10th edition, was published in 1991[2]. The current edition runs to 2650 pages and has over 6000 entries.

Because the Human Gene Mapping Workshops became complex, costly undertakings, a formal structure was developed to support the meetings and to generate the proceedings. There is an Executive Committee; there are permanent committees assigned responsibility for specific chromosomes, such as the "Committee on the Genetic Constitution of Chromosome 8"; and there are Committees devoted to specialized topics such as the "Comparative Committee for Human, Mouse and Other Rodents". The most important committee for symbols and terms is the "Nomenclature Committee", currently chaired by Dr. Phyllis McAlpine, University of Manitoba, Winnipeg, Canada. The Nomenclature Committee has become the *de facto*, "official" body establishing symbols for human genes, providing an extremely valuable service to the genetics community.

Since the Human Gene Mapping Workshops are intermittent affairs, whereas the production of new genetic information is continuous, and since clones and/or mapped human genes are being reported at an exponentially-increasing rate, there is a critical need for an online computer database, tied to the Workshops but continuously available and continuously updated. Such a database has been available for the past decade. The database was originally at Yale University[3], but in 1990 it moved to Johns Hopkins University in Baltimore, Maryland. The database is called the Genome Data Base or simply "GDB"[4]. GDB is sponsored by the National Institutes of Health and the US Department of Energy (operating through the NIH-DOE National Center for Human Genome Research.) GDB is available online, at no charge to the scientific community, and can be accessed by modem, by Internet or by other communication services[5]. At present there are over 15,000 human loci listed in GDB.

GDB has established a system of editors who chair committees devoted to specific chromosomes or subjects. These committees overlap with committees for the Human Gene Mapping Workshops. Thus the Nomenclature Committee is a component of both GDB and the Workshops. To establish gene symbols and terms at the time of a Workshop, interested individuals submit abstracts and participate in committee meetings. (The next Workshop is in Tokyo, Japan in 1993.) Between Workshops, requests for symbols, such as "D" numbers (e.g., "D8S165"), are submitted directly to GDB[5]. Trivial symbol assignments are made by the GDB staff but substantive decisions, such as symbols for degenerative retinal diseases, are made by the Nomenclature Committee.

Finally, the McKusick Catalog, like the HGM proceedings, has become extremely large and unwieldy, and needs constant updating. Thus this information source is also available as an online database, first established in 1985. The McKusick database, known as OMIM for "Online Mendelian Inheritance in Man", is physically at the same site as GDB (Johns Hopkins University), and the two databases are interconnected[4,5]. The main distinction between the databases is that GDB is an unannotated listing of human genes, symbols, map locations and other information, whereas OMIM contains text descriptions of each entry. Also, GDB is more global, including all reported loci, whereas OMIM focus mainly, though not exclusively, on inherited diseases. In addition, OMIM uses the same symbols and nomenclature as GDB (with occasional discrepancies) but OMIM also references each locus with a six-digit number referred to as the "McKusick number".

## GENE SYMBOLS FOR RETINAL DISEASES

GDB does not assign symbols for general categories of disease such as "autosomal dominant retinitis pigmentosa", instead, the GDB symbols are reserved for mapped or otherwise well-characterized loci. GDB uses uppercase letters only for human gene symbols and mixed upper-case, lower-case symbols for genes in other animals. With respect to retinitis pigmentosa, the Nomenclature Committee has specifically requested that autosomal dominant retinitis pigmentosa be referred to as "adRP" to indicate that this is not a gene symbol [personal communication, Dr. Phyllis

McAlpine]. Presumably this also applies to autosomal recessive and X-linked retinitis pigmentosa.

When a disease locus is mapped, or even tentatively mapped, to a specific chromosomal site, it is given a gene symbol. Thus the mapped forms of retinitis pigmentosa are RP1, RP2, etc., and the mapped forms of Usher Syndrome are USH1, USH2, etc.. If the map location changes (as was the case with RP1), then the symbol is retained, that is, the symbol belongs to the specific locus or to the defining patients. If a disease locus is subsequently found to be a protein entered earlier, such as rhodopsin, then the disease symbol is replaced by the protein symbol. On the other hand, if the protein is identified later, then the disease symbol is usually retained. Finally, if an apparently distinct locus is subsequently found to be the same as a previously named locus, then the symbol for the putative second locus is simply retired and not reused. This is likely to be the case for RP5.

The following table lists the mapped human genes known to cause retinal degeneration. Because new genes are being identified rapidly, the table will be out of date shortly. Also, some of these symbols may change as a result of new knowledge.

One additional note: the McKusick Catalog includes not only text descriptions of disease loci but, also, listings of specific mutations. Mutations are labeled with four-digit decimal values starting with ".0001". Thus the first described rhodopsin mutation, Pro23His[6], is McKusick entry 180380.001. How well this can be maintained with time remains to be seen.

**TABLE 1.** Cloned and/or Mapped Human Genes Causing Retinal Degeneration or Related Diseases (MD = macular dystrophy; RP = retinitis pigmentosa).

| Symbol | McKusick Number | Location | Protein; Disease | Comment [References] |
|---|---|---|---|---|
| BCP | 190900 | 7q31-q35 | blue cone pigment; rare retinal dystrophy | [7,8] |
| CHM | 303100 | Xq21.1-q21.3 | geranylgeranyl transferase A; choroideremia | [9,10] |
| CORD | 120970 | 18q21-q21.3 | cone-rod dystrophy 1 | deletion mapping [11] |
| CSNB1 | 310500 | Xp11.4-p11.23 | congenital stationary night blindness | deletion mapping [12-14] |
| DMD | 310200 | Xp21.3-p21.1 | dystrophin; may cause Oregon eye disease | [15] |
| FEVR | 133780 | 11q13-q23 | familial exudative vitreo-retinopathy | linkage mapping [16] |
| GCP | 303800 | Xq28 | green cone pigment; rare retinal dystrophy | [7,17] |
| KSS | 530000 | mitochondria | Kearns-Sayre syndrome | multiple large deletions [18] |
| MCDR1 | 136550 | 6q13-q16 | North Carolina MD | linkage mapping [19] |
| NDP | 310600 | Xp11.4-p11.3 | protein not named; Norrie Disease | gene cloned [20-22] |

(contd.)

Table 1 (Contd.)

| Symbol | McKusick Number | Location | Protein; Disease | Comment [References] |
|---|---|---|---|---|
| OAT | 258870 | 10q26 | ornithine aminotransferase; gyrate atrophy | [23] |
| PDEB | 180072 | 4p16.3 | cGMP phosphodiesterase $\beta$; recessive RP | mouse RD [24-27] |
| RCD1 | 180020 | 6q25-q26 | retinal-cone dystrophy 1 | deletion mapping [2] |
| RCP | 303900 | Xq28 | red cone pigment | [7] |
| RHO | 180380 | 3q21-q24 | rhodopsin; dominant and recessive RP | previously "RP4" [6,28,29] |
| RMCH | 216900 | 14 | autosomal rod monochromacy | mapping by uniparental isodisomy [30] |
| ROM1 | 180721 | 11q13 | rod outer membrane protein 1; possible dominant RP | [31,32] |
| RP1 | 180100 | 8p11-q21 | dominant RP | previously mapped to 1p [33] |
| RP2 | 312600 | Xp11.4-p11.23 | X-linked RP | [34] |
| RP3 | 312610 | Xp21.1 | X-linked RP | [35,36] |
| RP5 | 180102 | 3q | same as RHO | probably to be dropped [29] |
| RP6 | 312612 | Xp21.3-p21.2 | X-linked RP | [35,36] |
| RP7 | 179605 | 6p21.2-cen | peripherin-RDS; dominant MD and RP | mouse RDS [37-40] |
| RP9 | 180104 | 7p | dominant RP | linkage mapping [41] |
| RP10 | 180105 | 7q | dominant RP | linkage mapping [42] |
| USH1A | 276900 | 14q | Usher syndrome, French | linkage mapping; was "USH1" [43] |
| USH1B | 276903 | 11q14 | Usher syndrome type 1 | linkage mapping [44] |
| USH1C | 276904 | 11p15-p13 | Usher syndrome, Acadian | linkage mapping [45] |
| USH2 | 276901 | 1q | Usher syndrome type 2 | linkage mapping [46,47] |
| VMD1 | 153840 | 8q24 | atypical vitelliform MD | linked to GPT [48] |

Table 1 (Contd.)

| Symbol | McKusick Number | Location | Protein; Disease | Comment [References] |
|--------|----------------|----------|------------------|----------------------|
| VMD2 | 153700 | 11q13 | Best MD | linkage mapping [49] |
| VRNI | 193235 | 11q34 | neovascular inflammatory vitreoretinopathy | linkage mapping [50] |

## TERMS FOR MUTATIONS AND GENETIC VARIATION

There is no "Nomenclature Committee" for genetic terms, instead, usage is established by consensus. Of course it is notoriously difficult to establish consensus in science on anything, especially terminology. Therefore what follows is not "official" in any sense, and may or may not reflect consensus -- these comments are merely intended as suggestions for usage, recognizing that there are likely to be disagreements.

One problem with genetic terms such as *mutation*, or *allele*, or *polymorphism*, which had fixed definitions in classical genetics, is that the vast outpouring of new knowledge in molecular genetics has altered or confused their usage. There simply aren't rigid definitions for many of these terms any longer. What this implies is that the actual meaning of a genetic term is partially dependent on the context within which it is used. Thus, where a term may be ambiguous, an author should take care to define or delimit its meaning, even though to the author the meaning is transparent.

Some of the genetic terms of relevance to the medical genetics community, including those concerned with degenerative retinal diseases, are *locus, allele, mutation* and *polymorphism*. The definitions are superficially simple. A *locus* is a specific location in the human genome, or a specific DNA sequence or site, that codes for some observable trait such as the ABO blood group or the protein rhodopsin. An *allele* is one possible, specific form of a locus, e.g. the "A" or "B" alleles at the ABO locus, or the wild-type rhodopsin sequence. (A *gene* -- a trickey concept -- can be either a locus or an allele, depending on context.) A *mutation* is any heritable change in DNA, including nucleotide substitutions, insertions or deletions, or more complex rearrangements. Finally, a *polymorphism* is the existence in a population of more than one allele at a locus, where the alleles are in appreciable frequency. ("Appreciable" is usually taken to mean that the frequency of the 2nd most common allele, or the sum of the frequencies of the less common alleles, is at least 1%. A "population" is any substantial group ranging from Vietnamese in Houston to the entire human species.) Thus the ABO blood group is polymorphic but the amino acid sequence of rhodopsin is not, in spite of the dozens of reported mutations.

Note that *locus* applies to both chromosomes in an homologous pair, *allele* refers to one specific member of the pair, and *polymorphism* refers to the state of the locus in a population.

But nothing is ever this simple. The problem with *locus* is that the term treats genes as point-like entities, and did not, in its original usage, include non-coding DNA sequences. *Allele* suffers from similar difficulties. *Mutation* has become a real quagmire. The term should refer to a change in DNA but it often means any rare type detected in a family or in a patient population -- which may have arisen many generations previous. Also, *mutation* often implies a disease-causing or otherwise deleterious change. (A minor complication with *mutation* is that "heritable" implies from generation-to-generation, but the term is also applied to "somatic mutations" which pass from cell-to-cell.) Finally, *polymorphism* is occasionally misused to mean <u>any</u> allelic difference in a person or family, irrespective of its population distribution. The most egregious recent example of this is "SSCP" which is an acronym for "single-strand conformational polymorphism", a very powerful technique for detecting nucleotide substitutions in individuals, but which has little or nothing to do with populations.

Given these ambiguities, the following are suggestions on using genetic terms.

**Locus**. A locus is any identifiable stretch of DNA, as short as several nucleotides or as long as an entire gene including 5' regulatory sequences and 3' noncoding sequences, that has a detectable effect at a laboratory level or at a phenotypic level. A locus will typically have ill-defined ends in terms of DNA sequence. Thus rhodopsin and D8S165 (a polymorphic microsatellite on chromosome 8) <u>are</u> loci, since rhodopsin has a physiologic effect (phototransduction) and D8S165 has a laboratory effect (microsatellite variation), but a random DNA sequence is <u>not</u> a locus until it is mapped or a functional property is defined.

**Allele.** An allele is any specific expression of a locus detected at a phenotypic level, at a laboratory level or within a DNA sequence. Thus the Pro23His mutation in rhodopsin is an allele at the rhodopsin locus meeting all three criteria: it has a phenotypic effect, it can be detected by any of several laboratory techniques, and it is the result of a known nucleotide substitution. By contrast, alleles at the RH blood group locus are well defined by phenotypic effect and laboratory tests, but are not yet known at the sequence level. An important implication of the distinction between phenotypic versus sequence-defined alleles is that two phenotypically identical alleles may or may not be identical at the DNA level. In fact, at a DNA level any two sequences of more than several hundred base pairs from the same locus, even from within structural loci, are likely to differ from each other even though the differences may not lead to amino acid substitutions nor to observable phenotypic effects. It should also be noted that in common usage *allele* implies a relatively frequent expression of a locus, though this is not inherent in the definition.

**Mutation.** By now mutation has two distinct, though related, meanings, first, any very recent change in DNA and, second, a relatively rare, disease-causing change which may range from recent to ancient. Since the usage is ambiguous, it would be best to be more explicit: a deleterious mutation, a beneficial mutation, a neutral mutation, a silent mutation, a new mutation, an ancient mutation, etc. (A neutral mutation is a change in an amino acid which has no biologic consequence, that is, *neutral* refers to selectively neutral, not neutral pH. A silent mutation is a change in a codon which does not change an amino acid.) An alternative, to avoid the disease connotation, would be to use "variant". *Variant* has the advantage of having less implications regarding causes and consequences than does *allele* but it has the disadvantage of vagueness.

**Polymorphism.** A polymorphism is the presence in a population of two or more alleles at a locus where the alleles are in appreciable frequency, usually taken to mean that the frequency of the second most common allele or alleles is at least 1%. Polymorphism is such a useful word in its original genetic manifestation that the tendency to apply it to <u>any</u> allelic difference should be resisted. Also, "population" usually means a geographically or ethnographically defined population, not a patient-based population. Thus the Δ508 mutation causing cystic fibrosis is polymorphic among Caucasians[51], and the 4 bp insertion in hexoseaminodase A (which causes Tay Sachs disease) is polymorphic among Ashkenazi Jews[52], but the Pro23His mutation in rhodopsin, though frequent among adRP patients, is <u>not</u> polymorphic (as far as is known) in any geographic or ethnographic population.

## CONCLUSION

Though occasional confusion in the use of symbols and terms regarding retinal degeneration may be frustrating, it is the sign of a field with a rapidly growing body of practical knowledge. It is clear that all of us will have to rely more and more on computerized databases and text retrieval services to stay up to date.

## REFERENCES

1. Cytogenetics and Cell Genetics, Human Gene Mapping 11, Volume 58(3-4), London Conference 1991, S. Karger Publisher, Basel, Switzerland.
2. V. A. McKusick, "Mendelian Inheritance in Man", 10th edition, The Johns Hopkins University Press, May 1992.
3. R.K. Track, F.C. Ricciuti, K.K. Kidd. "Information on DNA Polymorphisms in the Human Gene Mapping Library (HGML)", Cold Spring Harbor Laboratory Press, Banbury Report 32:335-345, 1989.
4. P.L. Pearson, N.W. Matheson, D.C. Flescher, R.J. Robbins, The GDB$^{TM}$ Human Genome Data Base Anno 1992, Nucleic Acids Research 20:2201-2206(1992).
5. "NIH-DOE Award Supports Genome Data Base at Hopkins", Human Genome News 3(4):1-4, Nov. 1991.
6. T.P. Dryja, T.L. McGee, E.Reichel, L.B. Hahn, G.S. Cowley, D.W. Yandell, M.A. Sandberg, E.L. Berson, A point mutation of the rhodopsin gene in one form of retinitis pigmentosa, Nature 343(6256):364-366(1990).
7. J. Nathans, D. Thomas, D.S. Hogness, Molecular genetics of human color vision: the genes encoding blue, green and red pigments, Sci. 193: 202 (1986).
8. C.J. Weitz, Y. Miyake, K. Shinzato, E. Montag, E. Zrenner, L.N. Went, J. Nathans, Human tritanopia associated with two amino acid substitutions in the blue-sensitive opsin, Amer. J. Hum. Genet. 50:498-507 (1992).
9. F.P.M. Cremers, D.J.R. van de Pol, L.P.M. van Kerkhoff, B. Wieringa, H-H. Ropers, Cloning of a gene that is rearranged in patients with choroideraemia, Nature 347:674-677 (1990).
10. M.C. Seabra, M.S. Brown, J.L. Goldstein, Retinal degeneration in chorioderemia: deficiency of Rab geranylgeranyl transferase, Sci. 259:377-381 (1993).
11. M. Warburg, O. Sjö, L. Tranebjaerg, H.C. Fledelius, Deletion mapping of a retinal cone-rod dystrophy: Assignment to 18q211, Am.J.Med.Genet. 39:288-293 (1991).
12. M.A. Aldred, K.L. Dry, D.M. Sharp, D.B. Van Dorp, J. Brown, J.L. Hardwick, D.H. Lester, F.E. Pryde, P.W. Teague, M. Jay, A.C. Bird, B. Jay, A.F. Wright, Linkage analysis in X-linked congenital stationary night blindness, Genomics 144 (1992).
13. A. Gal, A. Schnizel, U. Orth, N.A. Fraser, F. Mollica, I.W. Craig, T. Kruse, M. Mächler, M. Neugebauer, L.M. Bleeker-Wagemakers, Gene of X-chromosomal conigenital stationary night blindness is closely linked to DXS7 on Xp, Hum. Genet. 81:315-318 (1989).
14. M.A. Musarella, R.G. Weleber, S.H. Murphey, R.S.L. Young, et al., Assignment of the gene for complete X-linked congenital stationary night blindness (CSNB1) to Xp11.3, Genomics 5:727-737 (1989).

15. P.N. Ray, D.E. Bulman, V.N. D'Souza, L.E. Becker, R.G. Worton, R.G. Weleber, D.M. Pillers, Dystrophin expression in the human retina is required for normal function, Amer. J. Hum. Genet. 51:A7 (1992).

16. Y. Li, B. Müller, C. Fuhrmann, C.E. van Nouhuys, H. Laqua, P. Humphries, E. Schwinger, A. Gal, The autosomal dominant familial exudative vitreoretinopathy locus maps on 11q and is closely linked to D11S533, Amer. J. Hum. Genet. 51:749-754 (1992).

17. J. Winderickx, E. Sanocki, D.T. Lindsey, D.Y. Teller, A.G. Motulsky, S.S. Deeb, Defective colour vision associated with a missense mutation in the human green visual pigment gene, Nat. Genetics 1:251-256 (1992).

18. D.C. Wallace, Mitochondrial genetics: a apradigm for aging and degenerative diseases? Sci. 628-632 (1992).

19. K.W. Small, J.L. Weber, A.Roses, F. Lennon, J.M. Vance, M.A. Pericak-Vance, North Carolina macular dystrophy is assigned to chromosome 6, Genomics 13:681-685 (1992).

20. W. Berger, A. Meindl, T.J.R. van de Pol, F.P.M. Cremers, H.H. Ropers, et al., Isolation of a candidate gene for Norrie disease by positional cloning, Nature Genetics 1:199-203 (1992).

21. Z-Y. Chen, R.W. Hendriks, M.A. Jobling, J.F. Powell, X.O. Breakefield, K.B. Sims, I.W. Craig, Isolation and characterization of a candidate gene for Norrie disease, Nature Genetics 1:204-208 (1992).

22. A. Meindl, W. Berger, T. Meitinger, D. van de Pol, H. Achatz, C. Dorner, M. Hassemann, H. Hellbrand, A. Gal, F. Cremers, H.-H. Ropers, Norrie disease is caused by mutations in an extracellular protein resembling C-terminal globular domain of mucins, Nat. Genet. 2:139-143 (1992).

23. D. Valle, O. Simell, The hyperornithinemias, in "The Metabolic Basis of INherited Disease", 6th Edition, C.R. Schriver, A.L. Beaudet, W. Sly, et al., Editors, McGraw-Hill, New York (1989).

24. J.B. Bateman, I. Klisak, T. Kojis, T. Mohandas, R.S. Sparkes, T.Li, M.L. Applebury, C. Bowes, D.B. Farber, Assignment of the $\beta$-subunit of rod photoreceptor cGMP phosphodiesterase gene PDEB (homolog of the mouse *rd* gene) to human chromosome 4p16, Genomics 12:601-603 (1992).

25. C. Bowes, T. Li, M. Danciger, L.C. Baxter, M.L. Applebury, D.B. Farber, Retinal degeneration in the *rd* mouse is caused by a defect in the $\beta$ subunit of rod cGMP-phosphodiesterase, Nature 347:677-680 (1990).

26. M.E. McLaughlin, E.L. Berson, T.P. Dryja, Mutations in the beta subunit of phosphodiesterase in patients with autosomal recessive retinitis pigmentosa, Invest. Ophthalmology Vis. Sci. 33:1148 (1993).

27. S.J. Pittler, W. Baehr, Identification of a nonsense mutation in the rod photoreceptor cGMP phosphodiesterase $\beta$-subunit gene of the *rd* mouse, Proc. Nat. Acad. Sci. USA 88:8322-8326 (1991).

28. G.J. Farrar, J.B.C. Findlay, R. Kumar-Singh, P. Kenna, M.M. Humphries, E. Sharpe, P. Humphries, Autosomal dominant retinitis pigmentosa: a novel mutation in the rhodopsin gene in the original 3q linked family, Hum. Mol. Genet. 1:769-771 (1992).

29. P. McWilliams, G.J. Farrar, P. Kenna, D.G. Bradley, M.M. Humphries, E.M. Sharp, et al. Autosomal dominant retinitis pigmentosa (ADRP): localization of an ADRP gene to the long arm of chromosome 3, Genomics 5:619-622 (1989).

30. L. Pentao, R.A. Lewis, D.H. Ledbetter, P.I. Patel, J.R. Lupski, Maternal uniparental isodisomy of chromosome 14: association with autosomal recessive rod monochromacy, Amer. J. Hum. Genet. 50:690-699 (1992).

31. R.A. Bascom, J. Garcia-Heras, C.L. Hsieh, D.S. Gerhard, C. Jones, U. Francke, H.F. Willard, D.H. Ledbetter, R.R. McInnes, Localization of the photoreceptor gene ROM1 to human chromosome 11 and mouse chromosome 19: sublocalization to human 11q13 between PGA and PYGM, Amer. J. Hum. Genet. 51:1028-1035 (1992).

32. R.A. Bascom, S. Manara, L. Collins, R.S. Molday, V.I. Kalnins, R.R. McInnes, Cloning of the cDNA for a novel photoreceptor membrane protein (rom-1) identifies a disk rim protein family implicated in human retinopathies, Neuron 8:1171-1184 (1992).

33. S.H. Blanton, J.R. Heckenlively, A.W. Cottingham, J. Friedman. L.A. Sadler, M. Wagner, L.H. Friedman, S.P. Daiger, Linkage mapping of autosomal dominant retinitis pigmentosa (RP1) to the pericentric region of human chromosome 8, Genomics 11:857-873 (1991).

34. S.S. Bhattacharya, A.F. Wright, J.F. Clayton, W.H. Price, C.I. Phillips, C.M.E. McKeown, M. Jay, A.C. Bird, P.L. Pearson, E.M. Southern, H.J. Evans, Close genetic linkage between X-linked retinitis pigmentosa and a restriction fragment length polymorphism identified by recombinant DNA probe L1.28, Nat. 309:253-255 (1984).

35. J. Ott, S. Bhattacharya, J.D. Chen, M.J. Denton, J. Donald, C. DuBay, G.J. Farrar, G.A. Fishman, et al., Localizing Multiple X chromosome-linked retinitis pigmentosa loci using multilocus homogeneity tests, Proc.Natl. Acad.Sci. USA 87:701-704 (1990).

36. M.A. Musarella, L. Anson-Cartwright, S.M. Leal, L.D. Gilbert, R.G. Worton, G.A. Fishman, J. Ott, Multipoint linkage analysis and heterogeneity testing in 20 X-linked retinitis pigmentosa families, Genomics 8:286-296 (1990).

37. G. Connell, R. Bascom, L. Molday, D. Reid, R.R. McInnes, R.S. Molday, Photoreceptor peripherin is the normal product of the gene responsible for retinal degeneration in the *rds* mouse, Proc. Natl. Acad. Sci. USA 723-726 (1991).

38. G.J. Farrar, S.A. Jordan, P.Kenna, M.M. Humphries, R.Kumar-Singh, P.McWilliam, V. Allamand, E. Sharp, P. Humphries, Autosomal dominant retinitis pigmentosa: localization of a disease gene (RP6) to the short arm of chromosome 6, Genomics 11:870-874 (1991).

39. S.A. Jordan, G.J. Farrar, R. Kumar-Singh, P. Kenna, M.M. Huphries, V. Allamand, E.M. Sharp, P. Humphries, Autosomal dominant retinitis pigmentosa (adRP; RP6): cosegregation of RP6 and the peripherin-RDS locus in a late-onset family of Irish origin, Amer. J. Hum. Genet. 50:634-639 (1992).

40. G. Travis, L. Christerson, P. Danielson, I. Klisak, R. Sparkes, L. Hahn, T. Dryja, J. Sutcliffe, The human retinal degeneration slow (RDS) gene: chromosome assignment and structure of the mRNA, Genomics 10:733-739 (1991).

41. C.F. Inglehearn, S.A. Carter, T.J. Keen, J. Lindsey, A.M. Stephenson, R. Bashir, M. Al-Magtheh, A.T. Moore, M. Jay, A.C. Bird, S.S. Bhattacharya. A new locus for autosoaml domimant retinitis pigmentosa (adRP) on chromosome 7p. Nat. Genet. 4:51-53 (1993).

42. S.A. Jordan, C.J. Farrar, P. Kenna, M.M. Humphries, D.M. Sheils, R. Kumar-Singh, E.M. Sharp, C. Ayuso, J. Benitez, P. Humphries. Localizatoin of an autosomal dominant retinitis pigmentosa gene to chromosome 7q. Nat. Genet. 4:54-58 (1993).

43. J. Kaplan, S. Gerber, D. Bonneau, J. Rozet, M. Briard, J. Dufier, A. Munnich, J. Frezal, Probable location of Usher type I gene on chromosome 14q by linkage with D14S13 (MLJ14 probe), Cyto.Genet.Cell.Genet. 58:1988 (1991).

44. W.J. Kimberling, C.G. Möller, S. Davenport, A. Priluck, P.H. Beighton, J. Greenberg, W. Reardon, M.D. Weston, J.B. Kenyon, J.A. Grunkemeyer, S.P. Dahl,

L.D. Overbeck, D.J. Blackwood, A.M. Brower, D.M. Hoover, P. Rowland, R.J.H. Smith, Linkage of Usher syndrome type I gene (USH1B) to the long arm of chromosome 11, Genomics 14:988-994 (1992).

45. R.J.H. Smith, E.C. Lee, W.J. Kimberling, S.P. Daiger, M.Z. Pelias, B.J.B. Keats, M.L. Jay, A. Bird, W. Reardon, M. Guest, R. Agyagri, F. Hejtmancik, Localization of two genes for Usher syndrome type I to chromosome 11, Genomics 14:995-1002 (1992).

46. W.J. Kimberling, M.D. Weston, C. Möller, S.L.H. Davenport, Y.Y. Shugart, I.A. Priluck, A. Martini, M. Milani, R.J. Smith, Localization of Usher syndrome type II to chromosome 1q, Genomics 7:245-249 (1990).

47. R.A. Lewis, B. Otterud, D. Stauffer, J-M. Lalouel, M. Leppert, Mapping recessive ophthalmic diseases: linkage of the locus for Usher syndrome type II to a DNA marker on chromosome 1q, Genomics 7:250-256 (1990).

48. R.E. Ferrell, H.M. Hittner, J.H. Antoszyk, Linkage of atypical vitelliform macular dystrophy (VMD-1) to the soluble glutamate pyruvate transaminase (GPT1) locus, Am.J.Hum.Genet. 35:78-84 (1983).

49. E.M. Stone, B.E. Nichols, L.M. Streb, A.E. Kimura, V.C. Sheffield, Genetic linkage of vitelliform macular degeneration (Best's disease) to chromosome 11q13, Nature Genetics 1:246-250 (1992).

50. E.M. Stone, A.E. Kimura, J.C. Folk, S.R. Bennett, B.E. Nichols, L.M. Streb, V.C. Sheffield, Genetic linkage of autosomal dominsnt neovascular inflammatory vitreoretinopathy to chrommomsome 11q13, Hum. Mol. Genet. 9:685-689 (1992).

51. E. Kerem, M. Corey, NB. Kerem, J. Rommens, D. Markiewicz, H. Levison, L.-C. Tsui, P. Durie, The relation between genotype and phenotype in cystic fibrosis - analysis of the most common mutation (delta-F508), New Eng. J. Med. 323:1517-1522 (1990).

52. R. Myerowitz, F.C Costigan, The major defect in Ashkenazi Jews with Tay-Sachs disease is an insertion in the gene for the $\alpha$-chain of $\beta$-hexosaminidase, J. Biol. Chem. 26:18587-18589 (1988).

# IDENTIFICATION OF CANDIDATE GENES FOR EYE DISEASES: STUDIES ON A NEURAL RETINA-SPECIFIC GENE ENCODING A PUTATIVE DNA BINDING PROTEIN OF LEUCINE ZIPPER FAMILY

Anne U. Jackson[1], Keqin Zheng[2], Teresa L. Yang-Feng[2], and Anand Swaroop[1,3]

[1]Department of Human Genetics
[3]Department of Ophthalmology, Kellogg Eye Center
University of Michigan
Ann Arbor, MI 48105
[2]Department of Genetics
Yale University School of Medicine
New Haven, CT 06510

## ABSTRACT

We have enriched human adult retina, fetal eye (11 wk) and retinal pigment epithelium (RPE) cell line cDNA libraries for tissue-specific genes using a novel biotin-based subtraction procedure[1,2]. At least 70% of the random cDNA clones from the subtracted retina and RPE libraries are not constitutively expressed. To identify candidate genes for eye diseases, several novel cDNA clones from subtracted libraries (as determined by sequence comparison) are being localized to human chromosomes using PCR-assays or *in situ* hybridization. Expression analysis of clones from a subtracted retinal library identified a cDNA, AS321, that showed neural retina-specific expression. AS321 encodes a putative DNA binding protein of the leucine zipper family and is named NRL[3]. We have now isolated and characterized the NRL gene and have localized it to human chromosome 14q11.1-q11.2[4]. Based on the gene structure, PCR primers have been designed and will be used to screen for mutations within the NRL gene in patients with retinal diseases. A polymorphic microsatellite $(CA)_n$ repeat has been identified within the NRL cosmid and should be valuable for linkage analysis.

## INTRODUCTION

Genetic abnormalities affecting eye function can be detected clinically due to their effects on vision. A large number of the over 4,000 catalogued human Mendelian disorders have a direct or indirect effect on visual function. Clinicians and geneticists, armed with sophisticated mathematical algorithms, an ever increasing number of polymorphic markers, and cooperative families, are making steady progress towards identifying the genetic loci for congenital and hereditary eye disorders. While a vast majority of inherited ocular diseases have not yet been mapped, chromosomal assignments do exist for several ocular disorders. Autosomal disease loci are shown in Table 1 (X linked loci are not included).

Once genetic disorders have been mapped by linkage analysis to polymorphic markers, several approaches may be taken to elucidate molecular defects underlying the disease phenotype. When little is known about the specific functional defect, positional cloning is often the best approach[7]. Positional cloning is a lengthy process whereby the genetic distances determined by linkage or, where possible, breakpoint analyses are translated into cloned DNA, which then serves as a repository for potential candidate disease genes. Positional cloning has successfully been used to isolate the genes for several ocular diseases, including aniridia, choroideremia, and Norrie's disease[8-11]. As more polymorphic markers are identified, the genetic distances should become smaller, making the positional cloning task less daunting.

An alternate approach is to directly search for mutations in genes whose functions are well characterized[7,12]. For ocular disease, this approach has been successful in about 40% of families with autosomal dominant retinitis pigmentosa[13]. Mutations segregating with the disease in independent families have been detected in the rhodopsin and peripherin-RDS genes[14,15]. Unfortunately, not many candidate eye genes have been discovered. Table 1 includes a comprehensive list of eye-specific genes.

A complementary approach to positional cloning and traditional candidate gene approaches is to increase the number of mapped candidate genes. For diseases specifically affecting the eye, one criterion for assigning candidate gene status is to assess gene expression in the ocular tissue. Once expression is verified, the chromosomal localization of novel genes must be determined to match potential candidate genes with particular disease loci. Mutational analyses may be then be initiated, even in families without large pedigrees. An additional benefit of chromosomal localization is that unique sequences derived from randomly isolated cDNA clones, called expressed sequence tags (ESTs)[16], can serve as molecular landmarks to be used for physical mapping of the human genome. Furthermore, novel polymorphic markers can be isolated from genomic clones containing ESTs by simply screening for microsatellites. Usefulness of ESTs is enhanced when they have been selected from eye- or developmental stage-specific libraries[17], for these genes may contribute more readily to our understanding of normal eye development.

## SUBTRACTION CLONING: MOLECULAR CHARACTERIZATION OF RANDOM CLONES, AND EXPRESSION ANALYSIS

To avoid repetitive isolation of constitutively expressed genes and to most efficiently select developmentally regulated and tissue-specific genes, our laboratory has generated several cDNA libraries from human ocular tissue, including fetal eye (11 weeks), RPE, and adult retina, and have enriched these libraries for retina-specific genes using a biotin-based subtraction scheme (Figure 1). A human adult retina cDNA library was enriched for tissue-specific genes by subtractive hybridization against a JY lymphoblastoid cell line library[1,2]. Likewise, a RPE cell line cDNA library was enriched for specific clones by subtraction against a mixture of RNA from JY and fetal eye cDNA libraries, and a fetal eye library was subtracted against RNA from a fetal brain cDNA library. The subtracted clones are in Bluescript KSM13(-) plasmid (Stratagene). DNA from approximately three hundred random cDNA clones from the enriched retina and RPE libraries was prepared and used for sequence and expression analysis to determine the subtraction efficiency. Expression was assessed using total RNA Northerns of total RNA or Southerns of DNA from cDNA libraries digested with EcoRI and HindIII to release the cloned inserts. More than 70% of

# Table 1. Mapped Autosomal Eye Diseases[1] and Genes

| Disease or Gene | Map Loc. | MIM # | Symbol |
|---|---|---|---|
| **Chromosome 1** | | | |
| Cataract,zon.pulverulent | 1q21-q25 | 116200 | CAE |
| Neur. ceroid-lipofuscin., Finnish(Santavuori) | 1p32 | 256730 | CLN1, INC1 |
| Phosducin, pineal gland | 1q | 171490 | PDC |
| Usher syndrome type II | 1q32 | 276901 | US2 |
| **Chromosome 2** | | | |
| Aniridia 1 | 2p25 | 106200 | AN1 |
| Cataract, anterior polar | 2p25 | 115650 | CAP |
| Coloboma of iris | 2p25.1 | 120200 | COI |
| Crystallin, γ polypep 1-6 cataract, Coppock | 2q33-q35 | 123660 | CRYG CCL |
| Optic atrophy, Kjer type | 2p | 165500 | OAK |
| Retinoic acid bind. prot., cellular type II | 2q | 180231 | CRABP2 |
| S-antigen (arrestin) | 2q24-q37 | 181031 | SAG |
| Waardenburg syn. 1 | 2q35 | 193500 | WS1 |
| **Chromosome 3** | | | |
| Cell. retinol bind. prot.1 | 3q21-q22 | 180260 | RBP1 |
| Cell. retinol bind. prot.2 | 3q21-qter | 180280 | RBP2 |
| Crystallin, γ polypept. 8 | Chr. 3 | 123730 | CRYG8 |
| Guanine nucleotide bind. prot., α transducin | 3p21 | 139330 | GNAT1 |
| Retinitis Pigmentosa-5 | 3q21-q23 | 180380 | RP5 |
| Rhodopsin | 3q21-q24 | 180280 | RHO |
| Von Hippel Lindau syn. | 3p26-p25 | 193300 | VHL |
| **Chromosome 4** | | | |
| Ant. seg. mesen.dysgen. | 4q28-q31 | 107250 | ASMD |
| Cyc. nucl. gated channel | 4p14-q13 | 123825 | CNCG |
| Rieger syndrome | 4q23-q27 | 180500 | RGS |
| Rod cAMP phosphodi-esterase, β ( rd) | 4p16.3 | 180072 | PDEB |
| **Chromosome 5** | | | |
| cGMP phosphodiest'se,α | 5q31.2-34 | 180071 | PDEA |
| Treacher-Collins syn. | 5q11 | 154500 | MFD1 |
| **Chromosome 6** | | | |
| Ocular albinism, AR | 6q13-q15 | 203310 | |
| Prog. foveal dystrophy, NC type macular | 6q13-q21 | 136550 | MCDR1 NCMD |
| Peripherin-rds | 6p21.1-cn | 170710 | PRPH |
| Retinal cone degenerat'n | 6q25-q26 | 180020 | |
| **Chromosome 7** | | | |
| Blue cone pigment | 7q22-qter | 190900 | BCP |
| GTP bind. prot., inhib. α | 7q21 | 139310 | GNAI1 |
| **Chromosome 8** | | | |
| Macular dyst, atyp vit. | 8q24 | 153840 | VMD1 |
| Retinitis Pigmentosa-1 | 8p11-q21 | 180100 | RP1 |
| **Chromosome 9** | | | |
| Gelsolin, amyloidosis V | 9q32-q34 | 137350 | GSN |
| Cockayne syn. A, late | 10q21.1 | 216400 | CS |
| Interst retinol bind prot3 | 10q11.2 | 180290 | IRBP |
| Ornith. aminotrans'ase, Gyrate atrophy | 10q26 | 258870 | OAT, OKT |
| **Chromosome 11** | | | |
| Aniridia 2 | 11p13 | 106210 | AN2 |
| Exud. vitreoretinopathy Criswick-Schepens | 11q13-q23 | 133780 | |
| Rod outer seg. prot.-1 | 11p13-q13 | 180721 | ROSP1 |
| Tyr-neg oculocuta-neous albinism 1 | 11q14-q21 | 203100 | OCA1 |
| Usher syndrome Ib | 11p | | |
| Usher syndrome Ic | 11q | | |
| Vitelliform macular degen. 2, Best's dis. | 11q13 | 153700 | VMD2 |
| **Chromosome 13** | | | |
| Retinoblastoma 1 | 13q14.1-14.2 | 180200 | RB1 |
| **Chromosome 14** | | | |
| Cataract, anter. polar | 14q24 | 115650 | CAP |
| NRL | 14q11.1-11.2 | | |
| Rod monochromatism | chr. 14 | 216900 | RMCH |
| Usher syndrome Ia | 14q | 276900 | |
| **Chromosome 15** | | | |
| Cohen syndrome | 15q11-q13 | 216550 | |
| Dyslexia-1 | 15q11 | 127700 | DYX1 |
| Ocular albinism, Tyr-positive | 15q11.2-q12 | 203200 | OCA2 |
| Retinoic acid bind. prot., cell. type I | 15q22-qter | 180230 | CRABP |
| **Chromosome 16** | | | |
| Cataract, Mariner type | 16q22.1 | 116800 | CTM |
| Fish-eye disease | 16q22.1 | 245900 | LCAT |
| Neur.ceroid lipofuscin., Batten disease | 16p12 | 204200 | CLN3, BTS |
| **Chromosome 17** | | | |
| Crystallin, β-B1 | 17q11.1-12 | 123610 | CRYB1 |
| Cyclic nucleotide phos-phodiesterase | 17q21 | 238300 | CNP |
| Leber's congenital amaurosis | 17q24-q25 | 204000 | LCA, CRB |
| Pigment epithithelium derived factor | 17p13.1 | | PEDF |
| Recoverin | 17p | | |
| Retinoic acid receptor | 17q21.1 | 180240 | RARA |
| Rod cGMP phospho-diesterase, γ | 17q21.1 | 180073 | PDEG |
| **Chromosome 18** | | | |
| Cone-rod dystrophy | 18q21.1-21.3 | 120970 | RCRD1 |
| **Chromosome 19** | | | |
| Green/blue eye color | chr. 19 | 227240 | GEY |
| **Chromosome 21** | | | |
| Crystallin, α-A | 21q22.3 | 123580 | CRYA1 |
| **Chromosome 22** | | | |
| Crystallin, β-B2 | 22q11.2-12.2 | 123630 | CRYB2 |
| Crystallin, β-B3 | 22q11.2-12.2 | 123630 | CRYB3 |
| Ocular coloboma, Cat Eye syndrome | 22q11 | 115470 | CES |

[1]Diseases primarily affecting retina and/or eye are listed. Other diseases with an ocular component , including renal, hepatic, skeletal, connective tissue, neurologic, dermatologic, craniofacial, hematologic, and multisystem syndromes, are not included. Table is derived from references 5 and 6.

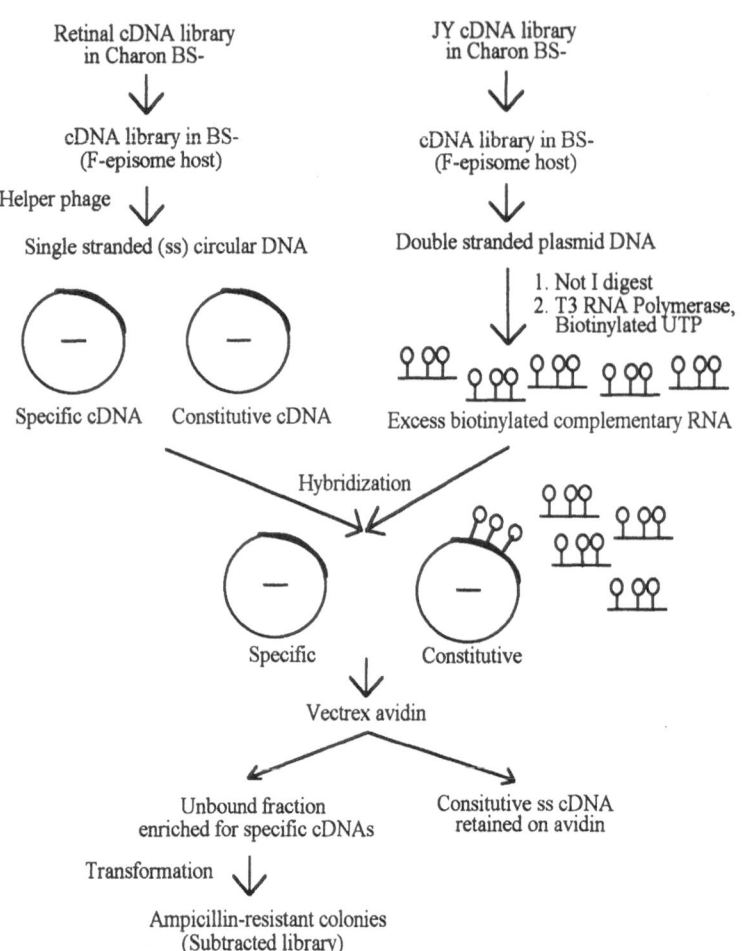

**Figure 1.** Subtraction cloning procedure. Directional cDNA libraries from human adult retina and JY lymphoblastoid cell line were constructed in Charon BS(-) vector and transferred to plasmid BS (-)[18]. Single stranded (ss) circular DNA was generated from the plasmid retina cDNA library with helper phage R408 (Stratagene). Double stranded plasmid from JY cDNA library was used to synthesize biotinylated run-off transcripts, complementary in orientation to the ss retinal cDNA. Typically, 0.2-1 ug ssDNA was annealed to 20-40 mg biotinylated RNA in a sealed capillary at 65°C for 16 hr. After hybridization, the mixture was incubated with 250 mg vectrex-avidin in two batches for 0.5-2 hr. at 22°C. The ssDNA in the unbound fraction, enriched for retina specific ss DNA, was precipitated and used directly to electrotransform XL1 Blue cells. Control experiment did not include any biotinylated RNA. Between 2-15% of clones were recovered after subtraction.

the clones are not expressed in JY cells, and several others are expressed only at low levels (Table 2). A large percentage of the randomly isolated clones showed a restricted pattern of expression, allowing these clones to be useful for gene mapping efforts.

Approximately 150-300 bases from the ends of subtracted clones with inserts were sequenced using T3 and T7 oligonucleotide primers by the dideoxy chain termination method[19]. These end sequences were compared to GenBank and EMBL nucleotide databases using the FASTA software program. The analyses revealed only a few clones of housekeeping function. Several others are brain or neuronal specific isoforms. Molecular characterization of some of the new genes identified has been published recently[20-22], and we are in the process of submitting new clones to GenBank for research use.

# Table 2. Identification and expression analysis of clones from subtracted retina libraries.

| Clone | Identification | Expression AR | JY | Clone | Identification | Expression AR | JY |
|-------|----------------|:---:|:---:|-------|----------------|:---:|:---:|
| 199 | Similar to mouse kinesin | + | - | 327 | | | |
| 200 | IGF binding protein 2 | + | - | 328 | | | |
| 201 | Apolipoprotein E | + | - | 329 | Ferritin H | + | + |
| 202 | Platelet glycoprotein Ib | + | - | 330 | | + | - |
| | β chain | | | 331 | | | |
| 208 | Similar to Human EST, | + | + | 333 | | | |
| | fetal hippocampus | | | 334 | | | |
| 211 | | | | 335 | | | |
| 220 | | + | - | 336 | | | |
| 301 | | + | - | 337 | | | |
| 301D | Opsin | + | - | 340 | | | |
| 302 | | + | - | 341 | | | |
| 304 | Human SINE | + | - | 342 | | | |
| 306 | | + | + | 343 | | | |
| 307 | | | | 345 | | | |
| 308 | similar to E. coli araE | | | 346 | | | |
| | proton symporter | | | 347 | | | |
| 310 | | | | 350 | | | |
| 312 | | | | 351 | | | |
| 313 | | | | 354 | | | |
| 317 | E. coli entF gene | | | 355 | | | |
| 318 | Protein kinase homology | | | 356 | | | |
| 319 | | + | - | 358 | | | |
| 320 | | | | 359 | rRNA 5' external | | |
| 321 | NRL | + | - | | transcribed spacer | | |
| 322 | | + | - | 361 | | | |
| 323 | Gs, a subunit | | | 366 | | | |
| 324 | | | | 370 | | | |
| 325 | | | | 372 | | | |
| 326 | Thyr. hormone bind.prot. | | | | | | |

Clones were identified by FASTA comparison of end sequences to GenBank/EMBL databases. Partial results are shown. Of the total clones analyzed[23], about 20% are known human genes, 10% are E. coli genes by homology to one end sequence, and 70% are novel unique sequences (shown as empty spaces in the columns). E. coli sequences are probably incorporated during the single-stranded phage step by illegitimate recombination[24]. Expression was assessed by probing total RNA blots from adult retina (AR) and JY lymphoblast cell line (JY) with nick-translated cDNA probes. +/- indicates low expression.

## CHROMOSOMAL LOCALIZATION OF SUBTRACTED CLONES

The sequence information of the ESTs was used to generate primer pairs for amplification by the polymerase chain reaction (PCR)[17]. PCR was performed on DNA from a panel of 18 human-rodent somatic cell hybrids (NIGMS repository, Camden, New Jersey). The presence or absence of a human-specific product was then correlated with the human chromosomal content within the particular hybrid. Chromosomal localization was assigned to those ESTs for which no discordancy was observed (Table 3)[17].

While PCR-based assays for chromosomal assignment are convenient, *in situ* hybridization to metaphase chromosomes provides much more specific and useful localization information. We have recently determined the regional assignment of several randomly isolated clones by *in situ* hybridization of [3]H-labeled cDNA to human chromosome spreads[25], as shown in Table 4. Based on this further localization, one clone, AA35, is being investigated for a potential role in Von Hippel Lindau syndrome[26].

**Table 3.** Chromosomal localization of ESTs by somatic cell hybrid analysis.

| Clone | Chr. | Clone | Chr. |
|-------|------|-------|------|
| AA36  | 1    | AA19  | 10   |
| AA35  | 3    | AA20  | 10   |
| AS324 | 3    | AA12  | 11   |
| AS259 | 7    | AA26  | 19   |
| AS266 | 7    | AA29  | 20   |
| AS269 | 7    | AA38  | 21   |
| AA32  | 7    |       |      |

Sequence from subtracted cDNAs was used to synthesize oligonucleotide primers on a DNA synthesizer (Applied Biosystems) for amplification of specific products from human genomic DNA. PCR primers were used to amplify DNA from human-rodent somatic cell hybrid panel DNAs (panels #1 and #2 from NIGMS repository, Camden, NJ) using a thermocycler from Perkin-Elmer/Cetus. Amplification products were visualized by ethidium bromide staining after agarose gel electrophoresis. The presence or absence of human-specific PCR product was correlated with the content of human chromosomes in the particular hybrid. ESTs were assigned to chromosomes based on no discordancy. Partial results have been previously published[17].

**Table 4.** Chromosomal localization by *in situ* hybridization (ISH).

| Clone | ISH | Clone | ISH |
|-------|-----|-------|-----|
| AA14  | 1p33-p35     | AS271 | 6p21.1-p12   |
| AA36  | 1q41-q42     | AA16  | 6q22-q23     |
| LG11  | 1q41-q42     | AA32  | 7q34-q36     |
| AS199 | 1q25-q31     | AA20  | 10q32-q24    |
| AS319 | 1q31         | AA12  | 11q23-q24    |
| AA36  | 1q41-q42     | AS322 | 11q23-q24    |
| AA1   | 3q11.2-q13.1 | AS321 | 14q11.1-q11.2|
| AA35  | 3p25-p24     | LG123 | 15q25-q26    |
| 6AR33 | 3p12         | LG110 | 19p13.3      |
| AA1   | 3q11.2-q13.1 | AA29  | 20p13        |
|       |              | AA38  | 21q11.2-q21  |

Subtracted cDNA clones were nick-translated with [$^3$H] deoxyribonucleotides to a specific activity of 2-4 X $10^7$ cpm/mg and hybridized to human chromosome spreads at a concentration of 25-35 ng/ml for 16-20 hours. After washing and emulsion coating, autoradiographs were exposed 10-12 days at 4°C. Chromosomes were G-banded using Wright's stain for silver grain analysis. Sites were assigned when labeling above background was detected. AA and LG clones were isolated from subtracted RPE library; AS clones and 6AR33 were isolated from subtracted retinal library. Partial results have been previously published[4,25].

## NRL, A GENE ENCODING A NEURAL RETINA LEUCINE ZIPPER PROTEIN

We have isolated a novel human cDNA[3], using subtraction cDNA cloning[1,2], which by Northern analysis is specifically expressed in the retina but not in any other tissue or cell line tested. Expression of this gene has been localized to all neuronal cell layers of the retina by *in situ* hybridization to RNA in baboon and mouse retina sections. Because the predicted polypeptide contains a basic motif and leucine zipper domain similar to DNA binding proteins of the *jun* and *fos* oncoprotein family, we have named this gene NRL (neural retina leucine zipper).

The NRL gene is conserved in vertebrates, and its expression has also been detected in bovine, mouse, and rat retina by Northern analysis[3]. Such evolutionary conservation suggests that NRL may play an important role in visual transduction and/or retinal development. Interestingly, an extended region of homology is observed between the DNA binding domain of the NRL polypeptide and that of v-maf[27], a transforming oncogene from the AS42 avian retrovirus.

The NRL gene was localized to chromosome 14 by Southern blot analysis of genomic DNA from a human-rodent somatic cell hybrid panel[4]. *In situ* hybridization to metaphase chromosomes (Figure 2) was performed to further localize the gene[4]. Of the 50 cells analyzed, 14 exhibited specific labeling at chromosome 14q11.1-q11.2. Silver grains over this region represented 16% of the total label; no other site was labeled above background.

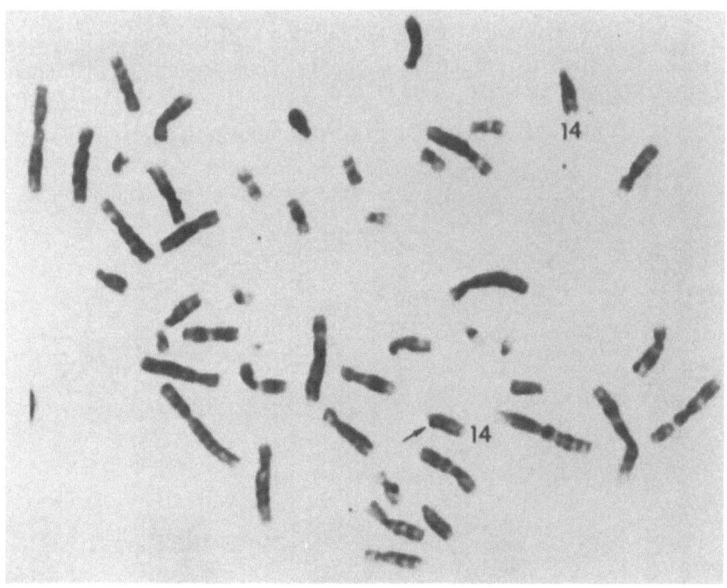

**Figure 2.** Metaphase chromosome spread. NRL cDNA was nick-translated with [³H]dCTP and [³H]TTP to a specific activity of $2.5 \times 10^7$ cpm/mg and hybridized to chromosome spreads at a concentration of 25 ng/ml.

The NRL gene consists of three exons, including a completely untranslated first exon[28]. The region of homology to v-maf is encoded by a single exon. To study expression and regulation of NRL in an animal model, multiple clones were obtained from mouse retinal cDNA and genomic DNA libraries[29]. DNA sequence analysis of the cDNA clones reveals that the murine and human genes are over 86% homologous, with even greater homology in amino acid sequence. The 5' and 3' untranslated regions display lesser, though still significant homology. Sequence of genomic mouse and human clones reveal that the gene structures of the human and murine genes are identical, with conservation of exon-intron junctions. Sequences of the human and mouse gene promoters, from +1 to approximately -250, show about 75% homology, perhaps indicating the presence of conserved regulatory elements. We are currently investigating the significance of 5' conserved elements in the tissue-specific regulation of this gene by gelshift and footprinting experiments using retinal and HeLa nuclear extracts. We have identified alternately polyadenylated forms of NRL in the human retina, accounting for the major 1.2 kb and less abundant 2.0 kb transcripts observed by Northern analysis. In the mouse, a polyadenylation signal of ATTAAA is present in close proximity to the poly-A tails of the cDNA corresponding to the major transcript of 1.9 kb. One curious difference between the mouse and human genes is that, in the mouse, the 3' untranslated region of the NRL gene contains a trinucleotide repeat consisting of $(AGG)_n$. Through PCR amplification of the region containing the repeat, it is evident that the region is polymorphic among different inbred mouse strains. Using other polymorphic markers and DNA from backcrosses of spretus and C57BL/6 mice, it appears that the gene is localized to mouse chromosome 14, in a region syntenic to human chromosome 14[28].

Given its chromosomal localization to 14q11.1-q11.2 and potential role in retinal development, the human NRL gene may serve as a candidate gene for ocular disorders which map to chromosome 14, including rod monochromacy[29]. To screen patients directly for mutations within the NRL gene, we are establishing a PCR-based approach (Figure 3). Primers have been designed to amplify overlapping DNA fragments that span the NRL coding region. Heteroduplex or single-strand conformational polymorphism (SSCP) analysis of the PCR products should enable detection of base composition differences.

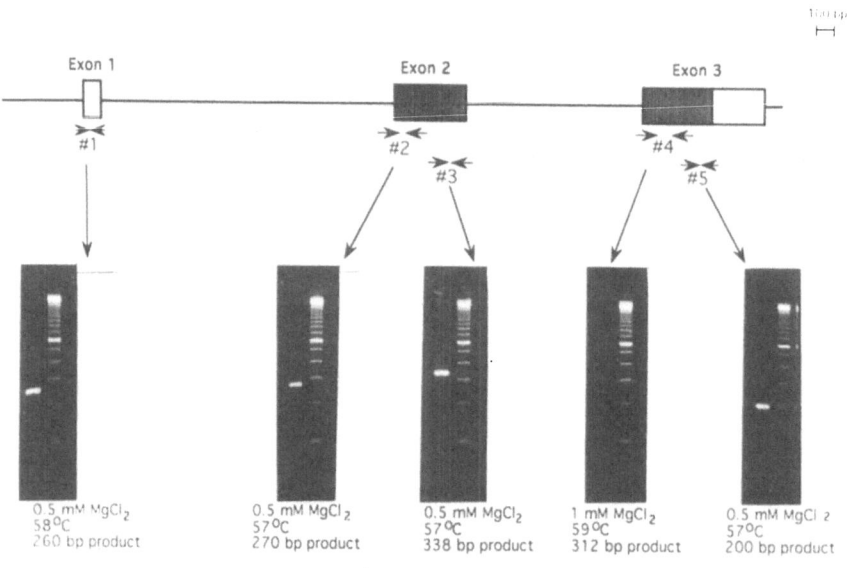

**Figure 3.** PCR based strategy for screening patients. PCR primers were designed to span the coding region of the NRL gene. PCR conditions are being optimized to generate a specific product, shown next to a 100 bp DNA marker. Products amplified from patient DNA will be analyzed for mutations in the coding region using heteroduplex or SSCP gels(MDE Hydrolink, J.T. Baker, Phillipsburg, NJ).

To enable genetic linkage analyses to be performed at the NRL locus, a polymorphic $(CA)_n$ repeat was isolated from a cosmid clone containing the NRL gene. Hybridization of Southern blots of digested cosmid DNA with a $^{32}P$-labeled $(CA)_{11}$ oligonucleotide detected a 400 bp RsaI fragment containing a $(CA)_n$ repeat. A library prepared from 400 bp RsaI fragments of cosmid DNA was screened with the microsatellite probe, and the isolated clones were sequenced using the dideoxy chain termination method. PCR primers were designed to amplify a 300 bp fragment containing the microsatellite. Recently, a new primer pair has been designed to amplify a 170 bp PCR product. When the primer pair for the larger fragment was used to amplify DNA from unrelated individuals, several alleles were detected (Figure 4). To determine the genetic distance between NRL and known chromosome 14 markers, genotyping of 155 individuals from 17 families was performed, The most likely position for NRL was calculated, using the LINKAGE 5.1 program, to lie between D14S54 and D14S50, with D14S54 the most centromeric locus[30].

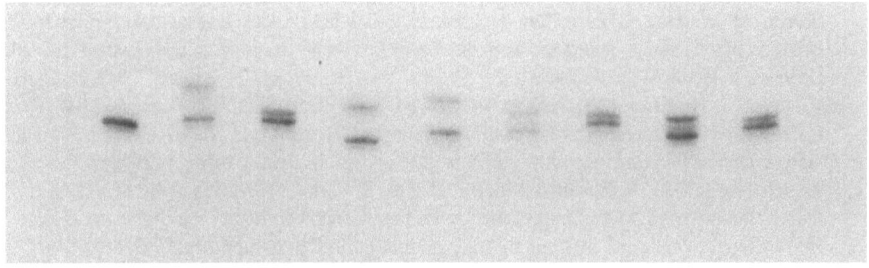

**Figure 4.** NRL $(CA)_n$ repeat is polymorphic. DNAs from 10 individuals were amplified by PCR using a primer pair designed to amplify a 300 bp fragment. One primer was end-labeled with $^{32}P$ using T4 polynucleotide kinase. Products were run on a 6% polyacrylamide/8M urea gel. The gel was dried and exposed to autoradiographic film overnight at -70°C.

## SUMMARY

A significant number of ocular disease loci have been localized to various chromosomes. The regionally-mapped cDNA clones from subtracted retinal and RPE libraries may provide a rich supply of candidate genes for inherited eye diseases. One potential candidate gene for diseases mapping to chromosome 14 is NRL which encodes a neural retina-specific, putative DNA binding protein. Identification of a polymorphic $(CA)_n$ repeat enables more precise mapping of and linkage analysis to this locus. This general scheme for interweaving positional and candidate gene approaches provides a way to more rapidly isolate bona fide candidate genes for ocular disorders.

## ACKNOWLEDGEMENTS

This work is supported in part by grants from the National Institutes of Health EY-07961 (A.S.), the George Gund Foundation, and the Retinitis Pigmentosa Foundation, Baltimore (to A.S. and T.L.Y.) and a National Science Foundation Graduate Fellowship (to A.U.J.)

## REFERENCES

1. A. Swaroop, J. Xu, N. Agarwal, and S. M. Weissman, A simple and efficient cDNA library subtraction procedure: isolation of human retina-specific cDNA clones, *Nucleic Acids Res.* 19:1954 (1991).
2. A. Swaroop, Construction of directional cDNA libraries from human retinal tissues/cells and their enrichment for specific genes using an efficient subtraction procedure, *in*:"Photoreceptor Cells", a volume of "Methods in Neurosciences", Paul Hargrave, editor, in press.
3. A. Swaroop, J. Xu, H. Pawar, A. Jackson, C. Skolnick, and N. Agarwal, A conserved retina-specific gene encodes a basic motif/leucine zipper domain, *Proc. Natl. Acad. Sci.* 89:266-270 (1992).

4. T. L. Yang-Feng and A. Swaroop, Neural retina-specific leucine zipper gene NRL (D14S46E) maps to human chromosome 14q11.1-q11.2, *Genomics* 14:491-492 (1992).

5. V. McKusick, Online "Mendelian Inheritance in Man", Johns Hopkins University (1992).

6. M. Musarella, Gene mapping of ocular diseases, *Surv. Ophthal.* 36:285-312 (1992).

7. F.S. Collins, Positional cloning: let's not call it reverse anymore, *Nature Genet.* 1:3-6 (1992).

8. C. C.T. Ton, H. Hirvonen, H. Miwa, M.M. Weil, P. Monaghan, T. Jordan, V. van Heyningen, N.D. Hastie, H. Meijers-Heijboer, M. Drechsler, B. Royer-Pokora, F. Collins, A. Swaroop, L.C. Strong, and G.F. Saunders, Positional cloning and characterization of a paired box and homeobox containing gene from the aniridia region, *Cell* 67:1059-1074 (1991).

9. F.P. Cremers, D. J. van de Pol, L.P. van Kerkhoff, B. Wieringa, and H.H. Ropers, Cloning of a gene that is rearranged in patients with choroideraemia, *Nature* 347:674-677 (1990).

10. W. Berger, A. Meindl, T.J.R. van de Pol, F. Cremers, H.H. Ropers, C. Doerner, A. Monaco, A.A.B. Bergen, M. Warburg, L. Zergollern, B. Lorenz, A. Gal, E.M. Bleeker-Wagemakers, and T. Meitinger, Isolation of a candidate gene for Norrie disease by positional cloning, *Nature Genet.* 1:199-203 (1992).

11. Z-Y. Chen, R.W. Hendriks, M.A. Jobling, J.F. Powell, X.O. Breadefield, K.B. Sims, and I.W. Craig, Isolation and characterization of a candidate gene for Norrie disease, *Nature Genet.* 1:204-208 (1992).

12. T. P. Dryja, Deficiencies in sight with candidate gene approach, *Nature* 347:614 (1990).

13. C.F. Inglehearn, J. Keen, R. Bashir, J. Lindsey, U. Atif, S. Carter, M. AlMaghtheh. M. Jay, A.C. Bird, and S. Bhattacharya, Molecular genetic studies in autosomal dominant retinitis pigmentosa, International Symposium on Retinal Degeneration, Porto Cervo, Sardinia (1992).

14. T.P. Dryja, L.B. Hahn, G.S. Cowley, T.L. McGee, and E.L. Berson, Mutation spectrum of the rhodopsin gene among patients with autosomal dominant retinitis pigmentosa, *Proc. Natl. Acad. Sci. USA* 88:9370-9374 (1991).

15. G.J. Farrar, P. Kenna, S.A. Jordan, R. Kumar-Singh, M.M. Humphries, E.M. Sharp, D.M. Sheils, and P. Humphries, A three-base-pair deletion in the peripherin-RDS gene in one form of retinitis pigmentosa, *Nature* 354:478-480 (1991).

16. M.D. Adams, J.M. Kelly, J.D. Gocayne, M. Dubnick, M.H. Polymeropoulos, H. Xiao, C.R. Merrill, A. Wu, B. Olde, R.F. Moreno, A.R. Kerlavage, W.R. McCombie, and J.C. Venter, Complementary DNA sequencing: expressed sequence tags and human genome project. *Science* 252:1651-1656 (1991).

17. L. Gieser and A. Swaroop, Expressed sequence tags and chromosomal localization of cDNA clones from a subtracted retinal pigment epithelium library, *Genomics* 13:873-876 (1992).

18. A. Swaroop and S. Weissman, Charon BS (+) and (-), versatile 1 phage vectors for constructing directional cDNA libraries and their efficient transfer to plasmids, *Nucleic Acids Res.* 16:8739 (1988).

19. F. Sanger, S. Nicklen, and A.R. Coulson, DNA sequencing with chain-terminating inhibitors, *Proc. Natl. Acad. Sci. USA* 74:5463-5467 (1977).

20. C.-L. Hsieh, A. Swaroop, and U. Francke, Chromosomal localization and cDNA sequence of human ralB, a GTP binding protein, *Som. Cell and Molec. Genet.*, 16:407-410 (1990).

21. N. Agarwal, C.-L. Hsieh, D. Sills, M. Swaroop, B. Desai, U. Francke, and A. Swaroop, Sequence analysis, expression, and chromosomal localization of a gene, isolated from a human subtracted retinal cDNA library, that encodes an insulin-like growth factor binding protein (IGFBP2), *Exp. Eye Res.* 52:549-561 (1991).

22. A. Swaroop, N. Agarwal, J.R. Gruen, D. Bick, and S. Weissman, Differential expression of novel Gsa signal transduction protein cDNA species, *Nucleic Acids Res.* 19:4725-4729 (1991).

23. A.U. Jackson, J. Xu, D. Sills, and A. Swaroop, manuscript in preparation.

24. B. Michel and S.D. Ehrlich, Illegitimate recombination at the replication origin of bacteriophage M13, *Proc. Acad. Natl. Sci. USA* 83:3386-3390 (1986).

25. K. Zheng, A. Swaroop, and T. L. Yang-Feng, manuscript in preparation.

26. A. Swaroop, D. Smith et al., manuscript in preparation.

27. M. Nishizawa, K. Kataoka, N. Goto, K.T Fujiwara, and S. Kawai, *V-maf*, a viral oncogene that encodes a "leucine zipper" motif, *Proc. Natl. Acad. Sci. USA* 86:7711-7715 (1989).

28. A. U. Jackson, unpublished results.

29. Q. A Farjo, A. U. Jackson, A. Swaroop et al,, manuscript in preparation.

30. M. Burmeister, A. Swaroop et al., manuscript in preparation.

31. L. Pentao, R.A. Lewis, D.H. Ledbetter, P.I. Patel, and J.R.Lupski, Maternal uniparental isodisomy of chromosome 14: association with autosomal recessive rod monochromacy, *Am. J. Hum. Genet.* 50:690-699 (1992).

32. S. Piecke-Dahl, A. Jackson, W.J. Kimberling, D. Blackwood, and A. Swaroop, Genetic mapping of NRL, a human retina-specific gene located on chromosome 14, *Am. J. Hum. Genet.* 51 (Suppl.):A185 (1992).

# NONRADIOACTIVE SINGLE STRAND CONFORMATION POLYMORPHISM (PCR-SSCP):
# A SIMPLIFIED METHOD APPLIED TO A MOLECULAR GENETIC SCREENING OF RETINITIS PIGMENTOSA

Mitsuru Nakazawa, Emi Kikawa, Yasushi Chida, Takashi Shiono, and
Makoto Tamai

Department of Ophthalmology
Tohoku University School of Medicine
1-1 Seiryo-machi, Aoba-ku
Sendai, 980, Japan

## INTRODUCTION

Retinitis pigmentosa (RP) regards to a group of pigmentary retinal degenerations that are clinically characterizwd by night blindness and a progressive loss of midperipheral visual field, with the eventual loss of central and far peripheral vision. With an incidence of about 1 in 4000 to 8000 people, RP is the third major cause of legal blindness in Japan's adult population.

Although most cases of RP are believed to be inherited, the pathogenesis of RP may be quite complicated. For example, the causes of RP have been thought to be heterogeneous because RP can be subdivided into X-linked recessive (XLRP), autosomal recessive (ARRP), autosomal dominant (ADRP), and putatively sporadic forms (SRP) according to its inheritance pattern. The results of pathological studies[1,2] have shown that degeneration occurs primarily in the photoreceptor and/or the retinal pigment epithelial cells. In most affected patients, RP occurs clinically without systemic disorders other than Usher or Refsum syndrome. Considering these pathological and clinical findings of RP, it has been thought that various kinds of genes coding for proteins that are specific for the function or structure of the photoreceptor or pigment epithelium have been generally altered.

Recent advances in molecualr genetics have provided some important clues to the understanding of RP at the molecular level. One of the most distinguished research breakthroughs in RP was the recent discovery of a mutation in codon 23 of the rhodopsin gene in families with ADRP.[3] This discovery prompted many investigators to search for mutations in the rhodopsin gene in patients with RP; thus far, over 40 mutations have been identified among patients with ADRP in North America, Europe, and Japan.[4, 5] These findings have revealed that mutations are almost randomly distributed in the rhodopsin gene and that different locations of mutations produce different clinical features of RP.[6-9] These data suggest that the clinical features of RP that are caused by mutations in the same gene, i.e, the rhodopsin gene, can show various patterns if the locations of the mutations are different. In addition to the rhodopsin gene, other genes such as the peripherin/RDS gene,[10, 11] chromosome 8 associated gene,[12] and ROM1 gene[13] have been shown to produce ADRP. Since these genes code for proteins specifically expressed in the retina, other genes have been proposed as additional candidates by considering their structure and metabolism specific to the retina.

To understand molecular genetic features of RP further, both clinical and basic researchers need to perform molecular genetic screening in a large number of patients with RP. Since the search for mutations in the rhodopsin gene has revealed this random distribution of mutations, other unknown kinds of mutations may be identified in future investigations. Accumulating such findings about the kind and location of mutations in the rhodopsin gene or in other candidate genes in affected patients will provide not only a new classification of RP at the molecular level but also some important clues to questions regarding the relationship between genotype and phenotype, the difference and similarity of mutations among different ethnic populations, and, finally, how to create an effective strategy to treat RP.

Methods of molecular genetic screening have been developed to detect not only known mutations[14, 15] but also randomely distributed unknown ones.[16-19] Methods of detecting unknown mutations commonly require the extensive use of radioisotopes or toxic agents, large gel apparatus, and/or temperature controlled systems, even if radioisotopes are not required. The analysis of single strand conformation polymorphism[16] of products of polymerase chain reaction (PCR-SSCP) is based on the fact that even a single base substitution in a stretch of DNA fragment can be detected as a mobility shift of single strand DNA by gel electrophoresis. Although this method has already been successfully used to detect mutations in the rhodopsin or peripherin/RDS gene in patients with RP,[11, 20] the use of radioisotopes makes this method cumbersome and limits its extensive use, particularly in clinical laborarories that should be strictly regulated for radiation safety.

For this reason, nonradioisotopic PCR-SSCP has recently been reported using mini-slab gel apparatus[21] or Pharmacia's Phast System.[22, 23] We report in this study the use of the nonradioisotopic PCR-SSCP as the screening method for the detection of mutation in the rhodopsin gene in patients with RP. In addition, we describe in a Japanese family with RP some clinical features due to the rhodopsin mutation confirmed by this method.


## MATERIALS AND METHODS

### Amplification of the Rhodopsin Gene by Polymerase Chain Reaction (PCR)

With the use of a thermocycler (Perkin Elmer Cetus, Norwalk, CT, U.S.A.), the PCR was carried out in 50 µl of reaction mixture containing 250 ng of genomic DNA, which was prepared from each patient's leukocytes as template; 20 µM of each primer; 200 µM each of dATP, dCTP, dGTP, and TTP; 1.25 units of Taq polymerase; and 0.5 unit of Perfect Match Enhancer (Stratagene, La Jolla, CA, U.S.A.). Buffer contained 50 mM KCl, 10 mM TrisCl (pH 8.3), and 1.5 mM $MgCl_2$. Reaction cycle was 30. The usual temperature settings for PCR were 94°C for 1 min of denaturation, 60°C for 2 min of anealing, and 72°C for 2 min of polymeration. Each pair for the amplification of the rhodopsin gene was determined in order that the size of each amplified fragment should not exceed 300 bp.

### Nonradioactive SSCP

A total of 2 µl of the PCR product was mixed with 2.5 µl of 95% deionized formamide containing 0.25% each of xylene cyanol and bromophenol blue. The mixture was heated at 95°C for 5 min, then cooled on ice before 4 µl of the mixture was loaded onto a 1.0 mm X 5 cm X 10 cm 10% polyacrylamide gel, or 0.5 X Hydrolink MDE gel (AT Biochem, Marvern, PA, U.S.A.), with or without 10% glycerol. Both the gel and the running buffer (1 X TBE) were precooled at 4°C. Electrophoresis was done at 4°C and at 100 V for 2 to 4 hours, depending on the size of the DNA. The gel was then silver stained with the use of Bio-rad silver stain kit. Glycerol was added and SSCP was redone when DNA band did not appear clearly enough to be identified but looked rather like a smear on gels without glycerol.

## RESULTS

### Putative Polymorphism in Exon 1

Figure 1 shows the SSCP pattern of the part of exon 1 prepared from unrelated patients with RP. SSCP was performed by using MDE gel without glycerol. Two different homozygous patterns were identified: one is shown in RP 222 and the other is shown in RPs 219, 223, 224, and 225. Also, one heterozygous pattern appeared in RPs 218, 220, and 221. The evidence that these different patterns of SSCP indicate polymorphism is shown in Fig. 2, which demonstrates the SSCP patterns in the family with ADRP. In this family, the heterozygous pattern occurs in both affected and non-affected family members. This pattern is not cosegragated with the disease.

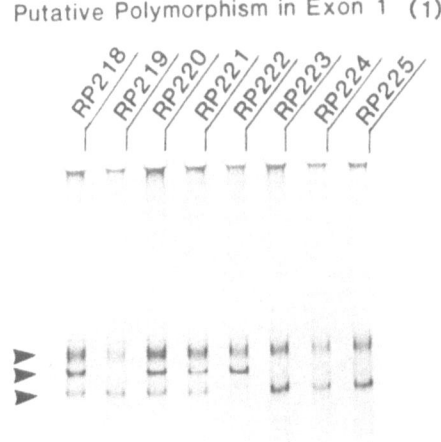

**Figure 1.** SSCP pattern of 257 bp fragment in Exon 1 in patients with ADRP. Arrow heads indicate the location of DNA bands. Note two different homozygous patterns and one heterozygouspattern (see the text).

### Mutation in Exon 5

In Fig. 3 are SSCP patterns of exon 5 on a polyacrylamide gel. Four patients (cases 1, 2, 3, and 5) were affected members in the same family with ADRP (Fig. 4). Genes of cases 1, 2, and 3 have been proved to have heterozygous C to T transition at the second nucleotide in codon 347 in the rhodopsin gene, resulting in amino acid change from proline to leucine in this codon.[24] In the present study, because DNA from the patient case 5 showed the same

## Putative Polymorphism in Exon 1 (2)

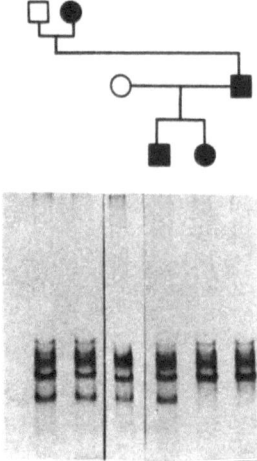

Figure 2. SSCP pattern of the same fragment as Fig. 1 in the family with ADRP (AD-27). The SSCP pattern is not cosegragated with the disease, suggesting that this pattern indicate polymorphism.

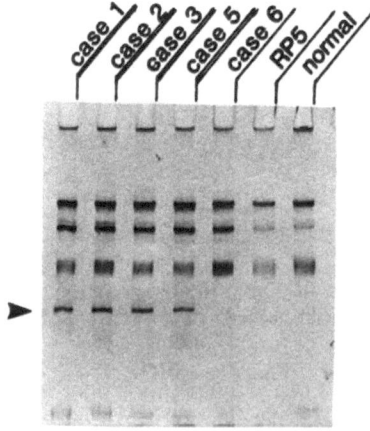

Figure 3. SSCP pattern of 251bp fragment in Exon 5 in the familt with ADRP (AD-3). Arrow head indicates mutant band seen in affected family members (cases 1, 2, 3, and 5). No variant band is seen in a non-affected family member (case 6).

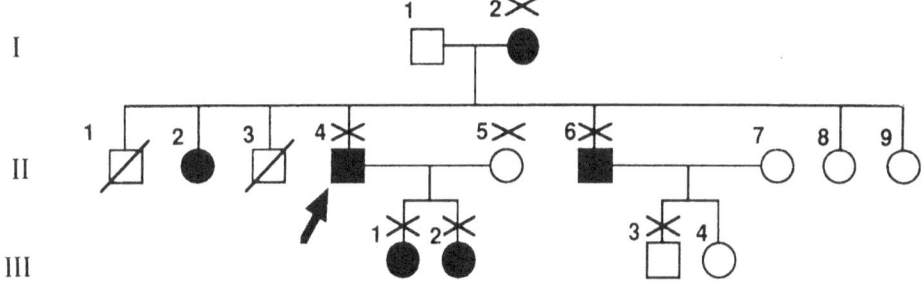

Figure 4. Family tree of AD-3. Family members of II-4 (proband), III-1, III-2, I-2, II-6, and III-3 were examined and they are referred to as cases 1, 2, 3, 4, 5, and 6, respectively.

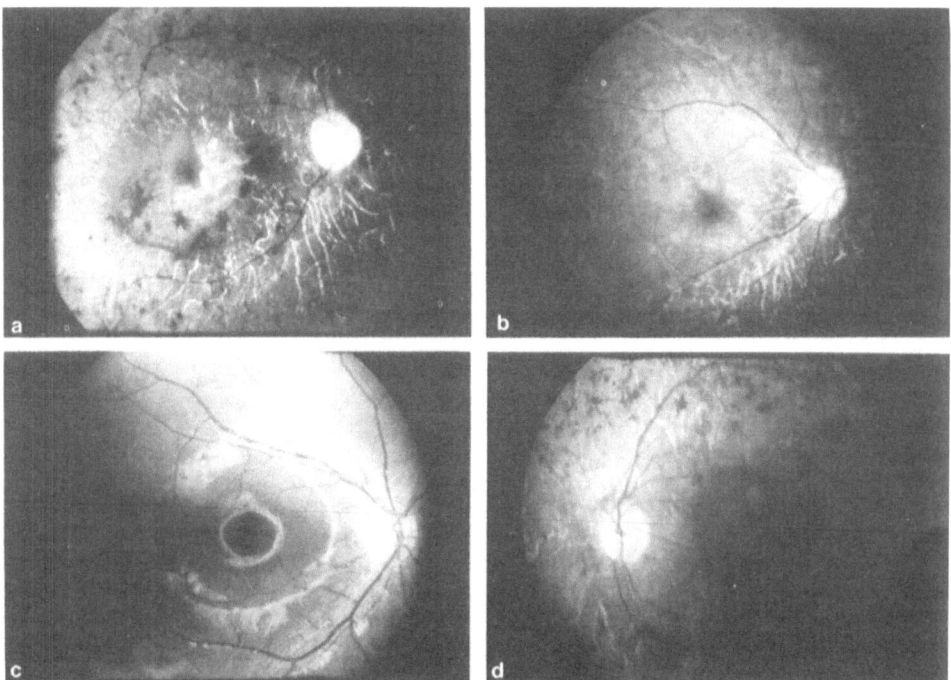

Figure 5. Fundi appearance of cases 1 (a), upper left, 2 (b), upper right, 3 (c), lower left, and 5 (d), lower right. Degeneration is seen in the fundi of cases 1, 2, and 5, and the fundus of case 3 shows attenuated retinal arteries without obvious degeneration. Marked similarity is observed between those of cases 1 and 5, suggesting intrafamilial concordance.

mobility as seen in cases 1, 2, and 3, case 5 was then confirmed to have the same mutation by SSCP. Also, case 6, who is a non-affected family member, showed no additional band, indicating that this mutation is cosegregated with the disease. These results have suggested that a single base substitution can be detected by the nonradioactive SSCP described in the present study.

Some clinical features of cases 1, 2, and 3 have been already reported in detail.[24] However, since then information has been obtained on case 5. In this study, we confirmed that case 5 has the same mutation as the other affected members. These additional clinical findings help to discern whether or not affected members with the same mutation show a similar clinical course. The fundi appearance of these four patients are shown in Fig. 5 a-d, and the clinical data of the affected members are summerized in the Table. Marked intrafamilial concordance is observed, particularly in the age of onset of night blindness. All had relatively good visual acuity until their 50s, onset of cataract in their 40s, pigmentary retinal degeneration in all four quadrants, and diminished ERG response, even in the early stage of RP.

**Table.** Summary of clinical features of affected members in AD-3 (Pro-347-Leu).

| Case | 1 (II-4) | 2 (III-1) | 3 (III-2) | 4 (I-2) | 5 (II-6) |
|---|---|---|---|---|---|
| Age | 49 | 20 | 12 | 75 | 48 |
| Sex | M | F | F | F | M |
| Visual acuity | 1.0 OD[1] 1.0 OS[1] | 1.0 OD 1.0 OS | 1.0 OD 1.0 OS | LS OD LS OS | 0.5 OD 0.2 OS |
| Cataract | + (IOL[2]) | – | – | +++ | + |
| Retinal degeneration | diffuse | diffuse | none | n.d.[3] | diffuse |
| ERG | extinguished | extinguished | decreased | extinguished | extinguished |
| EOG (L/D) | 1.0 | 1.0 | 1.0 | n.d.[4] | 1.0 |
| Night blindness[5] | 15 | 16 | – | 10s | 16 |

[1]Corrected visual acuity after intraocular lens implantation.
[2]Postoperative status after cataract extraction combined with intraocular lens implantation.
[3]Fundus can not be seen by the dense cataract.
[4]EOG was not performed.
[5]The age of onset of night blindness.

## DISCUSSION

To know the location and the kind of mutation causing a certain genetic disorder, the nucleotide sequence has to be determined. However, to perform sequencing on all affected patients is still quite laborious, even if an automatic sequencer is used. Therefore, recent advances in molecular biology have produced various kinds of diagnostic or screening methods for the detection of unknown mutations. These methods, in turn, have been helpful in reducing the laborious work of sequencing. One method, SSCP has been widely used to screen DNA samples and to detect putative mutations. In this study, we have described a nonradioisotopic version of SSCP using mini-slab gel apparatus to simplify the method and hydrolink MDE gel to increase resolution. As in the original SSCP method, in which sequencing gel apparatus is used, performing electrophoresis at a constant temperature is qute important to stabilize the conformation of single stranded DNA. This treatment is particularly true with mini-slab gel apparatus because of its relative thickness. The sensitivity of SSCP is not likely to be reduced by using mini-slab gel instead of sequencing gel. The reason for this speculation is the sensitivity of the heteroduplex method, in which mutation can be detected by the mobility shift formed by normal and mutant strands. The sensitivity of the heteroduplex method increased when 1-mm thick gel is used and lowers when 0.2- to 0.4-mm thick sequencing gel is used. Although the sensitivity of the SSCP or the heteroduplex method is about 80%, overall sensitivity can be raised to almost 100% by combining them. Nonradioactive SSCP may provide a new rapid, sensitive, and safe method of screening a large number of DNA specimens from patients with RP.

Moreover, the accumulation of clinical features of patients with RP due to a known mutation will provide important information regarding genotype-phenotype relationship. These data will help the physician in differentiating the clinical diagnosis of RP and counseling the patient and his or her family.

## REFFERENCES

1. J. G. Flannery, D. B. Farber, A. C. Bird, and D. Bok, Degenerative changes in a retina affected with autosomal dominant retinitis pigmentosa, Invest. Ophthalmol. Vis. Sci. 30: 191 (1989).
2. R. B. Szamier, E. L. Berson, R. Klein, and S. Meyers, Sex-linked retinitis pigmentosa: ultrastructure of photoreceptors and pigment epithelium, Invest. Ophthalmol. Vis. Sci. 18: 145 (1979).
3. T. P. Dryja, T. L. McGee, E. Reichel, L. B. Hahn, G. S. Cowley, D.W. Yandel, M. A. Sandberg, and E. L. Berson, A point mutation of the rhodopsin gene in one form of retinitis pigmentosa, Nature 343: 364 (1990).
4. T. P. Dryja, Doyne lecture. Rhodopsin and autosomal dominant retinitis pigmentosa, Eye 6: 1 (1992).
5. R. R. McInnes and R. A. Bascom, Retinal genetics: a nullifying effect for rhodopsin, Nature Genet. 1, 155 (1992).
6. E. L. Berson, B. Rosner, M. A. Sandberg, and T. P. Dryja, Ocular findings in patients with autosomal dominant retinitis pigmentosa and a rhodopsin gene defect (Pro23His), Arch. Ophthalmol. 109: 92 (1991).
7. E. L. Berson, B. Rosner, M. A. Sandberg, C. Weigel-DiFranco, and T. P. Dryja, Ocular findings in patients with autosomal dominant retinitis pigmentosa and rhodopsin, proline-347-leucin, Am. J. Ophthalmol. 111: 614 (1991).
8. G. A. Fishman, E. M. Stone, L. D. Gilbert, P. Kenna, and V. C. Sheffield, Ocular findings associated with a rhodopsin gene codon 58 transversion mutation in autosomal dominant retinitis pigmentosa, Arch. Ophthalmol. 109: 1387 (1991).
9. J. R. Heckenlively, J. A. Rodriguez, and S. P. Daiger, Autosomal dominant sectorial retinitis pimentosa. Two families with transversion mutation in codon 23 of rhodopsin, Arch. Ophthalmol. 109: 84 (1991).
10. G. J. Farrar, P. Kenna, S. A. Jordan, R. Kumar-Snigh, M. M. Hunphries, E. M. Sharp, D. M. Sheils, and P. Humphries, A three-base-pair deletion in the peripherin/RDS gene in one form of retinitis pigmentosa, Nature 354: 478 (1991).
11. K. Kajiwara, L. B. Hahn, S. Mukai, G. H. Travis, E. L. Berson, and T. P. Dryja, Mutations in the human retinal degeneration slow gene in autosomal dominant retinitis pigmentosa, Nature 354: 480 (1991).
12. S. H. Blanton, J. R. Heckenlively, A. W. Cottingham, J. Friedman, L. A. Sadler, M. Wagner, L. H. Friedman, and S. P. Daiger, Linkage mapping of autosomal dominant retinitis pigmentosa (RP1) to the pericentric region of human chromosoma 8, Genomics 11: 857 (1991).
13. R. A. Bascom, V. I. Kalmins, R. S. Molday, and R. R. McInnes, Molecular cloning of the cDNA for a novel photoreceptor-specific membrane protein (ROM-1) identified a disk rim protein family implicated in human degenerative retinopathies, Invest. Ophthalmol. Vis. Sci. 33 (suppl.): 946 (1992).

14. T. P. Dryja, T. L. McGee, L. B. Hahn, G. S. Cowley, J. E. Olsson, E. Reichel, M. A. Sandberg, and E. L. Berson, Mutations within the rhodopsin gene in patients with autosomal dominant retinitis pigmentosa, N. Eng. J. Med. 323: 1302 (1990).

15. M. Nakazawa, E. Kikawa-Araki, T. Shiono, and M. Tamai, Analysis of rhodopsin gene in patients with retinitis pigmentosa using allele-specific polymerase chain reaction, Jpn. J. Ophthalmol. 35: 386 (1991).

16. M. Orita, Y. Suzuki, T. Sekiya, and K. Hayashi, Rapid and sensitive detection of point mutations and DNA polymorphisms using the polymerase chain reaction, Genomics 5: 874 (1989).

17. V. C. Sheffield, D. R. Cox, L. S. Lerman, and R. M. Myers, Attachment of a 40-base-pair G+C rich sequence (GC-clamp) to genomic DNA fragments by the polymerase chain rection results in improved detection of single-base changes, Proc. Natl. Acad. Sci. USA 86: 232 (1989).

18. J. Keen, D. Lester, C. Inglehearn, A. Curtis, and S. Bhattacharya, Rapid detection of single base mismatches as heteroduplexes on hydrolink gels, Trend in Genet. 7: 5 (1991).

19. S. M. Forrest, H. H. Dahl, D. W. Howells, I. Dianzani, and R. G. H. Cotton, Mutation detection in phenylketouria by using chemical cleavage of mismatch: Importance of using probes from both normal and patient samples, Am. J. Hum. Genet. 49: 175 (1991).

20. P. J. Rosenfeld, G. S. Cowley, T. L. McGee, M. A. Sandberg, E. L. Berson, and T. P. Dryja, A null mutation in the rhodopsin gene causes rod photoreceptor dysfunction and autosomal recessive retinitis pigmentosa, Nature Genet. 1: 209 (1992).

21. P. J. Ainsworth, L. C. Surh, and M. B. Coulter-Mackie, Diagnostic single strand conformational polymorphism, (SSCP): a simplified non-radioisotopic method as applied to a Tay-Sachs B1 variant, Nucl. Acid Res. 19: 405 (1991).

22. B. Dockhorn-Dworniczak, B. Dworniczak, L. Brommelkamp, J. Buelles, J. Horst, and W. W. Boecker, Non-isotopic detection of single strand conformation polymorphism (PCR-SSCP): a rapid and sensitive technique in diagnosis of phenylketouria, Nucl. Acid Res. 19: 2500 (1991).

23. A. J. Mohabeer, A. L. Hiti, and W. J. Martin, Non-radioactive single strand conformation polymoephism (SSCP) using the Pharnacia "PhastSystem", Nucl. Acid Res. 19: 3154 (1991).

24. T. Shiono, Y. Hotta, M. Noro, T. Sakuma, M. Tamai, M Hayakawa, T. Hashimoto, K. Fujiki, A. Kanai, and A. Nakajima, Clinical features of Japanese family with autosomal dominant retinitis pigmentosa caused by point mutation in codon 347 of rhodopsin gene, Jpn. J. Ophthalmol. 36: 69 (1992).

# RETINOPATHIA PIGMENTOSA PLUS - THE VALUE OF ULTRA-STRUCTURAL EXAMINATION OF THE HUMAN RETINA

Hans H. Goebel

Division of Neuropathology
University of Mainz
Mainz/Germany

## INTRODUCTION

Retinopathia pigmentosa is more widely, but somewhat incorrectly known as Retinitis pigmentosa (RP). Its course as a primary exclusively retinal disease follows autosomal-dominant, autosomal-recessive, or X-linked recessive modes of inheritance, or it may be sporadic. However, a progressive retinopathy , also called tapeto-retinal degeneration, may also be associated with numerous disorders: retinopathia pigmentosa plus (RPP). Among these RPP are those which form part of certain syndromes, e.g. Laurence-Moon-Bardet-Biedl syndrome, the Hallgren syndrome, the Marinesco-Sjögren syndrome, to name a few. Other RPP are associated with disorders of different organs, the skin, e.g. Werner disease, the brain, e.g. Hallervorden-Spatz disease, the kidney, e g. nephronophthisis, the skeletal muscle, e.g. myotonic dystrophy; bone, e.g. osteopetrosis familiaris, and peripheral nerves, e.g. hereditary motor-sensory neuropathies III and VII. In some such forms of RPP, the generalized associated or underlying disorder may be marked by abnormalities of certain cytological organelles which are present, and therefore affected, in numerous tissues and organs, e.g. lysosomal diseases, mitochondrial diseases, and peroxysomal diseases. In other RPP, although comprising abnormalities in several organs and tissues, a common underlying structural abnormality is not apparent, e.g. adult Refsum disease (ARD). Studying the morphology of eyes in the two different categories of retinal degenerations, the RP and the RPP, entails certain differences, apart from morphological ones. In humans, morphology of the retina in RP and RPP can largely be studied only post mortem as human eyes enucleated from a living patient with RP or RPP only extremely seldom become available for study, usually when additional ocular or orbital lesions are present which require removal of the eye, such as trauma, tumour, or severe infection. As RP is a non-fatal disease RP patients, dying of other causes, may provide eyes at different stages of their RP, thus, occasionally enabling post mortem morphological, including electron microscopic examinations earlier in the course of RP. However, the underlying fatal, non RP-linked disease often prevents post mortem removal and study of the eye mostly based on lack of pertinent information of the RP at the time of autopsy. Conversely, RPP as part of the underlying fatal disorder, often presents itself for morphologic post mortem examination at its end-stage, but possibly specimens are more frequently available as the pathologist may be more aware of the RPP as part of the patient's disease at the time of autopsy, thus, securing eyes in greater numbers. This is actually reflected in a comparatively large number of recent published papers on the morphology of RPP compared to the scant morphological literature on RP, especially as it concerns the ultrastructure of the diseased retina.


MATERIAL AND METHODS


Eyes from patients of the following disorders have been studied by the author in neuronal ceroid-lipofuscinoses (NCL), infantile, late-infantile, and juvenile types, the Kearns-Sayre syndrome (KSS), dystrophic myotonia, Marfan syndrome, and ARD. Additional studies had comprised canine and ovine eyes in autosomal-recessive NCL.

Eyes retrieved at autopsy had been fixed in buffered glutaraldehyde by immersion, osmicated and subsequently embedded in epon using small strips of retina and underlying pigment epithelium from central, equatorial, and peripheral regions of the ocular fundi.

Non-specific retinal pathology like in RP (Fig. 1), Marfan syndrome (Fig. 2) and Refsum disease (Fig. 3) is marked by loss of photo-receptors, proliferation and displacement of pigmented cells into the atrophic retina and severe gliosis between the inner and outer limiting membranes.

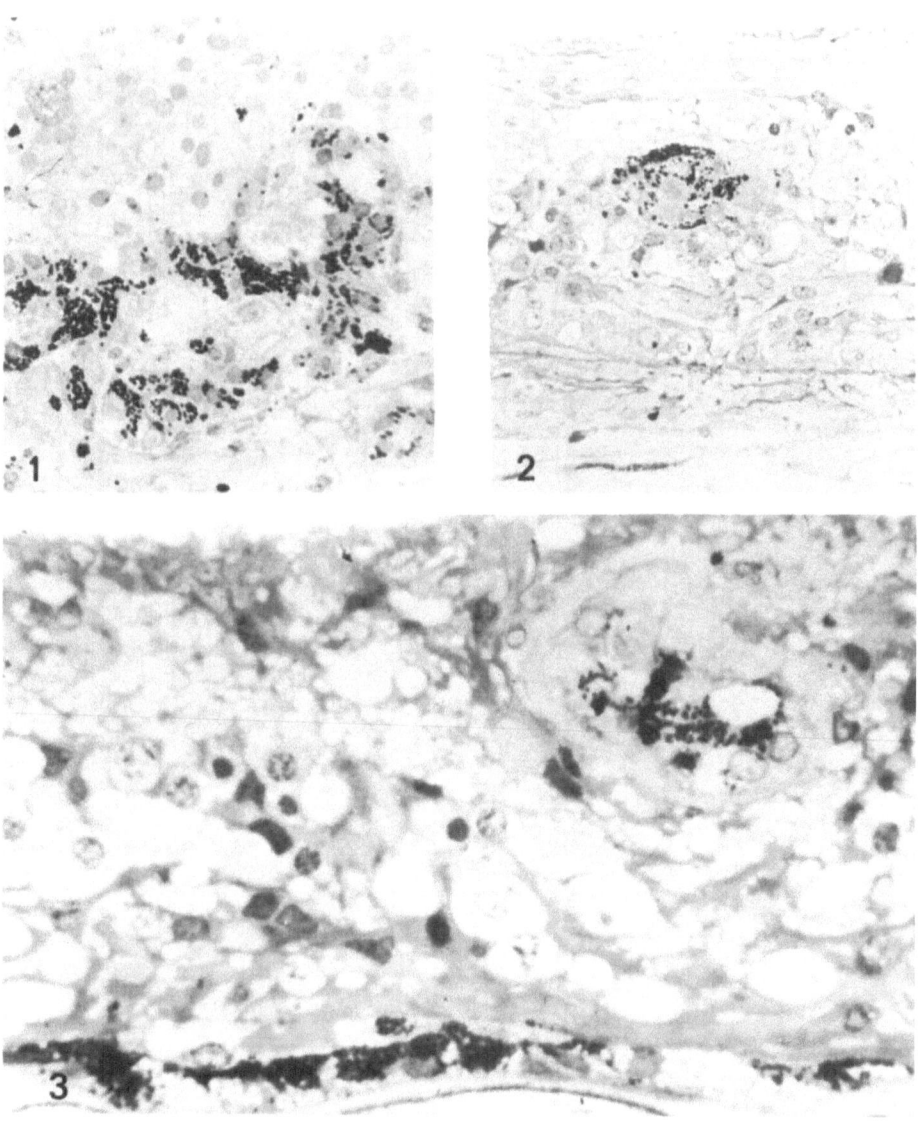

Fig. 1: Severe atrophy of the retina in primary retinopathia pigmentosa, peripheral region, 1 μm toluidine blue-stained semi-thin section, X 420

Fig. 2: Atrophic retina in Marfan's syndrome, equatorial region, 1μm toluidine blue-stained section, X 420

Fig. 3: Atrophic retina in ARD, equatorial region, 1μm-thick toluidine blue-stained section, X 678

In ARD, pigmented cells comprise three forms, those exclusively consisting of melanin (Fig. 4), of pure lipopigments (Fig. 5), of separate pigments and melanin inclusions (Fig. 6), and of both melanin and lipopigment components in the same inclusion body (Fig. 7).

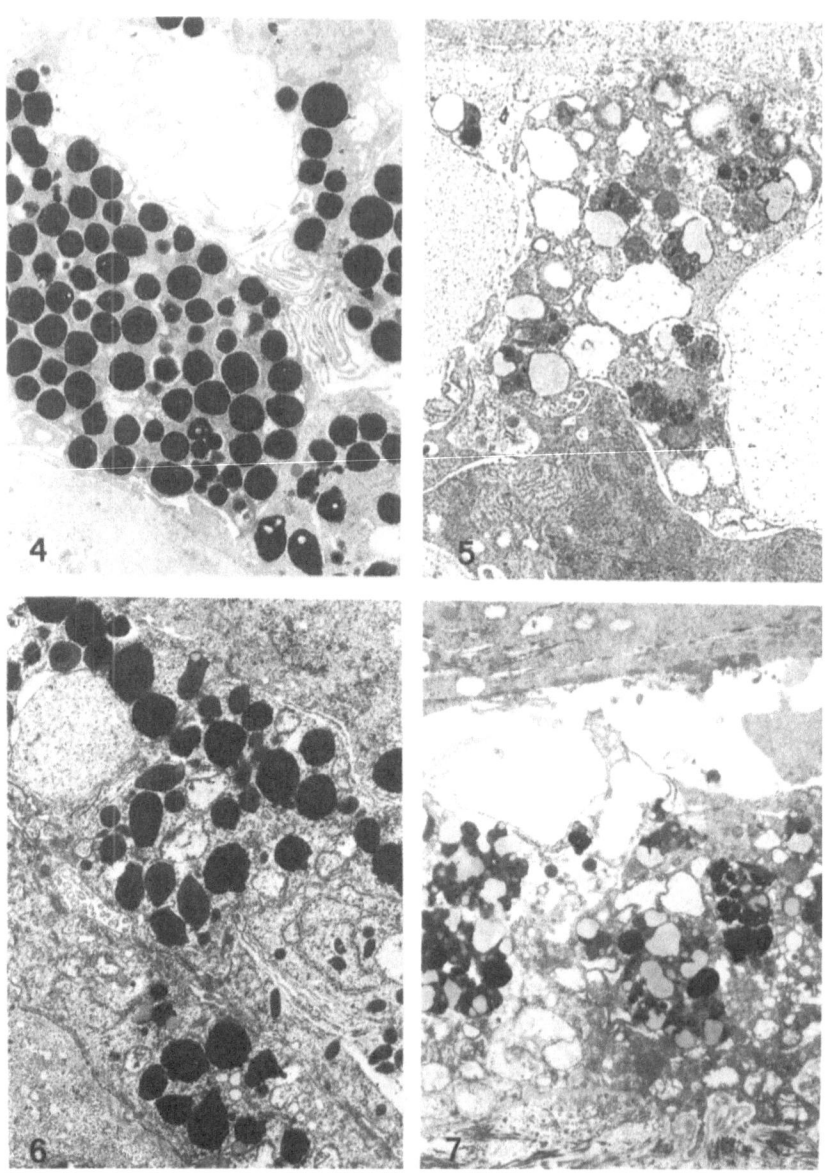

Fig. 4: RPE contain only pure melanin inclusions, ARD, X 6,475
Fig. 5: RPE contain exclusively lipopigment bodies with several lipid (L) droplets, ARD, X 14,245
Fig. 6: RPE contain both melanin inclusions and lipopigments (arrows), ARD, X 9,065
Fig. 7: RPE contain inclusions composed of both melanin granules and lipopigments with numerous lipid (L) droplets, ARD, X 7,238

Whether these four groups of pigmented cells represent different stages of evolution beginning with pure melanized cells or formation of lipopigments, followed by addition of the other component, and finally pigmented cells replete with fused melanin-lipopigment bodies, melanolysosomes, cannot be concluded from mere post mortem morphologic studies. As the external limiting membrane (ELM) was found re-established by glial cells after partial or complete drop-out of photoreceptors migration of pigmented cells from the largely obliterated subretinal space into the atrophic neural retina and its perivascular spaces seems to be a rather rapid process. This observation gives rise to the question whether discontinuities of the ELM may actually attract mobile pigmented cells from the subretinal space. Numerous lipid droplets of varying size in melanolysosomes of pigmented ARD cells (Fig. 7) may be a special feature of ARD and probably represent the equivalent of the droplets formerly seen by light microscopy.[1] In retinal pigment epithelial cells (RPE) of KSS, large mitochondria were encountered which may reflect the mitochondrial abnormalities that constitute this mitochondrial disorder affecting skeletal muscle, brain, heart, and retina However, crystalline inclusions as seen in muscle mitochondria of such patients were not encountered.[2,3] The frequent loss of RPE from their regular topographic site and mitochondrial abnormalities in RPP cells suggest a primary effect on the RPE rather than on the photoreceptors

## DISEASE-SPECIFIC ULTRASTRUCTURAL PATHOLOGY

As individual lysosomal diseases or groups of them are marked by ultrastructurally specific features, according to the accrued intra-lysosomal substrate, disease-specific lysosomes form a conspicuous ultrastructural hallmark These are seen in lysosomal disorders affecting the retina without associated retinopathy such as the sphingo-lipidoses,[4] metachromatic leukodystrophy,[5] infantile type II glyco-genosis[6] and lysosomal diseases associated with RPP Such lysosomal diseases with RPP are mucopolysaccharidoses types II,[7] III,[8] mucolipidosis IV,[9] and, especially, the childhood forms of NCL, where granular lipopigments mark the infantile type (Fig 8), curvilinear profiles the late infantile form (Fig 9), curvilinear-fingerprint inclusions the juvenile form (Fig 10). In adult NCL, the retina may actually be exempted from atrophy while NCL-specific cytosomes are present in retinal ganglion cells.[10] Clinically, retinal atrophy has been reported by Ikeda et al.[11] their patient being affected with adult NCL, deviating from the general experience that adult NCL does not affect the retina.[12] In this latter patient,[11] coincidence of adult NCL and sporadic RP can not be excluded. In acquired chloroquine intoxication, formation of abnormal lysosomes may be associated with degeneration of the retina.[13]

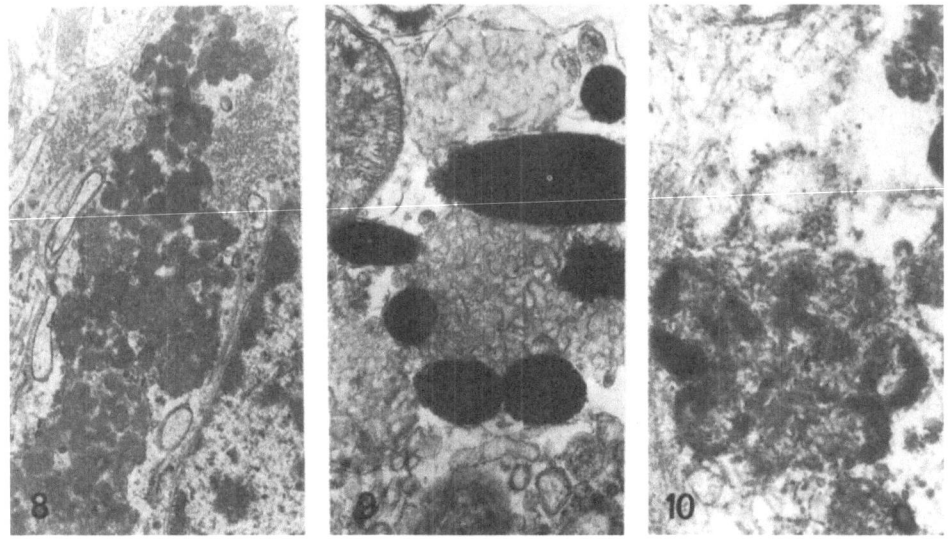

Fig.   8: Granular lipopigments in infantile NCL, X 21,840
Fig.   9: Curvilinear lipopigments in late-infantile NCL, X 26,400
Fig. 10: Curvilinear-fingerprint lipopigments in juvenile NCL, X 43,250

In canine and ovine NCL, disease-specific lipopigments abound (Fig. 11) while retinal atrophy is absent in canine NCL, but present in ovine NCL, though less than in human NCL. However, peculiar lamellar inclusions are a feature of RPE cell in canine (Fig. 12) and ovine (Fig. 13) NCL.

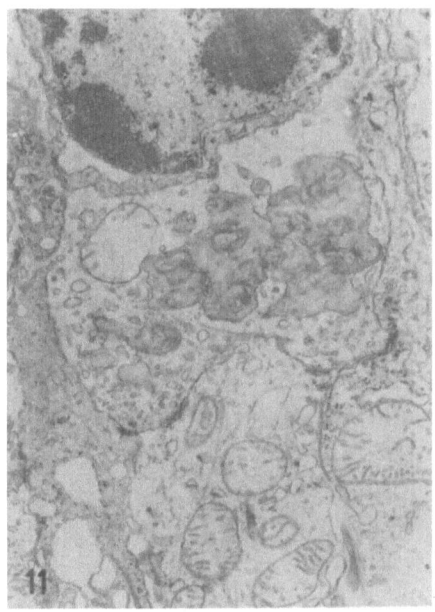

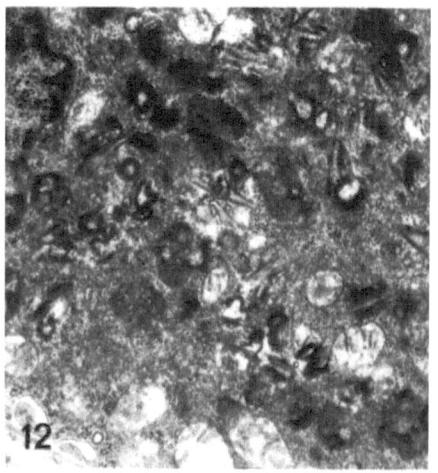

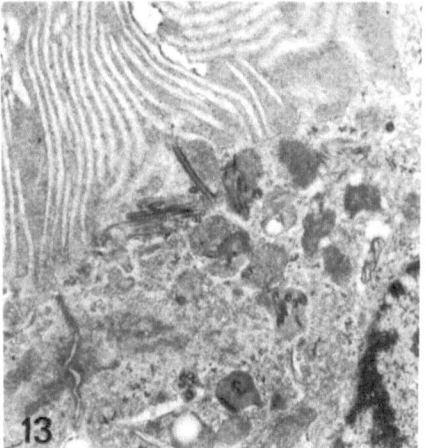

**Fig. 11:** Numerous lipopigments in photoreceptors of canine NCL, X 12,576
**Fig. 12:** Numerous lamellar inclusions in RPE, canine NCL, X 16,000
**Fig. 13:** Lamellar inclusions in RPE, ovine NCL, X 18,857

In certain peroxisomal disorders such as the cerebrohepatorenal or Zellweger syndrome, RPP cells contain bilamellar or bileaflet profiles,[14] and needle-like inclusions are also seen in retinal cells in neonatal adrenoleukodystrophy,[15] similar to those inclusions encountered in macrophages and Schwann cells.

## COMMENT

The value of studying the morphology of RPP is based on the possible elucidation of morphogenetic and pathogenetic factors derived from the underlying associated disorder, especially when they are of lysosomal, peroxisomal, or mitochondrial nature, and the comparative elucidation of etiological and pathogenetic similarities and dissimilarities between RP and RPP.

Disease-specific lipopigments which contain the subunit c of mito-chondrial ATP synthase[16] are present in NCL-affected patients, NCL-affected English setters, and NCL-affected sheep, but retinal degeneration is most marked in human NCL, less extensively present in ovine NCL, and absent in canine NCL.[17-21] Severe RPP is present both in infantile and juvenile human NCL the genes of which are located on different chromosomes, namely chromosome 1 in infantile NCL[22] and chromosome 16 in juvenile NCL.[23] Indicating that these two forms of human childhood NCL are different genetic diseases these observations and location of genes on different chromosomes in other RPP raise the question of common or uncommon denominators that link individual RPPs and their pathology to the respective individual generalized associated disorders. The ubiquitous involvement of lysosomes, peroxisomes and mitochondria in respective groups of disorders suggests a primary etiological association respective of the RPP with these diseased organelles, but the above-cited example of RPP in genetically different infantile and juvenile NCL emphasizes that other pathogenetic factors of RPP within the context of the underlying generalized disorder have to be considered. However, such factors are largely unknown and this is the more true of those RPP which are not marked by ultrastructural specificities such as ARD or Marfan syndrome.

## ABSTRACT

Retinopathia pigmentosa plus (RPP) denotes progressive retinal de-generation in diverse disorders, e.g. lysosomal, mitochondrial, peroxisomal, and other diseases. Concerning ultrastructural pathology of the retina in these RPP, two groups can be separated, one not showing any disease-specific pathology but one similar or identical to the ultrastructural pathology in primary retinopathia pigmentosa and another in which non-specific ultrastructural and light microscopic pathology is additionally marked by disease-specific features, usually only recognized at the electron microscopic level. Considering that many of these RPP are associated with disorders the genes of which are located on different chromosomes the pathogenetic factors responsible for RPP and the respective underlying disorders are still largely unknown. An investigative increase in post mortem studies of eyes from RPP patients may be useful for the elucidation of both RPP and RP.

**ACKNOWLEDGEMENTS**

Financially supported by the Deutsche Forschungsgemeinschaft, Schwerpunkt 'Erbliche Netzhautdegenerationen', Go 10/I. Technical work was performed by Mrs. M. Janocha and the photographs were prepared by Mr. W. Meffert. The manuscript was edited and typed by Mrs. A. Wöber.

## REFERENCES

1. D. Toussaint, and P. Danis, An ocular pathologic study of Refsum's syndrome, Am J Ophthalmol 72:342 (1971).

2. N.M. McKechnie, M. King, and W.R. Lee, Retinal pathology in the Kearns-Sayre syndrome, Br J Ophthalmol 69:63 (1985).

3. JR. Eagle, R.C. Hedges, and M. Yanoff, The atypical pigmentary retinopathy of Kearns-Sayre syndrome. A light and electron microscopic study, J Ophthalmol 89:1433 (1982).

4. D.G. Cogan, and T Kuwabara, The sphingolipidoses and the eye, Arch Ophthalmol 79:437 (1968).

5. H.H. Goebel, K. Shimokawa, A. Argyrakis, and H. Pilz, The ultrastructure of the retina in adult metachromatic leukodystrophy, Am J Ophthalmol 85:841 (1978).

6. H.H. Goebel, A Kohlschütter, and H. Pilz, Ultrastructural observations on the retina in type II glycogenosis (Pompe's disease) Ophthalmologica 176:61 (1978).

7. T.M. Topping, K.R. Kenyon, M.F Goldberg, and A.E. Maumenee, Ultrastructural ocular pathology of Hunter's syndrome. Systemic mucopolysaccharidosis type II, Arch Ophthalmol 86:164 (1971).

8. M.A. Lavery, W.R. Green, E.W. Jabs, M.W Luckenbach, and J.L. Cox, Ocular histopathology and ultrastructure of Sanfilippo's syndrome, type III-B, Arch Ophthalmol 101:1263 (1983).

9. K.G. Riedel, J. Zwaan, K.R. Kenyon, E.H. Kolodny, L. Hanninen, and D.M. Albert, Ocular abnormalities in mucolipidosis IV, Am J Ophthalmol 99:125 (1985).

10. J.J. Martin, J. Libert, and C. Ceuterick, Ultrastructure of brain and retina in Kufs' disease (adult type-ceroid-lipofuscinosis), Clin Neuropathol 6:231 (1987).

11. K. Ikeda, K. Kosaka, S. Oyanagi, and K. Yamada, Adult type of neuronal ceroid-lipofuscinosis with retinal involvement, Clin Neuropathol 3:237 (1984)

12. S.F Berkovic, S. Carpenter, F. Andermann, E. Andermann, L.S. Wolfe, Kufs' disease: a critical reappraisal, Brain 111:27 (1988).

13. M.S. Ramsey, and B.S. Fine, Chloroquine toxicity in the human eye. Histopathologic observations by electron microscopy, Am J Ophthalmol 73:229 (1972).

14. S.M.Z. Cohen, F.R. Brown III, L. Martyn, H.W. Moser, W. Chen, M. Kistenmacher, H. Punnett, W. Grover, Z.C. De La Cruz, N.R. Chan, and W.R. Green, Ocular histopathologic and biochemical studies of the cerebrohepatorenal syndrome (Zellweger's syndrome) and its relationship to neonatal adrenoleukodystrophy, Am J Ophthalmol 96:488 (1983).

15. S.M.Z. Cohen, W.R. Green, Z.C. De La Cruz, F.R. Brown III, H.W. Moser, M.W. Luckenbach, D.J. Dove, and I.H. Maumenee, Ocular histopathologic studies of neonatal and childhood adrenoleukodystrophy, Am J Ophthalmol 95:82 (1983).

16. D.N. Palmer, I.M. Fearnley, J.E. Walker, N.A. Hall, B.D. Lake, L.S. Wolfe, M. Haltia, R.D. Martinus, and R.D. Jolly, Mitochondrial ATP synthase subunit c storage in the ceroid-lipofuscinoses (Batten disease), Am J Med Genet 42:561 (1992).

17. H.H. Goebel, J.D. Fix, and W. Zeman, The fine structure of the retina in neuronal ceroid-lipofuscinosis, Am J Ophthalmol 77:25 (1974).

18. H.H. Goebel, W. Zeman, and E. Damaske, An ultrastructural study of the retina in the Jansky-Bielschowsky type of neuronal ceroid-lipofuscinosis, Am J Ophthalmol 83:70 (1977).

19. H.H. Goebel, N. Koppang, and W. Zeman, Ultrastructure of the retina in canine neuronal ceroid-lipofuscinosis, Ophthalmic Res 11:65 (1979).

20. H.H. Goebel, and E. Dahme, Ultrastructure of retinal pigment epithelial and neural cells in the neuronal ceroid-lipofuscinosis affected Dalmatian dog, Retina 6:179 (1986).

21. H.H. Goebel, and I. Dömpfmer, An ultrastructural study on retinal and pigment epithelial cells in ovine neuronal ceroid-lipofuscinosis, Ophthalmic Paediatr Genet 11:61 (1990).

22. I. Järvelä, P. Santavuori, L. Puhakka, M. Haltia, and L. Peltonen, Linkage map of the chromosomal region surrounding the infantile neuronal ceroid-lipofuscinosis on 1p, Am J Med Genet 42:546 (1992).

23. R.M. Gardiner, Mapping the gene for juvenile onset neuronal ceroid-lipofuscinosis to chromosome 16 by linkage analysis, Am J Med Genet 42:542 (1992).

## III. STUDIES OF RETINAL DEGENERATION USING TRANSGENIC MICE AND OTHER ANIMAL MODELS

Animal models of human diseases have been important tools in the study of retinal degenerations in numerous laboratories for the past three decades. Previous investigations were limited to the use of naturally occurring mutants as models for diseases which were thought to be similar to human disorders. This research has now come full circle with the ability to construct specific mutations in the laboratory and place mutant genes into the germ line of mice, thereby producing animal models with genetic defects which more closely approximate those causing human diseases. The first five chapters in this section deal with studies on transgenic mice that contain mutant genes. Another chapter describes nonsense mutations in the β-subunit of phosphodiesterase in the naturally occuring mutants of mouse and dog which cause photoreceptor degeneration. Another evaluates the possibility of systemic defects in fatty acid metabolism in inherited retinal degenerations. The final two chapters deal with more conventional descriptive studies using natural occurring mutants in mouse and dog models with degenerative retinal disorders.

# SIMULATION OF AUTOSOMAL DOMINANT RETINITIS PIGMENTOSA IN TRANSGENIC MICE

Muna I. Naash[*], Muayyad R. Al-Ubaidi[*], Joe G. Hollyfield[#], and Wolfgang Baehr[#]

[#]Department of Ophthalmology, Baylor College of Medicine, Houston, TX 77030
[*]*Present address:* UIC Eye Center, Department of Ophthalmology, University of Illinois College of Medicine, Chicago, IL 60612

"Retinitis Pigmentosa" (RP) comprises a group of hereditary retinal degenerative diseases characterized by photoreceptor cell degeneration and in most cases, eventual loss of vision. The typical clinical findings are pigmentary retinopathy, visual field loss, elevated dark-adapted thresholds and reduction of the ERG in the early stages of the disease (Heckenlively, 1988). In typical RP, three major modes of transmission have been identified: X-linked, autosomal dominant, and autosomal recessive (Heckenlively, 1988). There is extensive heterogeneity within each pattern of transmission.

Many laboratories have examined histological features of RP (Nizuno and Nishida, 1967; Kolb and Gouras, 1974; Szamier and Berson, 1977; Szamier et al., 1979; LaVail et al., 1985; Hollyfield et al., 1984). Due to the limitation in the availability of tissues from affected individuals, only a few biochemical studies have been carried out on human eyes with RP (Hollyfield et al., 1984; Rayborn et al., 1985; Farber et al., 1986; Flannery et al., 1989). Little is known of the pathogenesis of these retinal dystrophies, since most retinas acquired to date have been from donors with advanced RP. Since RP abnormalities are associated with photoreceptor cell degenerations, genes specifically expressed in photoreceptors are prime candidates for causing the disease. Humphries and his colleagues mapped the gene for ADRP in a large Irish pedigree to the long arm of chromosome 3 which contains the opsin locus (McWilliam et al., 1989). Dryja and colleagues, subsequently showed that a proline to histidine mutation at position 23 (P23H) near the N-terminus of human opsin is associated with a form of ADRP in 15 of 140 random patients (Dryja et al., 1990, Sung et al., 1991a, Berson et al., 1991). Patients with the P23H mutation exhibited a high degree of disease heterogeneity (Berson et al., 1991, Heckenlively et al., 1991; Sung et al, 1991b). Proline at position (23) is present in the mouse sequence and is conserved in all sequenced vertebrate and invertebrate opsin, and related G protein coupled receptors (Applebury and Hargrave, 1986). Apart from asparagine linked glycosylation, a precise function for the N-terminal end of opsin or for proline at position 23 has not been determined. Up to date over 50 different mutations in

*Retinal Degeneration*, Edited by J.G. Hollyfield *et al.*
Plenum Press, New York, 1993

rhodopsin affecting over 40 residues were found to segregate with ADRP (Farrar et al., 1993).

In a naturally occurring mouse strain with retinal degeneration, a null mutation was found in the *retinal degeneration slow (rds)* gene encoding the protein peripherin (Travis et al., 1989). This protein is thought to be important for structure of photoreceptor rod outer segment membrane discs. Both Humphries' group (Farrar et al., 1991) and Dryja's group (Kajiwara et al., 1991) have now identified specific point mutations or deletions in the coding region of the peripherin/RDS gene in some ADRP patients. An independent locus on chromosome 8 has also been implicated in some ADRP patients (Blanton et al., 1991).

Recently, transgenic mice were generated that carry either the normal or the mutant human opsin genes from a patient with P23H (Olsson et al., 1992). All the lines of transgenic mice with the mutated gene developed fast photoreceptor degeneration except for one with the lowest expression of the transgene, which showed less severe retinal degeneration (Olsson et al., 1992). One of the two transgenic lines for the normal allele showed severe retinal degeneration as well. Mouse mutants that exhibit characteristics similar to ADRP that could serve as a model for the human disorder are currently unavailable. To obtain such a model, we incorporated a mutated mouse opsin gene into the normal mouse genome. The consequences of the expression of the transgene were studied in one transgenic line and found to cause a retinal degeneration similar to ADRP.

## The Mutant Murine Opsin Gene

To distinguish between normal and mutant transcripts and to follow the fate of normal and mutant rhodopsins, two types of mutations were introduced. First, RFLPs which do not alter the amino acid sequence of opsin (addition of a HinfI site and removal of a NcoI site, Fig. 1A). Since no foreign DNA has been incorporated, the RFLPs are essential for identification of a founder mouse by PCR amplification of exon 1 and digestion of the amplified DNA with NcoI. Second, we generated a new epitope (Fig. 1B) which may be recognized by an anti-mutant peptide antibody by altering three amino acids near the N-terminus of rhodopsin (V20G, P23H, and P27L). One of the three altered residues is proline at position 23, a well-conserved residue in G protein linked receptors and the predominant point mutation in genes from human ADRP patients. The other two altered residues have not been observed in human ADRP and are much less conserved.

The transgene was constructed by oligonucleotide-directed mutagenesis, and insertion of a mutated fragment into λMOI clone comprising the complete murine opsin gene (Al-Ubaidi et al., 1990). Five oligonucleotide primers (Fig. 1A) were used for amplification of mutated opsin gene fragments by the Polymerase Chain Reaction (PCR). W75 (at position -245 to -209), MN1 (at position 177 to 201), MN2 (at position 211 to 142), MN3 (at position 446 to 377), and W11 (at position 1082 to 1062). The positions and sequences are as described by Al-Ubaidi et al., (1990). MN2 and MN3 are 70 meric antisense oligonucleotides which introduce the desired mutations. Amplified products W75/MN2 (446 bp) and MN1/MN3 (270 bp) were annealed, extended with Taq polymerase, and reamplified with W75/MN3. The product (681 bp) was reamplified in the presence of primer W11. This amplification gave a product of 1317 bp which was cut with XhoI, and the resulting XhoI fragment (bar in Fig. 1A, 349 bp) was cloned into MOPS2 (Baehr et al., 1988), a 5.1 kb EcoRI fragment containing the entire mouse opsin, from which the corresponding wild-type XhoI fragment had been removed. The defined clones were identified by the presence of the predicted Restriction Fragment Length Poly-

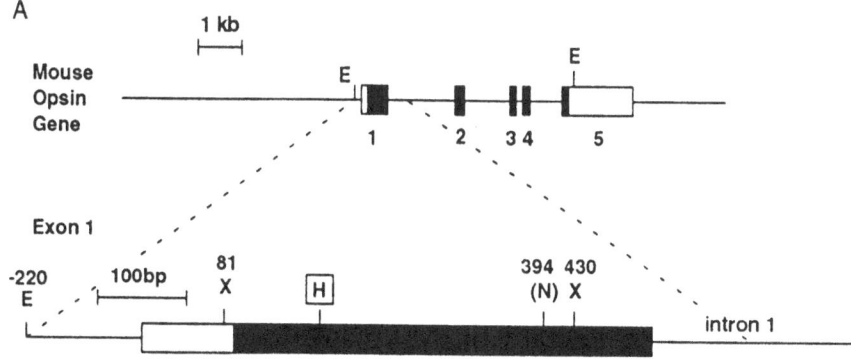

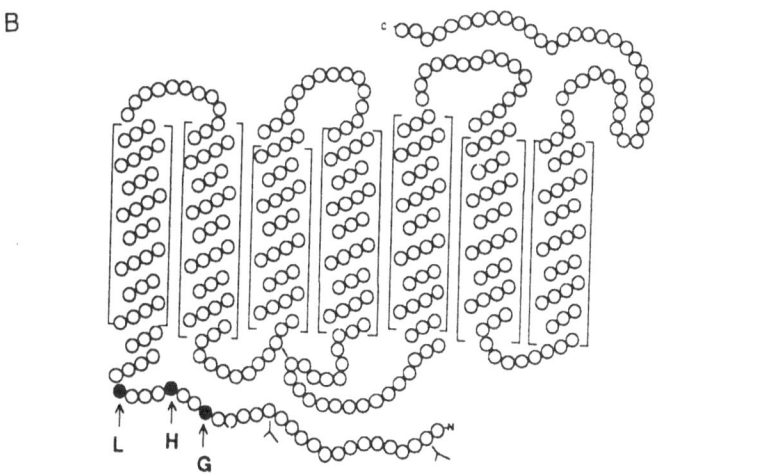

**Figure 1**. The mutated mouse opsin transgene and its product, mutated rhodopsin. (**A**) top, map of the mouse opsin gene (15 kb) containing the five exons encoding opsin (1-5, black boxes), lines connecting the exons depict the introns, the long line to the left represents the 5'-untranslated region containing the promoter (6 kb), while the right line represents the large 3'-untranslated region (3.5 kb) containing multiple polyadenylation signals (Al-Ubaidi et al., 1990). Middle drawing representing exon 1 of the transgene with relevant restriction sites and newly introduced RFLPs. E, EcoRI; X, XhoI; N, NcoI; boxed H, newly generated HinfI site; (N), deleted NcoI site. The nucleotide numbering system (Al-Ubaidi et al., 1990) starts with +1 at the transcription start site, depicted by the beginning of the open box (5' untranslated region). The filled box symbolizes the coding region of exon 1. W11, W75, MN1, MN2, and MN3 are primers used for PCR amplification and mutagenesis. The bar underneath symbolizes the mutated, 349 bp XhoI fragment which was used to replace the corresponding wild-type fragment of the opsin gene from λMOI (Al-Ubaidi et al., 1990). The transgene is delimited by SalI (S) sites deriving from the multiple cloning site of the vector. (**B**) Cartoon of mouse opsin molecule with three mutations at the N-terminal region. The three amino acids residues at position 20, 23, and 27 (Arrows) were altered in the transgene.

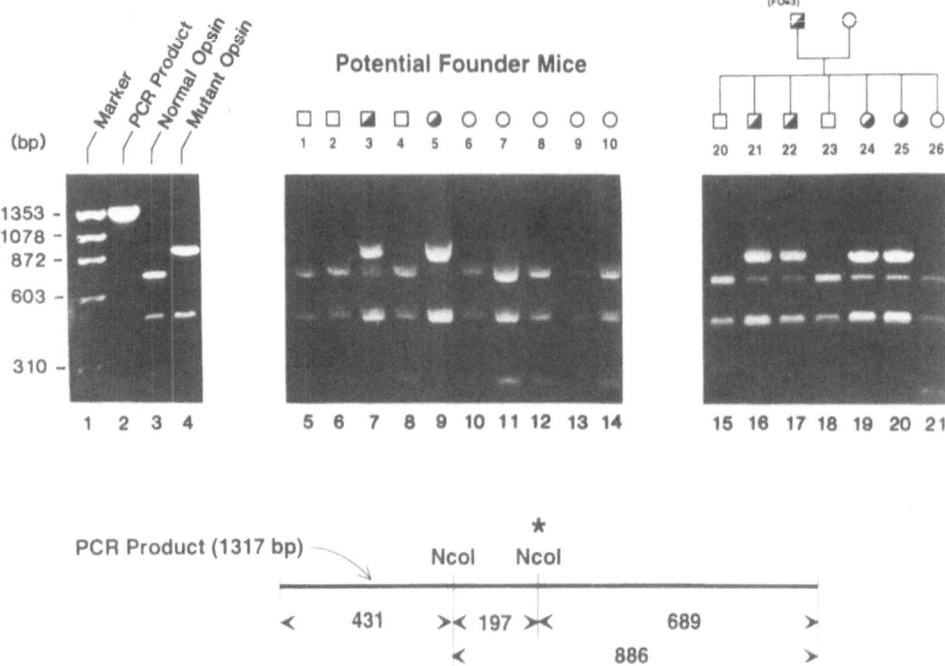

**Figure 2.** PCR amplification and NcoI digestion. F1 pedigree, mounted above a photograph of an ethidium bromide stained gel showing NcoI digests of PCR amplified product of exon 1 from DNA isolated from 11 mice born after the microinjection. PCR amplification with primers W11 and W75 produce a 1317 bp (lane 2) fragment which is normally cut twice with the restriction enzyme NcoI and give rise to 689, 431, and 197 bp bands (lane 3). However, deleting the NcoI site from the transgene results in generating 886 and 431 bp bands upon digestion with NcoI (lane 4). Of the 11 mice born, founder mice were identified by the absence of the NcoI site in genomic DNA (lanes 5-14). Lanes 15 to 21 represent the PCR amplification and NcoI digestion from one litter from founder mouse 3. The pedigree above the photograph to the right shows PCR amplified exon 1 from one littermate which has been cut with the restriction enzyme NcoI. These data indicate that founder mouse 3 passed the transgene to the offspring at 50% transmittance (Mendelian inheritance).

morphism's (RFLPs ) and checked for the correct orientation of the insert. Finally, one clone, which displayed the desired mutations and RFLPs, was verified by DNA sequencing. A 5.1 kb EcoRI fragment of this clone was ligated into λMOI, from which the wild-type EcoRI fragment had been removed, and packaged into viable EMBL3 phage. The resulting phage clones were screened for the correct orientation and sequence. The transgene, a 15 kb SalI fragment, was purified by agarose gel electrophoresis, diluted to 10 ng/ml, and injected into 1-day-old C57BL/6xSJL mouse embryos (DNX, New Jersey).

The final transgene (Fig. 1A) consisted of a 15 kb mouse opsin genomic fragment, which contained 6 kb upstream and 3.5 kb downstream sequences, and in which a part of the wild type exon 1 of the opsin gene was replaced by a fragment containing the mutations (Fig. 1A). Following microinjection of the transgene and reimplantation of embryos, 11 mice were born and two were identified as potential founder mice. Wild-

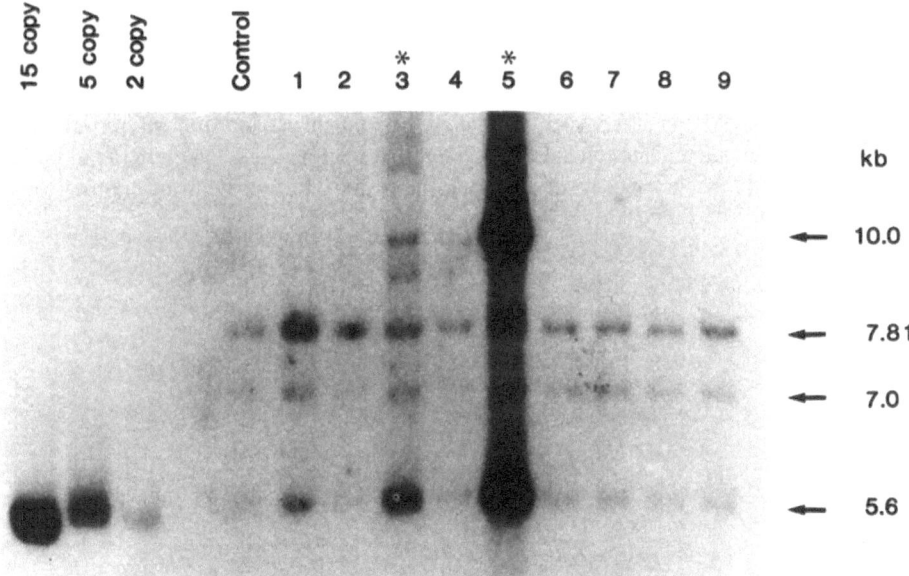

**Figure 3.** DNA blot analysis to determine the copy number of the transgene. EcoRI digest of genomic DNA isolated from tail clipping from all the potential founder mice. 16 μg of the digested genomic DNAs were run on 0.6% agarose in TAE buffer and transferred to a nylon membrane. The membranes were probed with radiolabeled mouse opsin genomic fragment (Mops2 that contains the 5 kb EcoRI fragment of the mouse opsin gene). For determination of transgene copy number, mouse opsin gene was digested with EcoRI and 15, 5, and 2 copy equivalent of the gene were run in the marked lanes.

type mice exhibit three fragments 689 bp, 431 bp, and 197 bp (Fig. 2). Transgenic (heterozygous) mice have an additional fragment (689 bp + 197 bp) due to the deletion of the NcoI site. One founder mouse (founder 3) passed the transgene to the offspring at 50% transmittance (Mendelian inheritance), while the other (founder 5) did not. Southern analysis of the potential founder mice indicated incorporation of 2 to 5 and more than 15 transgene copies per haploid genome into a single integration site of founders 3 and 5, respectively (Fig. 3).

**Simultaneous Expression of the Normal and Mutated Opsin Genes**

We previously showed (Naash et al., 1993) that expression of the mutated and normal opsin genes may be followed on the transcriptional level by two antisense 24-meric oligonucleotide probes, N and M. N was synthesized according to the normal opsin mRNA sequence (Figure 4). The introduced amino acid mutations (filled circles), including the HinfI RFLP (shaded), allowed for five non-identical nucleotides in the two probes (asterisks). We showed by RNA blotting that under stringent hybridization conditions, N will exclusively hybridize to the normal transcripts, while M will only hybridize to mutant transcripts (Fig. 2 in Naash et al.) . When polyA mRNA, isolated from transgenic retinas at postnatal day 20, was probed with N, the five normal transcripts of the murine opsin gene were seen, indicating that the mutant transgene did not integrate into the opsin locus on chromosome 6. Probing of this RNA with M also revealed five transcripts of the mutant gene. The mutant transcripts had the same mobility as the normal transcripts, indicating that the mutant transgene was transcribed and processed correctly. Moreover, the levels of mutant and normal transcripts were approximately identical. These results suggested that the transgene is a functional unit, is co-expressed with the normal opsin gene, and does not interfere with the normal opsin gene expression.

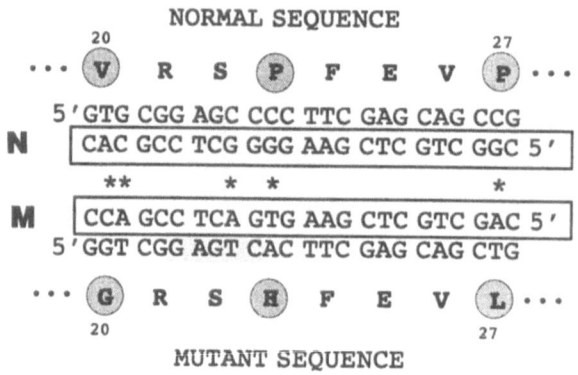

**Figure 4**. Sequences of normal and mutant antisense oligonucleotide probes N and M.

## Histological Evidence of the Slow Degeneration of the Photoreceptor Cells

Histological analyses of retinas from both normal and transgenic mice were indistinguishable at early stages of development (Fig. 5). At P13, shortly before eye opening (at P14), nascent outer segments are present in both normal and transgenic animals. By P20, when photoreceptor outer segments are near adult length in normal animals, outer segments in the transgenic animals are about one-half normal length (Naash et al., 1993). Outer segments in transgenic animals continue to shorten. This decrease in average length of the outer segments is accompanied by a decrease in the number of photoreceptors, as evidenced by the reduced thickness of the outer nuclear layer (Fig. 5). Throughout the postnatal period, the pigment epithelium in transgenic eyes appear to be normal, and no pigmentary changes typical for human RP are recognized (data not shown). Pyknotic photoreceptor nuclei, an indication of ensuing cell death, are frequently observed in the transgenic animals but only rarely in nontransgenic littermates. By P250, the oldest transgenic animal examined, extensive expanses of the central fundus are free of photoreceptors indicating that both rods and cones are affected by the degeneration (Naash et al., 1993). Photoreceptor survival at this age is more pronounced in the retinal periphery, with one to three rows of photoreceptor nuclei remaining. Analysis by electron microscopy indicates that both rod and cone cells are degenerated (data not shown).

Our results indicate that the transgene is transcribed concurrently with the endogenous murine opsin gene, but we have no evidence that the mutated rhodopsin is produced, transported to the outer segment and incorporated into disk membranes. In transgenic mice expressing a human rhodopsin gene containing the P23H mutation, the human mutant rhodopsin appeared to be incorporated into the rod outer segment without apparent gross structural distortions (Olsson et al., 1992). In tissue culture with embryonic kidney cell lines, however, P23H mutant rhodopsins were found to accumulate in the endoplasmic reticulum (Sung et al., 1991b). Preliminary EM studies in trangenic retinas using a monoclonal antibody that only recognizes normal rhodopsin show no accumulation of immunoreactivity in the inner segments or abnormal structures indicative of problems in the mutant rhodopsin transport.

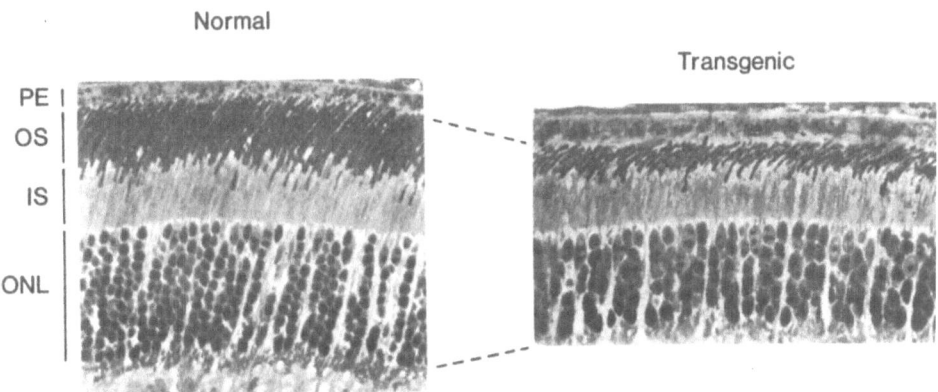

**Figure 5**. Temporal progressive changes in normal and transgenic littermates. Retinas shown were recovered 83 days postpartum. Pigment epithelium (PE), photoreceptor outer segments (OS), inner segments(IS), and outer nuclear layer (ONL) are indicated. The photoreceptor outer segment, located just below the level of the pigment epithelium at 83 days are shorter in transgenic than in the normal littermate. Additionally, the number of photoreceptor nuclei in the outer nuclear layer is reduced to 4 or 5 rows of nuclei in the transgenic as compared to the nontransgenic littermates. No differences were observed in the inner part of the retina between transgenic and normal littermates.

The transgenic mouse line presented here incorporated 2-5 copies of a transgene containing three mutations near the N-terminus of rhodopsin into a single site of the mouse genome. The transgene is expressed, and the generated phenotype displays many features characteristic of human ADRP. **First,** the effect of expression of the mutant opsin gene results in an autosomal dominant phenotypes similar to that in ADRP patients. **Second,** the subnormal light-evoked responses (ERGs) in mutant mice at P30, when no structural damages of photoreceptor cells except a shortening of the outer segment are evident, reflect a relatively early time of onset of the degeneration. In human ADRP patients, early ERG changes are the first clinical manifestations of the disease, with a median onset of night blindness in P23H patients at age 13-14 years (Berson et al., 1991). **Third,** the photoreceptor degeneration observed in this transgenic line progresses slowly, as evidenced by the gradual reduction of the a-wave, shortening of the photoreceptor outer segments and loss of photoreceptor cells (gradual decrease of the outer nuclear layer thickness). In human patients, ERGs and fundus analyses, the only effective means to follow the progression of the disease, slowly deteriorate with age, in most cases progressing over many decades. **Fourth,** the slow rod degeneration in the transgenic mice is accompanied by cone degeneration. Widespread loss of cone photoreceptors is characteristic of human ADRP at late stages. Thus, we have generated a transgenic mouse with a retinal (photoreceptor) degeneration which mimics human ADRP in many aspects. This mouse line will permit a variety of biochemical, molecular, and cell biological studies which are the main goals of this proposal.

Assuming that the mutant rhodopsin is produced, one of the challenging questions remaining to be answered is how mutations in the rhodopsin gene slowly disable and eventually disrupt photoreceptor cell function. To this end, the fate of the mutated rhodopsin in the living photoreceptor cell must be determined at the level of biosynthesis, posttranslational modifications, transport and incorporation into the disk membrane of outer segments. Eventually, we would like to understand the mechanistic link between mutations and retinitis pigmentosa, with the hope of designing rational approaches to retard or cure this group of diseases.

## REFERENCES

Al-Ubaidi, M. R., Pittler, S. J., Champagne, M. S., Triantafyllos, J. T., McGinnis, J. F., and Baehr, W. (1990) Mouse Opsin: Gene structure and molecular basis of multiple transcripts. *J. Biol. Chem.* 265: 20563-20569

Applebury, M. L. and Hargrave, P. A. (1986) Molecular biology of the visual pigments *Vision Res.* 26: 1881-1895

Baehr, W., Falk, I. D., Bugra, K., Triantafyllos, J. T., and McGinnis, J. F. (1988) Isolation and molecular analysis of the mouse opsin gene. *FEBS Lett.* 238: 253-256

Berson, E. L., Rosner, B., Sandberg, M. A., and Dryja, T. P. (1991) Ocular findings in patients with autosomal dominant retinitis pigmentosa and a rhodopsin gene defect (Pro-23-His). *Arch. Ophthalmol.* 109: 92-101

Blanton, S. H., Hekenlively J. R., Cottingham, A. W. Friedman, J., Sadler, L. A., Wagner, M., Friedman L. H., and Daiger, S. P. (1990) Linkage mapping of autosomal dominant retinitis pigmentosa (RPI) to the pericentric region of human chromosome 8. *Genomics* 11: 857-869

Dryja, T. P., McGee, T. L., Reichel E., Hahn, L. B., Cowley, G. S., Yandell, D. W., Sandberg, M. A., and Berson, E. L. (1990) A point mutation of the rhodopsin gene in one form of retinitis pigmentosa. *Nature* 343: 364-369

Farber, D. B., Flannery, J. G., Bok, D. and Bird, A. (1986) Abnormal distribution pattern of cyclic nucleotides in a retina affected with retinitis pigmentosa. *Invest. Ophthalmol. Vis. Sci.* 27: 56

Farrar, G. J., Jordan, S. A., Kumar-Singh, R., Inglehearn, C. F., Gal, A., Greggory, C., Al-Maghtheh, M., Kenna, P. F., Humphries, M. M., Sharp, E. M., Sheils, D. M., Bunge, S., Hargrave, P. A., Denton, M. J., Schwinger, E., Bhattacharya, S., and Humphries, P. (1993) Extensive genetic heterogeneity in autosomal dominant retinitis pigmentosa. In Retinal Degeneration: Experimental and Clinical Studies, ed. M. M. LaVail, J. G. Hollyfield, and R. E. Anderson. Alan R. Liss, Inc. (New York), submitted.

Farrar, G. J., Kenna, P. Redmond, R. McWilliam, P., Bradley, D. G. and Humphries, M. M. (1991) A three base pair deletion in the peripherin-*rds* gene in one form of retinitis pigmentosa. *Nature* 354: 478-480

Flannery, J. G., Farber, D. B., Bird, A. C and Bok, D. (1989) Degenerative changes in a retina affected with autosomal dominant retinitis pigmentosa. Invest. Ophthalmol. Vis. Sci. 30: 191-211

Hargrave, P. A., McDowell, J. H., Curtis, D. R., Wang, J. K., Juszczak, E., Fong, S-L.,

Mohanna Rao, J. K.., and Argos, P. (1983) The structure of bovine rhodopsin. *Biophys. Struct. Mech.* 9: 235-244

Heckenlively, J. R. (1988) Retinitis Pigmentosa. pp. 125-149, Lippincott, Philadelphia. Heckenlively, J. R., Rodriguez, J. A., and Daiger, S. P. (1991) Autosomal dominant sectoral retinitis pigmentosa. Two families with transversion mutation in codon 23 of rhodopsin. *Arch. Ophthalmol.* 109: 84-91

Hollyfield, J. G., Frederick, J. M., Tabor, G. A. and Ulshafer, R. J. (1984) Metabolic studies on retina tissue from a donor with a dominantly inherited chorioretinal degeneration resembling sectoral retinitis pigmentosa. *Ophthalmology* 91: 191-196

Kajiwara, K., Mukai, S., Travis, G., Berson, E. L., and Dryja, T. P. (1991) Mutations in the human retinal degeneration slow gene (rds) in autosomal dominant retinitis pigmentosa. *Nature* 354: 480-482

Kolb, H. and Gouras, P. (1974) Electron microscopy observations of human retinitis pigmentosa, dominantly inherited. *Invest. Ophthalmol. and Vis Sci.* 13: 487-498

LaVail, M. M., Yasumura, D. and Hollyfield, J. G. (1985) The interphotoreceptor matrix in retinitis pigmentosa: preliminary observations from a family with an autosomal dominant form of disease. *In: Retinal Degeneration, Experimental and Clinical Studies*, LaVail M. M., Hollyfield, J. G, and Anderson, R. E, editors. New York, Alan R. Liss, pp 51-62

McWilliam, P., Farrar, G. J., Kenna, P., Bradley, D. G., Humphries, M. M., Sharp, E. M., McConnel, D. J., Lawler, M., Sheils, D., Ryan, C., Stevens, K., Daiger, S. P. and Humphries, P. (1989) Autosomal dominant Retinitis pigmentosa (ADRP): Localization of an ADRP gene to the long arm of chromosome 3. *Genomics* 5: 619-622

Naash, M. I., Hollyfield, J., Al-Ubaidi, M. R., and Baehr, W. (1993) Simulation of human autosomal dominant retinatis pigmentosa in transgenic mice expressing a mutated murine opsin gene. *Proc. Natl. Acad. Sci. USA* in press

Nizuno, K. and Nishida, S. (1967) Part 1. Two cases of advanced retinitis pigmentosa. Electron microscopy studies of human retinitis pigmentosa. *Am. J. Ophthalmol.* 63: 791-803

Olsson, J., Gordon, J., Pawlyk, B., Roof, D., Hayes, A., Molday, R. Mukai, S., Cowley, G., Berson, E., Dryja, T. (1992). Transgenic Mice with Rhodopsin Mutation (Pro23His): A Mouse Model of Autosomal Dominant Retinitis Pigmentosa. *Neuron* 9: 815-830

Rayborn, M. E., Frederick, J. M., Ulshafer, R. J. Tabor, G. A., Moorhead, L. C. and Hollyfield, J. G (1985) Histopathology and in vitro metabolic studies on tissues from a family with an autosomal dominant form of retinitis pigmentosa. *In: Retinal Degeneration, Experimental and Clinical Studies*, LaVail M. M., Hollyfield, J. G, and Anderson, R. E, editors. New York, Alan R. Liss, pp 37-49

Sung, C. H., Davenport, C. M., Hennessey, J. C., Maumenee, I. H., Jacobson, S. G., Heckenlively, J. R., Nowakowski, R., Fishman, G., Gouras, P., and Nathans, J. (1991a) Rhodopsin mutations in autosomal dominant retinitis pigmentosa. *Proc. Natl. Acad. Sci. USA* 88: 6481-6485

Sung, C., Schneider, B. Agarwal, N., Papermaster, D., Nathans, J. (1991b) Functional Heterogeneity of Mutant Rhodopsins Responsible For Autosomal Dominant Retinitis Pigmentosa. Proc. Natl. Acad. Sci. USA 88: 8840-8844

Szamier, R. B. and Berson, E. L. (1977) Retinal ultrastructure in advanced retinitis pigmentosa. *Invest. Ophthalmol. and Vis. Sci.* 16: 947-962

Szamier, R. B. Berson, E. L., Klein, R. and Meyers, S. (1979) Sex-linked retinitis pigmentosa: ultrastructure of photoreceptor and pigment epithelium. *Invest. Ophthalmol. and Vis. Sci.* 18: 145-160

Travis, G. H., Bernnan, M. B., Dantelson, P. E., Kozak, C. A., and Sutcliffe, J. G. (1989) Identification of a photoreceptor-specific mRNA encoded by the gene responsible for retinal degeneration slow (rds). *Nature* 338: 70-73

# CREATING TRANSGENIC MOUSE MODELS OF PHOTORECEPTOR

# DEGENERATION CAUSED BY MUTATIONS IN THE RHODOPSIN GENE

Fulton Wong

Departments of Ophthalmology and Neurobiology
Duke University School of Medicine
Durham, NC 27710

## INTRODUCTION

Photoreceptor dysfunction is the major cause of vision loss in patients with retinitis pigmentosa (RP) -- a heterogeneous group of progressive hereditary retinal degenerative disorders.[1] Based on the various clinical attributes of these disorders, the rod photoreceptors have been identified as the primary focus of RP. This concept is strongly supported by the findings of the molecular genetic basis of one form of autosomal dominant RP (ADRP), which is a mutation in the rod photoreceptor-specific rhodopsin gene.[2] Recently, mutations in the *rds*/peripherin gene have been identified in some ADRP patients whose rhodopsin genes are normal,[3,4] thereby suggesting that a mutation in the *rds*/peripherin gene is the primary defect that causes this form of ADRP. (See chapters on the *rds*/peripherin gene in this volume.)

Once the mutations that cause RP are identified, the next step in understanding pathogenesis is to identify the relationship between the genetic defect and the mechanisms that cause the clinical phenotype. This is not always an easy task because the genetic defect may not cause the observed symptoms directly. In fact, the phenotype may be caused by mechanisms that are far removed from the genetic defect. For example, in the case of rhodopsin, it is believed that its major, or perhaps only, function is to capture photons and thus initiate visual transduction.[5,6] Since this is a highly specialized function, rhodopsin does not play a direct role in cell maintenance or cell viability; accordingly, its mutations probably only act as triggers that initiate other mechanisms leading to widespread rod photoreceptor degeneration.

Because the immediate effects of the rhodopsin mutations in ADRP are localized to the rod photoreceptors, it should be most productive to concentrate on analyzing the defective rod photoreceptors themselves to find the initial cause of the physiological deficits. Furthermore, it is important to study these cells early in the disease, before too many subsequent events are triggered by the primary defect. For this approach to succeed, an animal model using a rhodopsin mutation identical to that in humans would be helpful. Effects of the mutation on the rod photoreceptors can then be studied. Better still, if the animal model shows a phenotype similar to human RP, the chain of events leading to the clinically observable symptoms may be studied.

A variety of retinal degenerative diseases has been observed in animals. Rodent and canine species have been particularly well studied. (See other chapters in this volume.) Although these animal models have contributed significantly to our understanding of retinal function and diseases, to the best of my knowledge there is no available mammalian model with known mutations in the rhodopsin gene. Therefore, my research group has decided to evaluate the effects of mutations of the rhodopsin gene by creating and studying transgenic mice. We chose to study the mutations identified in codon 347 because not only are they among the most frequently observed mutations of the rhodopsin gene[7,8] but they also cause some of the most severe phenotypes among ADRP patients.[9]

We hypothesized that in ADRP patients who had a mutation in codon 347 of one of their two rhodopsin gene alleles, it is the presence of the mutant rhodopsin molecules, rather than a 50% reduction in functional rhodopsin, that causes the phenotype. According to this hypothesis, an extra rhodopsin gene (the transgene) with the appropriate mutation in codon 347 would simulate the human ADRP condition and cause rod photoreceptor degeneration in the transgenic mice. We were able to test this theory in our transgenic mice.

## MATERIALS AND METHODS

### Constructing Mutations in the Pig Rhodopsin Gene

The normal pig rhodopsin gene is highly similar to those of the mouse[10] and human.[11] Therefore, we assumed that the pig rhodopsin gene is compatible with the regulatory mechanisms of the mouse and that it would be properly expressed in the rod photoreceptors. We chose to express the pig rather than human, mouse, or bovine rhodopsin gene in the mouse because of our interest in creating a transgenic pig at some point in the future.

We have isolated the pig rhodopsin gene from a genomic library. A DNA fragment encompassing regulatory sequences that are required for proper expression of the rhodopsin gene has been cloned in a plasmid vector. We have constructed two point mutations in codon 347: proline$_{347}$ to leucine$_{347}$ (Pro347Leu) and proline$_{347}$ to serine$_{347}$ (Pro347Ser). Both of these mutations have been identified in ADRP patients.[7] In addition, we have constructed a mutation in codon 319 which changed threonine to STOP (TAA319). Both the TAA319 construct and the normal pig rhodopsin gene served as controls in our studies. The desired mutations were constructed using oligonucleotide-directed mutagenesis based on procedures[12] supplied in the Muta-Gene phagemid kit (BIO-RAD) and confirmed by sequence analysis.

### Creating Transgenic Mice

Transgenic mice were created by standard methods[13] of injecting DNA into fertilized eggs. Fertilized eggs were obtained either from inbred C57BL/6J or from F1 hybrids of C57BL/6J × SJL. The four DNA constructs were injected separately at concentrations of 1 ng/μl. The injected zygotes were transferred to pseudopregnant mothers who then gave birth to the transgenic candidates.

## Screening and Propagating Transgenic Mice

High-molecular-weight DNA were isolated from 1 - 2 cm of mouse tail, and these samples were screened by Southern blot analysis. The blots were probed with both a mouse-specific and a pig-specific rhodopsin gene probe. The pig-specific probe was designed to identify a novel band of 2.5 kb derived from the transgene. Positive founders were mated to inbred C57BL/6J mice in order to establish lines. Offspring were checked to screen against retinal degenerative traits such as *rd* which could be in the SJL background.[14]

## Histological Analysis

Enucleated eyes were fixed and processed by established methods for histological studies.[15] For light microscopy, 1-μm thick sections were taken through the vertical meridian at or near the optic disc.

## RESULTS

We performed several rounds of zygote-injection with the 4 different DNA constructs and screened over 500 offspring which yielded 31 transgene-positive animals. Examples of the screen by Southern blot analysis are shown in Figure 1, lanes A through D. Lane A was derived from a transgene-negative animal and shows only the mouse DNA band (black arrow). Lane B was derived from a transgene-positive animal and it shows, in addition to the mouse band, a novel band of 2.5 kb (arrow head) which was expected if the transgene had integrated properly. The relative intensities of the mouse and pig DNA bands show that 4 to 6 copies of the transgene had integrated. Lane C shows the results obtained from a founder similar to that shown in lane B except that the 2.5-kb band is more intense, indicating that more copies of the transgene (more than 20) had integrated. Lane D shows the results obtained from a transgene-positive mouse with a DNA band at 4 kb (white arrow) instead of at 2.5 kb suggesting that the injected DNA had been rearranged. Only transgene-positive mice that show the 2.5 kb band were used to establish lines. Of the 31 transgene-positive animals obtained, 20 lines representing the 4 different genotypes were established.

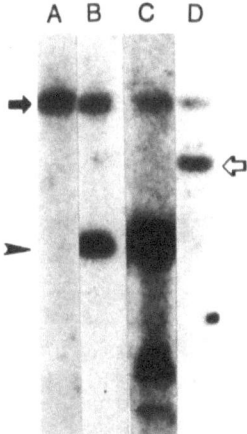

**Figure 1.** Southern blot analysis of transgenic mice. A) transgene-negative animal shows only mouse DNA band, B) novel band of 2.5 kb expected with proper integration of transgene, low copy number, C) novel band of 2.5 kb with higher copy number, D) novel band of 4 kb, suggesting rearrangement of DNA.

Mice from these 20 lines appeared physically normal. When the retinas of 3-week old animals were examined, many showed thinning of the outer nuclear layer (ONL) resulting from photoreceptor degeneration. The rest of the retina appeared normal. There was a spectrum of severity of the phenotype between mice from different lines that had no correlation with the genotype. However, the severity of the phenotype was highly correlated with the copy number. For instance, almost all the lines with copy numbers of 20 or more showed severe degeneration, that is, no remaining photoreceptor by postnatal 3 weeks. Those with copy numbers between 10 and 20 showed a more moderate phenotype, 2 to 3 rows of nuclei remaining at 3 weeks of age.

Among the group of mice with copy numbers between 2 and 6, there was one line, #6, which was of the Pro347Ser genotype, and 3 lines, #23, #25, and #3, in which the transgene was the normal pig rhodopsin gene. Mice of the #6 line showed slow photoreceptor degeneration: the ONL was reduced to about 50% by 7 weeks of age.[16] On the other hand, mice from lines #23, #25, and #3 showed normal retinas at a comparable age. In the case of line #23, we examined a 7-month old animal and observed no degeneration.

By 12 weeks of age, the retinas of mice from line #6 showed further degeneration;[16] the thickness of the ONL was about 25% of normal (Figure 2). In addition, "pigment-laden" cells (arrow) were visible in the layer of outer and inner segments of the remaining photoreceptors.

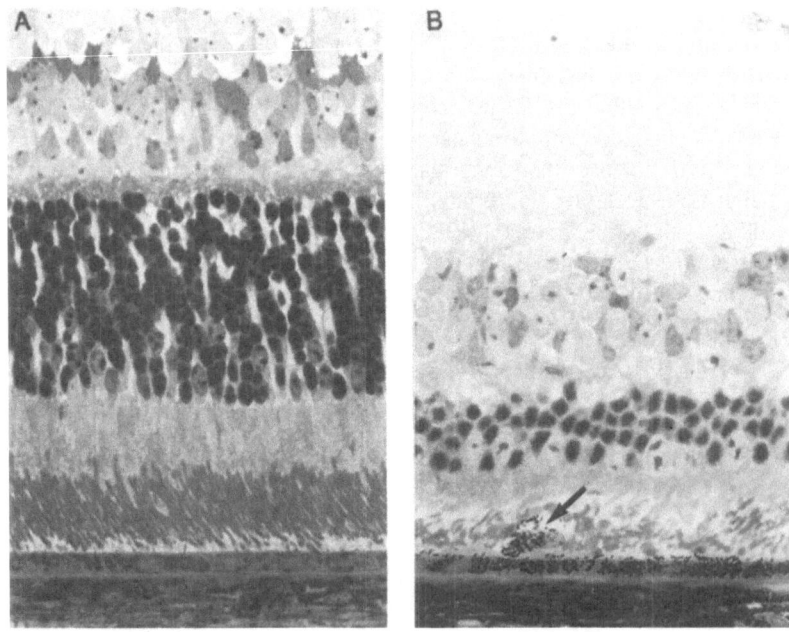

**Figure 2.** A. Retina of normal mouse. B. Retina of 12-week old transgenic mouse expressing the Pro347Ser mutant transgene.

## DISCUSSION

This project was carried out in order to create an animal model that simulates the human ADRP condition. We have succeeded in identifying line #6 which expresses the Pro347Ser mutant transgene and shows a phenotype of slow photoreceptor degeneration similar to human ADRP. We believe that these transgenic mice will prove useful for studying the mechanisms of pathogenesis and for developing methods of therapeutic intervention.

Results shown in Figure 1 demonstrated that the pig rhodopsin gene can integrate into the mouse genome and can be inherited. Most of the time, the injected DNA will integrate in a head-to-tail array[13], yielding internal bands of novel but predictable lengths. The 2.5-kb pig-specific DNA band was predicted and observed. We also observed a range of copy numbers reflecting the size of the array. Occasionally, the injected DNA are rearranged resulting in the appearance of bands other that the expected one. We were able to screen against those cases by Southern blot analysis (Figure 1, lane D).

Deficits in these transgenic mice were observed mainly in the ONL of the retina suggesting that the expression of the transgene is specific to photoreceptors, probably only rod photoreceptors. (However, this is difficult to ascertain because of the small number of cones in the mouse retina.) The tissue-specificity of rhodopsin transgenes has been demonstrated by several studies in which mouse[17,18], bovine[19], or human[20] rhodopsin gene sequences were used. As demonstrated in our studies, the pig rhodopsin gene showed the same tissue-specificity.

In these lines of transgenic mice, we observed severe photoreceptor degeneration in those with high copy number of transgene independent of the genotype. As high copy number could mean high levels of expression of the transgene, our results indicate that high levels of expression of mutant or normal rhodopsin may cause rapid photoreceptor degeneration with all photoreceptors disappearing by 3 weeks of age. This observation of rapid degeneration due to overexpression of the rhodopsin transgene was made independently by Olsson et al.[20] who studied transgenic mice expressing normal and mutant human rhodopsin genes. The mechanisms of photoreceptor degeneration due to overexpression of the rhodopsin gene are likely not to be the same as those due to specific mutations in the rhodopsin gene, as the time course of degeneration is different in each case. In the former case, degeneration is complete a few weeks after birth, whereas in the latter case, many photoreceptors remained after several months. While we will continue to characterize the group of transgenic mice showing rapid degeneration, they are clearly not good models for RP.

On the other hand, the group of mice having low copy numbers allows us to evaluate the effects of the transgene unconfounded by factors such as overexpression. Those mice expressing the normal pig transgene (#23, #25, and #3) showed no degeneration whereas #6 which expressed the Pro347Ser mutant transgene showed slow photoreceptor degeneration. While these observations suggest that the presence of normal pig rhodopsin molecules in the mouse photoreceptors does not cause degeneration, these molecules may or may not be functional in the mouse photoreceptors. Presence of the mutant rhodopsin molecules, though, did cause degeneration. If the mutant pig rhodopsin molecules were not made in #6 mice, there should be no observable degeneration since the normal amount of mouse rhodopsin was present in the photoreceptors. Accordingly, pathogenesis in these mice must be due to the mutant rhodopsin molecules. It is these defective molecules that trigger the events leading to the observable phenotype. These defective molecules may interfere with the normal synthesis, transport, or degradation of rhodopsin. Alternatively, they may interfere with the function of rhodopsin in the rod outer segment. Neither can other possible disease mechanisms[21] be ruled out at this time.

The most important feature of the phenotype of line #6 is the time course of photoreceptor degeneration, which is slow like that seen in ADRP patients.[1] Disappearance of photoreceptors, as indicated by the thinning of the ONL, is noticeable at post-natal 3 weeks. However, roughly 25% of photoreceptors continue to survive even at 3 months of age (Figure 2B). If mutant rhodopsin molecules are present in all rod photoreceptors and it is their presence that triggers the pathogenic events, what mechanisms might account for the large variance in longevity seen in these photoreceptors? Perhaps the mechanisms directly responsible for photoreceptor degeneration are not restricted to those immediately involved with the function of rhodopsin. The phenomena observed in these transgenic mice suggest that the photoreceptors disappear as though some retinal mechanisms are removing them. Once these retinal events are initiated, most likely triggered by the mutant rhodopsin molecules, they would run their course, causing the progressive disappearance of the photoreceptors. Whatever the mechanisms of degeneration may turn out to be, we believe that our transgenic mice will be useful for deciphering this missing link between mutation of the rhodopsin gene and photoreceptor degeneration.

## ACKNOWLEDGMENTS

The author acknowledges the contributions of Ying Hao, Cheryl Bock, and Alicia Gaitan to the work reported here. Zygote injections were performed in the Transgenic Mouse Shared Facility of the Duke University Cancer Center. This research was supported by NEI (grants EY06862 and EY09047) and RP Foundation of Baltimore, Maryland. Also thank you to Gail Atwater for editorial assistance.

## REFERENCES

1.  J.R. Heckenlively. "Retinitis Pigmentosa," Lippincott, Philadelphia (1988).
2.  T.P. Dryja, T.L. McGee, E. Reichel, L.B. Hahn, G.S. Cowley, D.W. Yandell, M.A. Sandberg, and E.L. Berson, A point mutation of the rhodopsin gene in one form of retinitis pigmentosa, *Nature*. 343:364 (1990).
3.  G.J. Farrar, P. Kenna, S.A. Jordan, R. Kumar-Singh, M.M. Humphries, E.M. Sharp, D.M. Sheils, and P. Humphries, A three-base-pair deletion in the peripherin-RDS gene in one form of retinitis pigmentosa, *Nature*. 354:478 (1991).
4.  K. Kajiwara, L.B. Hahn, S. Mukai, G.H. Travis, E.L. Berson, and T.P. Dryja, Mutations in the human retinal degeneration slow gene in autosomal dominant retinitis pigmentosa, *Nature*. 354:480 (1991).
5.  L. Stryer, Visual excitation and recovery, *J Biol Chem*. 266:10711 (1991).
6.  H.G. Khorana, Rhodopsin, photoreceptor of the rod cell. An emerging pattern for structure and function, *J Biol Chem*. 267:1 (1992).
7.  T.P. Dryja, T.L. McGee, L.B. Hahn, G.S. Cowley, J.E. Olsson, E. Reichel, M.A. Sandberg, and E.L. Berson, Mutations within the rhodopsin gene in patients with autosomal dominant retinitis pigmentosa, *N Engl J Med*. 323:1302 (1990).
8.  A. Gal, A. Artlich, M. Ludwig, G. Niemeyer, K. Olek, E. Schwinger, and A. Schinzel, Pro-347-Arg mutation of the rhodopsin gene in autosomal dominant retinitis pigmentosa, *Genomics*. 11:468 (1991)
9.  E.L. Berson, B. Rosner, M.A. Sandberg, C. Weigel-DiFranco, and T.P. Dryja, Ocular findings in patients with autosomal dominant retinitis pigmentosa and rhodopsin, Pro-347-Leucine, *Am J Ophthalmol*. 111:614 (1991).
10. M.R. Al-Ubaidi, S.J. Pittler, M.S. Champagne, J.T. Triantafyllos, J.F. McGinnis, and W. Baehr, Mouse opsin - gene structure and molecular basis of multiple transcripts, *J. Biol Chem*. 265:20563 (1990).
11. J. Nathans and D.S. Hogness, Isolation and nucleotide sequence of the gene encoding human rhodopsin, *Proc Natl Acad Sci USA*. 81:4851 (1984).
12. T.A. Kunkel, J.D. Roberts, and R.A. Zakour, Rapid and efficient site-specific mutagenesis without phenotypic selection, *Methods in Enzymol*. 154:367 (1987).
13. B. Hogan, F. Constantini, and E. Lacy, "Manipulating the mouse genome: A laboratory manual," Cold Spring Harbor Laboratory, Cold Spring Harbor (1986).
14. S.J. Pittler, and W. Baehr, Identification of a nonsense mutation in the rod photoreceptor cGMP phosphodiesterase ß-subunit gene of the rd mouse, *Proc. Natl. Acad. Sci. USA*. 88:8322 (1991).

15. M.M. LaVail, and B.-A. Battelle, Influence of eye pigmentation and light deprivation on inherited retinal dystrophy in the rat, *Exp Eye Res*. 21:167 (1975).

16. F. Wong and Y. Hao, Transgenic mouse model of retinal degeneration caused by mutations in proline 347 of rhodopsin, Invest Ophthalmol Vis Sci. 33 (Suppl):944 (1992).

17. J. Lem, M.L. Applebury, J.D. Falk, J.G. Flannery, and M.I. Simon, Tissue-specific and developmental regulation of rod opsin chimeric genes in transgenic mice, *Neuron*. 6:201 (1991).

18. M.R. Al-Ubaidi, J.G. Hollyfield, P.A. Overbeek, and W. Baehr, Photoreceptor degeneration induced by the expression of simian virus 40 large antigen in the retina of transgenic mice, *Proc Natl Acad Sci USA*. 89:1194 (1992).

19. D.J. Zack, J. Bennett, Y. Wang, C. Davenport, B. Klaunberg, J. Gearhart, and J. Nathans, Unusual topography of bovine rhodopsin promoter-lacZ fusion gene expression in transgenic mouse retinas, *Neuron*. 6:187 (1991).

20. J.E. Olsson, J.W. Gordon, B.S. Pawlyk, D. Roof, A. Hayes, R.S. Molday, S. Mukai, G.S. Cowley, E.L. Berson, and T.P. Dryja, Transgenic mice with a rhodopsin mutation (Pro23His): A mouse model of Autosomal dominant retinitis pigmentosa, *Neuron*. 9:815.

21. C.-H. Sung, B.G. Schneider, N. Agarwal, D.S. Papermaster, and J. Nathans, Functional heterogeneity of mutant rhodopsins responsible for autosomal dominant retinitis pigmentosa, *Proc Natl Acad Sci USA*. 88:8840 (1991).

# A MOLECULAR CHARACTERIZATION OF THE *RETINAL DEGENERATION SLOW* (*rds*) MOUSE MUTATION

Gabriel H. Travis[1] and Dean Bok[2]

[1]Department of Psychiatry, UT Southwestern Medical Center, Dallas
[2]Jules Stein Eye Institute, UCLA School of Medicine, Los Angeles

## INTRODUCTION

*Retinal degeneration slow* (*rds*) is a neurological mutation of mice that is characterized phenotypically by abnormal development of rod and cone photoreceptors, followed by their slow degeneration. No other cell types in the retina or the CNS are affected. In *rds/rds* homozygotes, the retina undergoes entirely normal development and differentiation of cells until the first postnatal week, the time at which photoreceptor outer segments normally appear. While other retinal cells continue their normal development, the photoreceptors fail to elaborate outer segments and rarely form outer segment discs (1). The photoreceptor inner segments, however, including the ciliary processes, are morphologically normal. The synaptic termini of photoreceptors with second-order retinal neurons also appear normal in these mutants.

The process of photoreceptor degeneration in *rds/rds* mice is first detectable histologically during the third postnatal week. The rate of loss of photoreceptors is greatest during the following few weeks and then becomes more gradual. Virtually all are gone by one year. Rods and cones are affected equally. Although the *rds* mutation was originally described as autosomal recessive, *rds/+* heterozygotes are phenotypically somewhat abnormal. In contrast to homozygotes, heterozygotes do form outer segments. However, these are reduced in length and contain oversized discs distorted into whorl structures. Very slow degeneration of photoreceptors is seen in *rds/+* heterozygotes (2).

## CLONING THE *rds* mRNA AND GENE

Prior to the work described below, nothing was known about the underlying biochemical defect in *rds*. The strategy that we used to clone the *rds* gene was based upon two observations: (*i*) patchy dysplasia and degeneration of photoreceptors were seen in the retinas of *rds/rds* ⟷ +/+ and *rds/+* ⟷ +/+ tetraparental mice (3,4); and (*ii*) photoreceptors are the only cell types affected in *rds* mutant retina. These observations suggested to us that in normal mice, the *rds* gene is expressed exclusively in photoreceptors in the retina. We employed subtractive cDNA cloning (5) to isolate a collection of photoreceptor-specific mRNAs (6). To this end, we took advantage of the unrelated mouse mutant, *retinal degeneration* (*rd*). Mice homozygous for this mutation have virtually no photoreceptors by the fourth postnatal week, while all other retinal cells are spared (7). We operationally defined an mRNA as photoreceptor-specific if it was present in wild-type retina, but absent from fully degenerate *rd* retina. A total of 12 distinct mRNAs were represented in a group of nearly 200 photoreceptor-specific cDNAs derived from the subtractive screen. Next, the chromosome-assignment for the genes encoding each of these photoreceptor-specific mRNAs was made using a panel of mouse x hamster hybrid cell line DNAs. Clone B9A mapped to mouse chromosome 17 with 100% concordance, and thus became a candidate for the *rds* mRNA.

Northern blot analysis was performed on retinal RNA from two month (pre-degenerate) *rds/rds* and wild-type mice, using clone B9A as a probe. In the wild-type lane, two bands at 1.6 and 2.7 kb were detected. In the *rds/rds* lane, these species were not detected, but a doublet band at 12-14 kb was visualized (6). These results suggested that B9A was a clone of *rds*, and that a significant disruption of the gene had occurred.

To define the mutation at the level of the gene, genomic libraries were prepared in phage lambda from *rds/rds* and wild-type mice, and clones of both loci were isolated. Fine structure analysis of these clones revealed that the mutation was due to the transposition of a 10-12 kb mouse repetitive element into a protein-coding exon in the *rds* gene (6). Due to this insertion, the putative protein encoded by the mutant *rds* gene is truncated at $Asn_{230}$, with loss of the carboxy terminal third of the molecule. This suggested two possible mechanisms for a phenotype in the heterozygote: (*i*) haploinsufficiency, where the 50% reduction in wild-type gene dosage results in the inadequate production of normal protein; or (*ii*) a direct dominant effect of the putative abnormal gene-product, such as the incorporation of a non-functional protein into an oligomeric complex.

## TRANSGENIC RESCUE OF *rds*: Generation of Transgenic Mice

To confirm that we have cloned the *rds* gene, and as a first step towards a genetic analysis of the mutation, we attempted an *in vivo* rescue of the phenotype (8). To this end we assembled a transgenic construct containing regulatory sequences upstream of the mouse opsin gene, transcriptionally fused to a wild-type *rds* minigene (Figure 1). A fragment of SV40 containing the early splice and polyadenylation sites was included downstream as a transcription terminator. DNA containing the construct was injected into the pronuclei of fertilized eggs from *rds* mutant mice. Eggs surviving micromanipulation were transferred to the oviducts of pseudopregnant foster mothers. A total of six transgenic lines were established. Three were chosen for study based upon their breeding productivity. We confirmed the *rds/rds* genotype of each $F_2$ mouse by checking for a restriction fragment length polymorphism on Southern blots that results from the insertional mutation within the *rds* gene (6). All mice were maintained on 12 hour light/dark cycles at low levels of illumination to minimize photic injury.

### Expression of the transgene

To quantify expression of the transgene in the three lines, RNA was prepared from the retinas of individual transgenic and non-transgenic $F_2$ mice for Northern blot analysis, using SV40 DNA and the mouse *rds* cDNA as probes. The SV40 probe hybridized to a 2.6 kb band in the lanes containing retinal RNA from the transgenic

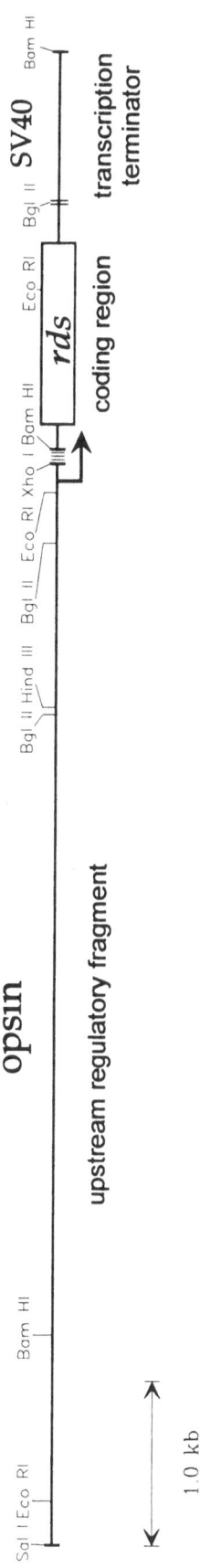

**Figure 1.** Physical map of the opsin-*rds*-SV40 fusion transgene. A segment of the mouse opsin gene from nucleotide -6,500 to +80 is followed by a central fragment of cDNA clone B9A (6) containing the complete wild-type *rds* coding block, followed by the SV40 t-intron and polyadenylation signal. The opsin transcription initiation site (25) is indicated by the arrow, and the *rds* coding block by the open box. Restriction sites are indicated by the labelled vertical lines.

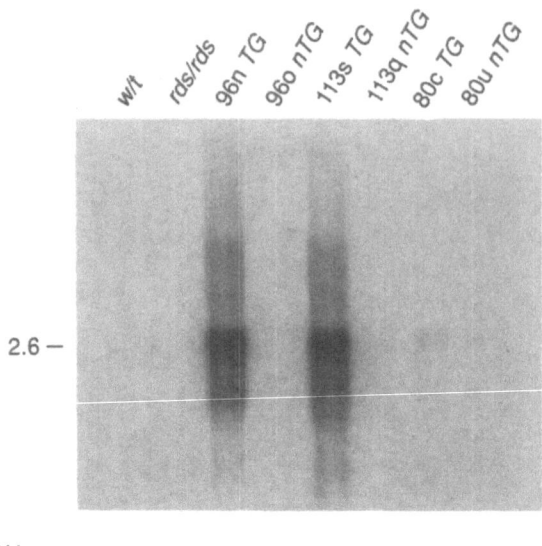

G3PDH

**Figure 2.** Northern blot analysis of retinal RNA showing expression of the transgene. RNA from line 96, 113 and 80 transgenic (*TG*) and non-transgenic (*nTG*) mice probed with the transcription-terminator fragment of SV40. RNA from non-transgenic C57BL/6 wild-type (*w/t*) and *rds/rds* homozygous mutant control retina are included as controls. Sizes of the major RNA bands in kb are indicated to the left. The insets labelled G3PDH shows the autoradiographic signal after the blots were stripped and re-probed with a cDNA of glyceraldehyde-3-phosphate dehydrogenase as a control for equal loading of lanes.

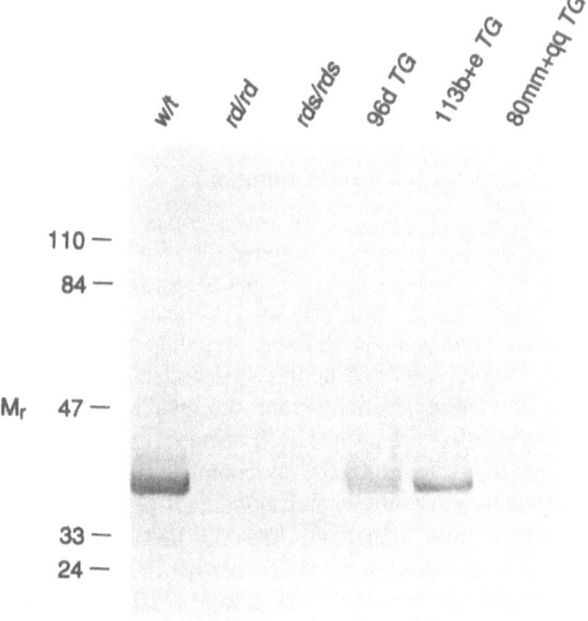

**Figure 3.** Protein expression in *rds* transgenics. Immunoblot analysis of retinal proteins showing expression of the transgene. Retinal extracts from line 96, 113 and 80 transgenics, each on an *rds/rds* genetic background, were reacted with an anti-β-gal-*rds* fusion protein serum. Retinal extracts from non-transgenic C57BL/6 wild-type (*w/t*), fully degenerate *rd* (*rd/rd*), and pre-degenerate *rds* (*rds/rds*) mice are included as controls. Positions of molecular weight size standards are shown on the left in kDA.

mice, but not the non-transgenic control mice (Figure 2). This was the predicted size for the transcript from the opsin-*rds*-SV40 transgene. The 2.6 kb transcript was abundant in RNA from lines 96 and 113, but rarer in RNA from line 80. By probing with *rds* clone B9A, we showed that the abundance of the 2.6 kb mRNA product of the transgene in lines 96 and 113 was similar to that of the 2.7 kb endogenous *rds* transcript in wild-type mice. In line 80, the 2.6 kb transcript was approximately 5% as abundant as the endogenous 2.7 kb transcript in wild-type mice.

The level of expression of the *rds* protein in the three transgenic lines was estimated by immunoblotting, using polyvalent antisera raised against a β-galactosidase-*rds* fusion protein. These antisera reacted with a protein species of approximately 39 kDA in retinal extracts from wild-type mice, but not from pre-degenerate *rds/rds* mutant mice. An immunoreactive band of 39 kDA was also present in lanes containing retinal extracts from the line 96 and 113 transgenic/*rds* mutants, but not from the non-transgenic littermate controls (Figure 3). This band was shifted downward by approximately 2 kDA after endoglycosidase F digestion, similar to the shift seen with the wild-type protein. Thus, an *rds*-immunoreactive glycoprotein of normal size is produced in the retinas of high-expressing transgenic lines 113 and 96. There was no detectable immunoreactivity in the lanes containing extracts from transgenic line 80. This was probably due to the low level of transgene expression in line 80.

**Phenotype in transgenic *rds/rds* homozygous mutants**

Retinas from line 96, 113 and 80 transgenic mice on an *rds/rds* mutant background at two and five months of age were studied by electron microscopy. Non-transgenic littermates were used as controls. All features of the *rds* phenotype for animals at these ages were present in the controls, including absence of outer segments, partial degeneration of photoreceptor cell bodies, and vesicular debris in the subretinal space (Figure 4A). In contrast, retinas from transgenic lines 96 and 113 were indistinguishable from wild-type, with normal outer segment morphology and no evidence of photoreceptor degeneration (Figure 4B,C). A complete rescue of the *rds* phenotype was thus effected in rod cells from line 96 and 113 mice.

Interestingly, retina from transgenic line 80 had an appearance that was intermediate between those of wild-type mice and *rds/rds* mutants (Figure 4D). There were fewer extracellular vesicles, and whorls of disorganized discs were present in the subretinal space, likely representing dysplastic outer segments. This pattern is similar to that seen in *rds/+* heterozygotes (9), except that in animals from line 80, the outer segment dysplasia was more severe. This difference in severity may be due to a dosage effect; in line 80, the level of the mRNA encoding the normal *rds* protein is only 5-10% of wild-type, whereas in heterozygotes it is approximately 50%.

In normal mice, the *rds* gene is expressed in both rods and cones (10). If we assume that the *rds*-containing transgene, transcriptionally regulated by the opsin promoter, has the same pattern of expression as the endogenous opsin gene, then we would expect the *rds* phenotype to have been rescued in rod but not cone photoreceptors. We are currently unable to confirm rod-specific correction however, since only 3% of photoreceptors in mice are cones (7), and since the oldest transgenics that have been studied are five months, and hence should have only lost about half of their cones.

In two previous reports, where smaller upstream genomic segments of opsin were used to read a *lacZ* reporter gene, non-uniform patterns of expression of the *lacZ* product were observed in the retinas from several transgenic lines (11,12). In the present study, rescue of the *rds* phenotype in photoreceptors was uniform across the retina in all three lines, with no patchy or gradient changes in the morphology of outer segments. Here, more uniform expression of the transgene may be due to our use of a larger opsin promoter segment.

Complete rescue of the phenotype in *rds/rds* mutant mice by transgenic complementation constitutes formal proof that we have cloned the *rds* gene. Since rescue was effected despite the presence of the mutant transcription products, it is unlikely that the putative abnormal *rds* protein is having any dominant effect. Preliminary *in situ* hybridization analysis with cRNA probes derived from *rds* clone B9A showed a predominantly nuclear hybridization signal in retina from *rds/rds* mutant mice (Travis and Bok, unpublished observations). In wild-type mice, the signal was predominantly over the inner segments, and in rescued transgenic *rds/rds* mice, there was a signal over both the inner segments and the photoreceptor nuclei. These results suggest that the 12-14 kb transcript of the mutant gene is not transported out of the nuclei, and hence cannot be translated into protein. This is consistent with our observation of no signal on Western blots in retinal extracts from mutant mice with antibodies against the *rds* protein (Figure 3). From these observations, it appears likely that the phenotype in *rds/+* heterozygotes is due to haploinsufficiency, and that *rds* is a null allele in mice.

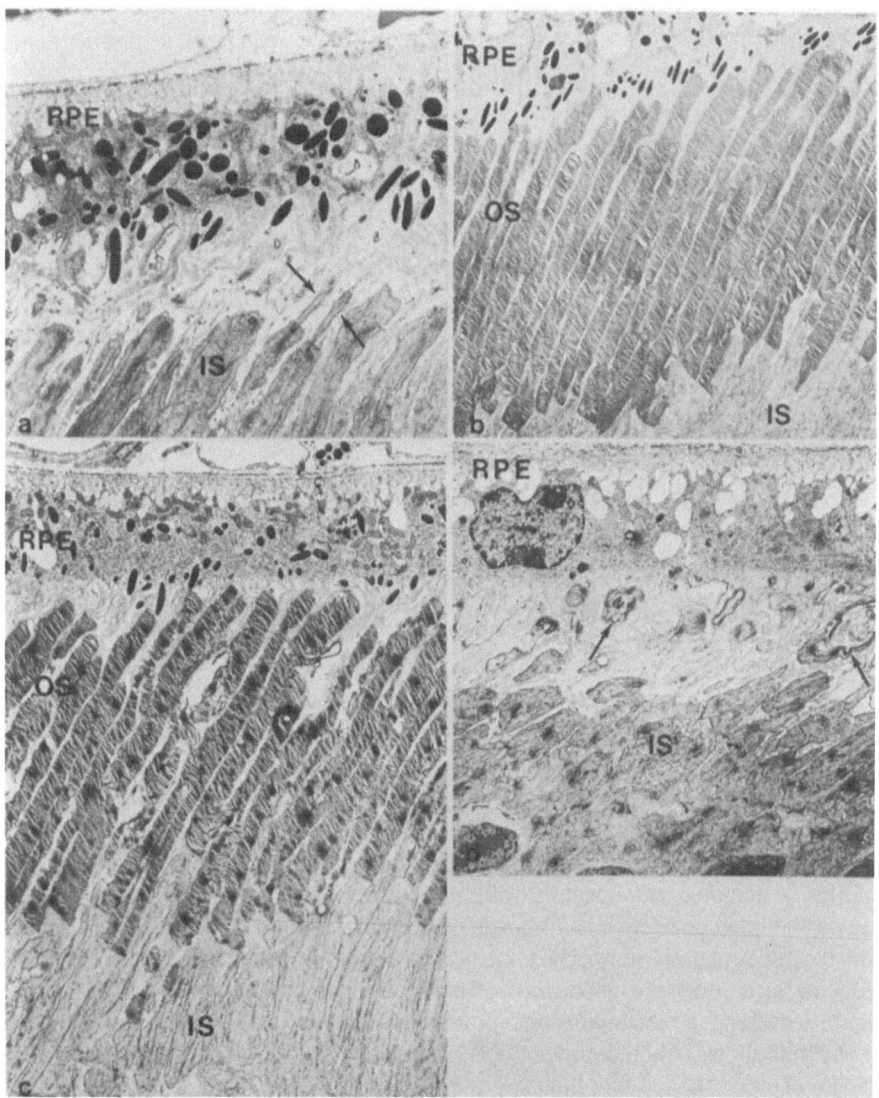

**Figure 4.** Electron microscopic analysis of retinas from transgenic lines 96, 113 and 80 on an *rds/rds* mutant background. In all cases, the retinal pigment epithelium (**RPE**) is at the top of the figure. (a) Two month-old, non-transgenic control mouse 96j. Rod inner segments (**IS**) and connecting cilia (arrows) are present, but outer segments are absent. x 3,300. (b) Two month-old transgenic mouse 96q (littermate of 96j) superior nasal retinal quadrant. Fully developed outer segments (**OS**) containing regularly arranged disc structures are visible between the retinal pigment epithelium (**RPE**) and inner segments (**IS**). x 2,900. (c) Five month-old transgenic mouse 113h superior temporal quadrant. As with 96q, the outer segments (**OS**) appear morphologically normal and there is no evidence of photoreceptor degeneration. x 3,800. (d) Two month-old, low-expressing transgenic mouse 80q inferior temporal quadrant. Partial development of outer segment discs is evident from the membranous whorls (arrows) between the retinal pigment epithelium (**RPE**) and inner segments (**IS**). x 3,800.

## THE HUMAN *rds* GENE (*RDS*)

The histopathological changes of autosomal dominant retinitis pigmentosa (ADRP) include photoreceptor degeneration with shortening and disorganization of the outer segments (13). This pattern is similar to the phenotype in *rds/*+ heterozygotes (9). To test the hypothesis that some patients with ADRP have mutations in the *rds* gene, we cloned the human homolog of *rds* (*RDS*) (14). This was found to encode a protein that was 91% identical to the mouse protein, containing the same number of residues and with the same predicted secondary and tertiary structural features (see below). We mapped the *RDS* gene to proximal human chromosome 6p (14). The sequence of the *RDS* mRNA was used to define the intron/exon junctions in the *RDS* gene. Oligonucleotide primers were designed to amplify *RDS* exons in human DNA. In two studies on patients with ADRP, point mutations and small deletions were detected in the *RDS* gene, resulting in non-conservative single amino acid substitutions and deletions (15,16) in the D2 loop and the third membrane-spanning segment. Thus, mutations in *RDS* are responsible for a subset of ADRP in humans.

## THE *rds* PROTEIN

From the mRNA sequence, the predicted mouse *rds* protein contains 346 residues. Initial protein database comparisons revealed no similarity to other known proteins. Computerized secondary structure analysis showed four uncharged $\alpha$-helical regions of 23-26 residues, suggesting four membrane-spanning segments. In addition, there were two potential sites at residues 53 and 229 for N-linked glycosylation.

To begin a biochemical characterization of the *rds* protein, we raised antibodies against a synthetic peptide corresponding to residues 311-323, near the carboxy terminus (peptide J97). We also raised antibodies against a $\beta$-galactosidase-*rds* fusion protein, using the prokaryotic expression vector, pUR292 (17). These were used for immunoblot analysis of retinal protein extracts. In wild-type extracts, these reagents reacted with a broad band of approximately 39 kDA (18). No reactivity was observed in fully degenerate *rd/rd* retinal extracts, consistent with the original observation that the *rds* mRNA is photoreceptor-specific. Endoglycosidase F digestion of the retinal extract caused the immunoreactive band to shift downward by approximately 2 kDA, suggesting that the *rds* protein is normally glycosylated. When extracts were prepared in a non-reducing medium, the immunoreactive band migrated at approximately 78 kDA. This result suggests that the native *rds* protein is covalently coupled to another protein, possibly itself as a homo-dimer. Extraction of proteins with Triton X-114 prior to immunoblotting showed that the *rds* protein was confined to the membrane-rich fraction. This confirmed that the *rds* protein was associated with outer segment membranes. No immunoreactivity was observed in retinal extracts from *rds/rds* mice with the antipeptide or the anti-fusion protein sera.

The distribution of the *rds* protein in photoreceptors was determined by immunocytochemistry using antisera against the J97 peptide and the $\beta$-galactosidase-*rds* fusion protein. Under the light microscope, immunoreactivity was confined to the outer segments, precisely the structure absent in *rds/rds* mutants. Under the electron microscope, we observed two different patterns of immunoreactivity in the outer segments. When we used antisera raised against the J97 peptide, we saw relatively uniform labelling across the width of the discs (18). However, when we used antisera against a $\beta$-galactosidase-*rds* fusion protein, we saw labelling restricted to the disc rims and the disc incisures. Robert Molday and coworkers have independently characterized the bovine homolog of the *rds* protein, which they named "peripherin" (19). With two monoclonal antibody reagents recognizing epitopes near the carboxy terminus, they also observed a disc rim distribution. Since there is now concordance with several reagents for localization to the disc rim, we ascertain that this represents the *rds* protein's true distribution. However, we are left without an explanation for the uniform disc labelling by the J97 peptide. An alternative explanation is that the epitopes detected by the anti-fusion protein sera and both of Robert Molday's monoclonal reagents are masked in the disk interior, while the epitopes detected by the J97 peptide are available throughout the discs.

**Figure 5.** Alignment of the rom-1 (20) and human *rds* (*HRDS*) (14) protein sequences predicted from the cloned mRNAs. Identical residues are indicated with double dots. The four membrane spanning α-helical segments for both proteins are indicated by the solid bars and the designation M1-M4. The two intradiscal loops are designated D1 and D2. The three cytoplasmic segments, including the amino and carboxy termini, are designated C1-C3.

Recently, Rod McInnes and co-workers have published the sequence of a rod-specific outer segment protein from humans, which they named rom-1 (20). This protein has important similarities to the *rds* protein (Figure 5). They are of comparable lengths (346 vs 351 residues), with an overall identity of 35%. They share similar predicted tertiary structures, each with four putative membrane spanning α-helical regions. Within the intradiscal D2 loop (between the third and fourth membrane spanning domains), there is a stretch of 29 residues with 79% identity. Interestingly, rom-1 was also found to be distributed in the periphery of outer segment discs (20). McInnes and co-workers showed by immunoaffinity chromatography that rom-1 and the *rds* protein are associated non-covalently in detergent extracts of outer segments (20).

Before making predictions about the possible function of the *rds* protein, it is useful to consider the *rds/rds* phenotype in more detail. In these mutants, absence of the normal *rds* protein results in failed development of the photoreceptor outer segments. Vesicular structures containing opsin have been observed budding from the ciliary projections of otherwise normal inner segments of *rds/rds* mice (21,22,23), suggesting that outer-segment membrane biogenesis is occurring in the mutants but that the membrane is unable to fold into disc structures. Since disc membranes in normal rods and cones contain tight, thermodynamically unfavorable folds at the disk periphery, a stabilizing interaction is required to hold the like-charged, extracellular disc faces in apposition. In the model proposed by Steinberg *et al.* (24) the outer segment discs are formed by evagination of the plasma membrane at the base of the outer segment, followed by the propagation or *zippering* of the disc rim to separate the disc from the plasma membrane.

It is our hypothesis that the *rds* protein, possibly in conjunction with rom-1, provides this zipper function. In the protein's absence in *rds/rds* mutant mice, no disc folding occurs, and hence the subretinal space is filled with vesicular debris but no outer segments. In *rds/+* heterozygotes, or in the partially rescued line 80 transgenic mice, the level of the *rds* protein is decreased. This may cause a reduction in the ratio of disc rim to disc surface area, resulting in the observed outer segment whorl structures. We have proposed that the *rds* protein is an adhesion molecule, participating in homophilic or heterophilic interactions across the disc space (18). We have suggested that this interaction involves conserved residues in the D2 loop, possibly including the N-linked glycan. The D2 loop is implicated as a critical functional segment of the *rds* protein by several observations: (*i*) in a phylogenetic comparison, the D2 segment is the most conserved segment of the molecule (14); (*ii*) in a comparison of the *rds* protein to the related rom-1 protein, the D2 loop contains a region of strikingly high similarity (Figure 5); (*iii*) the D2 segment contains the preponderance of mutations in humans with *RDS*-associated ADRP; and (*iv*) the D2 loop has the appropriate geometry for an interaction across the intradiscal space in the disc periphery.

An alternative hypothesis is that the *rds* protein stabilizes outer segment discs through interactions between the cytoplasmic face of the disc and the outer segment plasma membrane. This would require involvement of a cytoplasmic segment of the *rds* protein, most likely the long carboxy terminus. One difficulty with this is that the carboxy terminus is the least conserved segment, both in a phylogenetic comparison of the *rds* protein to itself and in a comparison of the *rds* protein to rom-1, and hence seems less likely to be involved in a critical interaction with another protein. Also, it is less easy to understand how the demonstrated *rds* protein/rom-1 non-covalent association (20) could result in disc stabilization if their interacting segments do not span the intradiscal space.          The degenerative phase of the *rds* phenotype is probably less correlated to the specific function of the *rds* protein than the dysplastic phase. Given the high degree of non-allelic heterogeneity in the inherited retinal degenerations of mammals, it is clear that there are many routes to photoreceptor death. In *rds* mutants, the photoreceptors may degenerate as a response to internal metabolic disruptions resulting from their failure to form stable outer segment structures. Alternatively, given the large gradients in $PO_2$ across the outer segments in normal retina, the absence of outer segments in *rds/rds* retina may result in the photoreceptor cell bodies being exposed to unusually high oxygen tensions. The mutant photoreceptors may thus be degenerating from prolonged $O_2$-toxicity. Finally, photoreceptor degeneration may result from loss of outer segment-dependent trophic influences provided by other retinal cells, such as those of the pigment epithelium, Müller cells or second-order neurons.

## SUMMARY

We have cloned the gene responsible for the photoreceptor dysplasia and degeneration of *rds* mutant mice, and have identified the genetic defect as an insertional mutation. We have completely rescued the *rds* phenotype by transgenic complementation. These results confirm our identification of the gene, and suggests that *rds* is a null allele in mice. It has been shown by us and others that mutations in the human homolog of the *rds* gene (*RDS*) are responsible for autosomal dominant retinitis pigmentosa in a subset of pedigrees. The wild-type *rds* gene encodes a novel, outer segment disc membrane-associated, glycoprotein. Based upon the phenotype, this protein is required for the normal folding or the stabilization of outer segment discs. We propose that the *rds* protein is functioning as an adhesion molecule, which acts through homophilic or heterophilic interactions across the intradiscal space in the disc periphery.

## ACKNOWLEDGEMENTS

We gratefully acknowledge Karen Groshan, Marcia Lloyd and Orna Yaron for their excellent technical assistance, and Jenny Price for her skillful manipulation of mouse ooctyes. This work was supported by grants from the National Eye Institute and the National Retinitis Pigmentosa Foundation. G.H.T. is a John Merck Fund Scholar. D.B. is the Dolly Green Professor of Ophthalmology.

## REFERENCES

1. Sanyal S., Chader G. and Aguirre G. Expression of retinal degeneration slow (*rds*) gene in retinal of the mouse. In "Retinal Degeneration: Experimental and Clinical Studies," (New York: Alan R. Liss, Inc.) pp 239-256 (1985).
2. Hawkins R.K., Jansen H.G. and Sanyal S. Development and degeneration of retina in *rds* mutant mice: photoreceptor abnormalities in the heterozygotes. *Exp. Eye Res.* **41**: 701-720 (1985).
3. Sanyal, S. and Zeilmaker, G.H. Development and degeneration of retina in *rds* mutant mice: light and electron microscopic observations in experimental chimeras. *Exp. Eye Res.* **39**: 231-246 (1984).
4. Sanyal, S., Dees, C. and Zeilmaker, G.H. Development and degeneration of retina in *rds* mutant mice: observations in chimeras of heterozygous mutant and normal genotype. *J. Embryol. exp. Morph.* **98**: 111-121 (1986).
5. Travis G.H. and Sutcliffe J.G. Phenol emulsion-enhanced DNA-driven subtractive cDNA cloning: isolation of low-abundance monkey cortex-specific mRNAs. *Proc. Natl. Acad. Sci. USA* **85**: 1696-1700 (1988).
6. Travis G.H., Brennan M.B., Danielson P.E., Kozak C.A. and Sutcliffe J.G. Identification of a photoreceptor-specific mRNA encoded by the gene responsible for retinal degeneration slow (*rds*). *Nature* **338**: 70-73 (1989).
7. Carter-Dawson, L.D., LaVail, M.M. and Sidman, R.L. (1978) Differential effect of the *rd* mutation on rods and cones in the mouse retina. *Invest. Ophthalmol. Visual Sci.* **17**: 489-498 (1978).
8. Travis G.H., Groshan K.R., Lloyd M. and Bok D. (1992) Complete rescue of photoreceptor dysplasia and degeneration in transgenic *retinal degeneration slow* (*rds*) mice. *Neuron* **9**, 113-119.

9. Hawkins R.K., Jansen H.G. and Sanyal S. Development and degeneration of retina in *rds* mutant mice: photoreceptor abnormalities in the heterozygotes. *Exp. Eye Res.* **41**: 701-720 (1985).

10. Arikawa, K., Molday, L.L., Molday, R.S., Williams, D.S. Localization of peripherin/*rds* in the disk membranes of cone and rod photoreceptors: relationship to disk membrane morphogenesis and retinal degeneration. *J. Cell Biol.* **116**: 659-667 (1992).

11. Lem J., Applebury M.L., Falk J.D., Flannery J.G. and Simon M.I. Tissue-specific and developmental regulation of rod opsin chimeric genes in transgenic mice. *Neuron* **6**: 201-210 (1991).

12. Zack D.J., Bennett J., Wang Y., Davenport C., Klaunberg B., Gearhart J. and Nathans J. Unusual topography of bovine rhodopsin promoter-*lacZ* fusion gene expression in transgenic mouse retinas. *Neuron* **6**: 187-199 (1991).

13. Flannery J.G., Farber D.B., Bird A.C. and Bok D. Degenerative changes in a retina affected with autosomal dominant retinitis pigmentosa. *Invest Ophthalmol Vis Sci* **30**: 191-211 (1989).

14. Travis G.H., Christerson L., Danielson P.E., Klisak I., Sparkes R.S., Hahn L.B., Dryja T.P., Sutcliffe J.G. The human *retinal degeneration slow* (*rds*) gene: Chromosome assignment and structure of the mRNA. *Genomics* **10**: 733-739 (1991).

15. Farrar, G.J., Kenna, P., Jordan, S.A., Kurar-Singh, R., Humphries, M.M., Sharp, E.M., Sheils, D.M., Humphries, P. A three-base-pair deletion in the peripherin-*RDS* gene in one form of retinitis pigmentosa. *Nature* **354**: 478-480 (1991).

16. Kajiwara, K., Mukai, S., Travis, G.H., Berson, E.L. and Dryja, T.P. Mutations in the human retinal degeneration slow gene (*RDS*) in autosomal dominant retinitis pigmentosa. *Nature* **354**: 480-483 (1991).

17. Rüther, U. and Müller-Hill, B. Easy identification of cDNA clones. *The EMBO Journal* **2**: 1791-1794 (1983).

18. Travis G.H., Sutcliffe J.G. and Bok D. The retinal degeneration slow (*rds*) gene product is a photoreceptor disc membrane-associated glycoprotein. *Neuron* **6**: 61-70 (1991).

19. Connell, G., Bascom, R., Molday, L., Reid, D., McInnes, R.R., and Molday, R.S. Photoreceptor peripherin is the normal product of the gene responsible for retinal degeneration in the *rds* mouse. *Proc. Natl. Acad. Sci. USA* **88**: 723-726 (1991).

20. Bascom, R.A., Manara, S. Collins, L., Molday, R.S., Kalnins V.I. and McInnes, R.R. Cloning of the cDNA for a novel photoreceptor membrane protein (rom-1) identifies a disk rim protein family implicated in human retinopathies. *Neuron* **8**: 1171-1184 (1992).

21. Nir, I. and Papermaster, D.S. Immunocytochemical localization of opsin in the inner segment and ciliary plasma membrane of photoreceptors in retinas of rds mutant mice. *Invest. Ophthalmol. Vis. Sci.* **27**: 836-840 (1986).

22. Usukura, J. and Bok, D. Changes in the localization and content of opsin during retinal development in the *rds* mutant mouse: immunocytochemistry and immunoassay. *Exp. Eye Res.* **45**: 501-515 (1987).

23. Jansen, H.G., Sanyal, S., DeGrip W.J. and Schalken, J.J. Development and degeneration of retina in rds mutant mice: ultraimmunohistochemical localization of opsin. *Exp. Eye Res.* **44**: 347-361 (1987).

24. Steinberg, R.H., Fisher, S.K. and Anderson, D.H. Disc morphogenesis in vertebrate photoreceptors. *J. Comp. Neurol.* **190**: 501-518 (1980).

25. Baehr, W., Falk, J.D., Bugra, K., Triantafyllos, J.T. and McGinnis, J.F. Isolation and analysis of the mouse opsin gene. *FEBS Letters* **238**, 253-256 (1988).

# TRANSGENIC MOUSE STUDIES OF RETINAL DEGENERATION: EXPRESSION OF THE β-SUBUNIT OF cGMP PHOSPHODIESTERASE AND TRANSDUCIN α-SUBUNITS

Janis Lem[1], John G. Flannery[2], Debora B. Farber[3], Meredithe L. Applebury[4], Carol Raport[5], James B. Hurley[5], and Melvin I. Simon[1]

[1]Division of Biology, California Institute of Technology, Pasadena, CA 91125
[2]Department of Ophthalmology, University of Florida College of Medicine, Gainesville, FL 32610
[3]Jules Stein Eye Institute, UCLA School of Medicine, Los Angeles, CA 90024
[4]Visual Sciences Center, University of Chicago, Chicago, IL 60637
[5]Department of Biochemistry and Howard Hughes Medical Institute, University of Washington, Seattle, WA 98195

## INTRODUCTION

A decade ago, little was known about the mechanisms leading to retinal degenerative disease. Since then, rapid advances in molecular biological techniques have unravelled the molecular basis of a number of retinal degenerations. The initial findings of Humphries and collaborators showing linkage to human chromosome 3 of an autosomal dominant form of retinitis pigmentosa (RP) in a human pedigree[1,2] led to the equally important finding by Dryja and collaborators of the Pro23His mutation in rhodopsin[3]. Since that time, numerous other mutations in the rhodopsin gene have been identified and linked to familial autosomal dominant[4-6] and recessive[7] forms of RP.

Naturally occuring animal models of retinal degeneration have also been important in characterizing the biochemical and morphological changes in the progressive disease process[8-11]. The *rd* [9](retinal degeneration) and the *rds* [10] (retinal degeneration slow) mouse models have been particularly useful for defining the molecular basis of retinal degenerations in the mouse. Using a subtractive hybridization strategy, the candidate genes for these degenerations have been identified, respectively, as the β-subunit of cGMP phosphodiesterase (PDE) and peripherin, encoding a disc membrane structural protein of rod photoreceptors[12-14]. The cloned cDNAs have been used to isolate the human homologues of these genes[15,16]. While mutations in the human β-PDE gene have yet to be linked to autosomal recessive retinal degenerations[17], examination of pedigrees using the human homologue of the peripherin gene as a probe has linked a number of mutations with a specific sub-class of autosomal dominant RP[18,19].

In spite of these important advances, the identified genes and their respective mutations account for less than 25% of inherited retinal degenerations in humans. Clearly, other genes are involved. It is likely that mutations in other genes encoding components of the phototransduction cascade will cause retinal degenerations. Likely candidates include genes encoding specific proteins of the rod and cone visual transduction cycle. A number of these genes have been cloned recently, including the rod and cone opsins[20-22], transducin subunits[23-28], cGMP phosphodiesterase subunits[15,29-31], arrestin[32], rhodopsin kinase[33], recoverin[34], phosducin[35] and the cGMP-gated channel[36]. Our ability to make defined mutations using molecular biological techniques combined with transgenic animal technology provides us with a powerful tool for understanding the role of these cloned genes in normal visual function and in retinal degenerations.

Thus, we have undertaken the study of the molecular basis of retinal degenerative disease by creating transgenic mice with specifically altered retina-specific genes. As a first priority, the study of retina-specific genes required the ability to appropriately direct their expression in the retina. The mouse rod opsin gene was selected because of its high levels of expression in rod photoreceptor cells. In the mouse retina, rod cells comprise approximately 95% of photoreceptors, with cones comprising the remaining 5%[37]. In addition, the rod opsin protein makes up approximately 90% of the disc membrane protein[38]. Described below are our studies characterizing mouse rod opsin 5' regulatory sequences[39]. We have also used this upstream region to drive the expression of two other retina-specific genes, the β-subunit of cGMP PDE[40] and the α-subunits of rod and cone transducin[41].

## EXPERIMENTAL RESULTS

**Characterization of the Mouse Rod Opsin 5' Regulatory Sequences.** To determine the specificity of expression of mouse rod opsin upstream 5' regulatory sequences, gene fusions between rod opsin upstream sequences and the diphtheria toxin A subunit (DTA) or β-galactosidase (Lac Z) structural genes were introduced into transgenic mice (figure 1A-C). The DTA subunit encodes an ADP ribosyltransferase which inhibits protein synthesis and leads to cell death. A single molecule of diphtheria toxin is sufficient to kill a cell[42], thus providing a sensitive assay for the analysis of tissue-specific gene expression. As the DTA subunit cannot be taken up by neighboring cells, its effects are restricted to expressing cells only. The Lac Z reporter gene encodes the bacterial *E. coli* β-galactosidase enzyme, whose activity is readily detectable using the chromogenic substrate X-Gal (5-bromo-4-chloro-3-indolyly-β-D-galactopyranoside). Expression of the Lac Z gene product in tissue appears as an insoluble blue precipitate visible by light microscopy, thus clearly localizing the site of expression.

Two promoter lengths were examined in these studies: a 4.4 kb and a 500 bp fragment upstream of the ATG start site. Transgenic animals were examined for appropriate cell-specific, temporal and spatial patterns of expression. The DTA transgene was tested only with the 4.4 kb opsin promoter. Six founder lines carrying the DTA transgene were identified. Five displayed no phenotypic evidence of diphtheria toxin expression. However, the sixth transgenic line, designated RDT9-1, exhibited a retinal degenerative phenotype similar to that reported in *rd* mice. Only photoreceptor cell layers were ablated in transgenic mice (figure 2B,C) as compared to a nontransgenic control sibling (figure 2A),

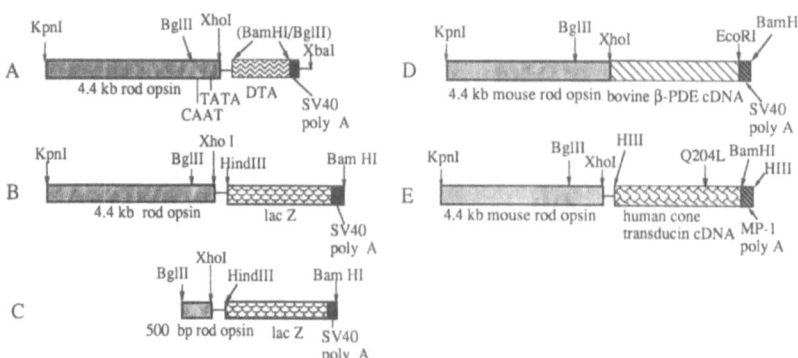

Figure 1. Mouse rod opsin fusion transgene constructions using 4.4 kb or 500 bp of 5' upstream regulatory sequences. A. 5.2 kb diphtheria toxin fusion gene; B. 8.2 kb Lac Z fusion gene; C. 4.3 kb lac Z fusion gene; D. 7.4 kb bovine β-PDE cDNA fusion gene; E. 6.2 kb human cone transducin cDNA fusion gene.

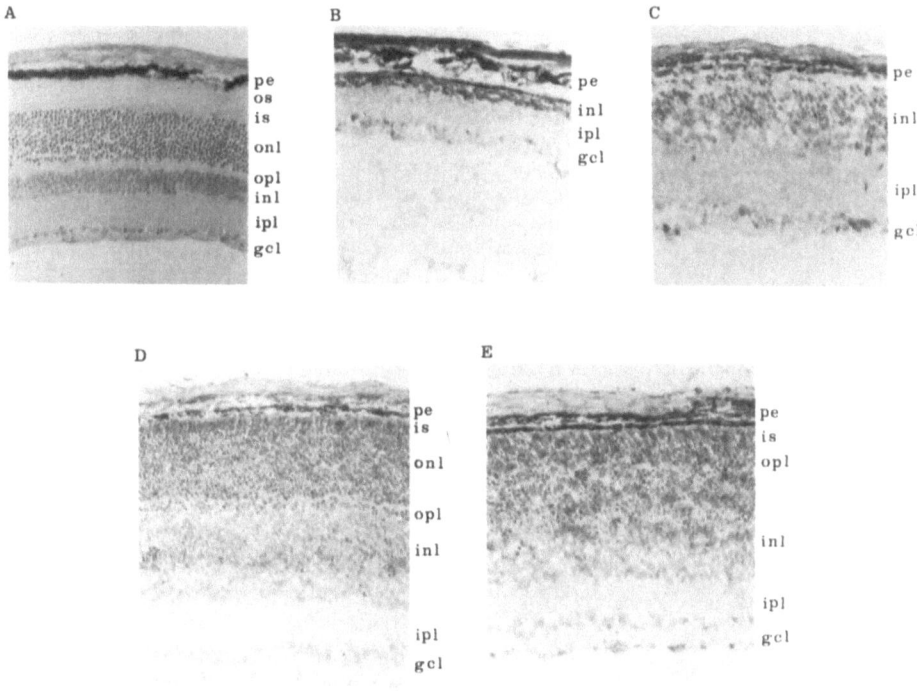

Figure 2. Frozen histological sections from diphtheria toxin transgenic mouse retinas. A. Normal control adult retina; B. Adult transgenic retina; C. Postnatal day 18 transgenic retina showing loss of majority of photoreceptor outer segments and nuclei; D. Postnatal day 7 control retina from non-transgenic mouse; E. Age-matched postnatal day 7 transgenic retina. Note the increased numbers of pyknotic cells; pe=pigment epithelium, os=outer segment, is=inner segment, onl=outer nuclear layer, opl=outer plexiform layer, inl=inner nuclear layer, ipl=inner plexiform layer, gcl=ganglion cell layer.

indicating the tight cell-specificity directed by the promoter. Time course studies first showed detectable expression of diphtheria toxin by postnatal day 7 (figure 2D, E). This is about the time that endogenous opsin protein is reported to be detected using immunohistochemical methods in wildtype mice[43,44]. Thus, 4.4 kb of mouse rod opsin upstream sequences are capable of directing photoreceptor specific expression following the appropriate time course.

With both the 4.4 kb and 500 bp promoter lengths, β-galactosidase enzyme activity was observed specifically in photoreceptor cells of adult animals. Animals expressing the 4.4 kb rod opsin/lac Z transgene stained predominantly in photoreceptor inner segments and the outer plexiform layer in a rod-like distribution (figure 3A). The 500 bp promoter similarly directed photoreceptor cell-specific expression. However, expression was nonuniform, with only sporadic photoreceptor cells staining for β-galactosidase activity (figure 3B). The developmental time course of expression was examined in animals from 5 days to one year of age. While endogenous opsin protein has been reported to be detectable at postnatal days 8 or 9, in fusion genes with 4.4 kb of opsin promoter, expression of β-galactosidase was observed at postnatal day 11-13 (figure 3C). This is much later than the time course of diphtheria toxin transgene expression. The difference in the time of onset most likely reflects a difference in assay sensitivity. While one molecule of diphtheria toxin is sufficient to kill a cell, approximately 1,000 molecules of β-galactosidase are required per cell. In addition, β-galactosidase must tetramerize to take on an active conformation. In transgenes using the 4.4 kb promoter, spatial patterns of β-galactosidase expression initially appeared in the superior/temporal aspect of the posterior retina. As the animal developed, expression spread peripherally to span the entire retina (figure 3C-E). In the adult mouse, spatial expression was uniform in one line of animals (figure 3G), but expressed as a characteristic, non-uniform gradient in another line (figure 3F). This gradient persisted to one year of age. The 500 bp promoter showed a very different spatial pattern of β-galactosidase expression, with expression strikingly restricted to a band across the retina (figure 3H). These studies showed that 500 bp of upstream flanking sequences were sufficient to confer photoreceptor-specific gene expression. However, regulatory sequences further upstream may be necessary for appropriate spatial expression of the transgene. The 4.4 kb upstream regulatory sequences were capable of directing appropriate uniform spatial expression. However, as will be described below, this regulatory sequence often produced the same characteristic spatial gradient in independently derived lines of transgenic mice driving expression of other retina-specific genes. This suggests the gradient might be a property of the upstream regulatory sequences rather than an effect of flanking sequences.

**Expression of the β-subunit of the cGMP PDE Gene.** The cGMP PDE β-subunit has been identified as a candidate gene for autosomal recessive retinal degeneration in the *rd* mouse[45,46]. To explore the role of the gene in retinal degeneration, the β-subunit of cGMP PDE was introduced into transgenic mice homozygous for the *rd* allele to test the prediction that expression of a functional β-PDE subunit would rescue photoreceptor

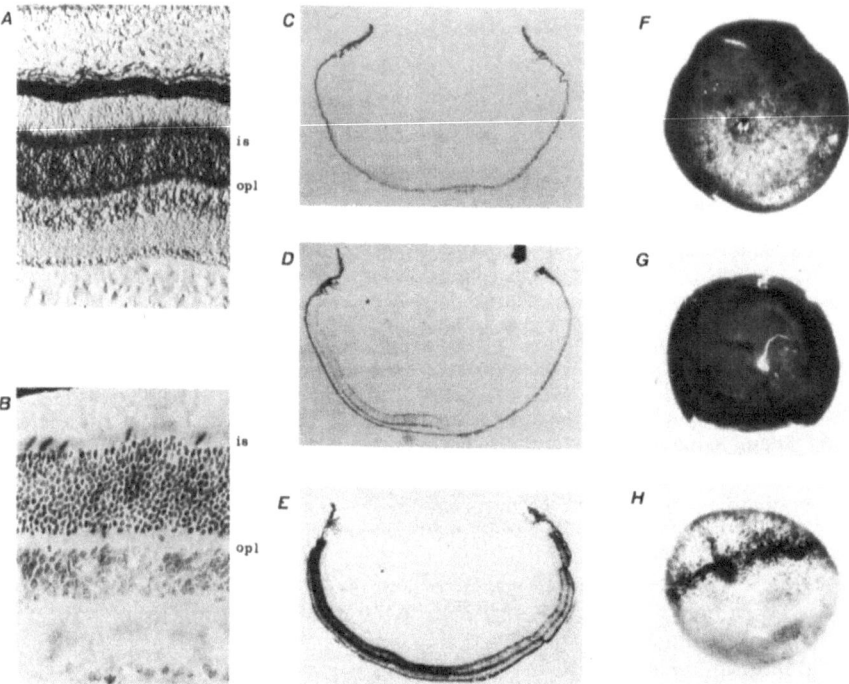

Figure 3. Histological sections and whole mounts from Lac Z expressing transgenic mouse retinas. Lac Z gene product appears as dark stain. A. Retinal section from transgenic mouse retina expressing lac Z gene from 4.4 kb of mouse rod opsin upstream sequences. Lac Z appears in the inner segment (is) and outer plexiform layer (opl); B. Histological section from transgenic mouse retina expressing lac Z gene from 500 bp of rod opsin regulatory sequences. Sporadic staining occurs in is and opl of some photoreceptor cells; C-E Developmental time course of lac Z gene expression in transgenic mice driven from 4.4 kb of rod opsin upsteam sequences; ages are P13, P15 and 6 months, respectively; expression first appears accentrically, spreading preripherally with age; nasal/temporal aspect is right to left, with sections cut along horizontal meridian; F. stained whole mount retina showing a gradient of expression driven from 4.4 kb of rod opsin sequences, with expression heaviest in the superior/temporal aspect of the retina; G. stained whole mount retina from transgenic line expressing the lac Z gene from 4.4 kb of rod opsin sequences and showing uniform expression across the entire retina; H. stained whole mount retina from from transgenic mouse line driving lac Z gene expression from 500 bp of rod opsin regulatory sequences. Nasal/temporal aspect is left to right and superior/inferior aspect from top to bottom.

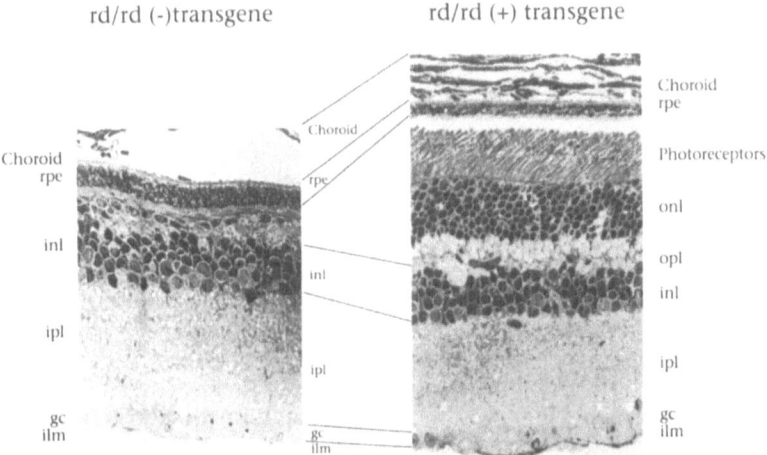

rd/rd (-)transgene        rd/rd (+) transgene

Figure 4. Histological sections from *rd/rd* control and β-PDE transgenic mouse retinas. Transgenic retina is morphologically normal in appearance, while photoreceptors have degenerated in control *rd/rd* retina. rpe= retinal pigment epithelium, gc=ganglion cells, ilm=inner limiting membrane.

degeneration (figure 1D). Four lines of mice were generated showing partial or complete rescue of photoreceptor degeneration. Two lines were studied in further detail. The RP4-28 line of transgenic animals was produced in a pigmented C57BL6/J *rdle* genetic background while the RP33 line was produced in an albino FVB/n genetic background. In control non-transgenic *rd* mice, degeneration was complete by 3 weeks of age, with the photoreceptor, outer nuclear and outer plexiform layers absent (figure 4). Albino RP33 transgenic mice showed rescue of photoreceptor cells across the entire retina. Photoreceptor cell morphology appeared normal by light (figure 4) and electron microscopy. The pigmented RP4-28 line of mice showed partial rescue of photoreceptor cells in a gradient of expression similar to that shown in the rod opsin/lac Z transgenic mice described above (figure 5a and b). Photoreceptors in the superior hemisphere of the retina were morphologically normal while those in the inferior hemisphere were degenerate. At the interface between the two hemispheres, intermediate degeneration was observed. In this region, photoreceptors with distorted inner segments were observed. In addition, there were a few photoreceptors which appeared essentially normal with the exception of vesiculation at the basal-most region of the outer segment. This abnormality at the site of new disc membrane synthesis is very puzzling in light of the fact that disc membranes synthesized earlier and lying further above in the outer segment appeared morphologically normal.

Biochemical assays indicated cGMP PDE activity was restored to near wildtype levels in the RP33 line of transgenic animals (see data presented by D. Farber *et al.*, published in this book). PDE activity in age-matched control *rd/rd* siblings showed minimal levels of activity. In RP4-28 transgenic retinas, higher levels of cGMP PDE activity were detected in the superior rescued hemisphere relative to the inferior degenerated hemisphere in these animals. Thus, lower cGMP PDE activity of the inferior hemispheres correlated with reduced numbers of rescued photoreceptor cells. Surprisingly, recent examination of older animals 4-7 months of age have revealed the degeneration to be progressive. While expression of the transgene has significantly delayed the degenerative process, it has not completely arrested it. The continued degeneration indicates a more complex mechanism of degeneration than the simple absence of the β-subunit of cGMP PDE. This will be discussed in further detail below in light of the results of our mutant cone transducin transgenic mouse studies.

**Expression of Wildtype or Mutant α-subunit Genes of Rod and Cone Transducin.** Next to rhodopsin, one of the best biochemically characterized retinal proteins is transducin [47,48]. It provides an excellent experimental model for analysis in transgenic animals. Both rod- and cone-specific forms of transducin have been cloned[23, 25]. Evidence suggests that the phototransduction pathway (figure 6) in rods and cones is very similar[47,48]. Light activation of opsin induces a conformational change in the retinal chromophore, which allows it to bind to the heterotrimeric transducin protein. In the inactive state, GDP is bound to the α-, β- and γ- transducin complex. Binding of activated opsin causes substitution of GTP for GDP, which results in the dissociation of the transducin β- and γ-subunits from the GTP bound α-transducin subunit. The activated transducin α-subunit in turn, activates cGMP PDE. The activated cGMP PDE hydrolyzes cGMP, causing closure of cGMP gated channels and a hyperpolarizing membrane response. Plasma membrane hyperpolarization leads to closure of the cGMP gated channel, producing a low internal calcium concentration and activation of guanylate cyclase in concert with recoverin (figure 6). Rhodopsin kinase and arrestin deactivate light activated rhodopsin. The role of the 33K phosducin protein, which has been shown to complex with the βγ subunits of transducin[49], remains to be elucidated.

In these studies, we have produced transgenic mouse lines expressing the wildtype human cone transducin α-subunit cDNA, or a mutant form with a point mutation in the GTPase activity region of the gene (Q204L, figure 1E). This mutation inactivates the intrinsic GTPase activity. Thus, the α-subunit of cone transducin is constitutively activated to hydrolyze cGMP by activation of cGMP PDE (figure 6) and would be predicted to increase cGMP hydrolysis. Mutations in the analogous site in other G proteins have been shown to decrease GTPase activity[50-53]. Transgenic animals have been analyzed morphologically for retinal degenerations, biochemically for changes in cGMP hydrolysis, and electrophysiologically by single cell recordings of electrical activity in response to light. No biochemical, morphological or electrophysiological changes were observed in 5 lines of control transgenic mice expressing the wildtype cone transducin. Although 4 lines of wildtype and one line of mutant rod transducin transgenic mice were also generated, we were unable to distinguish between the human transgene product and the endogenous mouse transducin protein. There were no obvious morphological, biochemical or electrophysiological differences in these rod transducin transgenic mice and their nontransgenic siblings. It will be necessary to produce other mutant rod transducin transgenic lines for further analysis.

However, transgenic mice expressing mutant cone transducin produced very surprising results. Fourteen transgenic founders were identified. Seven showed germline transmission and expression of the transgene. Two lines of animals expressing the mutant cone transducin designated RTcΔ47 and TRcΔ66, were analyzed in detail. Both lines

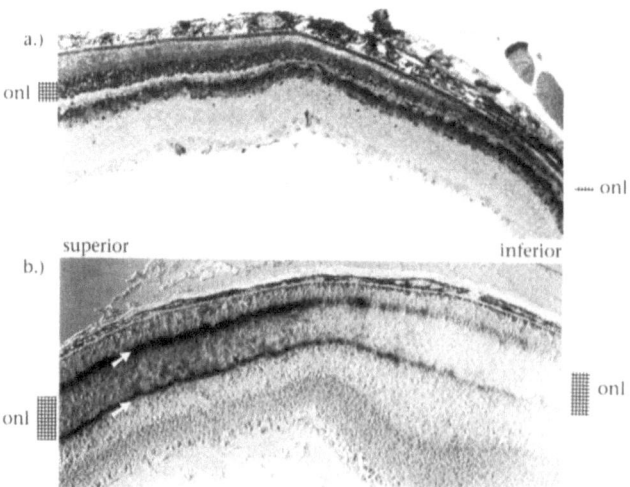

Figure 5. Retinal sections from transgenic mice showing gradients of transgene expression. a. β-PDE transgenic mouse RP4-28 showing a gradient of degeneration. b. Transgenic mouse RG2-2 showing gradient of β-galactosidase expression.

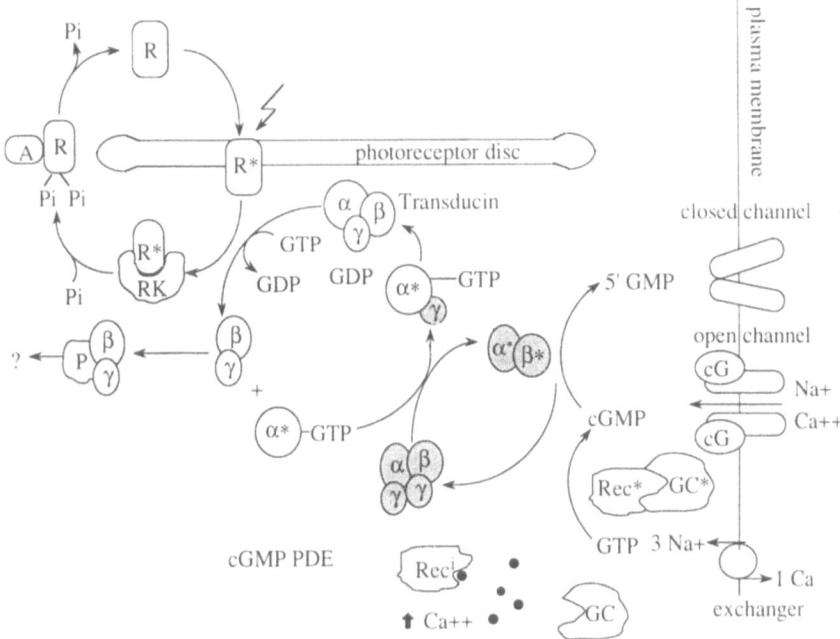

Figure 6. Phototransduction pathway. See text for details.

showed uniform expression of the transgene product across the entire span of the retina, as determined by immunohistochemical staining with antibody specific for human cone transducin. Expression of the transgene localized to photoreceptor cell outer segments. Although transgenic mouse lines showing the characteristic superior to inferior gradient of expression described above in transgenic animal lines expressing the rod opsin/lac Z and the rod opsin/β-PDE transgene were also observed, these lines were not examined in detail because of the technical difficulty of identifying expressing photoreceptor cells for single cell recording studies.

To assess the ability of the transducin protein to activate cGMP PDE , PDE activity was measured using two methods. First, PDE light sensitivity using different levels of light activated retinal homogenate was measured in the presence of $^3$H-cGMP as substrate. Instead of the expected increase in PDE activity from transgenic retinal homogenates, PDE activity was significantly diminished relative to nontransgenic sibling control animals. The RTcΔ47 and RTcΔ66 transgenic mice had a PDE activity that was 30% and less than 10% of

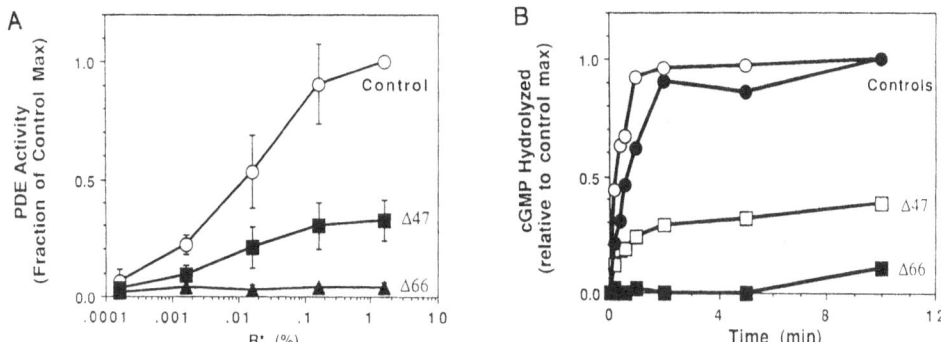

Figure 7. Assay of PDE activity in control and mutant cone transducin transgenic retinas. A. PDE light sensitivity was measured with increasing intensities of light flashes correlating with increased PDE activity. Transgenic retinal homogenates for RTcΔ47 and Δ66 have approximately 30 and 10% of control PDE activity. B. Time course of cGMP hydrolysis was measured following flash activation of .008% of rhodopsin. Levels of cGMP hydrolysis for mutant cone transducin transgenic homogenates is reduced relative to control levels.

that observed in non-transgenic control animals (figure 7A). In a second biochemical assay measuring the time course of cGMP hydrolysis following a light flash activating .008% of the rhodopsin, retinal homogenates from transgenic animals once again showed much reduced levels of cGMP hydrolysis relative to that of nontransgenic control animals (figure 7B). Transgenic mouse lines RTcΔ47 and RTcΔ66 hydrolyzed about 30% and 10%, respectively, of control levels.

Electrophysiological function of transgenic photoreceptors was assessed by single cell recordings from intact retinal cell clumps prepared from dark adapted transgenic and non-transgenic retinas[54]. A rod outer segment from intact photoreceptor cells was pulled into a suction electrode and photocurrents were recorded. The flash response times and amplitudes of transgenic RTcΔ47 and nontransgenic photoreceptors were essentially identical. The intensity-response relation was measured to determine the flash sensitivity of photoreceptors. As shown in figure 8, a plot of the intensity response to photons of light shows that mutant cone photoreceptors require more photons for half-maximal response than wildtype photoreceptor cells. The observed decrease in sensitivity in mutant transducin transgenic photoreceptor cells is consistent with decreased cGMP hydrolysis and PDE activity.

To examine the basis of the unexpected decrease in PDE activity and cGMP hydrolysis, levels of endogenous PDE in transgenic retinas were assayed. Western blot hybridization for the PDE α-, β- and γ-subunits indicated that levels of PDE α- and β-subunit protein were diminished in transgenic retinas relative to control retinas, while levels of γ-PDE appeared to be unaffected.

## DISCUSSION

If the β-subunit of PDE is truly the candidate gene for retinal degeneration in *rd* mice, the retinas of the mutant cone transducin mice would be expected to degenerate. Thus far, light microscopic examination of 4 week old transgenic mice expressing the mutant cone transducin transgene show no obvious evidence of retinal degeneration. Decreased levels of both PDE α- and β-subunits may be regulated at the transcriptional or the translational level. If regulation is at the transcriptional level, this would suggest a feedback mechanism of transducin on PDE α- and β-subunit message levels. Alternatively, if regulation is at the translational level, it is possible that α- and β-PDE protein stability has been diminished. Constitutively activated transducin may bind and stabilize γ-PDE, but prevent it from reassociating with and stabilizing α- and β-PDE. This would explain why γ-PDE protein levels are unaffected. This hypothesis can be tested by measuring PDE message levels in transgenic retinas.

Perplexingly, there appears to be no simple correlation between the rate of degeneration and level of PDE expression. Heterozygous *rd* mice have morphologically normal retinas. Thus, at half the normal level of PDE, degeneration is not observed. In our β-PDE transgenic animals, even though the level of PDE activity is close to that of wildtype, we see a progressive degeneration that is delayed for months. In the mutant cone transducin transgenic animals, there is no obvious degeneration at one month of age even though PDE

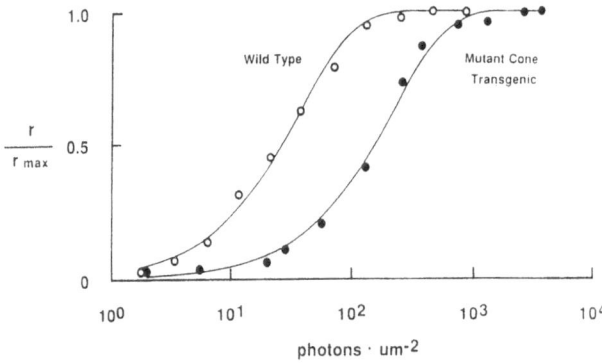

Figure 8. Intensity-response relation of control and mutant cone transducin transgenic mouse RTcΔ47 photoreceptor cells. Transgenic photoreceptors are almost one order of magnitude less sensitive than control photoreceptors.

activity is 10 to 30% of normal. Different assay methods were used to measure PDE activity in the β-PDE and the mutant cone transducin transgenic animals. Thus PDE activity levels are not strictly comparable.

The progressive degeneration seen in the RP4-28 and RP33 transgenic animals expressing the β-PDE subunit and the absence of an obvious degeneration in the RTcΔ47 transgenic animals with low levels of α- and β- PDE suggest that retinal degeneration in *rd* mice is not strictly the result of deficient β-PDE protein. Clearly, from the results of our mutant cone transducin experiments, a mutation in one gene (transducin) can affect the levels of other phototransduction proteins (α- and β-PDE). A mutation in another as yet unidentified gene may also be effecting the *rd* phenotype by down regulation of the PDE β-subunit. This seems unlikely, however, since mutations in the gene have been linked to the degeneration[55,56]. The connection between β-PDE deficiency and disc membrane morphogenesis is also extremely intriguing. The effect on disc membrane morphogenesis suggests that the stoichiometry of phototransduction components may be critical for maintaining the structural integrity of photoreceptor cells. Numerous studies have shown that varied rhodopsin levels produce differing susceptibility to light induced retinal degenerations in rats[57]. It is possible that varying susceptibility would also be seen with different levels of β-PDE. It will be essential to study animals that express not only appropriate levels of PDE, but also express in a temporally correct manner during development. Elucidating the molecular basis for the differences in time course and severity of these degenerations will be key to understanding the progression of retinal degenerations.

ACKNOWLEDGEMENTS

Special thanks to Clint Makino and Denis Baylor for single cell recordings measuring transgenic mouse photoreceptor responses. This work was supported in part by National Institutes of Health (NIH) Program Project Grant AG97687 (M.I.S), NIH EY08285 (D.B.F.), EY940801 (M.L.A.), and EY06641 (J.B.H.) George Gund/National Retinitis Pigmentosa Foundation grants (J.L., M.I.S., M.L.A., D.B.F.), American Cancer Society Grant #PF-3033 (J.L.). M.L.A. is a recipient of the Research to Prevent Blindness (RPB) Jules and Doris Stein Professorship. J.G.F. is a recipient of the RPB Research Career Development Award. D.B.F. is a recipient of the RPB Senior Scientific Investigator Award.

REFERENCES

1. P. McWilliam *et al.*, Autosomal Dominant Retinitis Pigmentosa (ADRP): Localization of an ADRP Gene to the Long Arm of Chromosome 3, *Genomics* 5:619 (1989).

2. G.J. Farrar *et al.*, Autosomal Dominant Retinitis Pigmentosa: Linkage to Rhodopsin and Evidence for Genetic Heterogeneity, *Genomics* 8:35 (1990).

3. T.P. Dryja *et al.*, A point mutation of the rhodopsin gene in one form of retinitis pigmentosa, *Nature* 343:364 (1990).

4. T.P. Dryja *et al.*, Mutation spectrum of the rhodopsin gene among patients with autsomal dominant retinitis pigmentosa. *Proc. Natl. Acad. Sci. USA* 88:9370 (1991).

5. C.-H. Sung *et al.*, Rhodopsin mutations in autosomal dominant retinitis pigmentosa, *Proc. Natl. Acad. Sci. USA* 88:6481 (1991).

6. C.F. Inglehearn *et al.*, A completed screen for mutations of the rhodopsin gene in a panel of patients with autosomal dominant retinitis pigmentosa. *Human Molec. Gen.* 1:41 (1992).

7. P.J. Rosenfeld *et al.*, A Null mutation in the rhodopsin gene causes rod photoreceptor dysfunction and autosomal recessive retinitis pigmentosa, *Nature Gen.* 1:209 (1992).

8. J.E. Dowling and R.L. Sidman, Inherited Retinal Dystrophy in the Rat, *J. Cell Biol.* 14:73 (1962).

9. A. Sorsby, P.C. Koller, M. Attfield, J.B. Davey, and D.R. Lucas. Retinal dystrophy in the Mouse: Histological and Genetic Aspects, *J. Exp. Zoology* 125:171 (1954).

10. R. Van Nie, D. Ivanyi, and P. Demant, A new H-2 linked mutation, *rds*, causing retinal degeneration in the mouse, *Tissue Antigens* 12:106 (1978).

11. G. Aguirre *et al.*, Rod-Cone Dysplasia in Irish Setters: A Defect in Cyclic GMP Metabolism in Visual Cells, *Science* 201:1132 (1978).

12. C. Bowes *et al.*, Isolation of a candidate cDNA for the gene causing retinal degeneration in the *rd* mouse, *Proc. Natl. Acad. Sci. USA* 86:9722 (1989).

13. C. Bowes *et al.*, Retinal degeneration in the *rd* mouse is caused by a defect in the β-subunit of rod cGMP-phosphodiesterase, *Nature* 347:677 (1990).

14. G.H. Travis *et al.*, Identification of a photoreceptor-specific mRNA encoded by the gene responsible for retinal degeneration slow (*rds*), *Nature* 338:70 (1989).

15. B. Weber *et al.*, Genomic organization and complete sequence of the human gene encoding the β-subunit of the cGMP phosphodiesterase and its localisation to 4p16.3, *Nucl. Acids Res.* 19:6263 (1991).

16. G.H. Travis *et al.*, The human retinal degeneration slow (*rds*) gene: chromosome assignment and structure of the mRNA, *Genomics* 10:733 (1991).

17. O. Riess *et al.*, The Search for Mutations in the Gene for the Beta Subunit of the cGMP Phosphodiesterase (PDEB) in Patients with Autosomal Recessive Retinitis Pigmentosa, *Am. J. Hum. Genet.* 51:755 (1992).

18. G.J. Farrar *et al.*, A three-base-pair deletion in the peripherin-RDS gene in one form of retinitis pigmentosa, *Nature* 354:478 (1991).

19. K. Kajiwara *et al.*, Mutations in the human retinal degeneration slow gene in autosomal dominant retinitis pigmentosa, *Nature* 354:480 (1991).

20. J. Nathans and D.S. Hogness, Isolation and nucleotide sequence of the gene encoding human rhodopsin, *Proc. Natl. Acad. Sci. USA* 81:4851 (1984).

21. W. Baehr, J.D. Falk, K. Bugra, J.T. Triantafyllos and J.F. McGinnis, Isolation and analysis of the mouse opsin gene, FEBS Letters 238:253 (1988).

22. J. Nathans, D. Thomas and D.S. Hogness, Molecular Genetics of Human Color Vision: The Genes Encoding Blue, Green, and Red Pigments, *Science* 232:193 (1986).

23. M.A. Lochrie, J.B. Hurley, and M.I. Simon, Sequence of the Alpha Subunit of Photoreceptor G Protein: Homologies Between Transducin, *ras*, and Elongation Factors, Science 228:96 (1985).

24. D.C. Medynski *et al.*, Amino Acid sequence of the α subunit of transducin deduced from the cDNA sequence, *Proc. Natl. Acad. Sci. USA* 82:4311 (1985).

25. C.L. Lerea *et al.*, Identification of Specific Transducin α Subunits in Retinal Rod and Cone Photoreceptors, *Science* 234:77 (1986).

26. H.K.W. Fong *et al.*, Repetitive segmental structure of the transducin β-subunit: Homology with the CDC4 gene and identification of related mRNAs, *Proc. Natl. Acad. Sci. USA* 83:2162 (1986).

27. K. Sugimoto et al., Primary structure of the β-subunit of bovine transducin deduced from the cDNA sequence *FEBS Lett.* 191:235 (1985).

28. K. Yatsunami *et al.*, cDNA-derived amino acid sequence of the γ subunit of GTPase from bovine rod outer segments, *Proc. Natl. Acad. Sci. USA* 82:1936 (1985).

29. N. Tuteja and D.B. Farber, γ-Subunit of mouse retinal cyclic GMP phosphodiesterase: cDNA and corresponding amino acid sequence, *FEBS Lett.* 232:182 (1988).

30. W. Baehr *et al.*, Complete cDNA sequences of mouse rod photoreceptor cGMP phosphodiesterase α- and β-subunits, and identification of β'-, a putative β-subunit isozyme produced by alternative splicing of the β-subunit gene, *FEBS Lett*. 278:107 (1991).

31. V.M. Lipkin *et al.*, β-Subunit of Bovine Rod Photoreceptor cGMP Phosphodiesterase, *J. Biol. Chem.* 265:12955 (1990).

32. T. Shinohara *et al.*, Primary and Secondary Structure of Bovine Retinal S Antigen (48-kDa Protein, *Proc. Natl. Acad. Sci. USA* 84:6975 (1987).

33. W. Lorenz *et al.*, The receptor kinase family: Primary structure of rhodopsin kinase reveals similarities to the β-adrenergic receptor kinase, *Proc. Natl. Acad. Sci. USA* 88:8715 (1991).

34. S. Ray *et al.*, cloning, expression, and crystallization of recoverin, a calcium sensor in vision, *Proc. Natl. Acad. Sci. USA* 89:5705 (1992).

35. C.M. Craft *et al.*, Rat Pineal Gland Phosducin: cDNA Isolation, Nucleotide Sequence, and Chromosomal Assignment in the Mouse, *Genomics* 10:400 (1991).

36. U.B. Kaupp *et al.*, Primary structure and functional expression from complementary DNA of the rod photoreceptor cyclic GMP-gated channel, *Nature* 342:762 (1989).

37. L.D. Carter-Dawson and M.M. LaVail, Rod and cones in the mouse retina. I. Structural analysis using light and electron microscopy, *J. Comp. Neurol.* 188:245 (1979).

38. J. Nathans, Molecular Biology of Visual Pigments, *Ann. Rev. Neurosci.* 10: 163 (1987).

39. J. Lem *et al.*, Tissue-Specific and Developmental Regulation of Rod Opsin Chimeric Genes in Transgenic Mice, *Neuron* 6:201 (1991).

40. J. Lem *et al.*, Retinal degeneration is rescued in transgenic *rd* mice by expression of the cGMP phosphodiesterase β subunit, *Proc. Natl. Acad. Sci. USA* 89:4422 (1992).

41. Manuscript in preparation.

42. M. Yamaizumi, E. Mekada, T. Uchida, and Y. Okada, One molecule of diphtheria toxin fragment A introduced into a cell can kill the cell, *Cell* 15:245 (1978).

43. S.-I. Ishiguro *et al.*, Accumulation of immunoreactive opsin on plasma membranes in degenerating rod cells of rd/rd mutant mice, *Cell Struct. Funct.* 123:141 (1987).

44. C. Bowes, T. van Veen, and D.B. Farber, Opsin, G-protein and 48 kDa protein in normal and *rd* mouse retinas: developmental expression of mRNAs and proteins and light/dark cycling of mRNAs, *Exp. Eye Res.* 47:369 (1988).

45. C. Bowes *et al.*, Isolation of a candidate cDNA for the gene causing retinal degeneration in the *rd* mouse, *Proc. Natl. Acad. Sci. USA* 86:9722 (1989).

46. C. Bowes *et al.*, Retinal degeneration in the *rd* mouse is caused by a defect in the b subunit of rod cGMP-phosphodiesterase, *Nature* 347:677 (1990).

47. L. Stryer, Visual Excitation and Recovery, *J. Biol. Chem.* 266:10711 (1991).

48. Y.-K. Ho *et al.*, Transducin: A Signaling Switch Regulated by Guanine Nucleotides, *Current Topics in Cellular Regulation* 30:171 (1989).

49. R.H. Lee, B.S. Lieberman, and R.N. Lolley, A Novel complex from Bovine Visual Cells of a 33,000-Dalton Phosphoprotein with $\beta$- and $\gamma$-Transducin: Purification and Subunit Structure, *Biochemistry* 26:3983 (1987).

50. S.B. Masters *et al.*, Mutations in the GTP-binding Site of $G_{s\alpha}$ Alter Stimulation of Adenylyl Cyclase, *J. Biol. Chem.* 264:15467 (1989).

51. M.P. Graziano and A.G. Gilman, Synthesis in Escherichia coli of GTPase-deficient Mutants of $G_{s\alpha}$*, *J. Biol. Chem.* 264:15475 (1989).

52. Y.H. Wong *et al.*, Mutant $\alpha$ subunits of $G_{i2}$ inhibit cyclic AMP accumulation, *Nature* 351:63 (1991).

53. Y.H. Wong, B.R. Conklin, and H.R. Bourne, $G_z$-Mediated Hormonal Inhibition of Cyclic AMP Accumulation, *Science* 255:339 (1992).

54. Makino, C. L., Baylor, D.A., Lem, J., Simon, M.I., Raport, C. and Hurley, J.B., Responses of Transgenic Mouse Rods Expressing Altered Transducins, *Invest. Ophth. Vis. Sci.* 33:1103 (1992).

55. C. Bowes *et al.*, Analysis of XMV-28 Proviral Integration within the *rd* $\beta$ PDE Gene. Invest. Ophthalmol. Vis. Sci. 33:945 (1992).

56. S. J. Pittler and W. Baehr, Identification of a nonsense mutation in the rod photoreceptor cGMP phosphodiesterase $\beta$-subunit gene of the *rd* mouse, *Proc. Natl. Acad. Sci. USA* 88:8322 (1991).

57. L.M. Rapp and S.C. Smith, Morphologic Comparisons Between Rhodopsin-Mediated and Short-Wavelength Classes of Retinal Light Damage, *Invest. Ophth. Vis. Sci.* 33:3367 (1992).

# RETINAL DEVELOPMENT UNDER THE INFLUENCE OF SV40 T-ANTIGEN IN TRANSGENIC MICE

Muayyad R. Al-Ubaidi *, Joe G. Hollyfield[#,] Ramon L. Font [#] and Wolfgang Baehr[#]

*University of Illinois at Chicago Eye Center, Chicago, IL
#Cullen Eye Institute, Baylor College of Medicine, Houston, TX

During the past several years a number of oncogenes, viral or cellular, have been expressed in various tissues of transgenic mice[1]. Generally, the expression of a transgenic oncogene results in the development of proliferative abnormalities, supporting the association of oncogenes with tumorigenesis. Many of the tumors induced in transgenic mice are unprecedented, since normal mice rarely or never develop such tumors.

Simian virus 40 (SV40) is a small DNA containing tumor virus. One of its gene products, the large tumor antigen (T antigen), is essential for both viral replication and cell transformation[2]. T antigen has been found to interact with several cellular proteins involved in tumor suppression[3,4], DNA replication[5], and regulation of transcription[6,7], but its precise role in oncogenesis is not well understood. Introduction of T antigen in the germline of mice induced the formation of tumors where allowed by the heterologous promoters. Tumors arose in the lens, where naturally occurring tumors are unknown, when the αA-crystalline promoter was used to direct the expression of T antigen[8]. T antigen expression in the lens was also achieved by the Moloney murine sarcoma virus (MSV) enhancer in conjunction with the SV40 T-antigen promoter[9,10]. Directed by the phenylethanolamine N-methyltransferase promoter[11,12], T antigen was expressed in amacrine neurons of the retinas of transgenic mice, producing a tumor consisting of a pleomorphic population of small cells with high mitotic index. A retinoblastoma-like tumor originating in the inner nuclear layer was generated in the retinas of mice, although in only one of 15 families, by expression of T antigen under the control of the luteinizing hormone β-subunit promoter[13].

In this article, we describe the results of employing two promoter fragments to direct expression of T antigen to photoreceptor cells. One fragment was derived from the mouse opsin gene[14] and the other from the human interphotoreceptor retinoid binding protein gene[15]. The consequences of expression of T antigen and the resulting phenotypes are discussed.

## A. PHOTORECEPTOR CELL DEATH INDUCED BY THE EXPRESSION OF T ANTIGEN DIRECTED BY THE MOUSE OPSIN PROMOTER

There are no known tumors that arise from the rod photoreceptor cells. Therefore,

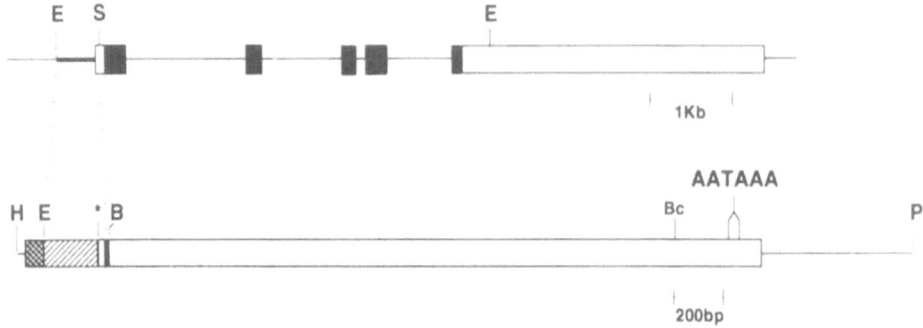

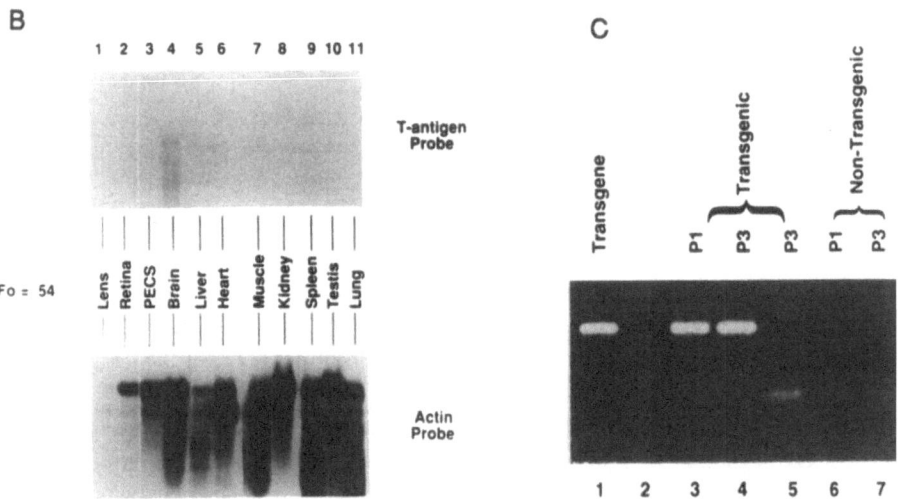

**Figure 1. A**. (Top) Mouse opsin gene. Filled boxes indicate coding exons; open boxes are untranslated regions; lines represent introns and flanking sequences. The promoter region used for the construct is flanked by EcoRI (E) and SacI (S) sites. (Bottom) Schematic representation of the transgene delimited by HaeII (H) and PvuI (P) sites of the vector. Crosshatched box, genomic sequences upstream of the aA-crystalline promoter; hatched box, opsin promoter depicting the transcription start site (asterisk); open box, opsin untranslated region; shaded box, SV40 T-antigen gene containing 64 bp of sequences upstream from the ATG, the coding sequences for both large T and small t antigens and the early region polyadenylation sites (AATAAA). BamHI (B) and BclI (Bc). **B**. Northern blot analysis of poly (A) RNA from different tissues of adult heterozygous transgenic mice. Five micrograms of RNA was loaded in each lane, and the blot was first probed with T-antigen-specific sequences, then stripped and reprobed with actin sequences. Note the presence of a band detected by the T-antigen probe only in the brain RNA sample. **C**. PCR analysis of expression of transgene at early postnatal stages of retinal development. Lane 1, PCR amplification using the transgene as template. Lane 2, empty. Lanes 3 through 5, PCR amplification of first strand cDNA prepared of RNA from transgenic retinas at P1 and P3. Lane 5, PCR amplification of first strand cDNA made of retinal RNA at P3 after treatment with DNase free RNase A, demonstrating that the amplification is RNA specific. Lanes 6 and 7, PCR amplification of first strand cDNA made of RNA from normal littermates' retinas at P3.

experiments were performed in which T antigen was introduced in the rod photoreceptors. The promoter chosen for this task was the mouse opsin promoter fragment, shown in Figure 1A. The promoter fragment (254 bp) was composed of 33 bp of 5' untranslated sequences and 221 bp upstream from the transcription initiation site[14]. The promoter fragment was introduced into the promoterless EcoRI/BamHI fragment of the plasmid pαA366$_a$-T[8]. The resultant construct was digested with HaeII/PvuI to yield a 3.7-kb fragment containing the mouse opsin promoter sequences, the entire coding sequences for the SV40 T antigen, and some flanking sequences from the plasmid pBR322. The 3.7-kb fragment also contained 80 bp of sequences upstream from the mouse αA-crystalline promoter which are irrelevant to promoter activity[8].

The tissue specificity of the promoter fragment was assessed by Northern analysis of poly (A) mRNA obtained from various tissues of heterozygous animals. T-antigen transcripts were detected at a considerable level in the brain of heterozygous transgenic animals (Fig. 1B), while no detectable expression of T antigen was observed in the retinas (Fig. 1B). In the developing transgenic retina, however, T-antigen transcripts were abundantly expressed at postnatal day 5 (P5), with significant decreases at P10 and P15[16]. Polymerase chain reaction (PCR) analysis of transcription levels detected the presence of T-antigen mRNA as early as P1 (Fig. 1C). The onset of transgene expression (days 1 through 5) correlates with the expression of opsin mRNA[14], indicating that the promoter fragment used contained the *cis*-acting elements required for tissue specificity.

Paralleling lack of T-antigen expression in the adult retina of heterozygous animals, electroretinograms showed no light-evoked response (not shown). To determine whether this functional deficit as detected by electroretinography is associated with morphologic changes, retinas from normal and transgenic littermates were examined histologically at successive stages of retinal development. At P5, the morphometric analysis of developing transgenic retinas was similar to that of the retinas of normal littermates (Fig. 2). At P10, degenerative changes became clearly evident, and no inner- or outer-segment elaboration was observed in the transgenic retina[15] (Fig. 2). Although pyknotic nuclei were present, there was no reduction in the thickness of the outer nuclear layer. At P15, the transgenic photoreceptor layer was only 30% of the thickness of the normal littermates' photoreceptor layer (Fig 2). At P21, there was more than 85% reduction in the thickness of the photoreceptor layer (Fig 2), where only one layer of photoreceptor nuclei was left. The surviving photoreceptor cells were shown to be cone nuclei, since they were not targeted by the mouse opsin promoter. Hence, while expression of T antigen in nonretinal cells typically promotes growth and division of the host cells and ultimately leads to tumorigenesis, its expression in photoreceptors leads to cell death.

Since some human retinal degenerations (cancer and melanoma associated retinopathies) are associated with high titers of antibody that may be causative of the disease[17-19], we explored the possibility that antibodies against T antigen may lead to photoreceptor cell death in the transgenic mice. Western analysis of retinal extracts from different developmental stages of both normal and transgenic retinas was performed. Results showed the absence of antibodies directed against T antigen in the serum of transgenic animals (data not shown), indicating that the photoreceptor cell death is not associated with the production of anti-T-antigen antibodies. Although macrophages were absent in sections of transgenic retinas, the photoreceptor death could be the result of a cellular rather than a humoral immune response.

Cyclosporin A (CsA) is an immunosuppressor that inhibits the cellular immune response through changes in the functions of T lymphocytes[20]. CsA also affects the functions of both T-independent B lymphocytes and macrophages. Treatment of transgenic mice with CsA at different developmental stages did not alter the onset or severity of the retinal degeneration caused by the expression of T antigen. Similar results were obtained from the

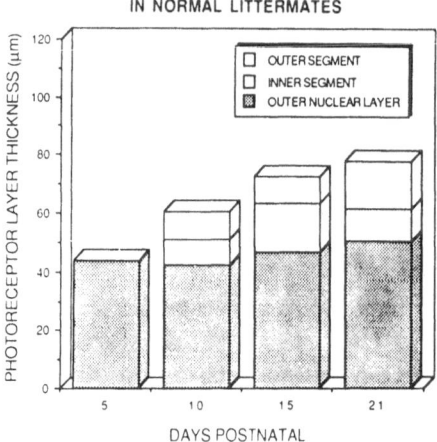
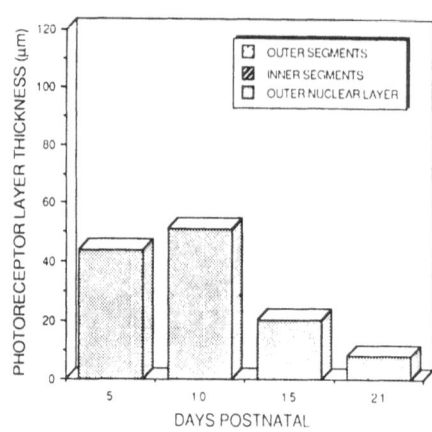

**Figure 2.** Developmental morphometric analysis of retinal sections from transgenic and normal littermates showing the reduction in thickness of the outer retina in transgenic animals.

treatment of transgenic mice with cortisone, which also suppresses the cellular immune response[21].

## B. RETINAL TUMORS IN TRANSGENIC MICE EXPRESSING T ANTIGEN UNDER CONTROL OF THE HUMAN INTERPHOTORECEPTOR RETINOID-BINDING PROTEIN PROMOTER

Opsin is expressed exclusively in rods starting at P1, while interphotoreceptor retinoid binding protein (IRBP) is known to be expressed in both rods and cones starting at early embryonic stages[22]. To study the effect of onset of expression of T antigen in the developing photoreceptor cell layer of transgenic mice, the mouse opsin promoter was replaced by the human IRBP promoter. In addition to an early onset of T-antigen expression, the IRBP promoter is expected to direct expression of the antigen to both rods and cones. All of the transgenic mice families produced from the introduction of hIRBP-T-antigen construct exhibited retinal and brain tumors, albeit at varying incidence levels[23]. The tumors were consistently trilateral and appeared to originate at embryonic or early postnatal stages. Most ocular and brain tumors were poorly differentiated with a variable degree of anaplasia and bizarre pleomorphic nuclei. The ocular tumors led to total retinal detachment (Fig. 3) with the tumor cells involving mostly the outer retinal layers[23]. Although most of the ocular tumors were poorly differentiated, some tumors were more differentiated and disclosed peculiar rosettes composed of a single layer of cuboidal cells that were centered around a small capillary lined by a single layer of flat endothelial cells with scattered erythrocytes observed in the lumens of the capillaries[23].

Although the retinal tumors were highly undifferentiated, the neoplastic cells reacted with antibodies to neuron-specific enolase, S antigen, and Leu 7. However, tumor cells were not reactive with either antibodies against neurofilament and glial fibrillary acidic protein or with photoreceptor-specific antibodies. The tumors exhibited immunoreactivity with antibodies to T antigen specifically restricted to cells that were mitotically active[23].

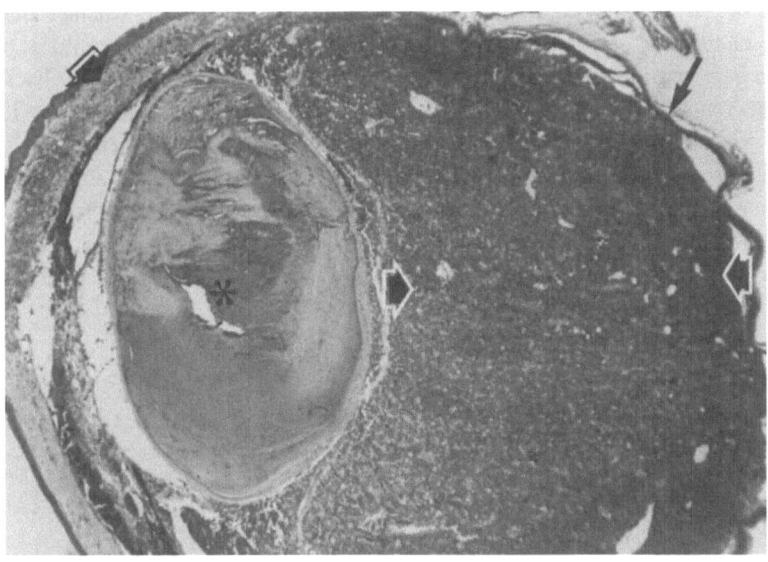

**Figure 3.** Section of an enucleated eye from 2 month old transgenic mouse expressing SV40 T antigen under control of the human IRBP promoter and displaying a large retinal tumor. The tumor originates from the outer retinal layers without involving the pigment epithelium and the schlera (long black arrow). Notice that the tumor, between the two white bordered black arrows, has filled the vitreal cavity and is directly opposed to the lens (asterisk). The black bordered black arrow points to the cornea.

## C. DISCUSSION

Although the mammalian retina is composed of several layers of different cell types, there is only one tumor that affects the retina in humans, namely retinoblastoma[24]. Retinoblastoma is a childhood retinal tumor that is rarely, if ever observed in adults. Some retinoblastoma tumors and cell lines express cone-specific antigens, suggesting the possible evolution of retinoblastoma along the line of cone cell lineage[25,26]. Therefore, no known retinal tumor expresses rod photoreceptor specific antigens. With that in mind we set out to test the effects of expression of an oncogene in the rod photoreceptor cells of transgenic mice. The SV40 T antigen, a viral oncogene[27], was chosen because its expression in all previously described tissues induced tumor formation[1]. The mouse opsin promoter[14] was used to direct the expression of the oncogene to the rod photoreceptor cells. Two transgenic mice families were established that expressed T antigen in the retina[15]. In one family, expression of SV40 T antigen in the retina of transgenic mice led to a slowly progressing retinal degeneration (family 2), whereas a rapidly progressing retinal degeneration was observed in members of the second family (family 54). Expression of the oncogene, in members of family 54, started at P1 concomitant with the expression of endogenous opsin, whereas the degeneration was observed after P10. The degeneration was completed by P21 to P25 and was associated with sustained DNA synthesis and mitotic cycles[15]. The continued cell-cycle activity was accompanied by increased thymidine kinase activity as a direct result of the expression of the oncogene[28]. In the normal retina, thymidine kinase activity decreases starting at P5 and reaches the basal adult level at P10. In the transgenic retina, however, thymidine kinase activity was consistently higher than normal until degeneration of the cells expressing T antigen was complete, when the activity dropped to normal adult levels[28]. Since the expression of the oncogene started at P1, we wished to investigate whether the onset of expression or the cell specificity of the promoter is responsible for the resulting phenotype. The human IRBP promoter has previously been used to direct expression of the bacterial CAT gene to the retina and pineal gland of transgenic mice[16]. Expression of the IRBP/CAT transgene, irrespective of the site of integration, starts at embryonic day 13 (E13), a time when developing retinal cells have just begun the final differentiation into rod and cone photoreceptor cells[23]. Since hIRBP/CAT transgene expression was shown to start as early as E13, this promoter was chosen to direct T-antigen expression to the retinas of transgenic mice. We found that expression of T antigen driven by the hIRBP promoter led to the formation of tumors in the brains and retinas of mice[23]. Tumors occurred at varying degrees of severity in different transgenic families, most likely because of the effect of site of integration on the levels of expression of the transgene.

The absence of immune response to the expression of T antigen in the transgenic mice that expressed T antigen under the control of the mouse opsin promoter suggests the involvement of another system, in situ, in the photoreceptor cell death. Normally, the photoreceptors enter a postmitotic state at the end of the first week of postnatal development[29-31], which is maintained as such for the entire life of the organism. In that state, the retinal cells are locked in $G_0$ phase and are prevented from DNA synthesis and cell division by internal cell cycle regulators[32]. Therefore, an endogenous (to the cell type) and/or exogenous (to the cell type or the entire retina), system must be responsible for maintaining the retina in that state. Since expression of an oncogene is generally associated with continued DNA synthesis and cell cycling[32], expression of T antigen, a potent viral oncoprotein, in the retinas of transgenic animals sustained the DNA synthesis (as late as P16) and cell cycling in the retinal cells[28]. Transgenic retinal cells may have overcome the effects of the internal blockers, such as Rb and P53, by the growth enhancement conferred by the oncogene. This is allowable as long as the retina has not entered the postmitotic state, which may help explain the absence of degeneration before P10. Once the retina enters the postmitotic state, the transgenic photoreceptor cells start to die. To determine whether transgenic photoreceptor cells can complete the mitotic cycle and proliferate when removed from their native environment, 10-day-old retinas from transgenic animals were excised, minced and placed in tissue culture. The cultured cells have undergone more than 30 passages and produced tumors on subcutaneous injection into nude mice[15]. Proliferation of photoreceptor cells from transgenic retinas in tissue culture suggests that factors, perhaps exogenous and antiproliferative, are present in the retina and are involved in inducing photoreceptor cell death in response to proliferative activities. Reintroduction of cultured transgenic cells into the vitreal cavity of animals failed to induce tumor formation, suggesting the existence of the exogenous factor. The growth of transgenic photoreceptor cells in culture also suggests that T antigen expression has overcome the effects of the endogenous antiproliferative factor(s).

SV40 T-antigen expression in another postmitotic system, Purkinje cells, led to cell ablation in transgenic mice rather than the expected tumors[33]. These mice displayed ataxia, a neurological disorder typical of cerebellar dysfunction, with onset reflective of the transgene copy number. This is the second postmitotic system in which cell death occurs as a result of sequential events initiated by reentry of postmitotic cells into the cell cycle. The reentry is followed by cell death caused by the inability of these cells to complete the cell cycle.

These results suggest a potential explanation as to why adult retinal tumors are rarely observed. They also suggest that instead of tumors, certain forms of retinal degenerations in humans could result from the untimely induction, by an oncogene, of proliferative activities in the postmitotic retina.

# References

1) D. Hanahan, *Dissecting multistep tumorigenesis in transgenic mice*. Ann. Rev. Genet. *22:* 479-519 (1988).

2) J.M. Bishop, *Molecular themes in oncogenesis*. Cell *64:* 235-248 (1991).

3) J.A. DeCaprio, J. W. Ludlow, J. Figge, J. Shew, C. Huang, W. Lee, E. Marsilio, E. Paucha and D. M. Livingston, *SV40 large tumor antigen forms a specific complex with the product with the retinoblastoma susceptibility gene*. Cell *54:* 275-283 (1988).

4) D.P. Lane and L. V. Crawford, *T antigen is bound to a host protein in SV40-transformed cells*. Nature *278:* 261-263 (1979).

5) I. Dornreiter, A. Hoss, A. K. Arthur and E. Fanning, *SV40 T antigen binds directly to the large subunit of purified DNA polymerase alpha*. EMBO J. *9:* 3329-3336 (1990).

6) M.E. Ewen, J. W. Ludlow, E. Marsilio, J. A. DeCaprio, R. C. Millikan, S. H. Cheng, E. Paucha and D. M. Livingston, *An N-terminal transformation-governing sequence of SV40 large T antigen contributes to the binding of both p110Rb and a second cellular protein, p120*. Cell *58:* 257-267 (1989).

7) S. Wagner and R. Knippers, *An SV40 large T antigen binding site in the cellular genome is part of acis-acting transcriptional element*. Oncogene *5:* 353-359 (1990).

8) P.A. Overbeek, A. B. Chepelinsky, J. S. Khillan, J. Piatigorsky and H. Westphal, *Lens-specific expression and developmental regulation of the bacterial chloramphenicol acetyltransferase gene driven by the murine αA-crystallin promoter in transgenic mice*. Proc. Natl. Acad. Sci. USA *82:* 7815-7819 (1985).

9) H.-W. Korf, W. Goetz, R. Herken, F. Theuring, P. Gruss and W. Schachenmayr, *S-antigen and rod-opsin immunoreactions in midline brain neoplasms of transgenic mice: similarities to pineal cell tumors and certain medullblastomas in man*. J. Neuropathol. Exp. Neurol. *49*: 424-437 (1990).

10) F. Theuring, W. Goetz, R. Balling, H. -W. Korf, F. Schulze, R. Herken and P. Gruss, *Tumorigenesis and eye abnormalities in transgenic mice expressing MSV-SV40 T-antigen*. Oncogene *5:* 225-232 (1990).

11) E.E. Baetge, R. R. Behringer, A. Messing, R. L. Brinster and R. D. Palmiter, *Transgenic mice express the human phenylethanolamine N-methyltransferase gene in adrenal medulla and retina*. Proc. Natl. Acad. Sci. USA *85:* 3648-3652 (1988).

12) J.P. Hammang, E. E. Baetge, R. R. Behringer, R. L. Brinster, R. D. Palmiter and A. Messing, *Immortalized retinal neurons derived from SV40 T-antigen-induced tumors in transgenic mice*. Neuron *4:* 775-782 (1990).

13) J.J. Windle, D. M. Albert, J. M. O'Brien, D. M. Marcus, C. M. Disteche, R. Bernards and P. Mellon, *Retinoblastoma in transgenic mice*. Nature *343:* 665-669 (1990).

14) M.R. Al-Ubaidi, S. J. Pittler, M. S. Champagne, J. T. Triantafyllos, J. F. McGinnis and W. Baehr, *Mouse opsin: gene structure and molecular basis of multiple transcripts*. J. Biol. Chem. *265:* 20563-20569 (1990).

15) M.R. Al-Ubaidi, J. G. Hollyfield, P. A. Overbeek and W. Baehr, *Photoreceptor degeneration induced by the expression of simian virus 40 large tumor antigen in the retina of transgenic mice*. Proc. Natl. Acad. Sci. USA *89*: 1194-1198 (1992).

16) G.I. Liou, L. Geng, M. R. Al-Ubaidi, S. Matragoon, G. Hanten, W. Baehr and P. A. Overbeek, *Tissue-specific expression in transgenic mice directed by the 5'-flanking sequences of the human gene encoding interphotoreceptor retinoid-binding protein*. J. Biol. Chem. *265:* 8373-8376 (1990).

17) C.E. Thirkill, A. M. Roth and J. L. Keltner, *Cancer associated retinopathy*. Arch. Ophthalmol. *105:* 372 (1987).

18) C.E. Thirkill, R. C. Tait, N. K. Tyler, A. M. Roth and J. L. Keltner, *The cancer-associated retinopathy antigen is a recoverin-like protein*. Invest. Ophthalmol. Vis. Sci. *33*: 2768-2772 (1992).

19) A.H. Milam, J. C. Saari, S. G. Jacobson, W. P. Lubinski, L. G. Feun and K. R. Alexander, *Autoantibodies against retinal bipolar cells in cutaneous melanoma-associated retinopathy.* Invest. Ophthalmol. Vis. Sci. 34: 91-100 (1993).

20) M.D. Chiara and F. Sobrino, *Modulation of the inhibition of respiratory burst in mouse macrophages by cyclosporin A: effect of in vivo treatment, glucocorticoids and the state of activation of cells.* Immunology 72: 133-137 (1991).

21) M.E. Gombert, L. B. Berkowitz, T. M. Aulicino and L. DuBouchet, *Therapy of pulmonary nocardiosis in immunocompromised mice.* Antimicrob. Agents Chemother. 34: 1766-1768 (1990).

22) L.D. Carter-Dawson, R. A. Alvarez, S. -L. Fong, G. I. Liou, H. G. Sperling and C. D. B. Bridges, *Rhodopsin, 11-cis-vitamin A, and interstitial retinol-binding protein (IRBP) during retinal development in normal and rd mutant mice.* Dev. Biol. 116: 431-438 (1986).

23) M.R. Al-Ubaidi, R. L. Font, A. B. Quiambao, M. J. Keener, G. I. Liou, P. A. Overbeek and W. Baehr, *Bilateral retinal and brain tumors in transgenic mice expressing simian virus 40 large T antigen under control of the human interphotoreceptor retinoid-binding protein promoter.* J. Cell Biol. 119: 1681-1687 (1992).

24) W.R. Green, *Retina. In Ophthalmic pathology: an atlas and textbook.* (Ed. W. H. Spencer) pp. 589-1351, W. B. Saunders Co., Philadelphia 1985.

25) E. Bogenmann, M. A. Lochrie and M. I. Simon, *Cone cell-specific genes expressed in retinoblastoma.* Science 240: 76-78 (1988).

26) R.L. Hurwitz, E. Bogenmann, R. L. Font, V. Holcombe and D. Clark, *Expression of the functional cone phototransduction cascade in retinoblastoma.* J. Clin. Invest. 85: 1872-1878 (1990).

27) J.S. Butel, *SV40 large T-antigen: dual oncogene.* Cancer Surv. 5: 343-365 (1986).

28) M.R. Al-Ubaidi, A. B. Quiambao, K. M. Myers, W. Baehr and J. G. Hollyfield, Death of photoreceptor cells expressing SV40 T-antigen is the result of reentry onto the cell cycle in a postmitotic environment. Dev. Biol. *in press* (1993).

29) L.D. Carter-Dawson and M. M. LaVail, *Rods and cones in the mouse retina.* J. Comp. Neuro. 188: 263-272 (1979).

30) L.D. Carter-Dawson and M. M. LaVail, *Rods and cones in the mouse retina.* J. Comp. Neuro. 188: 245-262 (1979).

31) R.W. Young, *Cell differentiation in the retina of the mouse.* Anat. Rec. 212: 199-205 (1985).

32) S. Travali, J. Koniecki, S. Petralia and R. Baserga, *Oncogenes in growth and development.* FASEB J. 4: 3209-3214 (1990).

33) M. Feddersen, R. Ehlenfeldt, W.S. Yunis, H.B. Clark and H.T. Orr, *Disrupted cerebellar cortical development and progressive degeneration of Purkinje cells in SV40 T antigen transgenic mice.* Neuron 9: 955-966 (1992).

# NONSENSE MUTATIONS IN THE ß SUBUNIT GENE OF THE ROD cGMP PHOSPHODIESTERASE THAT ARE ASSOCIATED WITH INHERITED RETINAL DEGENERATIVE DISEASE

Richard L. Hurwitz,[1,2,3] Steven J. Pittler,[5] Michael L. Suber,[1] Ning Qin,[4] Rehwa H. Lee,[6] Cheryl M. Craft,[7] Richard N. Lolley,[6] and Wolfgang Baehr[3,4]

Departments of [1]Pediatrics, [2]Cell Biology, [3]Ophthalmology, and [4]Biochemistry, Baylor College of Medicine, Houston, TX 77030
[5]Department of Biochemistry Molecular Biology and Ophthalmology, University of South Alabama College of Medicine, Mobile, AL 36688
[6]Department of Anatomy and Cell Biology, University of California School of Medicine, Los Angeles, CA 90024 and the Developmental Neurology Laboratories, Veterans Administration Medical Center, Sepulveda, CA 91343
[7]Department of Psychiatry, University of Texas Southwest Medical Center, and Veterans Administration Medical Center, Dallas, TX 75235

## INTRODUCTION

The phototransduction cascade is the series of biochemical reactions through which a photon of light results in decreased neurotransmitter release in the photoreceptor. Analogous but biochemically distinct pathways are responsible for both rod and cone phototransduction.[1] A photon of light results in the photoisomerization of rhodopsin from 11-cis retinal to all trans retinal in the rod. This allows transducin to exchange GTP for GDP. This activated transducin complex can then release the inhibitory γ subunit from the cGMP phosphodiesterase (PDE) resulting in a decreased concentration of free cGMP. A cGMP-gated cation channel is then closed decreasing sodium conductance and hyperpolarizing the photoreceptor.

The cGMP PDE consists of three separate subunits. The α and β subunits have apparent molecular weights of 88 and 84 kDa respectively.[2] The deduced amino acid sequences from cDNA clones indicate approximately 70% identity between the subunits and both contain intact cGMP binding and cGMP catalytic domains.[3,4] The carboxyl termini are posttranlationally modified by carboxymethylation and isoprenylation; however, the α subunit is farnesylated while the β subunit is geranylgeranylated.[5-7]

Cyclic GMP concentrations have been shown to be elevated in several animal models of inherited[8,9] and acquired[10] photoreceptor degeneration. These observations have led to investigations of possible defects in members of the phototransduction cascade as the potential genetic etiology of both the human and animal disease.

*Retinal Degeneration*, Edited by J.G. Hollyfield *et al.*
Plenum Press, New York, 1993

Retinal degeneration in the mouse, locus designation *rd*, was first identified as an autosomal recessive trait in wild mice caught in Switzerland, England, and France and later in inbred laboratory strains.[11] Homozygous animals have lost all rod photoreceptors by day 20. Cones survive much longer and all other retinal cells remain intact.[12] Cyclic GMP levels were found to be high[8] and there was an absence of histone activated PDE activity[13] suggesting that the genetic mutation could be present in this member of the phototransduction cascade. Using immunological methods, it was shown that at least one of the large subunits of the PDE was expressed at decreased levels in the homozygous *rd* mouse and the normal PDE complex that is necessary for regulation of cGMP levels in the rod was not formed.[14] Mice that are heterozygous for the mutation have histologically normal photoreceptors but show abnormal cGMP levels, altered PDE kinetics, abnormalities of cGMP binding and abnormal electroretinograms.[15-18]

The *rd* mouse mutation was genetically localized to chromosome 5.[11] The $\alpha$, $\beta$, and $\gamma$ subunits of the rod PDE were localized to chromosomes 18, 5, and 11 respectively,[19,20] a localization which excluded the $\alpha$ and $\gamma$ subunits as candidates for the *rd* locus. The $\beta$ subunit gene was mapped by interspecies backcross between *Afp* and *Gus*, near the *rd* locus.[21] The *rd* locus was also found to be linked to the xenotropic provirus *Xmv-28*.[22] Just prior to the onset of photoreceptor degeneration in the *rd* mouse, $\beta$ subunit mRNA levels decline.[23] Thus a mutation in the rod PDE $\beta$ subunit gene was implicated as a prime candidate in the etiology of photoreceptor degeneration by both biochemical and genetic parameters.[24]

To characterize the PDE $\beta$ subunit gene a strategy was employed that directly amplified normal and *rd* mouse (C57BL/6J) DNA fragments of the $\beta$ subunit from retinal cDNA and genomic templates using appropriate primers. In this way, overlapping segments of the complete $\beta$ subunit PDE cDNA were amplified from both normal and *rd* mice. No size differences in the paired amplified products were observed. The PCR fragments were then directly sequenced to avoid artifacts which sometime occur during PCR amplification. Nine differences were observed between the normal and *rd* mouse sequences. Three of the differences [position 2582 (G to C), 2612 (C to T), and 2657 (C to A)] occur in the 3' untranslated region, and five are silent changes within the coding region [nucleotide positions 86 (TTG to CTG), 331 (ACG to ACA), 826 (TAT to TAC), 1156 (GTC to GTT), 1834 CAA to CAG)]. The remaining difference (a C to A transversion at position 1048) resulted in a nonsense mutation converting codon 347, tyrosine (TAC) to a stop codon (TAA). The mutation of the $\beta$ subunit at codon 347 would predict a premature chain termination during translation which would result in a polypeptide of approximately 40 kDa. The mutation also introduces a new *Dde I* restriction site that can serve as an RFLP to identify the *rd* locus.[25]

The $\beta$ subunit PDE gene consists of at least 22 exons. The exact conservation of the lengths of exons within the $\alpha$ and $\beta$ subunits strongly suggest a common ancestral gene. Amplification and direct sequencing of the exons of the $\beta$ subunit *rd* gene revealed the same nine differences that were found in the *rd* cDNA.. In addition, several mutations in intron sequences that did not effect the amino acid sequence were identified. The nonsense mutation found at codon 347 was localized to exon 7. The heterozygote revealed both the TAC and TAA codon as expected.

The same TAA mutation at codon 347 was found in CBA/J (a mouse strain containing the *rd* phenotype and related to C57BL/6J); ST/bJ, WB/ReJ-W, PL/J, SWR/J, SJL/J, and BDP/J (six independent strains shown to contain a genetically identical defect); and FVB/N and Bub/J (two other retinal degeneration strains). Seven normal strains examined all contain the expected tyrosine codon TAC characteristic of normal $\beta$ subunit PDE.[25]

The rod/cone dysplasia found in Irish setter dogs is characterized by a rapidly progressing loss of photoreceptors[26] and is inherited as an autosomal recessive trait. In affected dogs, retinal and photoreceptor development appear normal until 13 days of age. By 1 month of age rod photoreceptor degeneration is evident; nearly all rod photoreceptors degenerate by 5 months. Cone photoreceptor degeneration is complete by 1 year.[27,28]

The first known biochemical aberration in the *rcd1* phenotype is a rapid accumulation of cGMP to levels approximately 10 fold greater than age-matched controls.[9,27,29,30] The $\alpha$ and $\gamma$ subunit mRNAs in affected animals are of normal size and abundance,[31] however, the $\beta$ subunit levels of mRNA appear to be reduced.[32] When the photoreceptor PDEs are examined by anion exchange HPLC, the cone isozyme appears to have normal activity while the rod activity is absent.[33,34] Western blot analysis of the rod PDE probed with a rod-specific antibody that recognizes both the $\alpha$ and $\beta$ subunits shows the presence of the $\alpha$ subunit but not the $\beta$ subunit in the *rcd1* homozygotes. The heterozygotes appear to have substoichiometric levels of the $\beta$ subunit.[33]

In order to determine an ISD PDE $\beta$ subunit cDNA sequence, the first strand cDNA produced from mRNA of heterozygous retinas was amplified. Sets of primers within the coding region were selected from a pool of primers used for amplification of the mouse PDE $\beta$ subunit or were derived from conserved sequences known for mammalian $\beta$ subunit.[25] The complete 3' untranslated sequence was amplified using a sequence specific primer and a universal 3' end primer.[35] A short 5' untranslated sequence was determined using the RACE PCR method[36] where the first strand cDNA was first polyadenylated at the 3' end and then amplified. Thus, a nearly complete cDNA sequence for the normal ISD PDE $\beta$ subunit could be assembled which includes the polyadenylation signal, a short poly(A) tract, the translation ATG start codon and 16 nucleotides upstream of the ATG start codon.[33] The 2746 nucleotide length of the sequenced cDNA (excluding the poly(A) tail) was in excellent agreement with the single 2.8 kb species detected on ISD Northern blots.[32] The composite cDNA sequence predicts that the ISD $\beta$ subunit contains 856 amino acid residues, exactly the same as other mammalian species (see Figure 1). When compared to other known mammalian $\beta$ subunit sequences, this sequence shows greater than 90% homology at both the amino acid and nucleotide levels and contains other features characteristic of all photoreceptor PDEs (cGMP and $\gamma$ subunit binding sites, CAAX box signaling posttranslational geranylgeranylation, and a homologous domain).[37-39]

To identify possible mutations in the affected ISD retinal PDE, the entire coding region of the $\beta$ subunit of affected ISD was amplified and directly sequenced. Comparison of the sequences generated from affected and heterozygous animals revealed differences in four nucleotides. Three of these substitutions did not change the amino acid sequence (C to T at position 267, G to A at position 606, and T to C at position 609). The other substitution, a G to A transversion at position 807, led to the generation of a stop codon (TGG to TAG) 49 codons upstream of the normal translation termination codon. The consequence of premature termination of the PDE $\beta$ subunit appears to be an unstable polypeptide since the predicted truncated protein does not accumulate at a detectable level.[33]

Putative exon 21 and a portion of exon 22 were amplified from genomic DNA isolated from normal, heterozygote, and affected homozygote samples in order to confirm the mutation. The length of the fragment generated indicated that the intron separating exons 21 and 22 is approximately the same size as that in the mouse[25] and human[40] $\beta$ subunit genes. Direct sequencing with a primer located 27 bp downstream of the nonsense mutation confirmed the presence of a TGG codon in (+/+), and a TAG codon in (-/-). As expected, the fragment amplified from heterozygous animals possessed a normal and a mutant allele that contained both the TGG and TAG codons.[33]

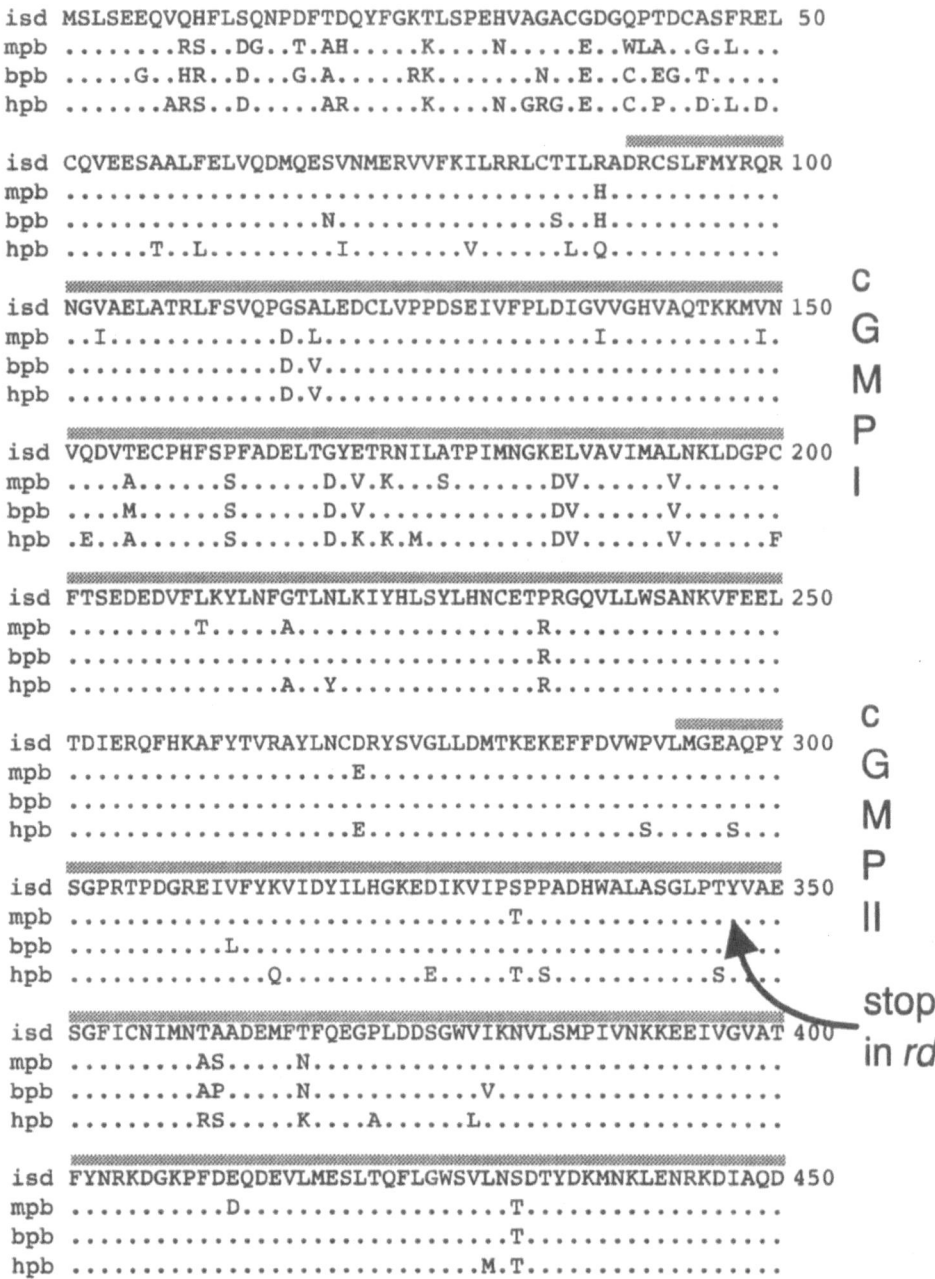

```
isd MSLSEEQVQHFLSQNPDFTDQYFGKTLSPEHVAGACGDGQPTDCASFREL 50
mpb ........RS..DG..T.AH.....K....N.....E..WLA..G.L...
bpb .....G..HR..D...G.A.....RK.......N.E..C.EG.T.....
hpb .......ARS..D.....AR.....K....N.GRG.E..C.P..D.L.D.

isd CQVEESAALFELVQDMQESVNMERVVFKILRRLCTILRADRCSLFMYRQR 100
mpb ......................................H...........
bpb .................N...................S..H..........
hpb ......T..L..........I.........V......L.Q...........

isd NGVAELATRLFSVQPGSALEDCLVPPDSEIVFPLDIGVVGHVAQTKKMVN 150
mpb ..I...........D.L...................I...........I.
bpb ..............D.V.................................
hpb ..............D.V.................................

isd VQDVTECPHFSPFADELTGYETRNILATPIMNGKELVAVIMALNKLDGPC 200
mpb ....A......S......D.V.K...S.......DV......V.......
bpb ....M......S......D.V.............DV......V.......
hpb .E..A......S......D.K.K.M.........DV......V......F

isd FTSEDEDVFLKYLNFGTLNLKIYHLSYLHNCETPRGQVLLWSANKVFEEL 250
mpb .........T.....A.................R................
bpb .................................R................
hpb ............A..Y.................R................

isd TDIERQFHKAFYTVRAYLNCDRYSVGLLDMTKEKEFFDVWPVLMGEAQPY 300
mpb .............................E....................
bpb ..................................................
hpb .............................E.............S.....S...

isd SGPRTPDGREIVFYKVIDYILHGKEDIKVIPSPPADHWALASGLPTYVAE 350
mpb ...............................T.................
bpb ..........L.......................................
hpb .............Q.........E.....T.S...........S.....

isd SGFICNIMNTAADEMFTFQEGPLDDSGWVIKNVLSMPIVNKKEEIVGVAT 400
mpb .........AS.....N................................
bpb .........AP.....N..........V.....................
hpb .........RS.....K...A......L.....................

isd FYNRKDGKPFDEQDEVLMESLTQFLGWSVLNSDTYDKMNKLENRKDIAQD 450
mpb ..........D...................T..................
bpb ..............................T..................
hpb ..............................M.T................
```

C G M P I

c G M P II

stop in *rd*

**Figure. 1** Alignment of the amino acid sequences of mammalian rod PDE β subunits. isd, Irish setter; mpb, mouse; bpb, bovine; hpb, human. The domain structure is indicated by bars above the aligned sequence, the CAAX box is shaded, and a variable domain between the putative catalytic domain and the C-terminus is boxed. The residues mutated in the *rd* β subunit (tyrosine, codon 347) and in the *rcd1* β subunit (tryptophan, codon 807) are identified by large black arrows.

```
isd MVLYHVRCDKDEIQLILPTRERLGKEPADCEEDELGILLKEVLPGPSKFD 500
mpb ............E....D..............KI...E...T...
bpb ........RE..............KI........A...
hpb ......K..R.........A.........D....EI...E...TT..

isd IYEFHFSDLECTELELVKCGIQMYYELGLVRKFQIPQEVLVRFLFSVSKG 550
mpb .........................V...............A
bpb .........................V...............
hpb ...........D.............V...............I...

isd YRRITYHNWRHGFNVAQTMFTLLTTGKLKSYYTDLEAFAMVTAGLCHDID 600
mpb .........G..........M...............
bpb .............R.....M...............
hpb ...............M...............

isd HRGTNNLYQMKSQNPLAKLHGSSILERHHLEFGKFLLSEETLNIYQNLNR 650
mpb ...............................A..S.........
bpb ...............................
hpb ...............................

isd RPSEHVIHLMDIAIIATDLALYFKKRTMFQKIVDESKNYEDRKSWVEYLS 700
mpb .QH..............................K.......
bpb .QH...............................
hpb .QH....................A.............Q.K....I..

isd LETTRKEIVMAMMMTACDLSAITKPWEVQSKVALLVAAEFWEQGDLERTV 750
mpb ...............................
bpb ...............................
hpb ...............................

isd LDQQPIPMMDRNKAAELPKLQVGFIDFVCTFVYKEFSRFHEEILPMFDRL 800
mpb ...............................
bpb ...............................
hpb ...............................

isd QNNRKEWKALADEYEAKLKALEEEKQQQEDRTTAKKAGTEICNGGPAPKS 850
mpb ......V.......KKE...VA...V...V.........
bpb ......V.....DQKKET---....V............R.
hpb ......V.....--KEE.E.VA...V............
```

**stop in *rcd1***

**Variable Domain**

856

```
isd ST CCIL
mpb .. ....
bpb .. .R..
hpb .. ....
```

**CAAX Box**

C a t a l y t i c   D o m a i n

## DISCUSSION

Both the *rd* mouse and the *rcd1* ISD share an autosomal recessive inherited degenerative disorder of rod photoreceptors that is characterized by a relatively early onset of rod degeneration followed by a later onset of cone degeneration. Both animals share elevated levels of cGMP in the retina, the absence of rod cGMP PDE activity prior to the onset of photoreceptor degeneration, and the presence of a nonsense mutation that signals premature translation termination of the β subunit of the PDE. In the mouse, degeneration of rod photoreceptors is complete by postnatal day 20. In the ISD, degeneration of rods begins by one month of age and progresses over the next four months.

The mouse mutation results in the termination of translation before the presumed catalytic domain of the PDE can be expressed. The ISD mutation allows the catalytic domain to be expressed but the CAAX box that is responsible for the geranylgeranylation site and the resultant membrane binding domain are absent. In both cases, the respective mutations apparently result in the unstable expression of rod PDE protein. Instability of the truncated polypeptides emphasizes the importance of an intact C-terminus and appropriate posttranslational processing. Both the α and γ subunits are apparently expressed, suggesting that the expression of an intact β subunit is important for stabilization of the PDE complex.

In human autosomal dominant Retinitis Pigmentosa (RP), over forty separate mutations of the opsin gene[41-43] and three mutations of the putative structural protein *rds*/peripherin[44,45] have been identified and associated with this disease. Another mutation in the rhodopsin gene appears to result in an autosomal recessive transmission of RP.[46] The β subunit of the PDE is the only other member of the phototransduction cascade which has been found to contain a mutation associated with recessive transmission of photoreceptor degeneration. A patient with autosomal dominant RP has been reported to have elevated levels of cGMP and depressed levels of photoreceptor cGMP PDE activity in a retinal biopsy,[47] suggesting that a defect in the processing or expression of the PDE may also be responsible for some human retinal degeneration. Indeed, defects in the β subunit PDE gene have recently been identified in patients with autosomal recessive RP.[48] Mutations causing retinal degeneration in the gene encoding the β subunit of the PDE may be found to be as diverse as the mutations found in the rhodopsin gene. These mutations will not only define the cause of photoreceptor degeneration but will help to further elucidate the mechanism of phototransduction. Now that mutations involving the phototransduction cascade have been identified as causing retinal degeneration, investigations will now be centered on understanding the pathogenesis of these disorders.

## ACKNOWLEDGMENTS

This research was funded by grants from the National Institutes of Health (EY06656, EY00395, EY08123, N528126), the National Retinitis Pigmentosa Foundation, the Gund Foundation, the RP Foundation Fighting Blindness (Baltimore), the Knights Templar Eye Foundation (Indianapolis), the Retina Research Foundation (Houston), the Ella G. McFadden Foundation, and a Merit Review from the V.A. Research Service. W.B. is the recipient of a Jules and Doris Stein Research to Prevent Blindness Professorship.

## REFERENCES

1. L. Stryer, Visual excitation and recovery, *J.Biol.Chem.* 266:10711 (1991).
2. W. Baehr, M.J. Devlin and M.L. Applebury, Isolation and characterization of cGMP phosphodiesterase from bovine rod outer segments, *J.Biol.Chem.* 254:11669 (1979).

3. V.M. Lipkin, N.V. Khramtsov, I.A. Vasilevskaya, N.V. Atabekova, K.G. Muradov, V.V. Gubanov, T. Li, J.P. Johnston, K.J. Volpp and M.L. Applebury, β-Subunit of bovine rod photoreceptor cGMP phosphodiesterase. Comparison with the phosphodiesterase family, *J.Biol.Chem.* 265:12955 (1990).

4. W. Baehr, M.S. Champagne, A.K. Lee and S.J. Pittler, Complete cDNA sequences of mouse rod photoreceptor cGMP phosphodiesterase α- and β-subunits, and identification of β'-, a putative β-subunit isozyme produced by alternative splicing of the β-subunit gene, *FEBS Lett.* 278:107 (1991).

5. O.C. Ong, I.M. Ota, S. Clarke and B.K.-K. Fung, The membrane binding domain of rod cGMP phosphodiesterase is posttranslationally modified by methyl esterification at a C-terminal cysteine, *Proc.Natl.Acad.Sci.USA* 86:9238 (1989).

6. P. Catty and P. Deterre, Activation and solubilization of the retinal cGMP-specific phosphodiesterase by limited proteolysis--Role of the C-terminal domain of the β-subunit, *Eur.J.Biochem.* 199:263 (1991).

7. N. Qin, S.J. Pittler and W. Baehr, *In vitro* isoprenylation and membrane association of mouse rod photoreceptor cGMP phosphodiesterase α and β subunits expressed in bacteria, *J.Biol.Chem.* 267:8458 (1992).

8. D.B. Farber and R.N. Lolley, Cyclic guanosine monophosphate: elevation in degenerating photoreceptor cells of the C3H mouse retina, *Science* 186:449 (1974).

9 G. Aguirre, D.B. Farber, R.N. Lolley and R.T. Fletcher, Rod-cone dysplasia in Irish setters: a defect in cyclic-GMP metabolism in visual cells, *Science* 201:1133 (1978).

10. D.A. Fox and D.B. Farber, Rods are selectively altered by lead: I. Electrophysiology and biochemistry, *Exp.Eye Res.* 46:597 (1988).

11. R.L. Sidman and M.C. Green, Retinal degeneration in the mouse; location of the *rd* locus in linkage group XVII, *J Hered* 56:23 (1965).

12. L.D. Carter-Dawson, M.M. LaVail and R.L. Sidman, Differential effect of the *rd* mutation on rods and cones in the mouse retina, *Invest.Ophthalmol.Vis.Sci.* 17:489 (1978).

13. R.H. Lee, B.S. Lieberman, R.L. Hurwitz and R.N. Lolley, Phosphodiesterase-probes show distinct defects in *rd* mice and Irish setter dog disorders, *Invest.Ophthalmol.Vis.Sci.* 26:1569 (1985).

14. R.H. Lee, S.E. Navon, B.M. Brown, B.K.-K. Fung and R.N. Lolley, Characterization of a phosphodiesterase-immunoreactive polypeptide from rod photoreceptors of developing *rd* mouse retinas, *Invest.Ophthalmol.Vis.Sci.* 29:1021 (1988).

15. J.A. Ferrendelli and A.I. Cohen, The effects of light and dark adaptation on the levels of cyclic nucleotides in retinas of mice heterozygous for a gene for photoreceptor dystrophy, *Biochem.Biophys.Res.Commun.* 73:421 (1976).

16. M. Doshi, M.J. Voaden and G.B. Arden, Cyclic GMP in the retinas of normal mice and those heterozygous for early-onset photoreceptor dystrophy, *Exp.Eye Res.* 41:61 (1985).

17. M.J. Voaden and N.J. Willmott, Evidence for reduced binding of cyclic GMP to cyclic GMP phosphodiesterase in photoreceptors of mice heterozygous for the *rd* gene, *Curr. Eye Res.* 9:643 (1990).

18. J.C. Low, The corneal ERG of the heterozygous retinal degeneration mouse, *Graefe's Arch. Clin. Exp. Ophthalmology* 225:413 (1987).

19. M. Danciger, C.A. Kozak, T. Li, M.L. Applebury and D.B. Farber, Genetic mapping demonstrates that the α-subunit of retinal cGMP-phosphodiesterase is not the site of the *rd* mutation, *Exp.Eye Res.* 51:185 (1990).

20. M. Danciger, N. Tuteja, C.A. Kozak and D.B. Farber, The gene for the gamma-subunit of retinal cGMP-phosphodiesterase is on mouse chromosome 11, *Exp.Eye Res.* 48:303 (1989).

21. M. Danciger, C. Bowes, C.A. Kozak, M.M. LaVail and D.B. Farber, Fine mapping of a putative *rd* cDNA and its co-segregation with *rd* expression, *Invest.Ophthalmol.Vis.Sci.* 31:1427 (1990).

22. W.N. Frankel, J.P. Stoye, B.A. Taylor and J.M. Coffin, Genetic analysis of endogenous xenotropic murine leukemia viruses: association with two common mouse mutations and the viral restriction locus Fv-1, *J.Virol.* 63:1763 (1989).

23. C. Bowes, M. Danciger, C.A. Kozak and D.B. Farber, Isolation of a candidate cDNA for the gene causing retinal degeneration in the *rd* mouse, *Proc.Natl.Acad.Sci.USA* 86:9722 (1989).

24. C. Bowes, T. Li, M. Danciger, L.C. Baxter, M.L. Applebury and D.B. Farber, Retinal degeneration in the *rd* mouse is caused by a defect in the β subunit of rod cGMP-phosphodiesterase, *Nature* 347:677 (1990).

25. S.J. Pittler and W. Baehr, Identification of a nonsense mutation in the rod photoreceptor cGMP phosphodiesterase β-subunit gene of the *rd* mouse, *Proc.Natl.Acad.Sci.USA* 88:8322 (1991).

26. H.B. Parry, Degenerations of the dog retina II: Generalized progressive atrophy of hereditary origin, *Br.J.Ophthal.* 37:487 (1953).

27. G. Aguirre, D.B. Farber, R.N. Lolley, P. O'Brien, J. Alligood, R.T. Fletcher and G.J. Chader, Retinal degeneration in the dog III: Abnormal cyclic nucleotide metabolism in rod-cone dysplasia, *Exp.Eye Res.* 35:625 (1982).

28. M.J. Voaden, Retinitis pigmentosa and its models, *Prog.Ret.Res.* 10:293 (1990).

29. Y.P. Liu, G. Krishna, G. Aguirre and G.J. Chader, Involvement of cGMP phosphodiesterase activator in an hereditary retinal degeneration, *Nature* 280:62 (1979).

30. G.J. Chader, R.T. Fletcher, S. Sanyal and G.D. Aguirre, A review of the role of cyclic GMP in neurological mutants with photoreceptor dysplasia, *Curr. Eye Res.* 4:811 (1985).

31. S.J. Pittler, R.H. Lee, R.N. Lolley, R.L. Hurwitz and W. Baehr, Molecular assessment of rod cGMP PDE alpha and gamma subunits in Irish Setters with progressive retinal atrophy (pra), *Invest.Ophthalmol.Vis.Sci.* 31:311 (1990).

32. D.B. Farber, J.S. Danciger and G. Aguirre, The β subunit of cyclic GMP phosphodiesterase mRNA is deficient in canine rod/cone dysplasia, *Neuron* 9:349 (1992).

33. M.L. Suber, S.J. Pittler, N. Qin, G.C. Wright, V. Holcombe, R.H. Lee, C.M. Craft, R.N. Lolley, W. Baehr and R.L. Hurwitz, Irish setter dogs affected with rod/cone dysplasia contain a nonsense mutation in the rod cGMP phosphodiesterase β subunit gene, *Proc.Natl.Acad.Sci.USA* (In Press).

34. R.L. Hurwitz, A.H. Bunt-Milam, M.L. Chang and J.A. Beavo, cGMP Phosphodiesterase in Rod and Cone Outer Segments of the Retina, *J.Biol.Chem.* 260:568 (1985).

35. M.R. Al-Ubaidi, S.J. Pittler, M.S. Champagne, J.T. Triantafyllos, J.F. McGinnis and W. Baehr, Mouse opsin. Gene structure and molecular basis of multiple transcripts, *J.Biol.Chem.* 265:20563 (1990).

36. M.A. Frohman and G.R. Martin, Rapid amplification of cDNA ends using nested primers, *Technique* 1:165 (1989).

37. W.A. Maltese, Posttranslational modification of proteins by isoprenoids in mammalian cells, *Faseb J.* 4:3319 (1990).

38. J.A. Beavo, Multiple isozymes of cyclic nucleotide phosphodiesterase, *Adv.Second Messenger Phosphoprotein Res.* 22:1 (1988).

39. S.J. Pittler and W. Baehr, The molecular genetics of retinal photoreceptor proteins involved in cGMP metabolism, *Prog.Clin.Biol.Res.* 362:33 (1991).

40. B. Weber, O. Riess, G. Hutchinson, C. Collins, B. Lin, D. Kowbel, S. Andrew, K. Schappert and M.R. Hayden, Genomic organization and complete sequence of the human gene encoding the β-subunit of the cGMP phosphodiesterase and its localisation to 4p16.3, *Nucleic Acids Res.* 19:6263 (1991).

41. C.F. Inglehearn, R. Bashir, D.H. Lester, M. Jay, A.C. Bird and S.S. Bhattacharya, A 3-bp deletion in the rhodopsin gene in a family with autosomal dominant retinitis pigmentosa, *Am.J.Hum.Genet.* 48:26 (1991).

42. C.-H. Sung, C.M. Davenport, J.C. Hennessey, I.H. Maumenee, S.G. Jacobson, J.R. Heckenlively, R. Nowakowski, G. Fishman, P. Gouras and J. Nathans, Rhodopsin mutations in autosomal dominant retinitis pigmentosa, *Proc.Natl.Acad.Sci.USA* 88:6481 (1991).

43. T.P. Dryja, L.B. Hahn, G.S. Cowley, T.L. McGee and E.L. Berson, Mutation spectrum of the rhodopsin gene among patients with autosomal dominant retinitis pigmentosa, *Proc.Natl.Acad.Sci.USA* 88:9370 (1991).

44. G.J. Farrar, P. Kenna, S.A. Jordan, R. Kumar-Singh, M.M. Humphries, E.M. Sharp, D.M. Sheils and P. Humphries, A three-base-pair deletion in the peripherin-*RDS* gene in one form of retinitis pigmentosa, *Nature* 354:478 (1991).

45. K. Kajiwara, L.B. Hahn, S. Mukai, G.H. Travis, E.L. Berson and T.P. Dryja, Mutations in the human retinal degeneration slow gene in autosomal dominant retinitis pigmentosa, *Nature* 354:480 (1991).

46. P.J. Rosenfeld, G.S. Cowley, T.M. McGee, M.A. Sandberg, E.L. Berson and T.P. Dryja, A null mutation in the rhodopsin gene causes rod photoreceptor dysfunction and autosomal recessive retinitis pigmentosa, *Nature Genetics* 1:209 (1992).

47. J.G. Flannery, D.B. Farber, A.C. Bird and D. Bok, Degenerative changes in a retina affected with autosomal dominant retinitis pigmentosa, *Invest.Ophthalmol.Vis.Sci.* 30:191 (1989).

48. M.E. McLaughlin, E.L. Berson, and T.P. Dryja, Mutations in the β subunit of phosphodiesterase in patients with autosomal recessive retinitis pigmentosa, *Invest.Ophthalmol.Vis.Sci.* (in press).

# SYSTEMIC ALTERATIONS IN DOCOSAHEXAENOIC ACID METABOLISM IN INHERITED RETINAL DEGENERATIONS

**Nicolas G. Bazan, Elena B. Rodriguez de Turco, William C. Gordon, Virginia C. Strand, and Rex E. Martin**

**LSU Eye Center and Neuroscience Center**
**Louisiana State University Medical Center School of Medicine**
**2020 Gravier Street, Suite B**
**New Orleans, LA   70112**

## INTRODUCTION

Photoreceptor membranes are highly enriched in phospholipids that contain docosahexaenoic acid (DHA, 22:6$n3$), a polyunsaturated fatty acid which is esterified mainly at the $C_2$ position of the glycerol backbone.  DHA represents 50% or more of the phospholipid acyl groups (i.e., phosphatidylethanolamine (PE) and phosphatidylserine (PS)) of disc membranes.[1-3]   Because animals cannot synthesize DHA *de novo*, either DHA or its precursor linolenic acid (18:3$n3$) must be obtained from the diet.[4-6]   This shorter chain fatty acid is then elongated and desaturated by the liver to form long chain polyunsaturated fatty acids (LC-PUFA) which, along with dietary DHA, are released to the plasma.[7-10]   Circulating lipoproteins and/or free DHA bound to plasma carrier proteins (*e.g.*, albumin) then deliver DHA to the retina.  Systemically-delivered [³H]DHA shows a time-dependent profile of organ labeling; uptake peaks early in liver [³H]DHA, and is followed by sustained uptake in retina (and brain) as labeling in liver [³H]DHA decreases.[11,12]   Uptake studies both *in vivo*[13,14] and *in vitro* [15,16]  involving analyses by light and electron microscope autoradiography have demonstrated that [³H]DHA is not evenly distributed among the cells of the retina.  In fact, retinal [³H]DHA accumulates preferentially in photoreceptor outer segments and synaptic terminals; comparisons of *in vitro* labeled frog, rabbit, monkey, and human retinas reveal similar patterns, with the majority of DHA (93% in frogs) accumulating within photoreceptor cells (Figure 1).

The functional significance and relevance of the unique enrichment of DHA-phospholipids in excitable membranes is largely unknown. Because membrane structure is linked to its function,[17] changes in the composition of membrane lipids may affect transmembrane events mediated by receptors, ion channels, and enzymes.[18-20] When diets are deprived of $n3$ fatty acids, both rats [21-23] and nonhuman primates[24,25] develop reduced levels of retinal DHA, as well as altered visual function.  Also, diets artificially low in $n3$ fatty acids given to very-low-birth-weight neonates have been correlated with impairments of retinal function.[26-28]  Collectively, efficient support of retinal DHA

*Retinal Degeneration*, Edited by J.G. Hollyfield *et al.*
Plenum Press, New York, 1993

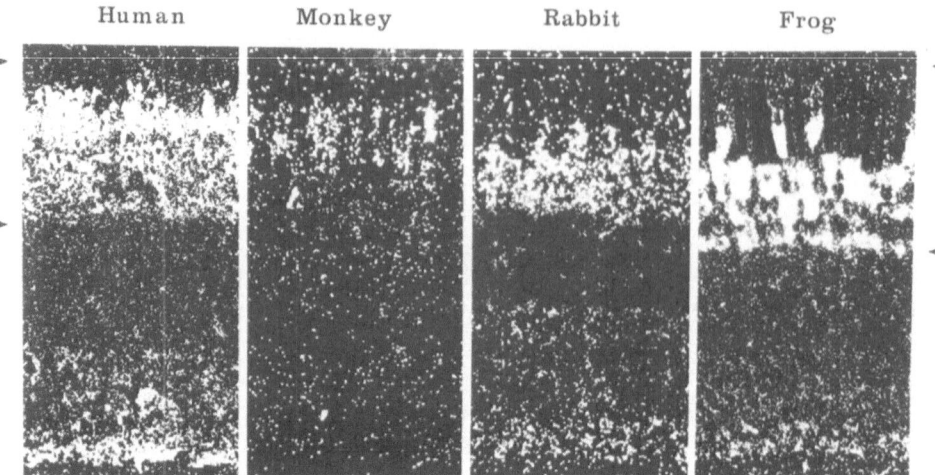

| Human | Monkey | Rabbit | Frog |

**Figure 1.** Light level autoradiograms of [³H]docosahexaenoic acid distribution in four representative retinas, following 4 hours of *in vitro* incubation. The majority of retinal label has accumulated in photoreceptors (93%, frog; 60% human). Photoreceptors are oriented upward and lie in the space denoted by the arrow heads. Each autoradiogram has been adjusted in size to match the lengths of the photoreceptors.

requirements by the liver, selective uptake of DHA by the retinal pigment epithelium (RPE), preferential delivery to photoreceptor cells through the interphotoreceptor matrix, and highly specific accumulation within photoreceptor membranes suggest a unique role for retinal DHA.

Because the retina depends largely upon a systemic supply of DHA for the synthesis of new disc membranes, it is relevant that patients with various forms of hereditary retinal degenerations have lower amounts of plasma DHA-lipids. A decreased DHA content in plasma phospholipids of Usher syndrome patients has been found.[29,30] Decreased plasma-DHA levels have also been observed in patients with retinitis pigmentosa (RP)[31-35] Furthermore, comparisons of total plasma DHA levels in 188 patients with RP have shown significant differences between some types of RP[36]; dominant and recessive RP DHA values remain relatively normal, while X-linked and isolate RP DHA values are reduced. Similarly, decreased levels of plasma DHA-phospholipids have been reported for poodles with progressive rod-cone degeneration (*prcd*).[37,38]

Experimental evidence implicates a deficient DHA supply to the retina in several retinal degenerative diseases, and artificial manipulation of dietary *n3* fatty acids induces changes in both the biochemistry and the function of photoreceptor cells. It has been suggested that, in inherited retinal degenerative diseases, there is a systemic alteration of DHA metabolism involving the liver.[7,29] However, the central issue is still whether or not decreases in plasma DHA are a cause or consequence of disease. Many specific questions are as follows: Is the synthesis of DHA and/or its packaging into lipoproteins altered in the liver? Is the uptake of DHA by retinal cells deficient? Is the mechanism of conservation/recycling affected? Are different pathways in the supply of DHA to the retina affected in different retinal degenerative diseases? Whether decreases in plasma DHA are causes or consequences, there is an important point to make: A limited supply of DHA to photoreceptor cells, as a result of impairments in a) synthesis in the liver, b) uptake and transport by RPE cells, or c) conservation by recycling from RPE phagosomal lipids, will contribute to deterioration of photoreceptor cell membranes and/or impairment in their function.

Because of the physiological relevance of supplying photoreceptor cells with adequate levels of DHA, we summarize in the following sections, a) the steps in DHA synthesis/trafficking that may contribute to retinal DHA deficiency during degenerative retinal diseases, and b) recent findings showing differences in DHA content and metabolism in peripheral and central regions of the retina correlated with histological alterations in the *prcd* dog.

## SYNTHESIS OF DOCOSAHEXAENOIC ACID IN THE LIVER

The dietary precursors linoleic acid (18:2*n6*) and 18:3*n3* are essential for the synthesis of the two most abundant long-chain polyunsaturated fatty acyl chains of cell membrane phospholipids, arachidonic acid (20:4*n6*), and DHA. Following intestinal digestion, the precursors incorporated into chylomicrons and very-low-density-lipoproteins (VLDL) go mainly to the liver through the lymphatics.[39] Desaturation-elongation reactions, similar for both *n3* and *n6* fatty acids, then lead to the synthesis of 20- and 22-carbon polyunsaturated fatty acids in the liver. The rate-limiting step in this process is the first desaturation, catalyzed by the Δ6-desaturase, and, because both precursors compete for this enzyme,[40] the dietary ratio of 18:2*n6*/18:3*n3* is an important factor.[41--43] Moreover, although 18:3*n3* is a better substrate for the Δ6-desaturase, it is less abundant in natural dietary sources (*e.g.*, vegetables, seeds, oils, and animal fats).[4,44]

Cellular DHA content, unlike 20:4*n6*, is extremely high in retina, sperm and brain as compared to other tissues.[4,5] The fatty acids of the *n3* family are found in high concentrations in seafoods (*i.e.*, fish oils),[44] and when present in the diet support tissue demands for DHA. Tissues can acquire 20:4*n6* or 18:2*n6* from plasma circulating lipoproteins where they are found in high concentrations. However, in contrast to fatty acids of the *n6* family, plasma levels of 18:3*n3* are very low, necessitating that organs rely on the liver for an adequate supply of DHA, especially during early postnatal development when a large requirement for DHA in photoreceptor biogenesis and synaptogenesis exists.[45]

The synthesis and delivery of DHA is orchestrated by the liver.[7,10] Its capacity for DHA synthesis is much higher than other tissues,[40,46] including retina.[47] The Δ6-desaturase peaks in the liver as it decreases in the brain during early postnatal murine life.[48,49] This coincides with photoreceptor outer segment biogenesis, neuronal cell proliferation and synaptogenesis in early postnatal murine development. Following administration of radio-labeled 18:3*n3* or DHA to 5- and 14-day-old mice,[10,50,51] to young adult rats,[12] and to frogs,[11] labeling first peaks in the liver; this is followed by a decrease in liver labeling that occurs parallel to a gradual increase of DHA label in brain and retina. The inter-organ, time-dependent labeling of liver and CNS with [³H]DHA is not observed in other organs. For example, frogs injected with [³H]DHA by way of the dorsal lymph sacs exhibit a very early, high DHA uptake in the lungs with the same order of magnitude or higher than in the liver. However, labeling in the lungs is reduced by 60% within the first 12-24 h postinjection (unpublished observations). In contrast, liver accumulates label for 5 days postinjection, with a gradual decrease occurring in its label during the subsequent 15 days.[11]

The liver plays a primary role in DHA synthesis and delivery to other tissues because it exhibits: a) a more efficient uptake of dietary 18:3*n3*, DHA, and other *n3* fatty acids; b) a more active conversion of precursors to DHA; c) the ability to accumulate DHA after its uptake unlike other organs such as lung (see above); and d) a sustained delivery of DHA to the CNS. The majority of dietarily-derived polyunsaturated fatty acids reach the liver through the lymph system and uptake is followed by activation to acyl-CoA, which, in turn, can be either esterified into triacylglycerols (TAG) or oxidized. Labeling by polyunsaturated fatty acids often results in a high and transient uptake in triacylglycerols in several tissues.[52] This may indicate that triacylglycerols are a transient reservoir of unsaturated fatty acids prior to their acylation in phospholipids or further metabolism (e.g. elongation, desaturation). When 18:3n3 is the polyunsaturated fatty acid taken up by the liver, then it is elongated and desaturated to DHA, and esterified into phospholipids.[10] This pathway is very active in hepatocytes.[53] Also, after i.p. administration of [$^{14}$C]18:3n3 to mouse pups,[10] a rapid decrease in liver labeling occurs within the first 24 hours (-70%) (Table 1), similar to that occurring after oral administration of [$^{14}$C]18:3n3 to suckling rats.[54] These authors showed that the rate of [$^{14}$C]18:3n3 oxidation during the first 24 hours is similar to that of oleic acid (18:1n9) (64% and 57%, respectively, of the administered dose) and higher than that of 18:2n6 (48%). This preferential utilization of 18:3n3 as an energy source may be the consequence of its selective activation to linolenoyl-CoA in hepatic mitochondrial outer membranes.[55] ß-oxidation will in turn lead to less 18:3n3 available for elongation and desaturation. It is interesting to note that 2-24 hours after i.p. injection of [$^{14}$C]18:3n3 in developing mice, there is a high accumulation of labeled palmitic acid (16:0) in brain and retina. This is contrasted with a low contribution of 16:0 to total liver and plasma fatty acid labeling (Table 1). High labeling of 16:0 was also observed in the brain of suckling rats when [$^{14}$C]18:3n3 was administered either i.p.[56] or per os.[57] In a more recent study, Li et al.[12] also reported high transient accumulation of [$^{14}$C]16:0 in retinas of adult rats after [$^{14}$C]18:3n3 gavage. These authors support the suggestion that retinal DHA and 16:0 may originate in the liver and are later taken up by the retinal pigment epithelium (RPE) via the choriocapillaris.

**Table 1.** Conversion of [$^{14}$C]18:3n3 to polyunsaturated fatty acids in the liver and transport through plasma to brain and retina in 3-day-old mouse pups.

| Organ | Time | | % distribution | | % of dose |
|-------|------|------|--------|-------------------------------|-----------|
| | (hrs) | 16:0 | 18:3n3 | Polyunsaturated Fatty acids | (total labeling) |
| Liver | 2 | 1.8 | 77.0 | 8.0 | 8.7 |
| | 24 | 23.2 | 48.2 | 39.3 | 2.8 |
| Plasma | 2 | 1.8 | 48.2 | 6.0 | 0.04 |
| | 24 | 3.6 | 11.4 | 66.0 | 0.04 |
| Brain | 2 | 50.0 | - | 9.0 | 0.22 |
| | 24 | 8.0 | - | 9.0 | 0.49 |
| Retina | 2 | 48.0 | - | 21.4 | 0.003 |
| | 24 | 32.1 | - | 50.0 | 0.005 |

3-day-old mouse pups were injected i.p. with 6 µCi [$^{14}$C]18:3 and killed 2 and 24 hrs later. Lipids were extracted from different organs and constituent fatty acids derivatized to n3 methylesters prior to HPLC. Peaks from different fatty acids, as identified by standards, were collected and radioactivity was determined. Polyunsaturated fatty acids = 20:5n3 + 22:4n3 + 22:5n3 + 22:6n3; % of dose: total labeling recovered in each organ with respect to the administered dose of [$^{14}$C]18:3n3 (6 µCi). % distribution: portion of labeled fatty acid with respect to total recovered radioactivity in each tissue. Values from Scott and Bazan.[10]

Either RPE cells and the blood-brain barrier take up circulating 16:0 polyunsaturated fatty acids (*e.g.*, 20:5*n3*, 22:5*n3*, and DHA), or the labeled 16:0 in brain or retina may be derived by recycling through partial ß-oxidation of [$^{14}$C]18:3*n3*. After injection of [$^{3}$H]16:0 or [$^{3}$H]DHA into the dorsal lymph sacs of frogs, marked differences in their RPE/retinal uptake takes place[58] further supporting the latter conclusion. RPE cells incorporate both fatty acids, with rapid, high labeling of cytoplasm and oil droplets. [$^{3}$H]DHA was actively transferred to the retina, where it continued to accumulate over the 34 days of this study. In contrast, the delivery of [$^{3}$H]16:0 from RPE to the retina reaches maximal values by 5 days postinjection, and only accounts for 38% of [$^{3}$H]DHA values.[58]

The liver-to-retina pathway is highly efficient, because oxidation of 18:3*n3*, as well as a decreased ability to convert 22:5*n3* to DHA,[47] results in retinal cells being more dependent on the systemic (hepatic) supply of DHA to support photoreceptor membrane biogenesis. Diminished output of DHA from the liver occurs during long term dietary deprivation of *n3* fatty acids, and results in decreased retinal DHA content and consequent abnormal retinal function.[5,21,24,25,59]

These and other impairments in the synthesis of DHA within the liver may also lead to the output of lipoproteins that are less enriched in this essential fatty acid. Deterioration of visual function during aging has been linked to a deficient liver supply of DHA as a consequence of a decreased Δ4-desaturation step.[60] In *prcd* dogs, lower plasma levels of DHA-phospholipids and increased ratios of 22:4*n6*/22:5*n6* and 22:5*n3*/22:6*n3* also point to alterations in the last step of long-chain polyunsaturated fatty acid synthesis.[37] The conversion of 22:5*n3* to 22:6*n3* involves the elongation of 22:5*n3* to 24:5*n3*, followed by a Δ6 desaturation to 24:6*n3*.[61] The initial two steps occur in microsomes, while the subsequent decarboxylation of 24:6*n3* to 22:6*n3* occurs in peroxisomes. In this context, it is of interest to mention that in patients with inherited peroxisomal disorders, and in animal models of this disease, alterations in *n3* fatty acid metabolism have been reported.[62] In Zellweger syndrome (cerebro-hepato-renal syndrome), levels of DHA and 22:5*n6* are greatly reduced in these three organs, as well as in the retina.[63] Retinas from English setters affected with hereditary canine lipofuscinosis (NCL), a disease similar to the juvenile form of neuronal ceroid lipofuscinosis, display a significant reduction of DHA-phospholipids.[64] Since reduced plasma levels of DHA have also been reported in patients with juvenile NCL,[62] it is possible that, in some retinal degenerative diseases, the function of liver peroxisomes may be abnormal, contributing to deficient hepatic output of DHA-phospholipids. In fact, in the liver of *prcd* dogs, increased ratios 22:4*n6*/22:5*n6* and 22:5*n3*/22:6*n3* have been observed (unpublished observations), yeilding similar changes in plasma lipids.[37]

## TRANSPORT AND METABOLISM OF DHA-LIPIDS IN PLASMA

Most plasma DHA is esterified in phospholipids (Figure 2). However, some is also carried by neutral lipids such as cholesterol esters (ChE) and TAG, and less than 3% (in free fatty acid form) is non-covalently bound to albumin and/or other carrier proteins.[12,45] The contribution of different lipids to total plasma DHA is very similar among the animal species studied except frogs, in which ChE and phospholipids accounts for 90% of the total, in a 1:1 molar relationship. These lipids are constituents, in various proportions, of the lipoproteins secreted by the liver and the intestine,[65] and are engaged in lipid metabolism of all organs. It is important to emphasize that, although DHA is highly enriched in photoreceptor cells and in other neuronal cells, all other organs contain DHA, although in smaller proportions. Whether or not cholesterol ester-containing DHA is

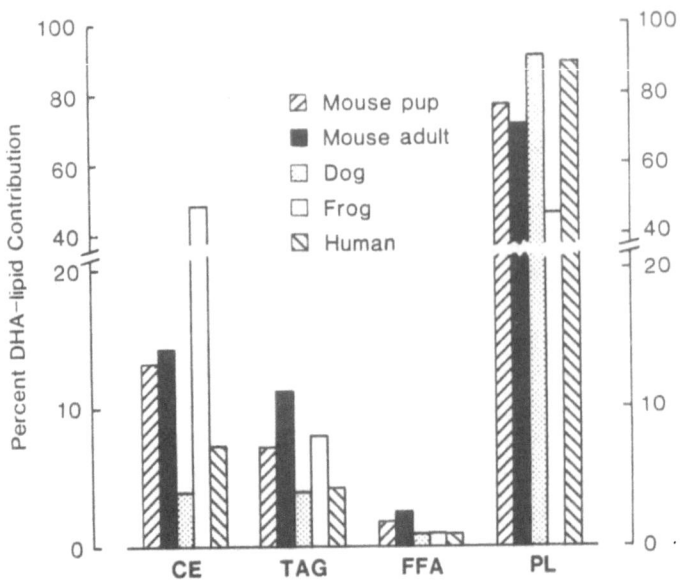

**Figure 2.** Distribution of plasma DHA among lipids in different animal species. Values are the average of 2-3 individual samples, and agree within 10%. Plasma lipids were extracted, isolated by TLC, and derivatized to FAME prior to GLC analysis.[66] Percent values represent the contribution of each individual lipid to total plasma DHA.

specifically related to the delivery of DHA (and of cholesterol) to photoreceptors remains to be studied.

Lipoproteins and/or carrier proteins utilized in the plasma for the delivery of DHA to the retina under normal conditions, and whether the transport of DHA is altered in retinal degenerations are not clearly defined. The mechanism(s) by which rod outer segments become highly enriched in DHA may involve efficient delivery by the liver, selective uptake by RPE, and retention by photoreceptors, including recycling by RPE cells.[6,7,10,67-69]

The distribution of DHA among different plasma lipids and lipoproteins is the combined result of DHA-lipid synthesis in liver and rate of clearance by different tissues. Whether or not there is subsequent metabolism of DHA within plasma, relevant to the supply of the fatty acid to the retina, has not been investigated. The liver acts as a primary source of plasma phospholipids, which are incorporated into nascent VLDL and high-density-lipoproteins (HDL), and released through the space of Disse into the circulation.[70-72]

Although it has been suggested that the RPE may be endowed with a highly efficient mechanism(s) for acquiring DHA from plasma lipoproteins through receptor-mediated endocytosis of lipoproteins[7] there is no direct experimental evidence as yet available. However, the presence of LDL (apo B/E) receptors in bovine RPE[73] and in bovine endothelium of brain capillaries[74] has been reported.

In summary, several mechanisms may contribute, with different degrees of selectivity, to the transport of DHA from the liver to retina, at different stages of development. Further studies are necessary to determine if a specific DHA-carrier exists in plasma and/or RPE. A unique DHA transporter within the RPE could tightly control, not only DHA uptake from plasma, but also DHA routing from phagosomal lipids (following shedding and phagosomal degradation), either out to the choriocapillaris, or back to the photoreceptors through the interphotoreceptor matrix.

# TRAFFICKING OF DHA WITHIN THE RETINA: RPE-PHOTORECEPTOR CELL INTERACTIONS

The RPE is actively involved in the uptake of nutrients from the bloodstream to support the metabolic requirements of photoreceptor cells. DHA also follows this route.[13] Following *in vivo* [³H]DHA administration to mouse pups[50,51] and frogs,[11] the trafficking of DHA from liver to RPE and retina, as a function of time, was traced. This pathway from liver to retina is a portion of a *long loop* that becomes closed when DHA, released from the retina/RPE-complex into the choriocapillaris, is returned to the liver. Recycling (*short loop*) involves the arrival of DHA to the RPE in newly-shed phagosomes, and the return of DHA to photoreceptors through the interphotoreceptor matrix.[7] The delivery of DHA to retinal photoreceptors and its recycling involves the photoreceptor matrix. Here there are proteins, including interphotoreceptor retinoid binding protein, that contain DHA. It has been suggested that they may participate in the routes of DHA movement between the RPE and the inner segment of photoreceptors.[7,8] [³H]DHA recycling in the retina was also followed in frogs as the [³H]DHA a) crossed the RPE, arriving at the photoreceptors after passing through the interphotoreceptor matrix,[75] b) was avidly taken up by photoreceptor cells,[13] c) gradually filled rod outer segments with densely labeled disc membranes, d) appeared in phagosomes after daily shedding, and e) was actively recycled from RPE back to photoreceptors.[69]

Frogs injected systemically with [³H]DHA indicate that RPE is rapidly saturated with label, while retina displays a slow, gradual increase.[11] This contrasts with early, transient, high labeling in other organs, suggesting that RPE very efficiently controls the DHA delivered through the interphotoreceptor matrix to photoreceptors. In fact, the profile of *in vivo* retinal labeling indicates that, as [³H]DHA is delivered to RPE and photoreceptors from plasma, it is actively utilized for the synthesis of new disc membranes, and appears autoradiographically as dense regions that expand from the base to the tip of outer segments.[13] Collectively this, with time, suggests the delivery of DHA from RPE to photoreceptors depends upon the metabolic requirements for DHA. Therefore, the gradually increasing content of DHA in lipids from plasma to RPE to outer segments reflects the ability of RPE to selectively take up DHA from plasma, and to selectively and efficiently make it available to photoreceptors.[76]

The equilibrium between photoreceptor shedding/phagocytosis and disc membrane synthesis produces outer segments of relatively constant length, and implies that RPE can control the balance between the amount of DHA that will be a) recycled from phagosomal lipids back to photoreceptors, b) released into the blood stream, and c) taken up from the circulation. Recent studies, done in rats that were fed diets containing different amounts of *n3* and *n6* fatty acids, indicate that the levels of 22-carbon polyunsaturated fatty acids (mainly 22:5n6) in plasma lipids may affect the rates of steps within these individual pathways, increasing recycling and decreasing the export of DHA.[22,23] As the ratio of 22:5n6/22:6n3 increases in plasma, and more 22:5n6 is incorporated into rod outer segment phospholipids, it is possible that more of 22:5n6 is exchanged between phagosomes/RPE and plasma, while the opposite occurs for DHA. In this way, as plasma DHA levels drop, the system must become less dependent on an external supply of DHA and compensates by stimulating the recycling pathway or *short loop*.

From these studies, it becomes evident that RPE plays a central role in DHA uptake from plasma and its recycling from phagosomal lipids back to photoreceptor cells.[7] Therefore, when alterations in the plasma content of DHA and/or the ratio of 22:5n6/DHA occur, the *short loop* may compensate for a decreased supply of DHA by conserving this essential fatty acid through recycling mechanisms. Alterations in this pathway, whether combined with an impaired systemic supply of DHA or not, may contribute to an inadequate availability of DHA for the support of photoreceptor cells.

## METABOLIC DIFFERENCES IN PERIPHERAL AND CENTRAL DOG RETINA: ALTERATIONS IN PROGRESSIVE ROD-CONE DEGENERATION

French poodles with a recessively inherited rod-cone degeneration are a model for human RP.[38] Degeneration is first observed after rod outer segment formation is complete and, as in RP, initial effects involve progressive loss of rod cells and a deterioration of retinal function.[77] Cone cells are less affected and can be seen even when rod nuclei are significantly decreased in number (Figure 3). At 150 days of age, a decrease in the rate of rod photoreceptor membrane renewal is observed and atrophy is detected in cells within the posterior pole.[78] However, as the disease progresses, in 250-day-old *prcd* dogs the periphery becomes severely affected, eventually showing the highest degree of cell loss.[79]

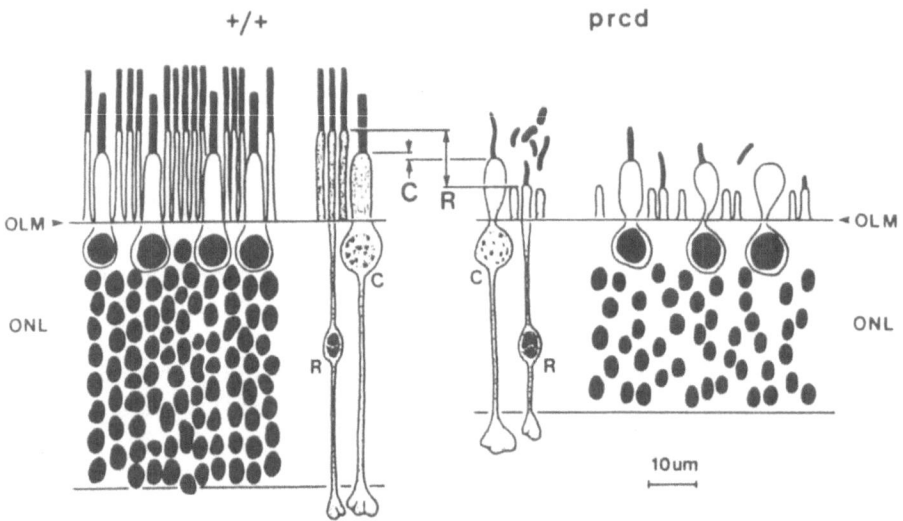

**Figure 3.** Diagrammatic representation comparing normal (+/+) and progressive rod-cone degeneration (*prcd*) dog retinas. Drawing is based upon histological sections obtained from the inferior-medial, posterior pole (central region) of the right eye. Differences in the length of rod (R) and cone (C) photoreceptor inner segments are indicated. OLM, outer limiting membrane; ONL, outer nuclear layer. Scale (10μm) is accurate in both dimensions.

Because the cause(s) of these regional differences in pathology are largely unknown, and because this deterioration is reminiscent of human RP, we have examined the lipid content and metabolism of [³H]DHA in retinas from normal and *prcd* dogs.[79] Analysis of acyl groups from total lipids reveals significant differences in the content of *n3* and *n6* fatty acids among different retinal areas. The periphery is more enriched in 20:4*n6* (20% more than central) and less enriched in 22:5*n6* (20% less than central), resulting in similar *n6* fatty acid content between these areas, with a ratio of 22:5*n6*/22:4*n6* of 1.59 in the periphery and 2.13 in the central area. While no differences are observed in the total content of DHA, 22:5*n3* content is 33% higher peripherally than centrally, with ratios of DHA/22:5*n3* of 34.5 and 27.0, respectively. Although the conversion of [¹⁴C]18:3 to DHA in the poodle retina is minimal 24 hours after intravitreal injection of the precursor,[80] the endogenous content of fatty acids suggests that the central region is more active than the periphery in the final steps involved in the conversion of shorter-chain *n3* and *n6* fatty acids to 22-carbon fatty acids. The same retinal areas were analyzed in three *prcd* dogs. These animals (age 258 days) showed marked differences in the progression of the disease, as measured both by the loss of rod nuclei and the content of DHA-phospholipids. Nevertheless, changes observed in endogenous fatty acids exhibited the same trends, with the patterns of deterioration being more accentuated in the periphery than in central areas. Also, in all animals observed, the central superior region is the least affected. The content of 20:4*n6* is increased in all *prcd* retinal areas from 70% centrally and 80% peripherally in the more affected animal, to 10% centrally and 40% peripherally in the least affected. The ratios of 22:5*n6*/22:4*n6* and DHA/22:5*n3* are significantly decreased. This is a result, not only of lower 22:5*n6* and DHA, but also of higher values for their immediate precursors. It is interesting that 22:5*n3* is increased by 60-70% in the central region of all *prcd* dogs (including the least affected), while in the periphery, 22:5*n3* is similar to controls. This suggests that the central retina has a more active 22:5*n3*-to-DHA pathway in normal animals, and that the impairment of this process in *prcd* dogs leads to the observed rise in 22:5*n3*.

When retinas were incubated with [³H]DHA (final concentration, 0.1 µM) for four hours, the uptake of the precursor was more avid peripherally than in central retina of normal animals. This was also observed within *prcd* retinas, which, in addition, have a similar or higher uptake than normal retinas. Most recovered label was equally distributed between phosphatidylcholine (PC) and phosphatidylethanolamine (PE) (more than 70% of the total) in both areas and for both normal and *prcd* dogs. Significant differences in phosphatidic acid (PA) and phosphatidylinositol (PI) labeling are observed between areas, with more [³H]PA and [³H]PI labelling seen centrally; These areas have a higher density of photoreceptor cells.[84] The percent of PI and PA labeling with [³H]DHA is much higher than their mass DHA contribution to total retinal DHA-lipids in both frog and human retinas.[51] These observations suggest that DHA-PA and/or DHA-PI may play a role in photoreceptor cells. DHA is mainly incorporated into retinal phospholipids through the *de novo* route leading initially to PA-DHA formation.[81-85] This route of DHA incorporation is also reflected in photoreceptor outer segments where PA displays relatively high endogenous DHA content.[82] In retinas from *prcd* dogs, PA and PI labeling was greatly decreased while TAG labeling increased. The marked decrease in DHA in PA and PI in the *prcd* retina may reflect a) diminished availability of DHA in the photoreceptor cells, or b) enhanced subsequent metabolism of DHA-PA. The latter may be brought about by an increase in the activity of phosphatidate phosphohydrolase, shifting the *de novo* pathway to TAG and decreasing the availability of PA for PI synthesis.

Anderson et al.[86] have reported that supplementing the *prcd* dog diet with fish oil restores circulating levels of DHA; however, no improvement in retinal DHA content, ERG abnormalities, or histological appearance of the retina occurs. Thus, an impairment in *prcd* dogs in the delivery of DHA, in agreement with previous suggestions, may take place.[85,87] Because total *in vitro* labeling, as well as *in vivo* labeling attained seven days after [³H]DHA injection into the hepatic portal vein, are similar for both normal and *prcd* dogs, it is possible that alterations in *prcd* dogs involve DHA synthesis, both in liver and in retina. A decreased [³H]DHA in plasma of *prcd* has been observed.[87] Because peripheral retinal areas, when compared with central regions in normal animals, display a) the highest labeling, b) similar DHA content, c) similar rates of membrane synthesis,[78] it is possible that the conservation (recycling) of DHA is less active in the periphery. Because plasma from *prcd* dogs contains reduced levels of DHA,[37] a subsequent lower availability of DHA to the retina occurs. Moreover, a less efficient recycling mechanism in the periphery could contribute to a rapid deterioration with loss of peripheral photoreceptor cells. Finally, impairments within the recycling pathway, in both retinal areas, could contribute to the net loss of retinal DHA, even while DHA uptake from plasma may be normal.

## ACKNOWLEDGMENTS

This work was supported by the National Institutes of Health (R01 EY04428).

## REFERENCES

1. M.I. Aveldaño de Caldironi and N.G. Bazan, Composition and biosynthesis of molecular species of retina phosphoglycerides. *Neurochem. Int.* 1:381 (1980).
2. K. Boesze-Battaglia and A.D. Albert, Fatty acid composition of bovine rod outer segment plasma membrane, *Exp. Eye Res.* 49:699 (1989).
3. S.J. Fliesler and R.E. Anderson, Chemistry and metabolism of lipids in the vertebrate retina, *Prog. Lipid Res.* 22:79 (1983).
4. A.P. Simopoulos, Omega-3 fatty acids in health and disease and in growth and development. *Am. J. Clin. Nutr.* 54:438 (1991).
5. J. Tinoco, Dietary requirements and functions of α-linolenic acid in animals, *Prog. Lipid Res.* 21:1 (1982).
6. N.G. Bazan, Supply of n-3 polyunsaturated fatty acids and their significance in the central nervous system, *in*: "Nutrition and the Brain, Vol. 8," R.J. Wortman and J.J. Wortman, eds., Raven Press Ltd, New York (1990)
7. N.G. Bazan, D.L. Birkle, and T.S. Reddy, Biochemical and Nutritional aspects of the metabolism of polyunsaturated fatty acids and phospholipids in experimental models of retinal degeneration, *in*: "Retinal Degeneration: Experimental and Clinical Studies," M.M. LaVail, G. Anderson, J. Hollyfield, eds., Alan R. Liss, New York (1985).
8. N.G. Bazan, T.S. Reddy, T.M. Redmond, B. Wiggert, and G.J. Chander, Endogenous fatty acids are covalently and non-covalently bound to interphotoreceptor retinoid-binding protein in the monkey retina, *J. Biol. Chem.* 260:13677 (1985).
9. A. Nouvelot, C. Delbart, and J.M. Bourre, Hepatic metabolism of dietary alpha-linolenic acid in suckling rats, and its possible importance in polyunsaturated fatty acid uptake by the brain. *Ann. Nutr. Metab.* 30:316 (1986).
10. B.L. Scott and N.G. Bazan, Membrane docosahexaenoate is supplied to the developing brain and retina by the liver, *Proc. Nat. Acad. Sci. USA.* 86:2903 (1989).
11. N.G. Bazan, W.C. Gordon, and E.B. Rodriguez de Turco, Delivery of docosahexaenoic acid ($^3$H-22:6) by the liver to the retina in the frog, *Invest. Ophthalmol. Vis. Sci.* 32(suppl):701 (1991).
12. J. Li, M.G. Wetzel, and P.J. O'Brien, Transport of n-3 fatty acids from the intestine to the retina in rats, *J. Lipid Res.* 33:539 (1992).
13. W.C. Gordon and N.G. Bazan, Docosahexaenoic acid utilization during rod photoreceptor cell renewal, *J. Neurosci.* 10:2190 (1990).
14. W.C. Gordon and N.G. Bazan, [$^3$H]Docosahexaenoic acid uptake and utilization by retinal pigment epithelium and photoreceptors, *Invest. Ophthalmol. Vis. Sci.* (1993, in press).
15. E.B. Rodriguez de Turco, W.C. Gordon, G.A. Peyman, and N.G. Bazan, Preferential uptake and metabolism of docosahexaenoic acid in membrane phospholipids from rod and cone photoreceptor cells of human and monkey retinas. *J. Neurosci. Res.* 27:522 (1990).
16. E.B. Rodriguez de Turco, W.C. Gordon, and N.G. Bazan, Rapid and selective uptake, metabolism, and differential distribution of docosahexaenoic acid among rod and cone photoreceptor cells in the frog retina, *J. Neurosci.* 11:3667 (1991).
17. M.T. Clandinin, S. Cheema, C.J. Field, M.L. Garg, J. Venkatraman, and T.R. Clandinin, Dietary fat; Exogenous determination of membrane structure and cell function, *FASEB J.* 5:2761 (1991).
18. J.M. Bourre, M. Francois, A. Youyou, O. Dumont, M. Piciotti, G. Pascal, and G. Durand, The effects of dietary α-linolenic acid on the composition of nerve membranes, enzymatic activity, amplitude of electrophysiological parameters, resistance to poisons and performance of learning tasks in rats, *J. Nutr.* 119:1880 (1989).

19. M.T. Clandinin, C.J. Field, K. Hargreaves, L. Morson, and E. Zsigmond, Role of diet fat in subcellular structure and function, *Can J. Physiol. Pharmacol.* 63:546 (1985).

20. A.A. Spector and M.A. Yorek, Membrane lipid composition and cellular function, *J. Lipid Res.* 26:1015 (1985).

21. T.G. Wheeler, R.M. Benolken, and R.E. Anderson, Visual membranes: Specificity of fatty acid precursors for the electrical response to illumination, *Science.* 188:1312 (1975).

22. R.E. Wiegand, C. A. Koutz, A. M. Stinson, and R. E. Anderson, Conservation os docosahexaenoic acid in rod outer segments of rat retina during n-3 and n-6 fatty acid deficiency, *J. Neurochem.* 57:1690 (1991).

23. A.M. Stinson, R. D. Wiegand, and R. E. Anderson, Recycling of docosahexaenoic acid in rat retinas during n-3 fatty acid deficiency, *J. Lipid Res.* 32:2009, (1991).

24. M. Neuringer, W.E. Connor, D.S. Lin, L. Barstad, and S. Luck, Biochemical and functional effects of prenatal and postnatal ω3 fatty acid deficiency on retina and brain in rhesus monkeys, *Proc. Natl. Acad. Sci. USA.* 83:4021 (1986).

25. M. Neuringer, W.E. Connor, D.S. Lin, G.J. Anderson, and L. Barstad, Dietary omega-3 fatty acids: Effects on retinal lipid composition and function in primates, *in:* "Retinal Degenerations," J.G. Hollyfield, R.E. Anderson, and M.M. LaVail, eds., CRC Press, Boca Raton (1991).

26. R.D. Uauy, D.G. Birch, E.E. Birch, J.E. Tyson, and D.R. Hoffman, Effect of dietary omega-3 fatty acids on retinal function of very-low-birth-weight neonates, *Pediatr. Res.* 28:485 (1990).

27. D.G. Birch, E.E. Birch, D.R. Hoffman, and R. Uauy, Retinal development in very-low-birth-weight infants fed diets differing in omega-3 fatty acids, *Invest. Ophthalmol. Vis. Sci.* 33:2365, (1992).

28. E.E. Birch, D.G. Birch, D.R. Hoffman, and R. Uauy, Dietary essential fatty acid supply and visual acuity development, *Invest. Ophthalmol. Vis. Sci.* 33:3242 (1992).

29. N.G. Bazan, B.L. Scott, T.S. Reddy, and M.Z. Pelias, Decreased content of docosahexaenoate and arachidonate in plasma phospholipids in Usher's syndrome, *Biochem. Biophys. Res. Comm.* 141:600 (1986).

30. L.L. Williams, L.A. Horrocks, L.E. Leguire, and B.T. Shannon, Serum fatty acid proportions in Retinitis Pigmentosa may be affected by a number of factors, *in:* "Inherited and Environmentally Induced Retinal Degenerations," M.M. LaVail, R.E. Anderson, and J. Hollyfield, eds., Alan R. Liss, Inc., New York, (1989).

31. R.E. Anderson, M.B. Maude, R.A. Lewis, D.A. Newsome, and G.A. Fishman, Fatty acid levels in retinitis pigmentosa, *Exp. Eye Res.* 44:155 (1974).

32. R.E. Anderson, M.B. Maude, R.A. Lewis, D.A. Newsome, and G.A. Fishman, Abnormal plasma levels of polyunsaturated fatty acids in autosomal dominant retinitis pigmentosa, *Exp. Eye Res.* 44:155 (1987).

33. C.A. Converse, H.M. Hammer, C.J. Packard, and J. Shepherd, Plasma lipid abnormalities in retinitis pigmentosa and related conditions, *Trans. Ophthalmol. Soc. UK.* 103:508 (1983).

34. C.A. Converse, T. McLachlan, C.J. Packard, and J. Shepherd, Lipid abnormalities in retinitis pigmentosa, *in:* "Advances in the Biosciences, Vol. 62: Research in Retinitis Pigmentosa," E. Zrenner, H. Krastel, and H.-H. Goebel, eds., Pergamon, Oxford (1987).

35. C.A. Converse, T. McLachlan, A.C. Bow, C.J. Packard, and J. Shepherd, Lipid metabolism in retinitis pigmentosa, *in:* "Degenerative Retinal Disorders: Clinical and Laboratory Investigations," J.G. Hollyfield, R.E. Anderson, and M.M. LaVail, M.M., eds., Alan R. Liss, New York (1987).

36. J. Gong, B. Rosner, D.G. Rees, E.L. Berson, C.A. Weigel-DiFranco, and E.J. Schaefer, Plasma docosahexaenoic acid levels in various genetic forms of retinitis pigmentosa, *Invest. Ophthalmol. Vis. Sci.* 33:2596 (1992).

37. R.E. Anderson, M.B. Maude, R.A. Alvarez, G.M. Acland, and G.D. Aguirre, Plasma lipid abnormalities in the miniature poodle with progressive rod-cone degeneration, *Exp. Eye Res.* 52:349 (1991).

38. R.E. Anderson, M.B. Maude, R. A. Alvarez, S. E. G. Nilsson, K Narfström, G. M. Acland, and G. Aguirre, Plasma lipid abnormalities in *prcd*-affected miniature poodles and abyssinian cats, *in:* "Retinal Degenerations," J.G. Hollyfield, R.E. Anderson, and M.M. LaVail, eds., CRC Press, Boca Raton (1991).

39. G.J. Nelson and R.G. Ackman, Absorption and transport of fat in mammals with emphasis on n-3 polyunsaturated fatty acids, *Lipids.* 23:1005 (1988)..

40. M.D. Rosenthal, Fatty acid metabolism of isolated mammalian cells. *Prog. Lipid Res.* 26:87 (1987).

41. P. Budowski and M.A. Crawford, α-linolenic acid as a regulator of the metabolism of arachidonic acid: Dietary implications of the ratio, n-6:n-3 fatty acids, *Proc. Nutrition Soc.* 44:221 (1985).

42. S.M. Innis, Essential fatty acids in growth and development, *Prog. Lipid Res.* 30:39 (1991).

43. K.J. Clark, M. Makrides, M.A. Neumann, and R.A. Gibson, Determination of the optimal ratio of linoleic acid to α-linolenic acid in infant formulas, *J. Pediatr.* 120:S151 (1992).

44. J.A. Nettleton, ω-3 fatty acids: Comparison of plant and seafood sources in human nutrition. *J. Am. Diet. Assoc.* 91:331 (1991).

45. R.E. Martin and N.G. Bazan, Changing fatty acid content of growth cone lipids prior to synaptogenesis, *J. Neurochem.* 59:318 (1992).

46. H. Sprecher, Long chain fatty Acid metabolism, *in* "Polyunsaturated Fatty Acids in Human Nutrition," R. Bracco and R.J. Deckelbaum, eds., Raven Pres, Ltd., New York (1992).

47. M.G. Wetzel, J. Li, R.A. Alvarez, R.E. Anderson, and P.J. O'Brien, Metabolism of linolenic acid and docosahexaenoic acid in rat retinas and rod outer segments, *Exp. Eye Res.* 53:437 (1991).

48. H.W. Cook, *In Vitro* formation of polyunsaturated fatty acids by desaturation in rat brain: Some properties of the enzymes in developing brain and comparisons with liver, *J. Neurochem.* 30:1327 (1978).

49. J.M. Bourre and M. Piciotti, Delta-6 desaturation of alpha-linolenic acid in brain and liver during development and aging in the mouse, *Neurosci. Lett.,* 141:65 (1992).

50. N.G. Bazan, E. B. Rodriguez de Turco, W. C. Gordon, Pathways for the uptake and conservation of docosahexaenoic acid in photoreceptors and synapses: biochemical and autoradiographic studies, *Can. J. Physiol. Pharmacol.* (submitted).

51. R.E. Martin, E.B. Rodriguez de Turco, and N.G. Bazan, Development and biochemical aspects of docosahexaenoic acid (DHA) accumulation in mouse retina and brain, (submitted).

52. H.E.P. Bazan and N.G. Bazan, Incorporation of [³H]-arachidonic acid into cattle retina lipids: High uptake in triacylglycerol, diacylglycerols, phosphatidylcholine and phosphatidylinositol, *Life Sciences* 17:1671 (1975).

53. T.-A. Hagve and B.O. Christophersen, Linolenic acid desaturation and chain elongation and rapid turnover of phospholipid n-3 fatty acids in isolated rat liver cells, *Biochim. Biophys. Acta.* 753:339 (1983).

54. J. Leyton, P.J. Drury, and M.A. Crawford, *In vivo* incorporation of labeled fatty acids in rat liver lipids after oral administration, *Lipids.* 22:553 (1987).

55. P. Clouet, I. Niot, and J. Bézard, Pathway of α-linolenic acid through the mitochondrial outer membrane in the rat liver and influence on the rate of oxidation: Comparison with linoleic and oleic acid, *Biochem. J.* 263:867 (1989).

56. G.A. Dhopeshwarkar, Metabolism of linolenic acid in developing brain: I. Incorporation of radioactivity from 1-¹⁴C-linolenic acid into brain fatty acids, *Lipids.* 10:238 (1975).

57. A.J. Sinclair, Incorporation of radioactive polyunsaturated fatty acids into liver and brain of developing rat. *Lipids.* 10:175 (1975).

58. W.C. Gordon, E.B. Rodriguez de Turco, F.O. Richardson, and N.G. Bazan, Marked differences in the liver supply and retinal/RPE uptake of palmitate (16:0) and docosahexaenoate (22:6), *Invest. Ophthalmol. Vis. Sci. (Suppl)* (1993, in press).

59. R.M. Benolken, R.E. Anderson, and T.G. Wheeler, Membrane fatty acids associated with electrical response in visual excitation, *Science.* 182:1253 (1973).

60. N.P. Rotstein, M.G. Ilincheta de Boschero, N.M. Giusto, and M.I. Aveldaño, Effects of aging on the composition and metabolism of docosahexaenoate-containing lipids of retina, *Lipids.* 22:253 (1987).

61. A. Voss, M. Reinhart, S. Sankarappa, H. Sprecher, The metabolism of 7, 10, 13, 16, 19-docosapentaenoic acid to 4, 7, 10, 14, 16, 19-docosahexaenoic acid in rat liver is independent of a 4-desaturase, *J. Biol. Chem.* 266:19995 (1991).

62. N.G. Bazan and E.B. Rodriguez de Turco, Supply, uptake, and retention of docosahexaenoic acid in the pathophysiology of Batten's disease, *J. Inher. Metab. Dis.* (1993, in press).

63. M. Martinez, Severe changes in polyunsaturated fatty acids in the brain, liver, kidney, and retina in patients with peroxisomal disorders, *in*: "Neurobiology of Essential Fatty Acids: Advances in Experimental Medicine and Biology," N.G. Bazan, M.G. Murphy, and G. Toffano, eds., Plenum, New York (1992).

64. T.S. Reddy, D.L. Birkle, D. Armstrong, and N.G. Bazan, Changes in content, incorporation, and lipoxygenation of docosahexaenoic acid in retina and retinal pigment epithelium in canine ceroid lipofuscinosis. *Neurosci. Let.* 59:67 (1985).

65. J.E. Vance and D.E. Vance, The assembly of lipids into lipoproteins during secretion, *Experientia.* 46:560 (1990).

66. V.L. Marcheselli and N.G. Bazan, Quantitative analysis of fatty acids in phospholipids, diacylglycerols, free fatty acids, and other lipies, *J. Nutr. Biochem.* 1:231 (1990).

67. R.E. Anderson, P.J. O'Brien, R.D. Wiegand, C.A. Koutz, and A.M. Stinson, Conservation of docosahexaenoic acid in the retina. *in*: "Neurobiology of Essential Fatty Acids. Advances in Experimental Medicine and Biology," N.G. Bazan, M.G. Murphy, and G. Toffano, eds., Plenum: New York. (1992).

68. N.G. Bazan, W.C. Gordon, and E.B. Rodriguez de Turco, Docosahexaenoic acid uptake and metabolism in photoreceptors: Retinal conservation by an efficient retinal pigment epithelial cell-mediated recycling process, *in*: "Neurobiology of Essential Fatty Acids. Advances in Experimental Medicine and Biology," N.G. Bazan, M.G. Murphy, and G. Toffano, eds., Plenum: New York. (1992).

69. W.C. Gordon, E.B. Rodriguez de Turco, and N.G. Bazan, Retinal pigment epithelial cells play a central role in the conservation of docosahexaenoic acid by photoreceptor cells after shedding and phagocytosis, *Current Eye Res.* 11:73 (1992).

70. J.E. Vance and D.E. Vance, Lipoprotein assembly and secretion by hepatocytes. Annu. Rev. Nutr. 10:337 (1990).

71. G.F. Gibbons, Assembly and secretion of hepatic very-low-density lipoprotein, *Biochem. J.* 268:1 (1990).

72. S. Eisenberg, High density lipoprotein metabolism. *J. Lipid Res.* 25:1017 (1984).

73. K.C. Hayes, S. Lindsey, Z.F. Stephan, D. Brecker, Retinal pigment epithelium possesses both LDL and scavenger receptor activity. *Invest. Ophthalmol. Vis.Sci.* 30:225 (1989).

74. S. Méresse, C. Delbart, J.-C. Fruchart, and R. Cecchelli, Low-density lipoprotein receptor on endothelium of brain capillaries, *J. Neurochem.* 53:340 (1989).

75. N.G. Bazan, E.B. Rodriguez de Turco, and W.C. Gordon, Docosahexaenoic acid and phospholipid metabolism in photoreceptor cells and in retinal degeneration *in*: "Retinal Degenerations," J.G. Hollyfield, R.E. Anderson, and M.M. LaVail, eds., CRC Press, Boca Raton (1991).

76. N. Wang, R. D. Wiegand, and R. E. Anderson, Uptake of 22-carbon fatty acids into rat retina and brain, *Exp. Eye Res.* 54:933 (1992).

77. G. Aguirre, J. Alligood, P.J. O'Brien, and N. Buyukmihci, Pathogenesis of progressive rod-cone degeneration in miniature poodles, *Invest. Ophthalmol. Vis. Sci.* 23:610 (1982).

78. G. Aguirre and P. O'Brien, Morphological and biochemical studies of canine progressive rod-cone degeneration: $^3$H-fucose autoradiography, *Invest. Ophthalmol. Vis. Sci.* 27:635 (1986).

79. E.B. Rodriguez de Turco, W.C. Gordon, W. K. Morgan, and N.G. Bazan, Heterogeneity in docosahexaenoic acid (22:6$\omega$3) metabolism in different retinal regions of control and prcd dogs, *Invest. Ophthalmol. Vis. Sci (Suppl.)* (1993, in press).

80. M.G. Wetzel, C. Fahlman, P.J. O'Brien, and G.D. Aguirre, Metabolic labeling of rod outer segment phospholipids in miniature poodles with progressive rod-cone degeneration (*prcd*) *Exp. Eye Res.* 50:89 (1990).

81. H.E.P. Bazan, M.M. Careaga, H. Sprecher, and N.G. Bazan, Chain elongation and desaturation of eicosapentaenoate to docosahexaenoate and phospholipid labeling in the rat retina in vivo, *Biochem. Biophys. Acta* 712:123 (1982).

82. N.G. Bazan, M.S. di Fazio de Excalante, M.M. Careaga, H.E.P. Bazan, and N.M. Giusto, High content of 22:6 (docosahexaenoate) and active [2-$^3$H]glycerol metabolism of phosphatidic acid from photoreceptor membranes, *Biochem. Biophys. Acta* 712:702 (1982).

83. H.E.P. Bazan, H. Sprecher, and N.G. Bazan, De novo biosynthesis of docosahexaenoyl phosphatidic acid in bovine retinal microsomes, *Biochem. Biophys. Acta* 796: 11-19 (1984).

84. H.E.P. Bazan, M.M. Careaga, and N.G. Bazan, Propranolol increases the biosynthesis of phosphatidic acid, phosphatidylinositol and phosphatidylserine in the toad retina. Studies in the entire retina and subcellular fractions, *Biochem. Biophys. Acta* 666:63 (1981)

85. E.B. Rodriguez de Turco, W.C. Gordon, G. M. Acland, G.V.M.D.Aguirre, and N.G. Bazan, Docosahexaenoic acid (DHA) metabolism in poodles with progressive rod-cone degenerations (PRCD), *Invest. Ophthal. Vis. Sci. Suppl.* 33:1066 (1992).

86. R.E. Anderson, M.B.Maude, R.A. Alvarez, M.G. Wetzel, P.J. O'Brien, G.M. Acland, and G. Aguirre, Role of docosahexaenoic acid in inherited retinal degenerations, *in:* "The Third International Congress on Essential Fatty Acids and Eicosanoids," A. Sinclair and R. Gibson, eds., American Oil Chemists' Society, Champaign (1993).

87. V.C. Strand, and N.G. Bazan, Supply of docosahexaenoic acid-containing lipoproteins to the retina: the role of the liver in the miniature poodle, *Invest. Ophthal. Vis. Sci. Suppl.* 32:1189 (1992).

# VARIABLE EXPRESSIVITY OF rd-3 RETINAL DEGENERATION DEPENDENT ON BACKGROUND STRAIN

John R. Heckenlively,[1] Bo Chang,[2] Chen Peng,[1] Norman L. Hawes,[2] Thomas H. Roderick[2]

[1]Jules Stein Eye Institute, Harbor-UCLA Medical Center, Torrance, California 90509, [2]The Jackson Laboratory, Bar Harbor, Maine 04609

## INTRODUCTION

Mouse models of retinal degeneration are proving to be useful for investigating analogous human conditions; while mice eyes are smaller, they are very similar to human, both morphologically and histologically. It is also possible to examine mouse retinas by indirect ophthalmoscopy and electroretinography, which provides a strong clinical basis for comparing mouse and human conditions.[1] Furthermore, because mouse genetics has been studied extensively for the last 60 years, much is known about the homology between mouse and human genes, and genetic discoveries in mouse can quickly lead to inquiries on the same gene in human.

The principle mouse retinal degenerations which have been studied are the retinal degeneration (gene symbol rd),[2] retinal degeneration slow (rds),[3] purkinje cell degeneration (pcd),[4] and nervous (nr).[5] Molecular genetic studies have determined that rd codes for ß-subunitt of cyclic GMP-phosphodiesterase which has been mapped to human chromosome 4p16,[6] and rds codes for a bridging protein called peripherin in the rod outer segment membrane which is found on human chromosome 6q.[7] Genes for various components of the phototransduction cascade and subunits of photoreceptors have been mapped in mouse and human, and mutations of these genes would be expected to lead to human retinal degenerations.[8]

We recently reported a new murine model for retinal degeneration with the gene symbol rd-3, which is inherited in the autosomal recessive manner and found in an albino mouse, the RBF/DnJ strain; these mice were originally derived from four wild mice captured in Valle de Poschiavo in Switzerland and crossed with "Swiss mice".[9] In the RBF/DnJ rd-3, photoreceptors develop normally through 2 weeks postnatally after which the retina degenerates leaving an extinguished ERG by six weeks of age. Crosses with rd, rds, nr, pcd revealed normal offspring by histology and electroretinographic testing. Linkage analysis places rd-3 on mouse chromosome 1, 10 ± 2.5 cM distal to Akp-1 and homology mapping suggests that the human homologous locus should be on chromosome 1q.

## METHODS

**Mice:** The _rd-3_ mice were originally trapped by Dr. Alfred Gropp in the Valle di Poschiavo in southeast Switzerland in 1969 and the mice were imported by Dr. T.H. Roderick to the Jackson Laboratory because they were known to have Robertsonian chromosomes. The mice were bred to several mouse strains, and then inbred to fix the lines for the Robertsonians (Figure 1). The simultaneous fixing of _rd-3_ was fortuitous and was discovered later. The lines with the _rd-3_ were RBF/DnJ, Meta-In(1)Rk, Rb(11.13)4Bnr, In(5)30Rk, and the lines without the rd-3 were In(13)31Rk, Rb4.6)2Bnr, Rb(16.17)7Bnr (Figure 1). The In30x4BnrF1 were produced by crossing In(5)30Rk with Rb(11.13)4Bnr. The _Rds_ mice, C3H+/+, _Rds/Rds_ and 020/A, were imported to the Jackson Laboratory from Dr. S. Sanyal of Erasmus University, Rotterdam, The Netherlands. The Jackson Laboratory _rd_ mice used for crosses were: CBA/J, C3H/HeJ, SWR/J, SJL/J, MOLD/Rk (inbred wild Japanese tree mice), and SF/CamRk (inbred wild mice trapped from San Francisco).

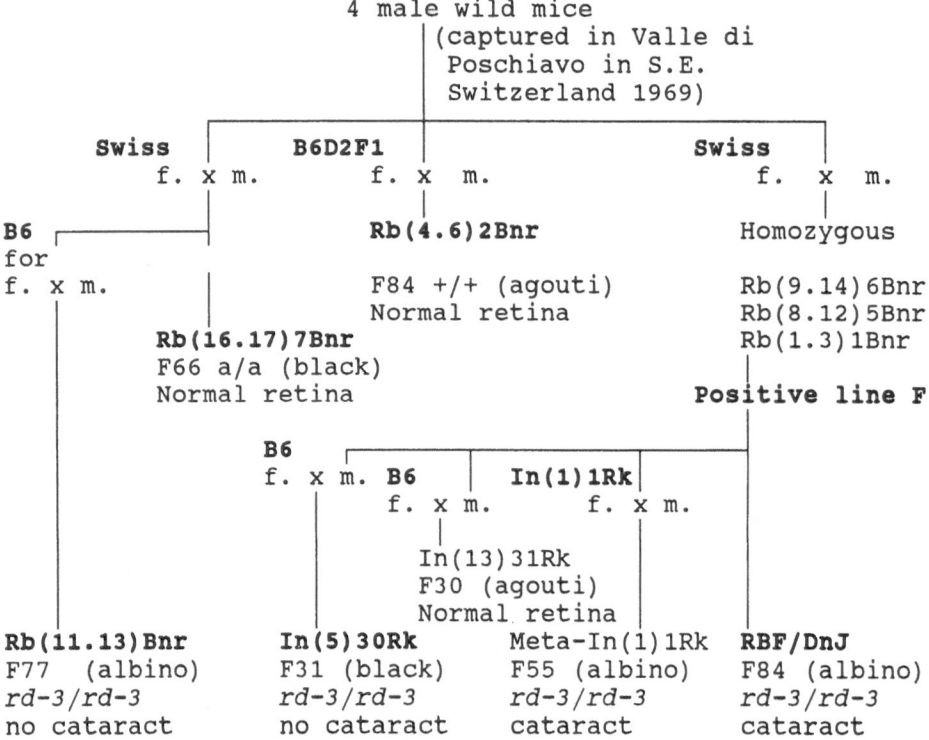

**Figure 1.** Pedigree chart showing origin and breeding strains used for inbreeding of mice carrying _rd-3_ gene (m.=male, f.=female, Rb.=Robertsonian Translocation, F=generation of inbreeding, B6=C57BL/6J, B6D2F1=first generation offspring of B57BL/6J x DBA/2J).

**Retina evaluation:**  The electroretinogram is performed using a 386 IBM compatible computer with a single 8-bit expansion card with a 8-channel, 12-bit analog to digital converter. The mice are dilated with 1% atropine drops, dark-adapted for at least 40 minutes, and then anesthetized with a mixture of 0.0153mg/gm xylazine and 0.024mg/gm ketamine.  Ground and reference electrodes are standard Grass subdermal electrode needs placed on either side of the mandibles, and the active lead is a Grass E-2 platinum wire lightly touching the superior cornea.  The light stimulus was from a Grass xenon lamp with neutral density filters and electronic controls changing the intensity under a standard protocol.  Normal control values have been established in several mice strains. Different rd-3 mice from the same inbred strain were tested on a weekly basis postnatally.

**Histology:**  After the ERG eyes were taken for histology, and fixed in a 1% paraformaldehyde and 2% glutaldehyde and 0.1M cacodylate buffer for 24 hours, and then transferred to a cold 0.1 M cacodylate buffer solution for 24 hours. Embedding was done with hydroxyethylmethacrylate and sections stained with H&E.  Photomicrographs were taken with 40x magnification using an Olympus microscope automated camera and Kodak ASA 100 color print film.  These prints were cut in one inch strips and lined up chronologically for comparison. Sections were not always in the same location in the eye and for this reason some variance occurs in thickness of retinal section.

**Clinical evaluation:**  Indirect ophthalmoscopy was performed with 40 and 60 diopter lenses, while fundus photographs were taken with a 90 diopter lens and a Nikon biomicroscope camera.

**RESULTS**

We found that there was varying expression of the rd-3 phenotype on different genetic backgrounds.  While all four of the strains investigated demonstrated abnormal photoreceptors at 3 weeks postnatally, there were marked differences in the time over which the retinal degeneration takes place on histology and the electroretinogram; a summary table below describes these changes.

Table 1.  Varying expression of rd-3 dependent on background strain.

| Background Strain | Cataract | Coat Color | ERG Extinguished | Histology ONL Degen. |
|---|---|---|---|---|
| RBF/DnJ | yes | albino | 6 wks | 14d-8 wks |
| Meta-In(1)Rk | yes | albino | 6 wks | 14d-8 wks |
| Rb(11,13)4Bnr | no | albino | 9 wks | 3-10 wks |
| (In-30x4BnrF1) | no | agouti | 12 wks | 3-12 wks |
| In-30 | no | black | 16 wks | 3-16 wks |

**Electroretinograms:** The electroretinograms are never normal in all strains with <u>rd-3</u> even though the photoreceptors appear normal on light histology at 2 weeks. However, even in the normal mouse, the ERG signal does not look similar to an adult waveform until four weeks of age. The ERG and photoreceptor loss correlates well in all strains of <u>rd-3</u> (Figures 1a-d, 2).

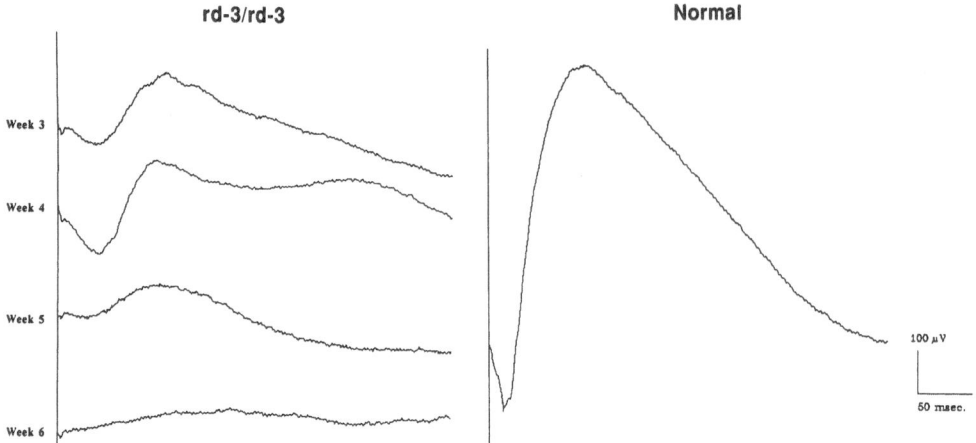

**Figure 1a.** RBF/DnJ <u>rd3/rd3</u> Electroretinogram, Scotopic Intensity 4, no neutral density filter

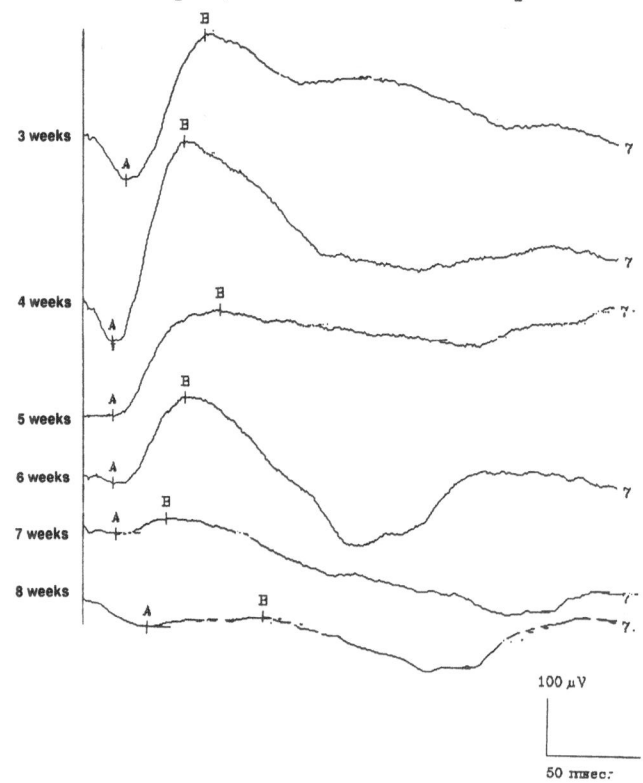

**Figure 1b.** Rb(11,13)4Bnr <u>rd3/rd3</u> Electroretinogram, Scotopic Intensity 4, no neutral density filter

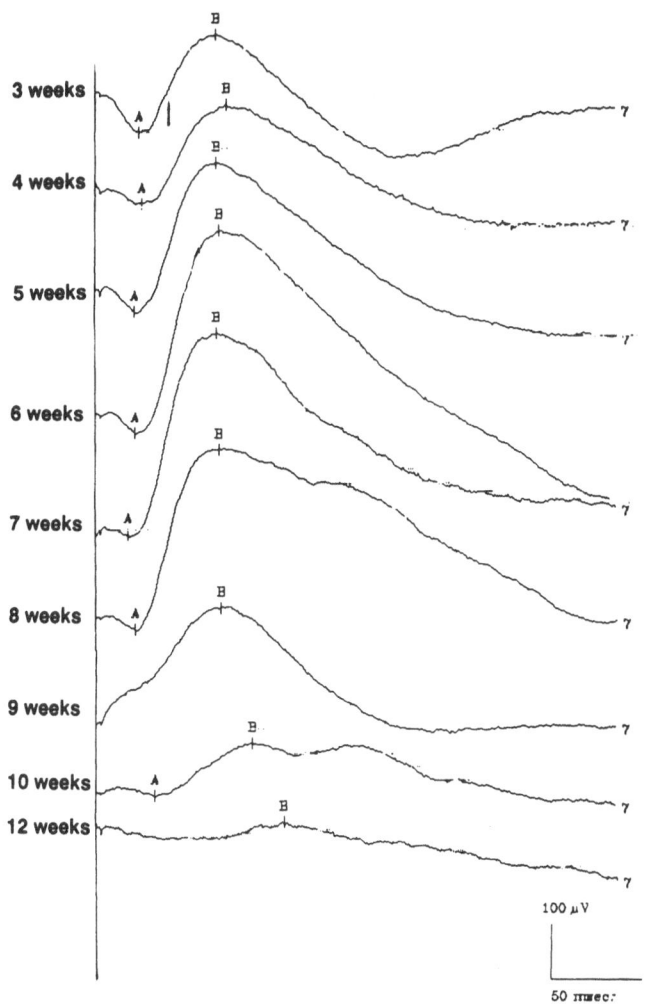

**Figure 1c.** First generation of cross In-30 x 4Bnr <u>rd3/rd3</u>
Electroretinogram, Scotopic, intensity 4, no
neutral density filter

     While the RBF/DnJ albino and the Meta-In(1)Rk have a
fast course with disorganized outer segments at 4 weeks, the
maximal ERG amplitudes still were seen at 4 weeks (Figure
1a&b). In the In(5)30Rk pigmented line, with the most
prolonged degeneration course, the maximal ERG amplitude was
at six weeks (Figure 1d); however it should be noted that the
a-wave was essentially lost or flattened at the higher
intensities by 6 weeks compared to normal mice which
typically have 200µv a-wave (negative) amplitudes.
     Of great interest were the mice derived by crossing the
In(5)30Rk pigmented line with the Rb(11.13)4Bnr albino line.
The F1 agouti offspring had ERG findings (Figure 1c) and
histological changes that were intermediate to that of the
parents. The ERGs extinguished at 12 weeks compared to 9
weeks for the 4Bnr albino and 15 weeks for the In(5)30Rk

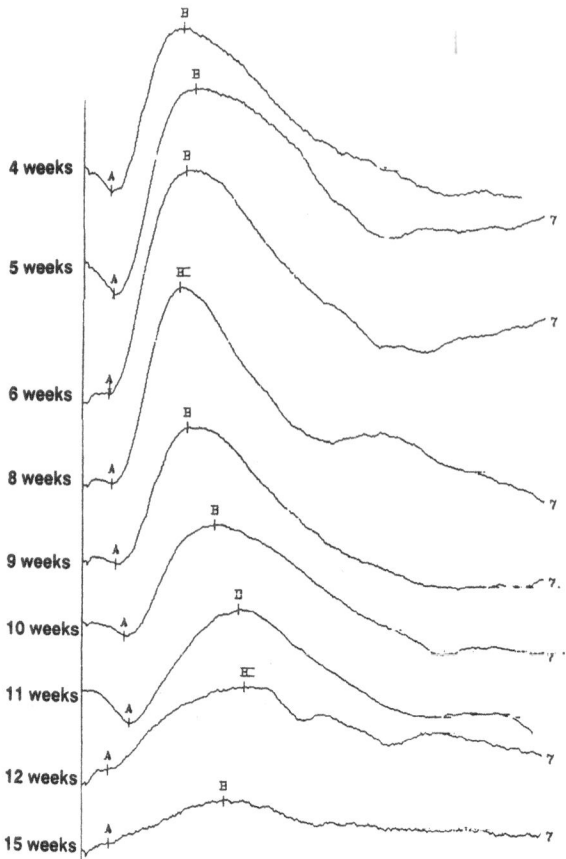

4 weeks

5 weeks

6 weeks

8 weeks

9 weeks

10 weeks

11 weeks

12 weeks

15 weeks

**Figure 1d.** In-30 <u>rd3/rd3</u> Electroretinogram, Scotopic Intensity 4, no neutral density filter

pigmented mice, which paralleled the histologic changes in each strain.

**Histology:** In comparing the photoreceptor outer and inner segment layers in the In(5)30Rk and 4Bnr backgrounds, formed photoreceptors are present in both strains at 3 weeks, after which disorganization and loss of the outer nuclear layer and photoreceptors occurs over 9 weeks in the 4Bnr and 16 weeks in the In(5)30Rk mice. These changes correlate well with the loss on the electroretinogram.

**Fundus appearance:** While retinal atrophy can be seen in albino mice as loss of the retinal vasculature, it is not possible to visualize the retinal pigment epithelium in any meaningful way in the albino mouse eye. However, the fundus changes which more closely resembles the human pathologic condition occurred in the In(5)30Rk black and In(5)30Rkx4BnrF1 agouti <u>rd-3</u> mice, where the retinal pigment epithelium can be visualized. In these mice, the retinal degeneration first appears as an increased generalized pigmented granularity to the retinal pigment epithelium, and by four months the degeneration manifests as a lobule dropout of the retinal pigment epithelium leaving bare sclera similar in appearance to the human condition called choroideremia.

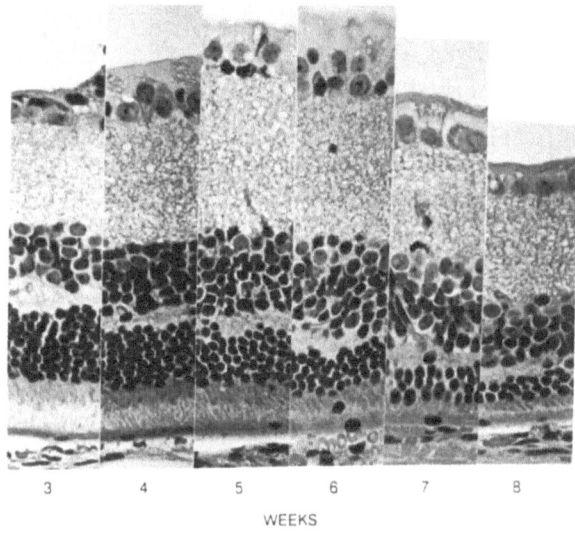

WEEKS

**Figure 2.** Light microscopy of Rb(11.13)4Bnr <u>rd3/rd3</u> retina from 3 to 8 weeks (magnification 40x)

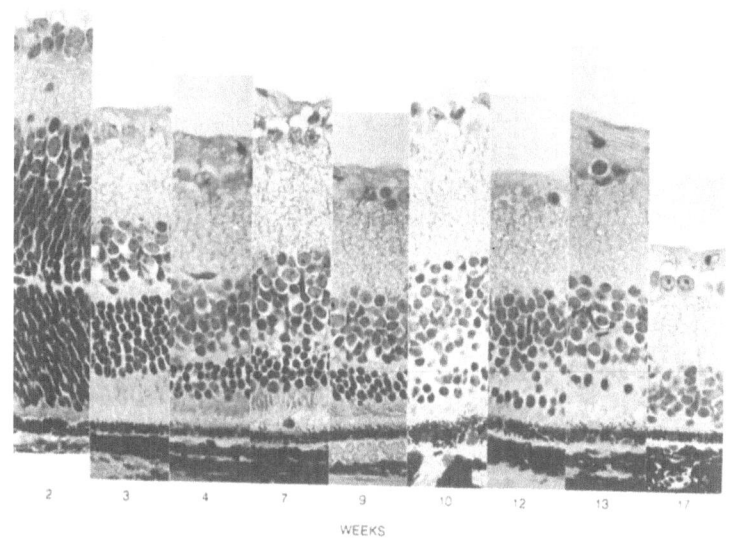

WEEKS

**Figure 3.** Light microscopy of In(5)30Rk <u>rd3/rd3</u> retina from 3 to 17 weeks

## DISCUSSION

The reason for the varying expression of <u>rd-3</u> in the various backgrounds is not known. Probably it is due to one or more secondary loci. Once the major gene is identified that is causing the retinal degeneration it may become obvious what secondary factor is altering the course of the retinal degeneration. It should be possible to do linkage analysis for this secondary factor which may help in its identification.

There are important ramifications to the discovery of varying expression of phenotype on different genetic backgrounds. First this finding validates the concept that a single gene causing a degeneration can have its effect altered by other genetic factors, and may help to explain the variance seen within families for genetic disease. It also implies that some single gene degenerations potentially can be altered or cured by secondary factors or therapeutic manipulation, and this line of investigation may be potentially fruitful in helping patients.

Much still has to be learned about rd-3, including isolating the gene, finding its product, understanding the mechanism of retinal degeneration and the secondary factor(s) which prolong the course of the degeneration.

**ACKNOWLEDGMENT**

Supported by NIH RO1 EY07758.

**REFERENCES**

1.  J.R. Heckenlively, J.V. Winston, T.H. Roderick, Screening for mouse retinal degeneration. I. Correlation of indirect ophthalmoscopy, electroretinograms, and histology. *Doc. Ophthalmol.* **71**:229 (1989).
2.  R.L. Sidman, M.C. Green, Retinal degeneration in the mouse: Location of the rd locus in linkage group XVII. *J. Hered.* **56**:23 (1965).
3.  R. Van Nie, D. Ivanyi, P. Demant, A new H-2 linked mutation, rds, causing retinal degeneration in the mouse. *Tissue Antigens* **12**:106, (1978).
4.  R.J. Mullen, E.M. Eicher, R.L. Sidman, Purkinje cell degeneration, a new neurological mutation in the mouse. Proc. Natl. Acad. Sci. USA **73**:208, (1976).
5.  R.L. Sidman, M.C. Green, Nervous, a new mutant mouse with cerebellar disease. In: "Les mutants pathologiques chez l'animal", M. Sabourdy, ed., Centre National de la Recherche Scientific, pp 69-79, Paris, (1970).
6.  J.B. Bateman, I. Klisak , T. Kojis, T. Mohandas, R.S. Sparkes, T. Li, M.L. Applebury, C. Bowes, and D.B. Farber, Assignment of the beta-subunit of rod photoreceptor cGMP phosphodiesterase gene PDEB (homolog of the mouse rd gene) to human chromosome 4p16. *Genomics* **12**:601 (1992).
7.  G.H. Travis, L. Christerson, P.E. Danielson, I. Klisak, R.S. Sparkes, L.B. Hahn, T.P. Dryja, J.G. Sutcliffe, The human retinal degeneration slow (RDS) gene: chromosome assignment and structure of the mRNA. *Genomics* **10**:733 (1991).
8.  B. Chang, J.R. Heckenlively, N.L. Hawes, T.H. Roderick, New Mouse Primary Retinal Degeneration (rd-3) linked to chromosome 1 distal to Akp-1. Genomics, Accepted for Publication.
9.  M.F.W. Festing "Inbred strains of mice", In: Genetic variants and strains of the laboratory mouse (eds M.F. Lyon and A.G. Searle), 2nd Ed. Oxford University Press, pp. 636-648, (1989).

# RETINAL DEGENERATIONS IN THE BRIARD DOG

P. Watson[1], A. Wrigstad[2], R.C. Riis[3], K. Narfstrom[2,4], P.G.C. Bedford[1], S.E.G. Nilsson [2]

[1] Department of Small Animal Studies, Royal Veterinary College. Hatfield. AL9 7TA. U.K.
[2] Department of Ophthalmology, University of Linkoping, S-581 85 Linkoping, Sweden
[3] Comparative Ophthalmology Service. College of Veterinary Medicine. Cornell University. Ithaca, NY 14853. U.S.A.
[4] Department of Surgery and Medicine, Faculty of Veterinary Medicine, Swedish University of Agricultural Sciences, S-750 07 Uppsala, Sweden

## INTRODUCTION

Retinal degeneration in the Briard dog has been described by a number of authors in different countries. Congenital night blindness has been reported in the USA[1] and in France[2]. In Sweden, the first Briards studied were congenitally night blind[3], but in successive generations of dogs day vision also became affected[4,5]. In contrast, a slowly progressive central retinal degeneration has been reported in the United Kingdom[6]. These dogs had apparently normal vision initially with no evidence of night blindness, but severe visual impairment developed with time. The problem in the USA has been referred to as the Briard lipid retinopathy, the problem in France simply as congenital night blindness and the problem in Sweden as congenital night blindness with severely reduced day vision. The condition reported in the UK has been referred to as central progressive retinal atrophy (cPRA) or retinal pigment epithelial dystrophy (RPED).

Retinal pigment epithelium (RPE) abnormalities are a prominent feature of each of these conditions in the Briard. The diseases, in particular those described in the USA, Sweden and France, have a number of features in common, and may represent manifestations of related disease processes. They may represent useful animal models for a number of human retinal degenerations. The purpose of this review is to compare these conditions more precisely.

In the studies discussed here all dogs underwent full clinical and ophthalmoscopic examinations. Fundus examination was performed by indirect ophthalmoscopy following pupil dilation. Behavioural included maze testing and falling cotton wool balls under conditions of dim and bright light illumination. Electroretinography was performed under specified conditions of neuroleptanalgesia or halothane anaesthesia, using corneal contact lens electrodes. Routine biochemistry was performed on fresh and frozen plasma along with detailed analysis of plasma lipids and fatty acids. Samples for histological examination were fixed immediately in 2-3% glutaraldehyde. Tissue for electron microscopy was post-fixed in osmium tetroxide. There were slight variations in methodology between studies but in each case groups of ophthalmoscopically normal Briards and crossbred dogs were examined as controls.

*Retinal Degeneration*, Edited by J.G. Hollyfield *et al.*
Plenum Press, New York, 1993

# NATURAL HISTORY / GENETICS

It has been reported in France that a very high proportion of the Briard population is affected and the results of a preliminary study of 62 affected dogs suggested an autosomal recessive mode of inheritance[7]. A study of three affected generations from one family of dogs in Sweden[3] also suggested an autosomal recessive mode of inheritance although this has yet to be confirmed. In the USA relatively few affected dogs have been identified.

An early study in the UK[6] established that 31% of Briards over 18 months of age were clinically affected by RPED. A recent survey (Bedford and Watson, Unpublished observations) has indicated however that currently the disease is much less prevalent. Initial studies suggested that this condition might be an autosomal recessive trait but further pedigree analysis does not entirely support this hypothesis.

## CLINICAL AND OPHTHALMOSCOPIC FEATURES

Dogs in the USA[1], and France[7] presented with congenital night blindness which could be detected readily by 5-6 weeks of age. At this stage there had been complete postnatal maturation of the retina. In Sweden most dogs presented with night blindness and severely reduced day vision[3,4]. Clinical behavioural testing clearly differentiated affected from normal dogs. Some affected dogs in the USA displayed an intermittent rapid vertical pendular nystagmus, consistent with a diagnosis of congenital retinal disease. In conditions of bright light most dogs had sufficient vision to permit normal behaviour. However the menace reflex could not be elicited in dogs in the USA. In Sweden some dogs showed marked visual deficits even in bright light. In all three studies pupillary light reflexes were normal although in general affected dogs had larger pupils than normal dogs in conditions of the same luminance.

In most affected dogs the fundus initially appeared ophthalmoscopically normal. Minor fundus abnormalities were detected in some dogs in the USA, including the development of small multifocal, linear non-tapetal depigmentations and subtle variations in the degree and nature of tapetal reflectivity. In Sweden the initial report described the condition as stationary, but later reports found it to be slowly progressive[8,9] with one dog (the oldest) which had been ophthalmoscopically normal when young developing minor fundoscopic abnormalities by 5 years of age. The principle abnormalities in this dog were progressive development of pale pigment spots in the tapetal fundus and slight vascular attenuation. All dogs remained otherwise clinically normal except in the USA where persistent unexplained problems of diarrhoea and pyrexia were reported[1].

Night blindness has not been reported in Briard dogs in the UK. The condition reported in this country, referred to herein as RPED, had a very variable age of onset and a variable rate of progression[6]. In early RPED no behavioural abnormalities were seen, however ultimately the condition caused severe visual impairment. Occasionally blindness was noted in affected dogs. Ophthalmoscopic abnormalities were first seen as early as 17 months but in other dogs not until 10 years of age. Most commonly abnormality was first detected between 2 and 5 years of age. The disease was characterised ophthalmoscopically by the bilateral development of small pale-brown circular patches of pigment in the temporal aspect of the tapetal fundus. With time the number and size of the patches increased, ultimately involving most of the tapetal fundus. Areas of hyper-reflectivity appeared between the pigment patches (Fig. 1). Vascular attenuation and optic atrophy were reported in some dogs. No ophthalmoscopic abnormalities were seen in the non-tapetal fundus.

## ELECTRORETINOGRAPHY

Comprehensive electroretinographic (ERG) testing has been performed on Briards in Sweden[4,10]. Single flash ERG on affected dogs at 7-12 months of age showed no definite a- and b-waves. DC ERG in these dogs showed no positive c-wave but instead a high amplitude, late appearing negative potential (Fig. 2).

In American dogs the ERG was extinguished at 6 months of age while in the affected dogs in France variable ERGs were recorded. In some French dogs a response could not be recorded, while in others the ERG was present but much reduced.

Studies in the UK on RPED affected Briards showed that abnormalities could be

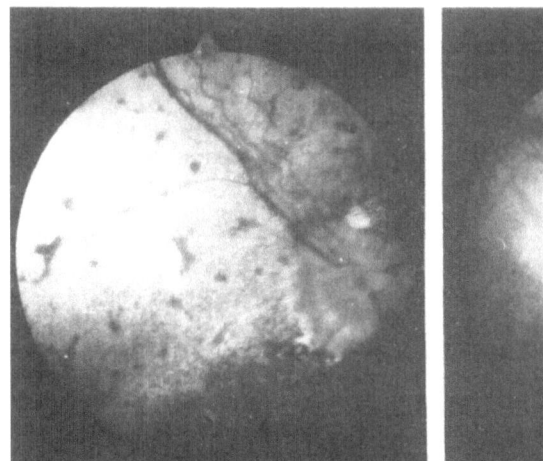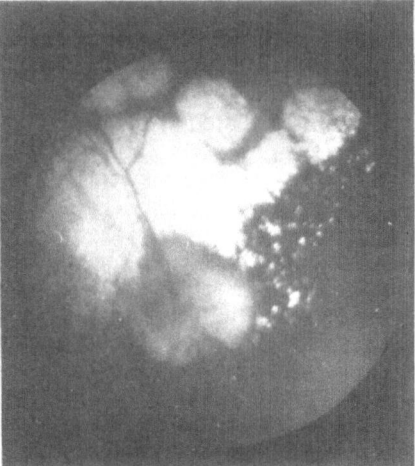

Fig.1. Fundus photographs of RPED affected Briards. **A.** Mid-stage disease with abnormal pigment patches in tapetal fundus. **B.** Late-stage disease with large patches of abnormal pigmentation, areas of hyper-reflectivity, attenuation of superficial retinal vasculature, optic nerve degeneration.

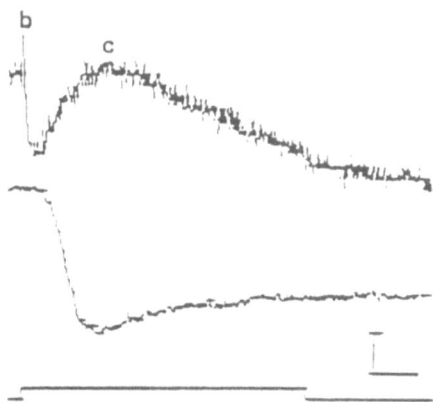

Fig.2. Top trace: The b- and c-waves of the DC ERG in a dark-adapted normal Briard dog, 2 years old. Stimulus intensity 4.5 log units above the normal b-wave threshold (approximately 24,000 lx at the level of the cornea). Amplitude and time calibrations 70μV and 1.5 sec., respectively. Bottom trace: A prominent (approximately 1,500μV) negative potential with a latency of about 1.5 sec. is seen in the DC ERG of a dark-adapted 7-month-old affected Swedish Briard dog. Stimulus intensity 4.5 log units above the normal b-wave threshold (approximately 24,000 lux at the level of the cornea). Amplitude and time calibrations 400μV and 1.5 sec., respectively.

detected several months before ophthalmoscopic abnormalities were evident[11]. These studies showed a marked reduction in b-wave amplitude with retention of the a-wave. The effects of the disease on c-wave response were not investigated.

## LIGHT (LM) AND ELECTRON (EM) MICROSCOPY

### 1. RPE

LM studies on affected dogs in the USA[1] showed vacuolation of the RPE. EM studies showed that the RPE was filled with osmiophilic bodies in all areas of the retina. These cytoplasmic inclusion bodies were non-membrane bound and appeared homogeneous. They ranged greatly in size and were more numerous in the central retina than in the periphery. They did not have a ceroid or lipofuscin profile. The RPE cytoplasm was rich in endoplasmic reticulum. In France lipid inclusions were reported to be present in RPE cells of affected dogs[2].

Ultrastructural studies on affected dogs in Sweden[4,5] described electron-lucent inclusions in cells of the RPE (Fig. 3A). These inclusions occurred in patches in both the tapetal and non-tapetal fundi but were most numerous centrally. They were located primarily in the basal parts of the RPE cells. They contained small amounts of an electron dense material near their periphery and were up to 5µm in diameter, sometimes even larger. Small, electron-dense, membrane bound inclusions were also seen throughout the RPE cytoplasm of affected dogs. These were present but much fewer in number in control dogs. Other cell organelles, including mitochondria and endoplasmic reticulum appeared normal. With increasing age the number of intracellular electron-lucent inclusions increased[9] until by 13 months of age there was massive vacuolation of the central RPE cells.

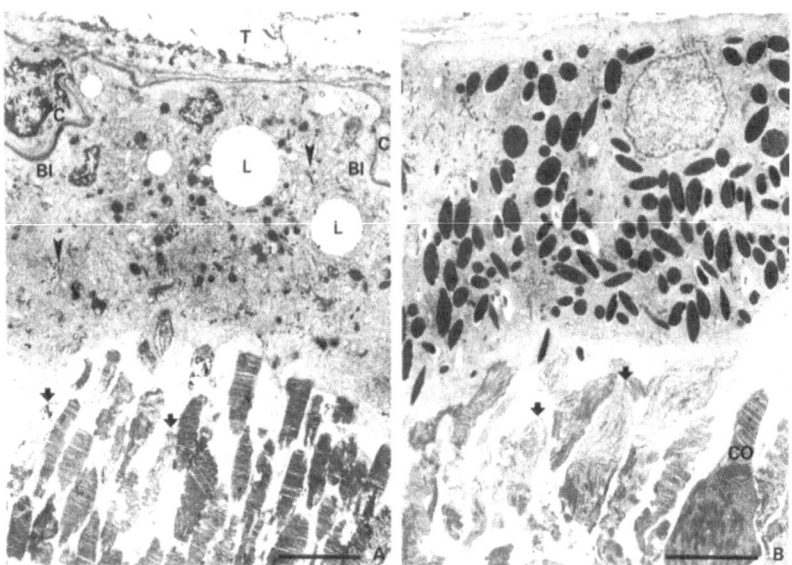

Fig. 3. Affected Swedish Briard (dam with reduced day vision in addition to night-blindness, sire with night-blindness but normal day vision), 8 months old. **A.** Central fundus. Electron-lucent inclusions (L) are located primarily in the basal parts of the RPE cells. Small, membrane-bound, electron-dense inclusions (arrowheads) are abundant throughout the cytoplasm. Basal infoldings (BI) are prominent along the capillary indentations (C). Rod outer segments with distorted disc membranes (arrows) are found together with normal-looking outer segments. T: tapetum. Bar: 5µm. **B.** Lower periphery of the non-tapetal area. Rod outer segment disc membranes (arrows) are severely disoriented. A cone (CO) seems better preserved. Bar: 5µm.

The principle histological characteristic of RPED in Briards was the development of abnormal lipofuscin-like pigment granules in cells of the RPE[11] (Fig. 4). Initially only RPE cells from the tapetal fundus (those lacking melanin pigment) were affected, but later in the disease similar granules were identified in melanin-containing RPE cells in the peripheral retina. The pigment granules within the RPE showed intense yellow-green autofluorescence with short wavelength light. Histological examination of eyes from aged, ophthalmoscopically normal Briards in the UK has revealed abnormal amounts of "age pigment" in the RPE of most eyes (Watson and Bedford, Unpublished observations). Electron microscopy of RPE cells of eyes affected by early disease demonstrated generally amorphous, but occasionally lamellar, lipofuscin-like particles within the RPE cytoplasm. As the disease progressed the affected cells became increasingly laden with such granules and cell organelles were scarce. In some cells lamellar "finger-print" type inclusion granules were seen at the basal surface.

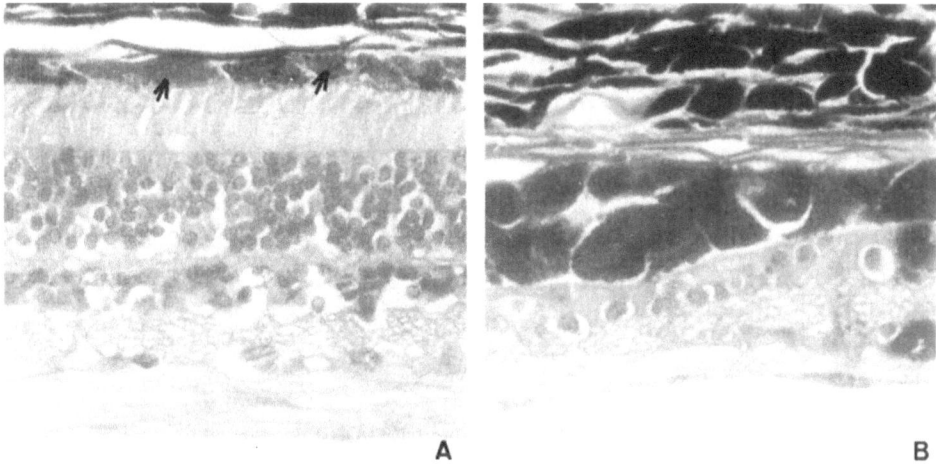

Fig. 4. Light microscopy of retina from RPED affected Briards. **A.** Early-stage disease with abnormal lipopigment accumulation in RPE cells (arrows). **B.** Late-stage disease with large amounts of disease pigment and severe degeneration of neuroretina. (PAS stained sections).

### 2. Photoreceptors and inner retina

Photoreceptor "thinning" was described in the central retina of young affected dogs in the USA while in older dogs more advanced photoreceptor degeneration was seen[1]. Throughout the retina there was evidence of photoreceptor outer segment vesiculation, disruption and degeneration but other layers of the neural retina were normal. In the French dogs the abnormality was summarised as a diminished photoreceptor population[2].

In an affected dog in Sweden examined at 5 weeks of age, rod outer segment disc membranes were disorientated throughout the retina with the abnormalities being most marked in the peripheral areas[9]. In dogs examined at 7-8 months of age the rod outer segments were disorientated and sometimes vesiculated over large areas, although normal outer segments could also be identified (Fig 3A). At this stage, phagocytes were seen in the subretinal space. Cones were less severely affected. The areas of most severe progressive rod outer segment degeneration (peripheral areas) (Fig. 3B) did not correlate anatomically with the areas of prominent RPE abnormality (central areas).

In Briards affected by RPED the development of the abnormal lipofuscin-like granules in RPE cells was followed by degeneration of adjacent neuroretina[11]. This involved shortening and then loss of outer and later inner segments with subsequent disappearance of the outer nuclear layer and finally degeneration of the inner retinal layers. Geographically these changes mirrored those in the RPE so that the most severe changes occurred initially in the central, tapetal areas of the retina. In some dogs outer segment degeneration was also identified in the most peripheral retina. Ultimately the majority of the retina became severely atrophic.

### PLASMA BIOCHEMISTRY

Plasma from affected dogs in the USA was variably lipaemic and affected dogs had a mild hypercholesterolaemia, with elevated beta-lipoprotein (low density lipoprotein or LDL) levels and a hypoproteinaemia[12]. The results of thyroid, pancreas and liver function tests were normal. Plasma arachidonic acid levels were significantly higher in affected dogs than in normals and were identified as a useful diagnostic indicator of the disease. Plasma vitamin A and vitamin E levels were normal for all dogs sampled. No evidence of abnormalities in plasma lipids have been identified in affected dogs in Sweden and no information on plasma biochemistry is available for affected dogs in France.

Hypercholesterolaemia and hyperlipoproteinaemia has been identified in Briards in the United Kingdom[13]. Mean fasting plasma cholesterol concentrations were approximately double the concentrations obtained for control dogs of other breeds. The samples were not visibly lipaemic and plasma triglyceride concentrations were normal. The results of other routine plasma biochemistry and of thyroid and adrenal gland function tests were normal. Preliminary investigations of the plasma lipoproteins suggested an elevation in the high density lipoprotein fraction referred to as HDLc and an increase in plasma levels of apoprotein E. No significant differences were observed however between Briards affected by RPED and ophthalmoscopically normal Briards. Preliminary studies of plasma fatty acids from affected and ophthalmoscopically normal Briards and control dogs of other breeds have not identified any significant differences. Fasting plasma vitamin E concentrations have been found to be significantly lower in RPED affected Briards than in ophthalmoscopically normal Briards or control dogs[14]. Fasting plasma taurine concentrations were significantly lower in all Briards compared to control dogs, but no differences were found between whole blood taurine concentrations for the different groups.

## DISCUSSION

The striking similarities between the retinal degenerations described in Briards in the USA, Sweden and France suggest that these three conditions may be closely related. A number of differences were seen however. Most of the Swedish dogs had severely reduced day vision, in addition to night blindness. Elevated plasma lipids were found in the dogs in the USA but not in Sweden. There were many marked contrasts between these conditions and RPED seen in Briards in the UK. It is significant however that in all conditions prominent abnormalities were described in cells of the retinal pigment epithelium.

The three conditions which include congenital night blindness all appear to be inherited as an autosomal recessive trait. Expressivity varied in the Swedish dogs however, since the initial dogs studied were night blind whereas in dogs produced by inbreeding, day vision was also severely affected. It is of interest that the dogs described in the USA[1] had common ancestors with affected Swedish dogs. RPED however has a very variable age of onset and rate of progression and does not appear to be a true autosomal recessive trait. Most ophthalmoscopically normal eyes from UK Briards examined post mortem showed some degree of abnormality. These factors have lead to the suggestion that in the Briard the expression of RPED might be being modified by environmental, possibly dietary, factors.

The absence of a- and b- waves in the Swedish dog ERGs was first thought to result from blocked transduction. When the prominent slow negative potential of long latency was discovered however it appeared that transduction was not entirely blocked but severely delayed[10]. It is suggested that the negative potential, replacing the positive c-wave reflects the photoreceptor potential (fast PIII) and the Muller cell hyperpolarisation in response to light (slow PIII), whereas the positive RPE hyperpolarisation (PI) is severely reduced or lacking.

It is noteworthy that affected dogs in the USA retained useful day vision despite the absence of any ERG response. Studies by Lightfoot[11] of RPED affected Briards in the UK showed a marked reduction in b-wave amplitude with retention of the a-wave while Parry[15] reported loss of the a-wave before the b-wave in studies of dogs of other breeds affected by RPED. The reason for this difference is not as yet clear.

Forms of night blindness have been reported in humans and in the horse[16]. In man there is a group of such diseases mainly characterised by normal fundus appearance, loss of scotopic vision and abnormal ERG. In one form the a-wave is prominent but the b-wave small or absent indicating a transmission defect. In the second form, both the a- and b- waves are minor or absent, indicating a transduction defect[17]. In the Appaloosa horse there was an a-wave but no b-wave. In the few human CSNB eyes which have been examined histologically few abnormalities have been found. Three genetic forms have been described - dominant, autosomal recessive and sex-linked recessive.

A characteristic histological finding in eyes from affected dogs in the USA, Sweden and France was the presence of abnormal electron-lucent inclusions in RPE cells. For the dogs in the USA these inclusions were also osmiophilic and were shown to be lipid in nature[1]. Such droplets have not been reported previously as inclusions in the RPE in clinical specimens. Experimentally they have been produced in rats by treatment with zinc-chelating agents[18] but the EM characteristics of these droplets varied in a number of ways from those described in the Briards. Plasma zinc levels were normal in the Briards in the USA. It is interesting that for the Swedish dogs the RPE abnormalities were found to be most marked in the central retina and were generally not directly correlated to those in the photoreceptor outer segments, which were most prominent in the peripheral retina. Peripheral photoreceptor degeneration was also found in some affected Briards in the UK[11]. In affected dogs in the USA both the RPE and the photoreceptor abnormalities were found to be most severe in the central retina. It is noteworthy that the rod-cone degeneration in the Swedish Briards was shown to be slowly progressive. This may also be true of the problem described in the USA.

The clinical and histological features of RPED in Briards are similar to those seen in a number of other dog breeds[19]. The condition is characterised primarily by the accumulation of abnormal lipofuscin-like pigment granules within cells of the RPE. Lipofuscin is a non-specific generic term which is used to describe a wide range of abnormal, complex and autofluorescent pigments which accumulate in a range of post-mitotic cells[20,21]. Lipofuscin accumulation in the RPE is thought to result from the incomplete degradation of photoreceptor outer segment membrane material[22]. Such pigments have been shown to develop gradually during normal ageing in a number of species[23,24] however the reasons for their development are not as yet understood. Considerable evidence has been assimilated to show that the pigments may arise through free radical-mediated peroxidation of rod outer segment membrane lipids although as yet this has not been specifically proven[25].

It is significant that lipofuscin accumulation in Briards occurs most rapidly in RPE cells in the central fundus which do not contain melanin. Melanin has been shown to be a potent free radical scavenger[26] and thus might protect adjacent retina against free radical-mediated damage[24]. The features of RPED closely resemble those which have been described in vitamin E deficiency retinopathy in dogs[27] and also in rats[28]. Low levels of this plasma antioxidant were found in RPED affected Briards in the UK and in addition, plasma levels of another potent antioxidant , taurine, were found to be lower than normal in all the UK Briards sampled[14]. These findings in the UK are consistent with the suggestion that the lipofuscin-like pigment might be accumulating due to defective protection of membrane lipids against free radical-mediated damage.

Plasma biochemistry analyses in the USA suggested that the condition described there may relate to abnormality in fatty acid metabolism. Similar relationships have been described for a number of retinal degenerations in man and animals. Low plasma levels of docosahexaenoic acid (DHA) have been identified in a number of common forms of human retinal degeneration, including Usher's syndrome[29], X-linked retinitis pigmentosa (RP)[30] and autosomal dominant RP[31]. It has been suggested[31] that the low plasma levels of DHA in some RP patients might be the result of a defective elongation and desaturation system for converting short-chain polyunsaturated fatty acids into long-chain polyunsaturated fatty acids. Gong et al[32] reported reduced plasma DHA levels in X-linked and in isolate forms of RP but not in autosomal dominant RP. There were no deficiencies in the plasma precursors of DHA which suggests that there may be a block at some stage of the pathway to DHA. This possibility was supported by a significantly higher ratio of 20:5(n-3) to DHA in X-linked RP patients compared to control groups. Reduced plasma DHA levels have also been reported in miniature poodles with progressive rod-cone degeneration[31,33]. These studies suggest a genetic defect in the delta-4 desaturase step that converts 22:5(n-3) and 22:4(n-6) to 22:5(n-6) in these dogs. In Abyssinian cats with progressive rod-cone degeneration the ratio of 22:5(n-3) to 22:6(n-3) was higher in affected animals than in controls, which is also consistent with a delta-4 desaturase deficiency[34].

Elevations in plasma phytanic acid levels have been described in Refsums disease in man, a condition in which accumulation of lipids in RPE is thought to cause the retinal degeneration, which is of RP type[35]. The initial visual symptom of this disease is night blindness. Plasma phytanic acid abnormalities were not seen however in affected Briards in the USA.

Any disturbance in the synthesis or availability of the lipids required for membrane biogenesis may lead to photoreceptor outer segment degeneration[36]. It was demonstrated that in the retina of dogs with canine ceroid lipofuscinosis (CCL) there was a significant decrease in docosahexaenoic acid content in all phospholipids. This was compensated by an increase in the content of arachidonic acid. In these animals there was an accumulation of lipopigments in the RPE but the photoreceptors themselves were normal in appearance.

Plasma biochemistry analysis for Briards in the UK affected by RPED is not a simple matter since the variable expression of the disease makes it difficult to define a normal group. This may explain the absence of a significant difference between fasting plasma cholesterol concentrations in normal and RPED affected Briards. Hypercholesterolaemia and abnormal plasma lipoprotein profiles were described in affected Briards in the USA and in both normal and affected Briards in the UK, as compared to control dogs of other breeds. Normal plasma lipoprotein profiles for the dog are well established[37]. It is significant that when separated by lipoprotein electrophoresis affected dogs in the USA had an elevation in the LDL fraction while Briards in the UK had an elevation in the fraction referred to as HDLc. The underlying causes of the hypercholesterolaemia and hyperlipoproteinaemia in the two countries are currently under investigation.

A sizeable proportion of human RP patients are hypercholesterolaemic[38,39], while a few, in one type of autosomal-dominant RP are hypocholesterolaemic[40]. Jahn et al[39] studied the apolipoprotein E isoforms of RP patients in Germany and found that apoE2 was present in RP patients more commonly than in normal individuals. The study also showed a widespread hypercholesterolaemia amongst the RP patients. Interestingly however, amongst the RP patients, mean serum cholesterol concentrations tended to be lower in those with the E2 isoform than in those without. Thus the presence of the E2 isoform itself is apparently not the cause of the hypercholesterolaemia. The full nature of the relationship between them in this group of patients is still under investigation.

## CONCLUSIONS

Various types of retinal degeneration have been described in Briard dogs. In each case the characteristic findings relate to an accumulation of abnormal material within cells of the RPE and a photoreceptor degeneration. Differences between the conditions are also seen. The RPE inclusions have been described as lipofuscin-like (UK) or as containing lipids (USA, France). For the Swedish Briards the content of the inclusions remains unknown. The retinal changes have been reported as mainly outer segment degeneration (Sweden, USA) or loss of photoreceptors (France, UK) and in advanced RPED, inner retinal degeneration (UK). Degeneration may start centrally in both RPE and neuroretina (UK) or centrally in the RPE but peripherally in the neuroretina (Sweden). Night blindness alone was found in the American and French dogs whereas most of the Swedish dogs showed both night blindness and reduced day vision. Night blindness was not described in Briards in the UK. Differences were also reported in plasma lipids. It thus seems that these diseases are all related but not identical. Future studies will relate to a more detailed comparison of the reported conditions, in particular to the specific analysis of the abnormal RPE material and the further characterisation of the anomalies in plasma biochemistry.

## REFERENCES

1. R.C. Riis and G.D. Aguirre, The Briard problem, *Procedings of the American College of Veterinary Ophthalmology*. Chicago (1983).
2. M. Roze, A. Lucciana, J. Capodano, S. Gambarelli, and M. Auphan, The Briard problem in France, the retina, light microscopy, electron microscopy, retinal antigen, *Procedings of the International Society for Veterinary Ophthalmology*. Vienna (1991).
3. K. Narfstrom, A. Wrigstad, and S.E.G. Nilsson, The Briard dog: A new animal model of congenital stationary night blindness, *Br. J. Ophthalmol.* 73:750-756 (1989).
4. S.E.G. Nilsson, A. Wrigstad, and K. Narfstrom, Congenital night blindness and partial day blindness in the Briard dog. *In:* "Retinal Degenerations", R.E. Anderson, J.G. Hollyfield, M.M. LaVail, Eds. CRC Press, Inc. Boca Raton, Ann Arbor, Boston, London (1991).
5. A. Wrigstad, S.E.G. Nilsson and K. Narfstrom, Ultrastructural changes of the retina and the retinal pigment epithelium in Briard dogs with hereditary congenital night blindness and partial day blindness, *Exp. Eye Res.* (In Press).
6. P.G.C. Bedford, Retinal pigment epithelial dystrophy (cPRA): a study of the disease in the Briard. *J. Small Anim. Pract.* 25:129-138 (1984).
7. M. Roze, The Briard problem in France, clinical and breed study. *Procedings of the International Society for Veterinary Ophthalmology*. Vienna (1991).
8. K. Narfstrom, A. Wrigstad and S.E.G. Nilsson, Congenital night blindness and partially reduced day vision in the Briard dog: Clinical and laboratory findings. *Procedings of the International Society for Veterinary Ophthalmology*. Vienna (1991).

9. A. Wrigstad, K. Narfstrom and S.E.G. Nilsson, Morphological changes in the retina and RPE in Briard dogs with congenital night blindness and partial day blindness are slowly progressive. *Invest. Ophthalmol. Vis. Sci.* 33: 1065 (1992).
10. S.E.G. Nilsson, A. Wrigstad and K. Narfstrom, Changes in the DC electroretinogram in Briard dogs with hereditary congenital night blindness and partial day blindness. *Exp. Eye Res.* 54: 291-296 (1992).
11. R. Lightfoot, Retinal pigment epithelial dystrophy in the Briard. *Thesis*. University of London (1988).
12. R.C. Riis and A.B. Siakotos, Inherited lipid retinopathy in a dog breed. *Procdings of the American College of Veterinary Ophthalmology*. (1989).
13. P. Watson, K.W. Simpson and P.G.C. Bedford, Hypercholesterolaemia in Briards in the United Kingdom. *Res. Vet. Sci.* In Press.
14. P. Watson, G. Anantharaman, J. Vichoud and P.G.C. Bedford, Antioxidant status of Briards with retinal pigment epithelial dystrophy. *Procdings of the World Small Animal Veterinary Association..* Rome (1992).
15. H.B. Parry, Degenerations of the dog retina. VI. Central progressive atrophy with pigment epithelial dystrophy. *Brit. J. Ophthalmol.* 38: 653-668 (1954).
16. D.A. Witzel, E.L. Smith, R.D. Wilson and G.D. Aguirre, Congenital stationary night blindness: an animal model. *Invest. Ophthalmol. Vis. Sci.* 17 : 788-795.1978.
17. R.E. Carr, H. Ripps, I.M. Siegel, and R.A.Weale, Rhodopsin and the electrical activity of the retina in congenital night blindness. *Invest. Ophthalmol.* 5: 497-507 (1966).
18. A.E. Leure-duPree, Electron-opaque inclusions in the rat retinal pigment epithelium after treatment with chelators of zinc. *Invest. Ophthalmol. Vis. Sci.* 21:1-9 (1981).
19. G.D. Aguirre and A. Laties, Pigment epithelial dystrophy in the dog. *Exp. Eye Res.* 23:247-256. (1976).
20. R.W. Young, A theory of central retinal disease. *In:* "New directions in ophthalmic research". M.L. Sears, Ed., Yale University Press. New Haven (1981).
21. R.D. Jolly and R.R. Dalefield, Lipopigments in veterinary pathology: Pathogenesis and terminology. *Adv. Exp. Med. Biol.* 266: 157-167 (1989).
22. L. Feeney-Burns and G.E. Eldred, The fate of the phagosome: Conversion to "age pigment" and impact in human retinal pigment epithelium. *Trans. Ophthalmol. Soc. UK.* 103: 416-421 (1984).
23. M.L. Katz, and W.G. Robison, Age-related changes in the retinal pigment epithelium of pigmented rats. *Exp. Eye Res.* 38: 137-151 (1984).
24. M. Boulton, Ageing of the retinal pigment epithelium. *In:* "Progress in Retinal Research, Vol. 11". N.N. Osborne and G.J. Chader, Eds. Pergamon Press. Oxford (1991).
25. G.E. Eldred and M.L. Katz, The lipid peroxidation theory of lipofuscinogenesis cannot yet be confirmed. *Free Rad. Biol. Med.* 10: 445-447 (1991).
26. M. Ostrovsky, N.L. Sakina and A.E. Dontsov, An antioxidative role for ocular screening pigments. *Vision Res.* 27: 893-899 (1987).
27. R.C. Riis, B.E. Sheffy, E. Loew, T.J. Kern and J.S. Smith, Vitamin E retinopathy in dogs. *Am. J. Vet. Res.* 42: 74-86 (1981).
28. M.L. Katz, W.L. Stone and E.A. Dratz, Fluorescent pigment accumulation in retinal pigment epithelium of antioxidant deficient rats. *Invest. Ophthalmol.Vis. Sci.* 17: 1049-1057 (1978).
29. N.G. Bazan, B.L. Scott, T.S. Reddy and M.Z. Pelias, Decreased content of docosahexaenoate and arachidonate in plasma phospholipids in Usher's syndrome. *Biochem. Biophys. Res. Comm.* 141: 600-604 (1986).
30. C.A. Converse, T. McLachlan, A.C. Bow, C.J. Packard and J. Shepherd, Lipid metabolism in retinitis pigmentosa. *In:* "Degenerative Retinal Disorders: Clinical and Laboratory Investigations." J.G. Hollyfield, R.E. Anderson and M.M. LaVail, Eds. Alan R. Liss, Inc. New York. (1987).
31. R.E. Anderson, M.B. Maude, R.A. Alvarez, G.M. Acland and G.D. Aguirre, Plasma levels of docosahexaenoic acid in miniature poodles with an inherited retinal degeneration. *Invest. Ophthalmol. Vis. Sci.* 29 (Supplement) : 169 (1988).
32. J. Gong, B. Rosner, D.G. Rees, E.L. Berson, C.A. Weigel-DiFranco and E.J. Schaefer, Plasma docosahexanoic acid levels in various genetic forms of retinitis pigmentosa. *Invest. Ophthalmol. Vis. Sci.* 33: 2596-2602 (1992).

33. M.G. Wetzel, C. Fahlman, M.B. Maude, R.A. Alvarez, P.J. O'Brien, G.M. Acland, G.D.Aguirre and R.E. Anderson,  Fatty acid metabolism in normal miniature poodles and those affected with progressive rod cone degeneration (prcd). *In:* "Inherited and Environmentally Induced Retinal Degenerations". M.M. LaVail, R.E. Anderson and J.G. Hollyfield, Eds. Alan R. Liss, Inc. New York (1989).

34. R.E. Anderson, M.B. Maude, S.E.G. Nilsson and K. Narfstrom,  Plasma lipid abnormalities in the Abyssinian cat with a hereditary rod-cone degeneration. *Exp. Eye Res.* 53: 415-417 (1991).

35. D. Toussaint and P. Danis,  An ocular pathologic study of Refsum's syndrome. *Am. J. Ophthalmol.* 72: 342-347 (1971).

36. N.G. Bazan, D.L. Birkle and T.S. Reddy,  Biochemical and nutritional aspects of the metabolism of polyunsaturated fatty acids and phospholipids in experimental models of retinal degeneration. *In :* "Retinal Degenerations: Experimental and Clinical Studies".  R.E. Anderson, J.G. Hollyfield and M.M. LaVail, Eds. Alan R. Liss, Inc., New York (1985).

37. R.W. Mahley and K.H. Weisgraber,  Canine lipoproteins and atherosclerosis. I. Isolation and characterisation of plasma lipoproteins from control dogs. *Circ. Res.* 35: 713-721 (1974).

38. C.A. Converse, T. McLachlan, H.M. Hammer, C.J. Packard and J. Shepherd,  Hyperlipidaemia in retinitis pigmentosa. *In:* "Retinal Degenerations: Experimental and Clinical Studies." R.E. Anderson, J.G. Hollyfield and M.M. LaVail. Eds. Alan R. Liss, Inc. New York (1985).

39. C.E. Jahn, K. Oette, A. Esser, K. v. Bergman and O. Leiss,  Increased prevalence of apolipoprotein E2 in patients with Retinitis Pigmentosa. *Ophthalmic Res.* 19: 285-288 (1987).

40. C.A. Converse, H.M. Hammer, C.J. Packard and J. Shepherd,  Plasma lipid abnormalities in retinitis pigmentosa and related conditions. *Trans. Ophthalmol. Soc. UK* 103: 508-512 (1983).

## IV.  AGENTS WHICH CAUSE OR PREVENT RETINAL DEGENERATION

In all research on retinal degeneration, one overriding concern is to ultimately be able to provide a treatment which might slow or retard the degeneration process.  The chapters in this section deal with a variety of issues related to possible therapeutic effects of various agents on retinal degeneration.  One chapter addresses the effects of growth factors on prevention of light damage to photoreceptors.  One uses laser therapy to prevent retinal degeneration and a third relates the production of heat shock proteins to the survival of photoreceptors.  The final chapters describe agents which cause photoreceptor degeneration including light and naphthalene on photoreceptor degeneration.  These studies may provide important information for designing therapies that alter the natural history of inherited degenerations and may, ultimately, provide a treatment for some of these disorders.

# GROWTH FACTORS AS POSSIBLE THERAPEUTIC AGENTS FOR RETINAL DEGENERATIONS

Matthew M. LaVail,[1,2,4] Kazuhiko Unoki,[2] Douglas Yasumura,[1]
Michael T. Matthes,[2] and Roy H. Steinberg[2,3,4]

[1]Department of Anatomy
[2]Department of Ophthalmology
[3]Department of Physiology
[4]Neuroscience Program
University of California, San Francisco
San Francisco, CA 94143-0730

## INTRODUCTION

In 1990, Faktorovich and co-workers[1] demonstrated that intraocularly injected basic fibroblast growth factor (bFGF) dramatically delayed photoreceptor degeneration in Royal College of Surgeons (RCS) rats with inherited retinal dystrophy. This showed that at least one growth factor, bFGF, could act as a survival-promoting agent in one form of hereditary retinal degeneration, and was consistent with its action as a survival-promoting neurotrophic agent in some other regions of the central and peripheral nervous system.[2-7]

The finding in RCS rats raised numerous questions about the mechanism(s) of the survival-promoting action of this peptide growth factor in the retina.[1] Although the RCS rat is an important model for the study of retinal degenerations and photoreceptor-retinal pigment epithelial cell interactions,[8] several features of this mutant and its retinal dystrophy make it less than optimal for analyzing the mechanism(s) of bFGF survival-promoting action. These disadvantages include 1) small litter size, 2) relatively poor breeding performance, 3) age-dependant experiments, and perhaps most significantly, 4) the 1- to 2-month post-injection survival time required to distinguish clearly photoreceptor "rescue" from degenerating retina. Thus, a need existed for a better, or at least more rapid, model for studying the survival-promoting action of bFGF and other possible agents.

The excitement of finding a pharmacologic agent that delayed photoreceptor degeneration in an inherited retinal degeneration[1] was tempered by the well-known fact that bFGF is a powerful mitogen and angiogenic factor, and that its injection increased the incidence of retinal macrophages and, occasionally produced proliferative vitreoretinopathy.[1] Thus, it was cautioned that therapeutic use of bFGF should wait until more was known of its potentially harmful side effects.[1] We then asked whether there were other survival-promoting or survival-potentiating neurotrophic agents that did not have side effects in the eye. Several agents were known to show survival-promoting activity in other parts of the nervous system without apparent side effects.[9-20]

The present chapter briefly summarizes our search for a "better" model with which to study survival-promoting factors and for potentially "better" survival-promoting factors than bFGF.

## THE LIGHT DAMAGE MODEL

It is widely known that after 1 or 2 weeks of constant fluorescent light, most photoreceptors of cyclic light-reared albino rats will degenerate and disappear without the loss of the retinal pigment epithelium.[21-26] Thus, when we exposed either F344 or Sprague-Dawley albino rats to 1 week of fluorescent light at an illuminance level of 115-200 ft-c, the retinas showed the expected loss of photoreceptor nuclei from the outer nuclear layer (ONL). Whereas the ONL of normal retinas contained 8-10 rows of photoreceptor nuclei (Fig. 1A), the ONL from rats exposed to constant light showed only 1-3 rows of nuclei in the most severely damaged region in the posterior superior part of the retina (Fig. 1B). In this region, only a few fragments of photoreceptor inner and outer segments remained (Fig. 1B). In the peripheral retina, fewer photoreceptor nuclei were lost and the ONL was correspondingly thicker.

If bFGF was injected intravitreally 2 days before the 1 week of constant light, then 2-4 times more photoreceptor nuclei survived (Fig. 1C) compared to phosphate buffered saline (PBS)-injected or uninjected eyes (Fig. 1B). Clearly, bFGF prevented degeneration and loss of photoreceptor cells. Further details of these experiments and methods to quantify degree of photoreceptor rescue by bFGF are presented elsewhere.[27]

The light damage model offers several advantages over the RCS rat for studying the survival-promoting activity of growth factors *in vivo*. The degree and kinetics of the "lesion" and photoreceptor degeneration can be experimentally controlled by varying the intensity and duration of constant light; the period of degeneration can be as short as a few days to 1-2 weeks; the animals can be used at almost any age; one eye of each animal can be used as a control (which avoids the inter-animal variability seen in light-damage studies, particularly in some strains of rats[27]); and the experiments can be done with abundant, commercially available rats. In addition, the fact that the "lesion" is produced non-invasively and growth factors are administered relatively simply within the eye, make light damage perhaps the most efficaceous CNS model for *in vivo* study of the survival-promoting activity of presumptive neurotrophic agents.

## SURVIVAL-PROMOTING FACTORS

Using the same paradigm described above for bFGF, we recently found that 8 different growth factors, cytokines and neurotrophins significantly protect photoreceptors from the damaging effects of constant light, although not all to the same degree.[28] As shown in Fig. 2, those factors that provide a high degree of photoreceptor rescue include brain-derived neurotrophic factor (BDNF), ciliary neurotrophic factor (CNTF), interleukin-1β, bFGF (either with or without heparin) and acidic fibroblast growth factor (aFGF) (either with or without heparin). Those that show some, but less, survival-promoting activity include neurotrophin-3, insulin-like growth factor-II and tumor necrosis factor-α; those that show little or no protective effect include nerve growth factor, insulin-like growth factor-I, insulin, laminin, heparin, epidermal growth factor, and platelet-derived growth factor.[28] Thus, various agents that typically act through 4 or more distinct receptor families are capable of protecting photoreceptors from the damaging effects of constant light.

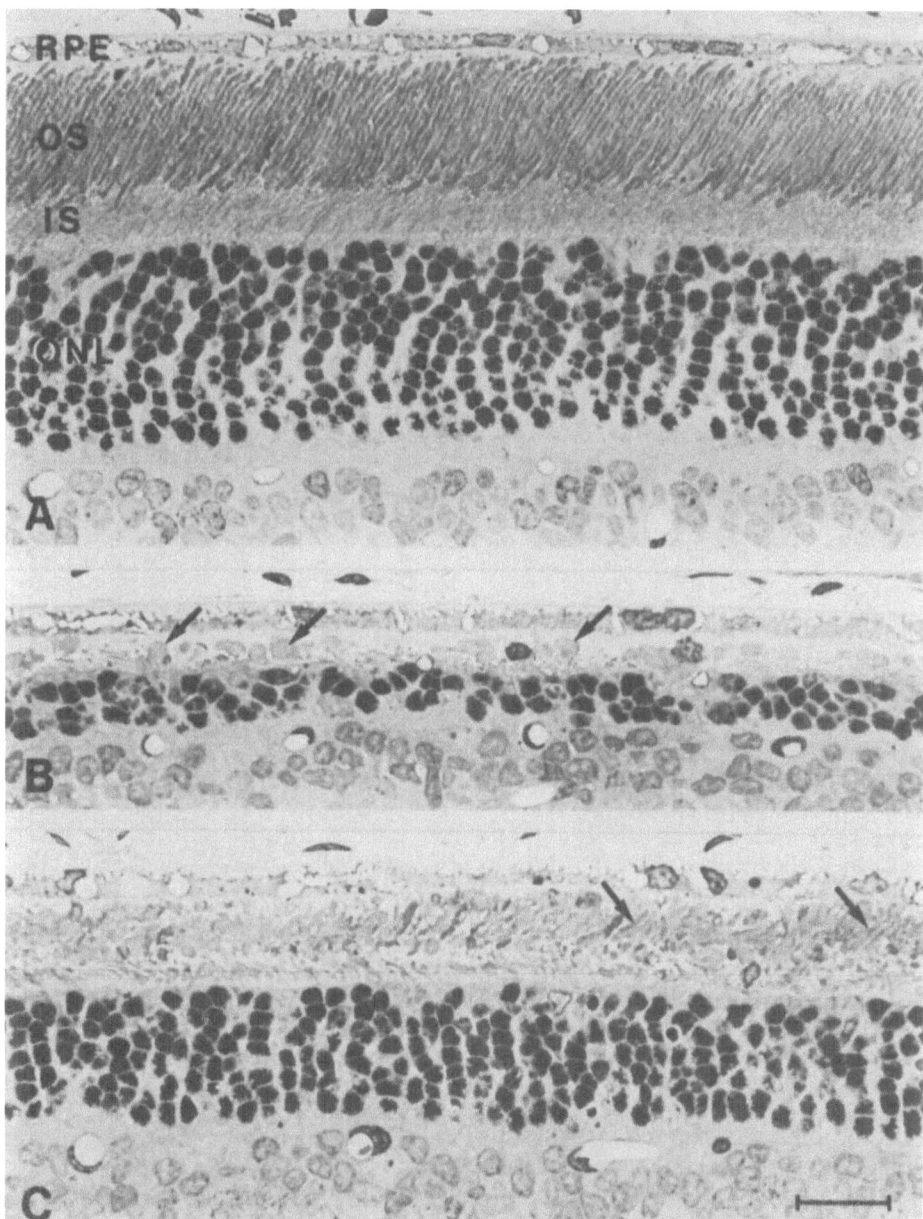

**Figure 1.** Light micrographs of retinas from F344 albino rats taken from the posterior superior region of the eye that is most sensitive to the damaging effects of constant light. (A) Normal retina from an uninjected rat kept in cyclic light. (B) Retina from an uninjected rat exposed to constant light for 1 week. The outer nuclear layer has been reduced to 1-3 rows of photoreceptor nuclei; no photoreceptor inner segments are present; and the remaining outer segments are in the form of large rounded or oblong profiles (arrows) that are as large as photoreceptor nuclei. (C) Retina from a rat injected intravitreally with bFGF 2 days before a 1-week exposure to constant light. The outer nuclear layer shows 6-7 rows of nuclei; some inner segments are present, albeit shorter than normal (cf Fig. 1A, above); and photoreceptor outer segments are present. Although most of the outer segments are somewhat disorganized, they are mostly of normal diameter (arrows). IS, inner segments; ONL, outer nuclear layer; OS, outer segments; RPE, retinal pigment epithelium. Toluidine blue stain. Bar = 20 μm. Reprinted with permission from the *Journal of Neuroscience* and Oxford University Press.[27]

295

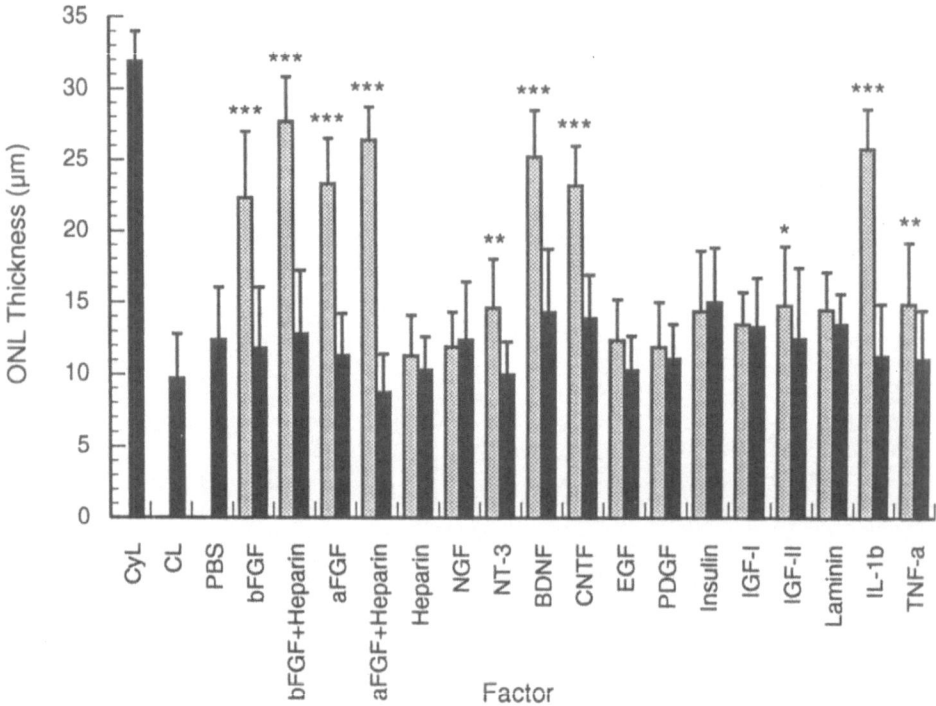

**Figure 2.** Measurements of ONL thickness in the retinas of Sprague-Dawley rats exposed to 1 week of constant light or maintained in cyclic light (CyL). Two days before the onset of light exposure, each rat received an intravitreal injection of one of the listed factors in one eye, and a buffer control (usually PBS) injection in the other eye. Uninjected controls were also exposed to light (CL). In each case, the bars represent the mean value, and the error bars are the standard deviation. ONL thickness of eyes injected with bFGF, aFGF, bFGF or aFGF with heparin, BDNF, CNTF and IL-1b showed the greatest significant difference in numbers of photoreceptor nuclei surviving (shaded bars) compared to their control eyes (solid black bars). NT-3, IGF-2 and TNF-a produced less photoreceptor rescue, but still significantly greater numbers of photoreceptor nuclei survived than in control eyes (* $P<.05$; ** $P<.01$; *** $P\leq.0001$). The remaining agents failed to show significant differences from controls. The number of rats injected (with an equal number of control eyes) was: bFGF, n=17; bFGF + heparin, n=13; aFGF, n=14; aFGF + heparin , n=11; heparin, n=14; NGF, n=12; NT-3, n=6; BDNF, n=16; CNTF, n=11; EGF, n=6; PDGF, n=14; insulin, n=6; IGF-I, n=10; IGF-II, n=5; laminin, n=5; IL-1b, n=8; TNF-a, n=10. Although we did not compare agents in the same animal, the bFGF + heparin value is significantly greater than bFGF alone ($P<.0025$), as is aFGF + heparin over aFGF alone ($P<.05$). Reprinted with permission from the *Proceedings of the National Academy of Sciences*. For further details see LaVail, et al.[28]

## GROWTH FACTORS AS POSSIBLE THERAPEUTIC AGENTS

There are many questions that must be answered before growth factors can be applied to human ocular diseases. A major question is the efficacy of any of the survival-promoting agents in other forms of retinal degeneration or ocular lesions. Although we have shown bFGF to delay photoreceptor degeneration in RCS rats[1] and 8 different agents, including bFGF, to protect photoreceptors from the damaging effects of constant light,[28] it remains to be shown whether any or all of the agents are effective in other ocular disorders.

Another question concerns the safety of the agents. The potential problems with bFGF were discussed above, and similar problems may exist with aFGF. In addition, interleukin-1b produced severe ocular and retinal inflammation, in addition to the protection of photoreceptor cells.[28] Thus, several of the agents may have side effects that prevent their use in ocular disorders, although some of the most potent agents, particularly BDNF and CNTF, showed no apparent side effects.[28]

If survival-promoting agents do prove capable of rescuing photoreceptors or other retinal cells in human ocular disorders, then many experimental questions still remain to be answered. These include the determination of dose-response relationships, biological and pharmacological half-lives in the retina, and optimal times in the degenerative process(es) to administer the agent(s). In addition, since several of the potentially useful agents are peptides that do not cross the blood-retinal barrier, effective intraocular delivery mechanisms need to be developed for these agents.

In the future, it will be essential to answer these questions, as well as to determine the mechanisms and cellular location of growth factor actions, which are not presently understood for any of the factors. For light damage, it will also be important to understand how mechanisms of rescue by growth factors are related to those produced by other agents or procedures.[29-36] Overall, our present findings offer the hope for eventual pharmacologic therapeutic approaches to several classes of ophthalmic disorders, including different forms of retinal degeneration.

## REFERENCES

1. E.G. Faktorovich, R.H. Steinberg, D. Yasumura, M.T. Matthes and M.M. LaVail, Photoreceptor degeneration in inherited retinal dystrophy delayed by basic fibroblast growth factor, *Nature* 347:83 (1990).
2. J. Sievers, B. Hausmann, K. Unsicker and M. Berry, Fibroblast growth factors promote the survival of adult rat retinal ganglion cells after transection of the optic nerve, *Neurosci. Let.* 76:157 (1987).
3. K. Unsicker, H. Reichert-Preibsch, R. Schmidt, B. Pettmann, G. Labourdette and M. Sensenbrenner, Astroglial and fibroblast growth factors have neurotrophic functions for cultured peripheral and central nervous system neurons, *Proc. Natl. Acad. Sci. USA* 84:5459 (1987).
4. M.E. Hatten, M. Lynch, R.E. Rydel, J. Sanchez, J. Joseph-Silverstein, D. Moscatelli and D.B. Rifkin, In vitro neurite extension by granule neurons is dependent upon astroglial-derived fibroblast growth factor, *Dev. Biol.* 125:280 (1988).
5. D. Dreyer, A. Lagrange, C. Grothe and K. Unsicker, Basic fibroblast growth factor prevents ontogenetic neuron death in vivo, *Neurosci. Let.* 99:35 (1989).
6. D. Otto, M. Frotscher and K. Unsicker, Basic fibroblast growth factor and nerve growth factor administered in gel foam rescue medial septal neurons after fimbria fornix transection, *J. Neurosci. Res.* 22:83 (1989).
7. D. Hicks, K. Bugra, B. Faucheux, J.-C. Jeanny, M. Laurent, F. Malecaze, F. Mascarelli, D. Raulais, Y. Cohen and Y. Courtois, Fibroblast growth factors in the retina, *in* "Progress in Retinal Research, Vol. 11," N. Osborne and G. Chader, ed., p. 333. Pergamon Press, Oxford (1991).
8. M.M. LaVail, Analysis of neurological mutants with inherited retinal degeneration, *Invest. Ophthalmol. Vis. Res.* 21:638 (1981).
9. R. Adler, K.B. Landa, M. Manthorpe and S. Varon, Cholinergic neuronotrophic factors: intraocular distribution of trophic activity for ciliary neurons, *Science* 204:1434 (1979).
10. J.E. Johnson, Y.-A. Barde, M. Schwab and H. Thoenen, Brain-derived neurotrophic factor supports the survival of cultured rat retinal ganglion cells, *J. Neurosci.* 6:3031 (1986).
11. C. Kalcheim, Y.A. Barde, H. Thoenen and N.M. Le Douarin, In vivo effect of brain-derived neurotrophic factor on the survival of developing dorsal root ganglion cells, *EMBO J.* 6:2871 (1987).
12. A. Rodriguez-Tébar, P.L. Jeffrey, H. Thoenen and Y.A. Barde, The survival of chick retinal ganglion cells in response to brain-derived neurotrophic factor depends on their embryonic age, *Dev. Biol.* 136:296 (1989).
13. Y. Arakawa, M. Sendtner and H. Thoenen, Survival effect of ciliary neurotrophic factor (CNTF) on chick embryonic motoneurons in culture: comparison with other neurotrophic factors and cytokines, *J. Neurosci.* 10:3507 (1990).

14. A. Hohn, J. Leibrock, K. Bailey and Y.-A. Barde, Identification and characterization of a novel member of the nerve growth factor/brain-derived neurotrophic factor family, *Nature* 344:339 (1990).

15. P.C. Maisonpierre, L. Belluscio, S. Squinto, N.Y. Ip, M.E. Furth, R.M. Lindsay and G.D. Yancopoulos, Neurotrophin-3: a neurotrophic factor related to NGF and BDNF, *Science* 247:1446 (1990).

16. M. Sendtner, G.W. Kreutzberg and H. Thoenen, Ciliary neurotrophic factor prevents the degeneration of motor neurons after axotomy, *Nature* 345:440 (1990).

17. M.H. Tuszynski, H.S. U, D.G. Amaral and F.H. Gage, Nerve growth factor infusion in the primate brain reduces lesion-induced cholinergic neuronal degeneration, *J. Neurosci.* 10:3604 (1990).

18. C. Hyman, M. Hofer, Y.A. Barde, M. Juhasz, G.D. Yancopoulos, S.P. Squinto and R.M. Lindsay, BDNF is a neurotrophic factor for dopaminergic neurons of the substantia nigra, *Nature* 350:230 (1991).

19. N.Y. Ip, Y. Li, I. Stadt, N. Panayotatos, R.F. Alderson and R.M. Lindsay, Ciliary neurotrophic factor enhances neuronal survival in embryonic rat hippocampal cultures, *J. Neurosci.* 11:3124 (1991).

20. R.W. Oppenheim, D. Prevette, Y. Qin-Wei, F. Collins and J. MacDonald, Control of embryonic motoneuron survival in vivo by ciliary neurotrophic factor, *Science* 251:1616 (1991).

21. W.K. Noell, V.S. Walker, B.S. Kang and S. Berman, Retinal damage by light in rats, *Invest. Ophthalmol.* 5:450 (1966).

22. W.K. O'Steen, C.R. Shear and K.V. Anderson, Retinal damage after prolonged exposure to visible light. A light and electron microscopic study, *Am. J. Anat.* 134:5 (1972).

23. W.K. Noell, Effects of environmental lighting and dietary vitamin A on the vulnerability of the retina to light damage, *Photochem. Photobiol.* 29:717 (1979).

24. M.M. LaVail, Eye pigmentation and constant light damage in the rat retina, *in* "The Effects of Constant Light on Visual Processes," T.P. Williams and B. B. Baker, ed., p. 357. Plenum Press, New York (1980).

25. L.M. Rapp and T.P. Williams, A parametric study of retinal light damage in albino and pigmented rats, *in* "The Effects of Constant Light on Visual Processes," T.P. Williams and B.N. Baker, ed., p. 135. Plenum Press, New York (1980).

26. M.M. LaVail, G.M. Gorrin, M.A. Repaci and D. Yasumura, Light-induced retinal degeneration in albino mice and rats: strain and species differences., *in* "Degenerative Retinal Disorders: Clinical and Laboratory Investigations," J.G. Hollyfield, R.E. Anderson and M.M. LaVail, ed., p. 439. Alan R. Liss, Inc., New York (1987).

27. E.G. Faktorovich, R.H. Steinberg, D. Yasumura, M.T. Matthes and M.M. LaVail, Basic fibroblast growth factor and local injury protect photoreceptors from light damage in the rat, *J. Neurosci.* 12:3554 (1992).

28. M.M. LaVail, K. Unoki, D. Yasumura, M.T. Matthes, G.D. Yancopoulos and R.H. Steinberg, Multiple growth factors, cytokines and neurotrophins rescue photoreceptors from the damaging effects of constant light, *Proc. Natl. Acad. Sci. USA* 89:11249 (1992).

29. W.K. Noell and R. Albrecht, Irreversible effects of visible light on the retina: role of vitamin A, *Science* 172:76 (1971).

30. W.K. O'Steen, K.V. Anderson and C.R. Shear, Photoreceptor degeneration in albino rats: dependency on age, *Invest. Ophthalmol. Vis. Sci.* 13:334 (1974).

31. W.K. Noell, Possible mechanisms of photoreceptor damage by light in mammalian eyes, *Vision Res.* 20:1163 (1980).

32. D.T. Organisciak, H.M. Wang, Z.Y. Li and M.O.M. Tso, The protective effect of ascorbate in retinal light damage of rats, *Invest. Ophthalmol. Vis. Sci.* 26:1580 (1985).

33. J.S. Penn, A.G. Howard and T.P. Williams, Light damage as a function of "light history" in the albino rat, *in* "Retinal Degeneration: Experimental and Clinical Studies," M.M. LaVail, J.G. Hollyfield and R.E. Anderson, ed., p. 439. Alan R. Liss, Inc., New York (1985).

34. M.M. LaVail, G.M. Gorrin, M.A. Repaci, L.A. Thomas and H.M. Ginsberg, Genetic regulation of light damage to photoreceptors, *Invest. Ophthalmol. Vis. Sci.* 28:1043 (1987).

35. M.F. Barbe, M. Tytell and D.J. Gower, Hyperthermia protects against light damage in the rat retina, *Science* 241:1817 (1988).

36. D.P. Edward, T.T. Lam, S. Shahinfar, J. Li and M.O. Tso, Amelioration of light-induced retinal degeneration by a calcium overload blocker. Flunarizine, *Arch. Ophthalmol.* 109:554 (1991).

# LASER EFFECTS ON PHOTORECEPTOR DEGENERATION IN THE RCS RAT

Martin F Humphrey, Yi Chu, Carol Parker and Ian J Constable

WA Retinitis Pigmentosa Research Centre,
Lions Eye Institute, University of Western Australia,
Nedlands, Western Australia 6009

## INTRODUCTION

The Royal College of Surgeons (RCS) rat has an inherited progressive photoreceptor degeneration which has been well characterised (1). The primary defect in the RCS rat seems to be a failure of the pigment epithelial cells to properly ingest and dispose of the rod photoreceptor outer segments (2). Although the prime defect is not in the photoreceptor cells the RCS rat has been used to assess procedures for improving photoreceptor survival in inherited retinal degenerations. In recent years several procedures have been found which alter the rate at which photoreceptors degenerate in the RCS rat. The most effective procedure, in terms of number of cells surviving and the period of survival, involves the sub-retinal transplantation of pigment epithelial cells from normal animals (3,4). In this case substantial numbers of photoreceptors survive into adult life maintaining apparently normal inner and outer segments as well as synaptic contact with the inner retina (4). By replacing the defective pigment cells the genetic defect was therefore essentially corrected.

Somewhat surprisingly it was found that even in control experiments, in which saline injections were done but no cells transplanted, there was a transient delay in the loss of photoreceptors (4,5). Maintained retinal detachment appeared to enhance this increased survival (5). The mechanisms of this transient rescuing effect are not well understood but may be of particular relevance to maintenance of photoreceptors either when the defect is primarily in the photoreceptor cell, and pigment cell transplantation is therefore not an effective option, or when photoreceptor function is compromised by other factors. Recent studies have shown that a similar transient photoreceptor rescue can be produced by intravitreal injections of basic fibroblast growth factor (bFGF) at the start of degeneration (6). Levels of bFGF mRNA are lower in RCS retinas although the pigment cells have similar levels to controls, therefore a lack of retinal bFGF may explain the effect of exogenous bFGF (7). Release of bFGF or other growth factors by injured retina could therefore explain the traumatic photoreceptor rescue but other explanations are also possible. In support of a growth factor role the rescue following pigment epithelial cell transplantation always extended beyond the region with transplanted cells (3,4). Certainly very little is currently known about the localisation and possible actions of growth factors within the retina in either development or degeneration.

One difficulty with studies of the traumatic rescue of photoreceptors is that they require the insertion of a needle into the eye at a stage when the eye is not yet fully developed. Such a procedure, even when done very carefully, has the potential to cause widespread and unreproducible disruption of the eye. This may be the basis for some of the differences in the effect of trauma on photoreceptor rescue between different published studies (3-6). Argon laser irradiation, by contrast, has well defined lesioning effects on the pigment cells, photoreceptors and choroidal cells and is relatively reproducible but does not require physical penetration of the eye (8,9). In particular, unlike previously used procedures, laser irradiation produces outer retinal damage without transient retinal detachment or bleeding and the ability to make a focal lesion allows for ready assessment of lateral spread of the effect. Laser irradiation is also an established clinical procedure and it is therefore important to know of possible effects on inherited dystrophies. To date, only one study has addressed the effect of laser irradiation on RCS rat retina (10) and this did not examine the effect on photoreceptor survival. We have therefore initiated a series of studies into the effects of argon laser irradiation on photoreceptor degeneration in the RCS rat with particular emphasis on quantifying cell survival and correlating this with other cellular changes.

In order to make our data comparable to that of previous studies in this field the lesions were done at post natal day 23 and the analysis done at 2 weeks, 1 month and 2 months after irradiation.


## METHODS

All procedures conformed to the Declaration of Helsinki and were approved by the University of Western Australia Animal Welfare Committee. Post-natal day 23 (P23) pink-eyed RCS and RCS-rdy+ rats were anaesthetized with ketamine, dilated with 0.5% tropicamide and 2.5% phenylephrine 30 minutes prior to the procedure. The fundus was observed with a conventional slit lamp by neutralising the corneal refraction with a coverslip placed gently on the eye. Unilaterally, a region superior to the optic disk was irradiated with a grid pattern (6 horizontal rows with variable lesion number per row) of 40-60 non-overlapping argon blue/green laser lesions of 50 μm diameter. Preliminary experiments showed that 500mW lesions often produced disruption of Bruch's membrane and damage to the choroid. At 400mW the retina blanched and there was damage to pigment cells and loss of photoreceptors in the irradiated region. At 250-300mW there was disruption to the pigment epithelium and debris layer but often no obvious loss of photoreceptors. These levels are higher than previous studies because the majority of pigment cells in these rats do not contain melanin granules and therefore absorb less energy. Due to the variation with energy we examined the effect of irradiation in two ranges, one above 350mW for 0.5 sec and the other below 300mW for 0.5 sec. At 7, 14, 30 and 60 days after the laser lesions the animals were deeply anaesthetised with Nembutal and the eyes enucleated. A small opening was made in the cornea and the eyes placed in 3% paraformaldehyde, 1.25% glutaraldehyde in 0.1M phosphate buffer pH7.4. Thirty minutes later the eyes were bisected along the vertical meridian, the lens and vitreous removed and the eye cup immersion fixed overnight at 4°C. The eyecups were then dehydrated and embedded in Epon/araldite using a Lynx tissue processor. The eyecups were then sectioned at 1-2 μm in the superior-inferior plane until sections including the optic nerve head, the lesion site and both superior and inferior ora serrata were obtained. The sections were stained with 0.2% toluidine blue and drawn at 400X magnification. The drawing of each section was divided into 100 μm zones centred on the centre of the lesion. Within each zone the number of photoreceptors (both normal and pyknotic) and macrophage-like cells within the debris layer and outer nuclear layer (ONL) was recorded. The average thickness of the debris layer in each zone was also measured and the condition of the pigment layer recorded. The data were plotted graphically to assess correlations between the various parameters. Comparisions were done with the same region of the non-irradiated eye from the same animal or from an unoperated RCS animal at an equivalent stage. A total of 88 animals were examined in this study, including preliminary series for testing energy levels. These comprised 14 at 7 days, 14 at 14 days, 32 at one month and 28 at two months.

## RESULTS

### Qualitative Description

#### One to two weeks after irradiation (P30-P37)

At seven days following irradiation the lesion site could readily be identified. In all cases with higher energy irradiation, and in some of the lower energy cases, the lesion centre was characterised by a complete loss of pigment cells, debris and photoreceptors with some underlying choroidal cell damage, although the extent of loss generally varied throughout the lesion area. In all cases the inner retinal layers were preserved. In some of the lower energy cases there was a slight loss of debris with disruption of the pigment layer but the photoreceptor layer was not markedly reduced.

In the zone flanking the lesion there was a progressive increase in the thickness of the debris layer and an increase in the thickness of the photoreceptor cell layer. On the flank of the lesion the photoreceptor cell layer was generally the same thickness as in controls, or sometimes slightly thinner due to lesion induced loss of cells. However, the appearance of the layer was quite different to that of both controls and adjacent, non-irradiated, retinal regions. In the control retinas and non-irradiated regions of the experimental retinas at 7 and 14 days there was a disorganized debris layer in the outer

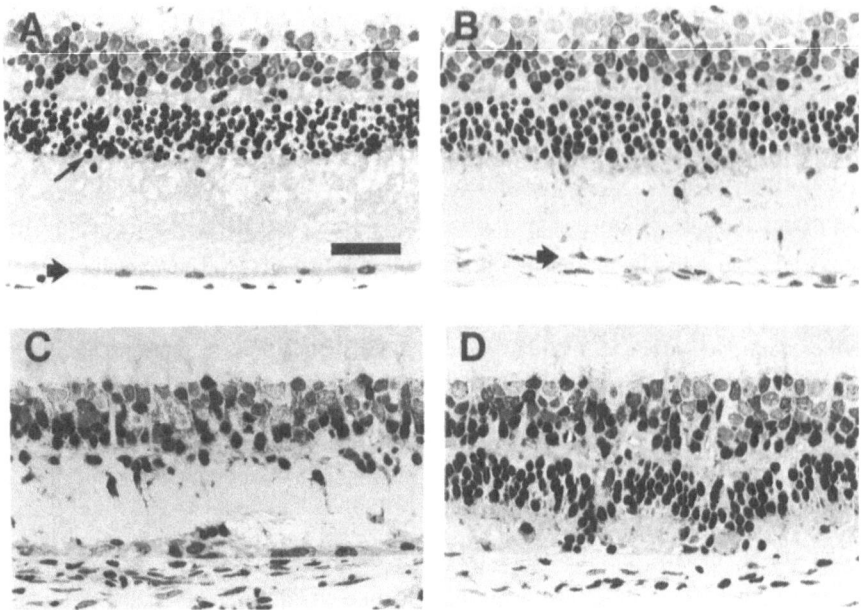

**Figure 1.** At 14 days after laser (P37) the OPL in non-irradiated regions contained many pyknotic cells (A, small arrow) and had an intact pigment epithelium (A, large arrow). On the flanks of lesions the photoreceptors had a normal appearance but the pigment epithelium was disrupted (B, arrow). At 30 days (P58) the non irradiated regions had one row of cells in the ONL (C), whereas the irradiated regions had 5 or 6 rows (D). Scale bar in A = 200 μm; and applies to all micrographs.

retina and the pigment cells formed an intact epithelium. Despite a very thick outer segment/debris layer there were very few macrophage/microglia-like cells in the debris layer (Figure 1A). Where pale staining cells with heterochromatic nuclei characteristic of this cell type were found they were generally adjacent to the inner segments (which were often darkly stained) or within the ONL. The thickness of the ONL was in the normal range but was largely composed of overtly pyknotic cells and densely stained cell fragments. This pyknosis was so strong that large aggregations of intensely toluidine blue stained cellular debris formed.

On the lesion flanks the ONL contained very few pyknotic cells and the majority of cell profiles resembled normal photoreceptors with intact inner segments (Figure 1B). There were many cells with heterochromatic nuclei present in the debris region on the lesion flank. Cells with similar morphology were located partly attached to Bruch's membrane or spanning the debris layer, while many others were found close to the ONL or even bridging the outer limiting membrane. In the non-irradiated regions similar cells were found in the ONL but none in the debris layer. There were less cells in the debris layer at 14 days than at 7 days and therefore we presume that most of these cells were in the process of migrating inwards to the ONL.

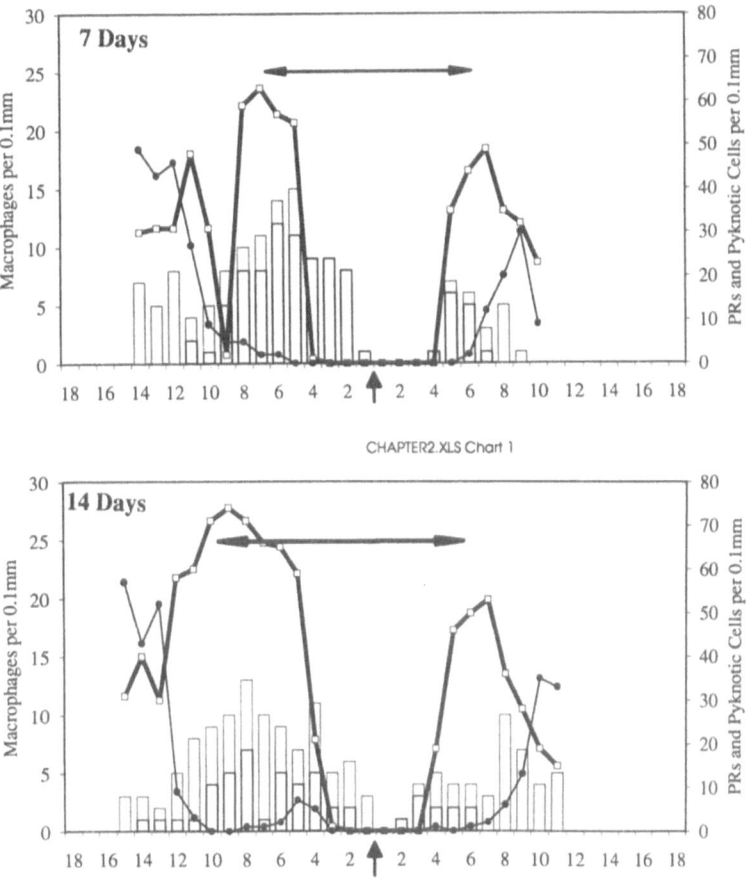

**Figure 2.** Graphs of the number of photoreceptors (thick lines), pyknotic cells (thin lines), macrophages in the debris layer (open bars) and macrophages in the OPL (speckled bars) versus distance from the lesion center at 7, 14, 30 and 60 days after laser irradiation. The bars are stacked and therefore total macrophage number is indicated by the height of the combined bars.

## One and two months after irradiation (P53 and P83)

Thirty days after irradiation the contrast between irradiated and non-irradiated regions was most dramatic. The non-irradiated posterior retina had a one to two cell thick ONL containing many cells with heterochromatic nuclei or pyknotic cells (Figure 1C). The irradiated regions contained an ONL four to five rows thick which largely contained photoreceptor cell somas of normal appearance (Figure 1D). In cases where the photoreceptors had been destroyed by the irradiation much of the inner nuclear layer (INL) had also degenerated and the choroid was hypertrophied. Where less energy was absorbed there was generally no pigment epithelium or debris layer but many photoreceptors at the centre of the irradiated region (Figure 1C). By 60 days after irradiation the number of

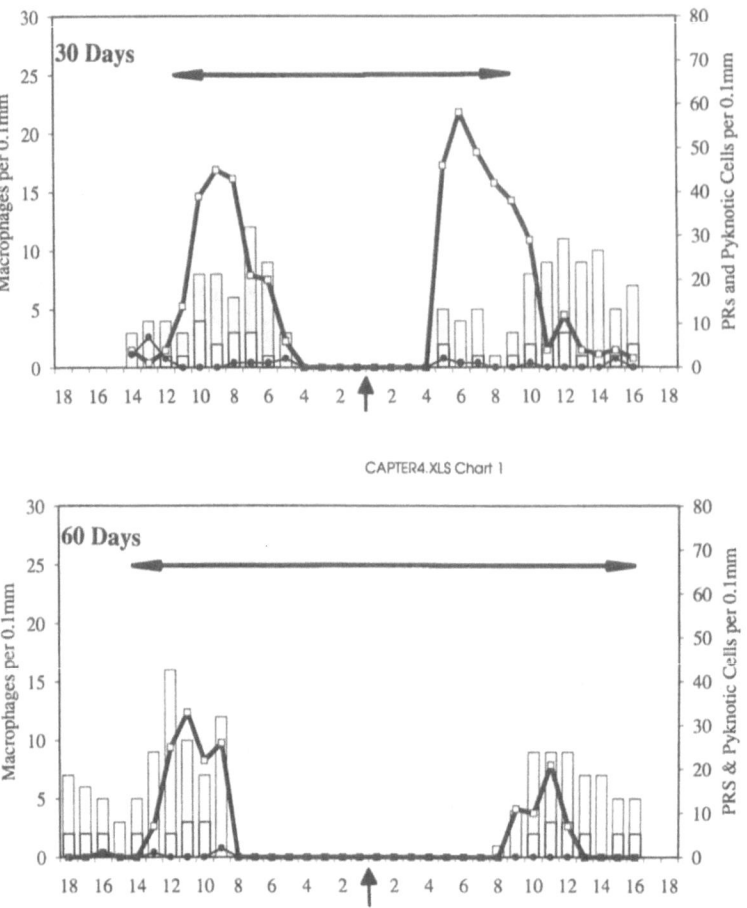

**Figure 2 (contd).** The values indicated are absolute numbers per 100μm zone measured from the lesion center. The horizontal arrows indicate the region in which the pigment epithelium was disrupted. The vertical arrows indicate the lesion center.

photoreceptors surviving on the flanks was reduced but was still increased over control levels. Generally very little debris remained and the pigment epithelium was not intact although occasional pigment cells were found on Bruch's membrane even in regions of elevated photoreceptor number.

## Quantitative Analysis

The cell counts confirmed the qualitative description above. There was a dramatic increase in normal photoreceptor soma number on the lesion flanks at all stages examined (Figure 2). At 7 and 14 days there were almost no pyknotic cells in the lesion flank zones although they comprised more than 50% of cells in the adjacent ONL. This correlated with an increased macrophage number, particularly in the debris layer. The increased photoreceptor number also correlated with an absence of an intact pigment epithelial layer. However, note that the rescue zone extended 200-300 μm beyond the pigment disruption. At 30 and 60 days there were still elevated macrophage numbers in the region of increased photoreceptor survival but a greater proportion were in the ONL rather than debris layer. At 7 and 14 days the rescue zone did not correlate very well with a decrease in the debris layer thickness although there was always a decreased debris thickness in the lesion centre. However, by 30 to 60 days the debris layer was reduced in the rescue zone as well as the lesion centre (data not shown). In some of the lower energy cases there was no major loss of photoreceptors at the center of the lesion. This pattern of rescue suggests that preservation could be obtained over wider areas of retina without destruction of photoreceptors.

## DISCUSSION

This study has demonstrated that a single application of argon laser irradiation, shortly after the onset of photoreceptor degeneration in the RCS rat, significantly delays the loss of photoreceptor cells from the outer retina. The time course of this delayed degeneration is similar to that following a needle stab or bFGF injection. The laser lesions did not produce any general trauma to the eye, no bleeding occurred and the effect was localised to the flanks of the irradiated region.. This therefore confirms that events local to the trauma site are sufficient to delay degeneration. When a moderate energy photocoagulation was used the saving was confined to the lesion flanks. However, if the energy level was low enough, cell rescue occurred throughout the irradiated retina. It may therefore be possible to do multiple irradiations over a period of time which may prolong the rescue period. Therefore laser irradiation provides a simple, reproducible method for producing trauma associated transient photoreceptor rescue.

Laser lesions are known to have a wide variety of effects on the retina. These include the release of injury and inflammatory related molecules, changes in oxygen and ionic flows, activation of Müller and pigment cells etc. (8-15). Any of these factors could potentially influence photoreceptor survival. In addition, alterations in cellular state, such as the induction of heat-shock proteins may play a role (16). Our histology shows that the major events which correlate with ONL changes are (i) migration of cells into the outer segment/debris layer, (ii) lack of an intact pigment epithelial layer and, at later stages (iii) a reduction in debris layer thickness. Further associated events may also be important such as alterations in Müller or photoreceptor cell function. Preliminary data with GFAP immunocytochemistry demonstrates that the Müller cells sprout processes in the laser region, possibly also responding to high growth factor levels. Immunocytochemistry for bFGF shows no major sustained alterations in levels around the lesion but a transient alteration triggering further changes as suggested for retinal detachment (17) cannot yet be excluded.

Silverman and Hughes (5) demonstrated somewhat longer lasting transient rescue of photoreceptors in the RCS rat by insertion of a gelatin slab in the subretinal space. In our study there was no physical detachment of the retina but rather activation of the pigment epithelial cells and loss of the intact pigment epithelium. However, in many respects laser irradiation may produce effects similar to a retinal detachment. For example, many cells infiltrate the sub-retinal space after detachment and many of these appear to derive from the pigment epithelium (18) just as we have found in the present study. It has been shown that photoreceptor protein synthesis may drop by 30-40% following photocoagulation (19) and Müller cell enzyme expression is changed following laser irradiation (20). Similarly outer segment production is reduced after detachment (23).

The majority of cells which migrate into the debris layer are probably of pigment epithelial origin, as has previously been reported to occur following argon laser irradiation (8-10). Normally in RCS rats the cells which migrate into the debris layer have been presumed to be macrophages (21), particularly as they have been found adjacent to apparently intact pigment epithelium. However it is often difficult to distinguish macrophages from reactive pigment cells. In the study of Ansell and Marshall (10) where pigmented rats were used, the cells which entered the debris layer contained primary melanosomes and were therefore classified as of pigment epithelial origin. Therefore, the activation produced by argon irradiation may be similar to the effect of transplantation of dystrophic pigment cells which produces greater transient rescue of photoreceptors but fails to sustain the cells in the long term (4). Therefore our results would suggest that the dystrophic pigment cells can either phagocytose more effectively or release beneficial factors, when removed from their usual epithelial constraints. Pigment epithelial cells have been shown to produce bFGF (22,23) as well as other growth factors (24,25) in tissue culture and their proximity to the ONL after laser may produce increased bFGF levels around the photoreceptors. It is also possible that some of the cells in the debris layer are macrophages or microglial cells which migrated from the choroid or ONL. Use of specific cell markers may resolve this issue.

It would appear to be unlikely that the sole mechanism of this transient rescue is a reduction in debris layer thickness. At 7 and 14 days after irradiation the thickness of the debris layer on the flanks where the photoreceptors had a more healthy appearance was only slightly less than adjacent regions. Nevertheless, at later stages (30 and 60 days) there was a reduction of debris thickness at sites of increased photoreceptor survival. Since there was no intact pigment epithelium the migrated cells must have removed part of the debris. This removal may also have been aided by a slowing of outer segment production in the irradiated region. However, Li and colleagues (26) have shown that sub-retinal injection of peritoneal macrophages, which was effective in reducing the debris layer did not have a marked effect on photoreceptor survival, although they did not examine early stages very thoroughly and a transient effect may be possible.

A beneficial role of macrophage invasion is controversial since although they do secrete substances which may be beneficial to cell survival activated macrophages may also release many cytotoxic substances which can destroy normal cells. Examples of active photoreceptor and ganglion cell destruction by macrophages have been documented (27,28). In our analysis of either non-irradiated regions or control RCS retinas the period of P30 - P53 was characterised by regions in which there was virtually no macrophage or pigment cell entry to the debris layer despite the active degeneration in the ONL. This suggests that the debris lacked any components to stimulate such a migration. Even after the laser irradiation the cells with heterochromatic nuclei in the debris layer tended to cluster around the inner segments asthough the prime stimulus for their migration was in the ONL and not the debris layer. At later stages in the non-irradiated regions macrophages or microglia did enter the debris layer in some areas. In these regions they were often located directly adjacent to the pigment cells or within the ONL. However, in no case was this associated with increased photoreceptor survival.

Usually following argon laser photocoagulation the pigment epithelial cells around the lesion divide and fill the gap in the epithelium (8,9,29) within a few weeks. In this study the lesioned areas showed very little sign of re-establishment of the pigment epithelium. This may be due to the pigment cell defect in the RCS rat or to the absence of intact outer segments which may act as a stimulus for this process.

In conclusion, the results of this and earlier studies suggests that there are several critical factors involved in the transient photoreceptor rescue following injury. These are likely to include an alteration in retinal state which reduces photoreceptor turnover, activation of pigment cells to increase phagocytosis and possibly growth factor release and a breakdown of the blood retinal barrier. We are currently examining whether the laser irradiation does release endogenous bFGF or other growth factors and also the extent to which continuous low level laser may maintain a favourable environment for photoreceptor survival.

## ACKNOWLEDGEMENTS

We would like to acknowledge the assistance of Dr M. LaVail in the establishment of an RCS rat colony and the Western Australian Retinitis Pigmentosa Foundation, the Australian Retinitis Pigmentosa Association and the Lions Save-Sight Foundation (Western Australia) for the provision of funds. We would also like to acknowledge Ms K. Salter for expert assistance with text and graph formatting.

## REFERENCES

1. R.J. Mullen and M.M. LaVail, Inherited retinal dystrophy: Primary defect in pigment epithelium determined with experimental rat chimeras. *Science* 192:799 (1976).
2. M.H. Chaitin and M.O. Hall, Defective ingestion of rod outer segments by cultured dystrophic rat pigment epithelial cells. *Invest. Ophthalmol. Vis. Sci.* 24:812 (1983).
3. L. Li and J.E. Turner, Inherited retinal dystrophy in the RCS rat: Prevention of photoreceptor degeneration by pigment epithelial cell transplantation. *Exp. Eye Res.* 47:911 (1988).
4. L. Li and J.E. Turner, Optimal conditions for long-term photoreceptor cell rescue in RCS rats: The necessity for healthy RPE transplants. *Exp. Eye Res.* 52:669 (1991).
5. M.S. Silverman and S.E. Hughes, Photoreceptor rescue in the RCS rat without pigment epithelium transplantation. *Curr. Eye Res.* 9:183 (1990).

6.  E.G. Faktorovich, R.H. Steinberg, D. Yasumura, M.T. Matthes and M.M. LaVail, Photoreceptor degeneration in inherited retinal dystrophy delayed by basic fibroblast growth factor. *Nature* 347:83 (1990).

7.  P.E. Rakoczy, M.F. Humphrey, D.M. Cavaney, Y. Chu and I.J. Constable, Expression of basic fibroblast growth factor (b-FGF) and b-FGF receptor mRNA in the retina of royal college of surgeons rats. A comparative study. *Invest. Ophthalmol. Vis. Sci.* In press (1993).

8.  J. Marshall and J. Mellerio, Pathological development of retinal laser photocoagulations. *Exptl. Eye Res.* 6:4 (1967).

9.  J. Marshall and J. Mellerio, Laser irradiation of retinal tissue. *Br. Med. Bull.* 26:156 (1970).

10. P.L. Ansell and J. Marshall, Laser induced phagocytosis in the pigment epithelium of the Hunter dystrophic rat. *Brit. J. Ophthal.* 60:819 (1976).

11. V.A. Alder, S.J. Cringle and M. Brown, The effect of retinal photocoagulation on vitreal oxygen tension. *Invest. Ophthalmol. Vis. Sci.* 28:1078 (1987).

12. R.L. Novack, E. Stefansson and D.L. Hatchell, The effect of photocoagulation on the oxygenation and ultrastructure of avascular retina. *Exp. Eye Res.* 50:289 (1990).

13. M.S. Burns and M. Robles, Müller cell GFAP expression exhibits gradient from focus of photoreceptor light damage. *Curr. Eye Res.* 9:479 (1990).

14. N. Naveh, E. Bartov and C. Weissman, Subthreshold argon-laser irradiation elicits a pronounced vitreal prostaglandin E2 response. *Graefe's Arch. Clin. Exp. Ophthalmol.* 229:178 (1991).

15. G.A. Peyman and D. Bok, Peroxidase diffusion in the normal and laser-coagulated primate retina. *Invest. Ophthalmol. Vis. Sci.* 11:35 (1972).

16. K. Yamaguchi, V.P. Gaur, M. Tytell and J.E. Turner, Heat shock protein (HSP70) in ocular tissues of rats with normal and dystrophic retinas. *Invest. Ophthalmol. Vis. Sci.* 32:1207 (1991).

17. G.P. Lewis, P.A. Erickson, C.J. Guérin, D.H. Anderson and S.K. Fisher, Basic fibroblast growth factor: A potential regulator of proliferation and intermediate filament expression in the retina. *J. Neurosci.* 12:3966 (1992).

18. N.F. Johnson and W.S. Foulds, Observations on the retinal pigment epithelium and retinal macrophages in experimental retinal detachment. *Bri. J. Ophthalmol.* 61:564 (1977).

19. J. Marshall, Thermal and mechanical mechanisms in laser damage to the retina. *Invest. Ophthalmol. Vis. Sci.* 9:97 (1970).

20. H. Ishigooka, A. Hirata, T. Kitaoka and S. Ueno, Cytochemical studies on pathological Müller cells after argon laser photocoagulation. *Invest. Ophthalmol. Vis. Sci.* 30:509 (1989).

21. E. Essner and G. Gorrin, An electron microscopic study of macrophages in rats with inherited retinal dystrophy. *Invest. Ophthalmol. Vis. Sci.* 18:11 (1979).

22. L. Schweigerer, B. Malerstein, G. Neufeld and D. Gospodarowicz, Basic fibroblast growth factor is synthesized in cultured retinal pigment epithelium. *Biochem. Biophys. Res. Comm.* 143:934 (1987).

23. M.D. Sternfeld, J.E. Robertson, G.D. Shipley, J. Tsai and J.T. Rosenbaum, Cultured human retinal pigment epithelial cells express basic fibroblast growth factor and its receptor. *Curr. Eye Res.* 8:1029 (1989).

24. H.C. Wong, M. Boulton, D. McCleod, M. Bayly, P. Clark and J. Marshall, Retinal pigment epithelial cells in culture produce retinal vascular mitogens. *Arch. Ophthalmol.* 106:1439 (1988).

25. P.A. Campochiaro, R.A. Sugg, G. Grotendorst and L.M. Hjelmeland, Retinal pigment epithelial cells produce PDGF-like proteins and secrete them into their media. *Exp. Eye Res.* 49:217 (1989).

26. L. Li, H. Sheedlo, V. Gaur and J.E. Turner, (1991) Effects of macrophage and retinal pigment epithelial cell transplants on photoreceptor cell rescue in RCS rats. *Curr. Eye Res.* 10:947 (1991).

27. R.A. Cuthbertson, R.A. Lang and J.P. Coghlan, Macrophage products IL-1$\alpha$, TNF$\alpha$ and bFGF may mediate multiple cytopathic effects in the developing eyes of GM-CSF transgenic mice. *Exp. Eye Res.* 51:335 (1990).

28. S. Thanos, Adult retinofugal axons regenerating through peripheral nerve grafts can restore the light-induced pupilloconstriction reflex. *Exp. Eye Res.* 55 (Suppl. 1):S37,112.

29. W.E. Smiddy, S.L. Fine, H.A. Quigley, G. Dunkelberger, R.M. Hohman and E.M. Addicks, Cell proliferation after laser photocoagulation in primate retina. An autoradiographic study. *Arch. Ophthalmol.* 104:1065 (1986).

# ROLE OF HEAT SHOCK PROTEIN 70 (HSP70) IN PHOTORECEPTOR CELL SURVIVAL IN THE AGED RAT

Michael Tytell,[1,2] Keiko Yamaguchi,[3] and Katsuhiro Yamaguchi[3]

[1]Department of Neurobiology and Anatomy
[2]Department of Ophthalmology
 Program in Neuroscience
 Bowman Gray School of Medicine
 Wake Forest University
 Winston-Salem, NC 27157-1010
[3]Department of Ophthalmology
 Tohoku University School of Medicine
 Sendai 980, Japan

## INTRODUCTION

The toll aging takes on the retina the rat and human is well-documented. In both organisms, there is a gradual loss of photoreceptors, especially in the peripheral retina[1-8]. Additionally, in the aged human, there is often some degree of macular degeneration or cone loss[9,10], as well as a loss of retinal ganglion cells[11]. Although many factors probably contribute to the loss of cells and subsequent comprise of function, we suggest that one important factor is likely to be the capacity of the cell to defend itself against metabolic stress.

The retina is subject to toxic products daily as a result of photochemical side reactions, especially lipid peroxidation[12-14]. Additionally, the photoreceptors are located in an avascular zone that depends solely on diffusion for oxygen and other nutrients; they may be bordering on hypoxia because of their high metabolic rate[15,16]. Therefore, a decrease in the photoreceptors' capacity to respond to such stress might exacerbate the loss of cell function or of cells themselves in response to the typical metabolic stress which they experience. Such a depressed stress response might be one consequence of aging and it is that possibility which we have investigated in the work described here.

One general feature of the response of cells to metabolic stress is the increased synthesis and accumulation of the 70 kDa heat shock protein (hsp70). A large number of observations have accumulated over the past decade which show that the hsp70 and other hsps play major roles in the capacity of most cells to tolerate periods of metabolic stress[17]. Hsp70 includes two very similar isoforms, the constitutive form (hsp70c; also called hsp73 to indicate its higher molecular weight) and the inducible form (hsp70i; also called hsp72 or hsp68 in reference to its lower molecular weight)[18]. Hsp70c is the form

*Retinal Degeneration*, Edited by J.G. Hollyfield *et al.*
Plenum Press, New York, 1993

that is most abundant in normal, unstressed cells, whereas hsp70i refers to the form whose synthesis typically increases more in response to metabolic stress than hsp70c. These two forms can be discriminated using two monoclonal antibodies as described in the Methods. However, whether or not these two forms have different functions in the metabolic stress response remains unclear.

In the rat retina, earlier work has shown that its content of hsp70i can be increased by raising the anesthetized rat's body temperature to 41-42° C[19] and that during the time of the elevated hsp70i, the retina is resistant to light damage[20,21]. On the basis of this correlation, we evaluated the levels of both forms of hsp70 in each of the layers of the retina of young (3 month) and old (20 month) Fischer 344 rats under normal conditions and after being challenged by a brief period of hyperthermia. The aim was to determine if a disturbance in retinal hsp70 might correlate with the well-documented gradual loss of photoreceptors in the Fischer rat retina as a function of increasing age[1,2,22,23].

## METHODS

Two sets of Fischer 344 male rats were used, one three months of age, the other twenty months. Some of each age were anesthetized with pentobarbital and made hyperthermic by placing them in a plexiglass covered chamber with circulating, humidified air at 40-41° C. Rectal temperature of each rat was monitored and each was kept in the chamber until its temperature reached 41° C. Subsequently, they were placed in another chamber maintained at 30° C until they recovered from anesthesia. The purpose of the warm recovery chamber was to prevent the rats from becoming hypothermic while under the influence of the barbiturate. Control rats of the two ages were treated identically in terms of anesthesia and recovery, but were not placed in the 41° C chamber.

To prepare the eyes for immunohistochemical detection of hsp70, two or three rats from the two age groups were sacrificed at 4, 18, or 50 hours after the heat treatment and two control rats were sacrificed at 50 hours after being anesthetized. The eyes were removed and fixed in Carnoy's solution, which is greatly superior to standard formaldehyde-based fixatives in preserving hsp70 immunoreactivity (Tytell and Hollman, unpublished observations). The details of the fixation and subsequent paraffin embedding, sectioning and processing for immunohistochemistry were the same as described previously[24], except for the primary antibody concentrations. The antibody that recognizes both hsp70c and hsp70i, which will be referred to as anti-hsp70ci, was used at 1:5000 dilution and the antibody specific for hsp70i, called anti-hsp70i, was used at 1:500 dilution. Both monoclonal antibodies were obtained from StressGen (Victoria, British Columbia, Canada). Bound antibody was detected using an avidin-biotin-horseradish peroxidase second antibody system (Vector Laboratories, Burlingame, CA) and diaminobenzidene (DAB) as the chromagen. Background controls for nonspecific immunoreaction product consisted of tissue sections treated with all the reagents except for the primary antibody.

Retinal sections through the optic nerve in the superior-inferior plane were used to quantify the amount of immunoreaction product in each layer except the retinal ganglion cell (RGC) layer. The irregular distribution of RGCs in the retina precluded them from being measured. A Leitz Laborlux 5 trinocular microscope with an MTI #65 video camera (DAGE-MTI, Michigan City, IN) and the Jandel Video Analysis software (JAVA version 4.0, Jandel Sciences, Corte Madera, CA) were used to capture images viewed through the 40X objective. To make sure that measurements from different sessions were comparable, the light intensity and condenser settings of the microscope were always kept the same. Beginning at the central superior retina, right next to the optic nerve head, the layer in which the staining density was to be measured was defined by drawing a rectangle 100 $\mu$m in width, with the height defined by the borders of the layer, as shown in Figure 1. Within this area, the software was used to calculate the average gray level. This process was repeated every 250 $\mu$m around the circumference of the superior and inferior retina, yielding a total of 30 gray level measurements for each retinal layer of every eye. Those 30 values were used to calculate a mean density for each layer. Similar measurements made of background control eyes were subtracted from the hsp70 stained eyes so that the net density values reflected only that due to the specific immunoreaction product and not the inherent density of the tissue or any nonspecific DAB reaction product that may have been present in the tissue.

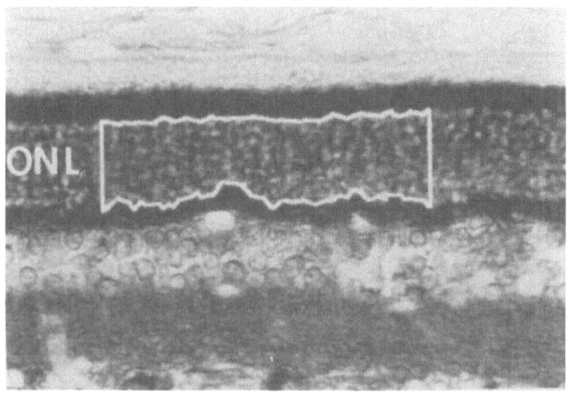

**Figure 1.** An example of an outlined layer, the outer nuclear layer (ONL) in this case, after capturing an image using the video morphometry system. The horizontal width of the area was calibrated to be 100 $\mu$. Within that area, the average gray level can be determined and, after subtracting from that value the average gray level of a retinal section not treated with the primary antibody (a minus antibody control), the resulting net average gray level is proportional to the density of the anti-hsp70 immunoreaction product. This procedure was repeated for each of the retinal layers.

## RESULTS

### Hsp70 Distribution In Retinas of Control Rats

The typical pattern of hsp70 staining in the retina is illustrated for a three month old rat in Figure 2. It shows that the normal retina has a considerable amount of anti-hsp70ci immunoreactive material, especially in the layers that contain the cytoplasmic regions of the photoreceptors, the inner segments (IS) and outer plexiform layer (OPL) (Figure 2A). Additionally, the retinal ganglion cell (RGC) bodies typically contain a lot of immunoreactive material, often being as darkly stained as the IS (Figure 2A). A section of the same eye, shown in Figure 2B, probed with anti-hsp70i, showed that this form of hsp70 is present in the retina as well, though in lower amounts, and has essentially the same relative distribution as the anti-hsp70ci immunoreactivity. Figure 2C shows that when the primary antibodies were omitted, very little nonspecific reaction product was deposited in the retinal section.

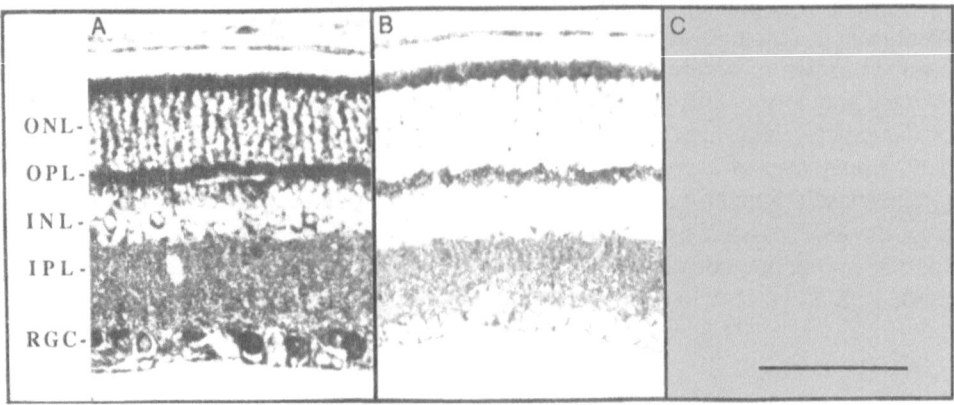

**Figure 2.** Examples of the distribution of immunoreaction product for hsp70ci (A) and hsp70i (B) in the retina of a three month old control rat. Panel (A) shows that high levels of anti-hsp70ci reaction product were present in the inner segments (IS), outer plexiform layer (OPL), and in the cytoplasm of the retinal ganglion cells (RGC). The outer nuclear layer also had considerable staining with this antibody, which appeared to be in cytoplasmic processes surrounding the photoreceptor nuclei; the cytoplasm of scattered cells in the inner nuclear layer (INL) were similarly stained. That staining pattern could correspond to Müller cell cytoplasmic processes and cell bodies, but that identification could not be made in this type of preparation. Immunostaining was prominent as well in the inner plexiform layer (IPL) and pigment epithelium (PE; the latter was not analyzed in this study). Panel (B) is another section from the retina of the same rat as in (A) treated with anti-hsp70i antibody. The overall distribution of immunoreactivity was similar to that seen in (A), but was much less intense, indicating that there was less hsp70i immunoreactive material in the retina compared to that for anti-hsp70ci. Panel (C) illustrates the very low background staining obtained when the primary antibody was omitted from the immunohistochemical procedure. Bar = 100 $\mu$.

Quantitative analysis by computer densitometry of hsp70ci and hsp70i immuno-reactivity in young and old control retinas confirmed qualitative impressions (Figure 3). The two antibodies yielded similar distributions of immunoreaction product in the six retinal layers of both young and old rats, with the anti-hsp70ci positive material always being greater than that for anti-hsp70i, except in the outer segments (OS). In that layer, though there was only a small amount of either form of hsp70 present, the young rats showed a trend toward more hsp70i than hsp70ci and the old rats had equivalent amounts of the two. One other apparent trend confirmed by this analysis was that the three layers which consisted of cytoplasmic portions of cells, the IS, the OPL, and the inner plexiform layer (IPL), had more immunoreactivity than the two nuclear layers, the outer nuclear layer (ONL) and inner nuclear layer (INL).

**Hyperthermia Differentially Affects Retinal Hsp70 Levels In Young And Old rats**

The rectal temperatures of each of the rats was individually monitored after they were anesthetized and placed in a chamber maintained at 40-41° C. They were kept in the heated chamber until each reached between 41-42° C. After that, they were transferred to a chamber maintained at about 30° C until they began to recover from the anesthesia in order to prevent hypothermia, a typical side effect of barbiturate anesthesia. The mean temperature records of the young and old rats are plotted in Figure 4 and show that the old rats responded more slowly than the young ones to the

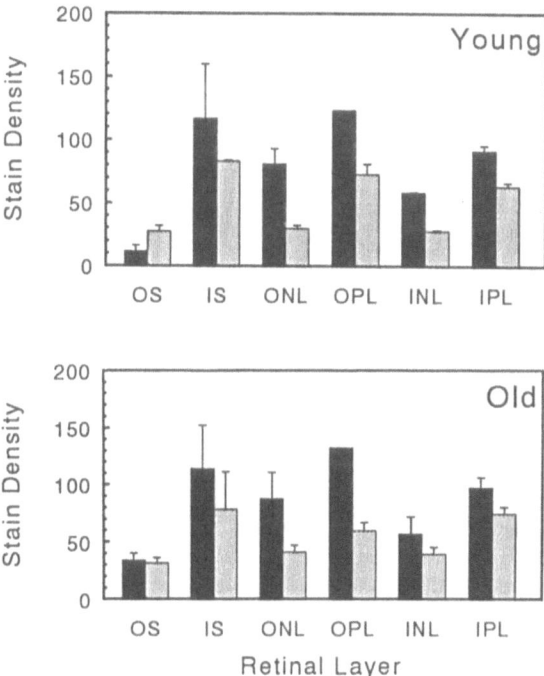

**Figure 3.** The relative amounts of hsp70ci (black bars) and hsp70i (stippled bars) immunoreactive material in each of the six layers of the retinas of young and old control rats. As described in the Methods, a computer-assisted video analysis system was used to convert the density of immunoreaction product in 30 sectors of each retinal layer to a gray level on a scale of 0-255, where 0 is white (complete absence of immunoreaction product) and 255 is black (saturating amount of immunoreaction product). The 255 step scale is defined by the resolution of the computer. In other work using the same system, a linear relationship was shown between a series of increasing hsp70 standards and the measured gray level in western blots probed with the same antibody[25]. The mean and standard error of the mean for each layer and the two antibodies are based on measurements in two retinas from different rats, except for the hsp70ci values in both young and old rat OPL and the INL in the young rats, which were based on readings from one retina only because of missing data.

heat treatment. The mean rectal temperature of the young rats peaked at 41.6 ± 0.1° C after 25 minutes in the heated chamber, whereas that of the old rats reached a similar level, 41.3 ± 0.3° C, but not until 35 minutes of exposure. Furthermore, as a consequence of their slower responses, the old rats' rectal temperatures remained elevated above 40° C for a longer time, about 29 minutes, in comparison to 24 minutes for the young rats. This slowed responsiveness of older rats has been observed previously[26].

Visual inspection of the staining of the retinas with the two antibodies at 4, 18, and 50 hours after the heat treatment suggested that the retinas of the old rats tended to have slightly less immunoreactive hsp70 than those of the young animals with either antibody. Some examples of these differences are shown in Figure 5 for retinas collected at 18 hours after heat treatment. However, the amount of immunoreactivity varied noticeably in different sectors of the same retina, so it was necessary to evaluate

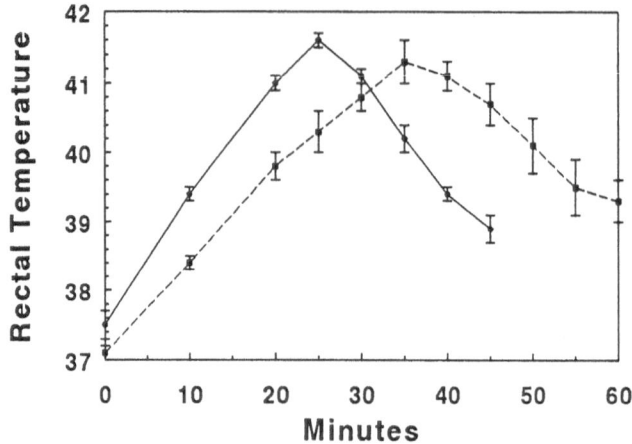

**Figure 4.** Average rectal temperatures during heat treatment as a function of time in the young (solid line) and old (dotted line) rats. For the young rats, each point is based on 9 animals, except at 35 minutes, which is a mean of 6. For the old rats, 8 animals were followed at each point except at 60 minutes, for which only 5 rats were evaluated. Monitoring of the young rats' temperatures was stopped after they had fallen below 39° C and for the old rats, it was stopped after one hour.

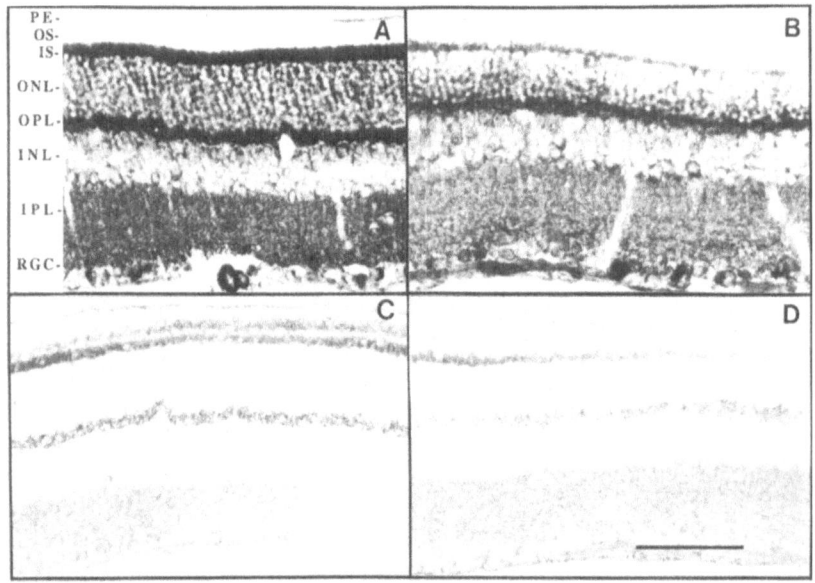

**Figure 5.** Examples of hsp70 immunostaining in retinas of young and old rats collected 18 hours after the hyperthermia. The upper pair of micrographs (A,B) show anti-hsp70ci staining in young (A) and old (B) retinas and the lower pair of micrographs (C,D) show anti-hsp70i staining in sections from the same young (C) and old (D) retinas used for the upper micrographs. Note that the overall staining of the young rat retinas with either antibody was slightly darker than that for the old rat retinas. Bar = 100 $\mu$

the retinas by computer-assisted video densitometry to confirm whether or not there were significant differences in the post-hyperthermia retinal hsp70 levels of the two groups. This analysis showed that there were significant distinctions in the responses of the retinas of the groups, but only in the layers that included the various portions of the photoreceptor cells. Figure 6 shows the results of the analysis of anti-hsp70ci immunoreactivity. Whereas the hsp70ci immunoreactivity in the outer retina of the young rats increased or stayed about the same after hyperthermia, that of old rats declined significantly, either in relation to the control old rats or in comparison to the corresponding group of young rats. This difference in response was most dramatic in the OS. Although the old rats started out with significantly more hsp70ci immunoreactivity before heating (Figure 6, OS, 0 hrs, $p<0.05$), by 18 hours after heating, the young rats showed almost a 4-fold rise ($p<0.01$, 0 hrs versus 18 hrs), whereas the old rats demonstrated more than a 2.5-fold decrease ($p<0.05$, 0 hrs versus 18 hrs), so that the hsp70ci immunoreactivity in the young rat OS was almost 4-fold higher than in the old rats ($p<0.01$, young 18 hr versus old 18 hr). The same trends were seen in the IS and OPL (Figure 6, lower two graphs), but was statistically significant only in the OPL at 4 and 50 hours ($p<0.05$).

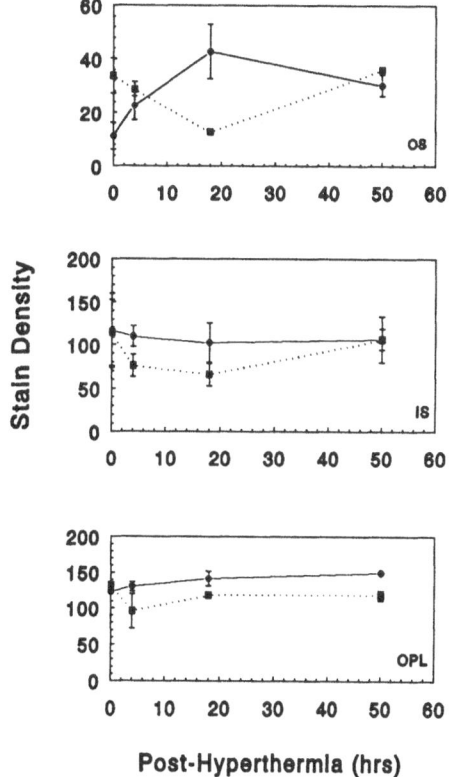

**Post-Hyperthermia (hrs)**

**Figure 6.** Quantitative analysis of the density of hsp70ci immunostaining in the outer layers (OS, IS, OPL) of the retinas of young (solid lines) and old (dotted lines) rats as a function of time after hyperthermic treatment. The data were analyzed by two-way analysis of variance, which showed a significant interaction effect ($p = 0.0036$); individual means were then compared by protected t-tests, the results of which are mentioned in the text. Each point was based on analyses of three retinas from separate rats, with the exception of the following for which two retinas were analyzed: 0 hour young and old for all layers; OS old 18 hour; OPL old 4 and 18 hour.

For hsp70i immunoreactivity, a similar tendency toward a decrease in the retinas of the old rats was observed, but it achieved significance only in the IS and ONL (Figure 7). In the IS, although the hsp70i immunoreaction product started out the same in the zero hour controls, by 50 hours after hyperthermia that of the older rats became significantly lower than the level in the IS of the young rat retinas (Figure 7, upper

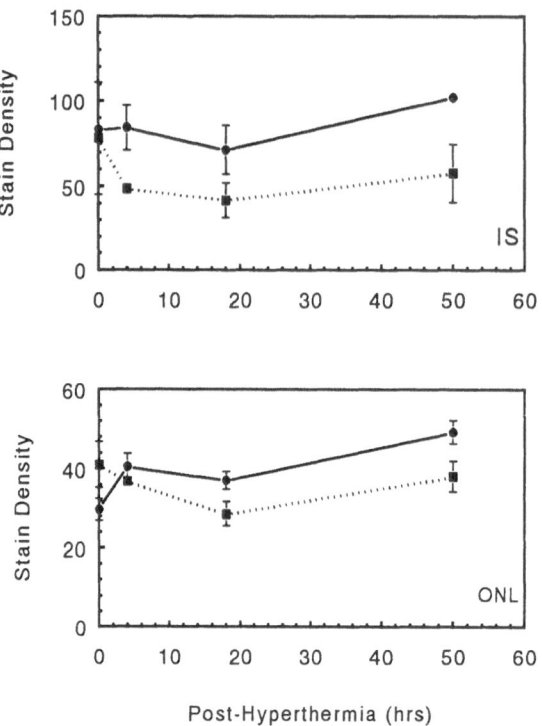

**Figure 7.** Quantitative analysis of the amount of hsp70i immunoreaction product in the IS and ONL of retinas from young (solid lines) and old (dotted lines) rats. Two-way analyses of variance showed in the IS a significant main effect of age ($p = 0.027$) and in the ONL, a significant main effect of time after hyperthermia ($p = 0.037$) and a significant interaction effect ($p = 0.0211$). Comparison of individual means were made by protected t-tests and significant differences are discussed in the text. The IS analysis was based on two retinas from separate rats, except for the 4 and 18 hour intervals in the young rats and the 50 hour interval in the old rats, which were based on three retinas. The ONL analysis was based on two retinas from separate rats for the zero and 50 hour points in the young group and zero and 18 hours in the old group and three retinas at all the other points for both groups.

graph). The ONL hsp70i immunoreactivity was initially higher in the old rat retinas ($p < 0.05$), similar to what was seen for hsp70ci in the OS (Figure 6, upper graph) and, after hyperthermia, it decreased ( zero versus 18 hours, $p < 0.05$) while that of the young rats rose (zero versus 4 and 18 hours, $p < 0.05$; 0 versus 50 hours, $p < 0.01$). At 50 hours post-hyperthermia, the ONL of the young rats had significantly more hsp70i immunoreactive material ($p < 0.05$).

## DISCUSSION

The general pattern of distribution of hsp70 detected in the Fischer rat retina with the anti-hsp70ci and -hsp70i antibodies (Figures 2 and 3) agrees with that which we have described previously in the Sprague-Dawley rat and in the Royal College of Surgeons rat prior to the onset of photoreceptor degeneration, being most prominent in the IS, OPL, and RGC cytoplasm, and lower in the IPL, INL, and ONL[24,27]. Thus, it is likely that this pattern is characteristic of the retina; in fact, preliminary studies of human retinal hsp70 appear qualitatively similar (Tytell and Hollman, unpublished observations). We can suggest three reasons to explain that pattern. First, the forms of hsp70 detected by the antibodies used here are primarily cytoplasmic proteins, except under conditions of severe metabolic stress, when hsp70i becomes concentrated in the nucleus[28]. In this context, it is logical that the layers of the retina that include most of the cytoplasmic processes of the cells, notably the IS, OPL, and IPL, contain more immunoreaction product than the nuclear layers. Second, the fact that hsp70c is involved in membrane biosynthetic events[29] provides a reason for its apparent abundance in the cytoplasm of the photoreceptors, which makes up much of the material in the IS and OPL, and in the cytoplasm of the RGCs. Both of those cell types have high rates of membrane production, the photoreceptors to replenish the OS and the RGCs to support their long axons. Third, the presence of the highest levels of hsp70 in the IS and OPL may also be a reflection of the greater degree of metabolic stress to which the photoreceptors are subject, since that part of the retina is avascular and is known to have very low oxygen tension[15,16] and is prone to photochemical damage as a side reaction of the absorption of light energy[12-14].

The responses of the young and old rats to the hyperthermic stress led to one unexpected observation. Although we used the same conditions that were shown previously to produce a sufficient increase in body temperature to elevate hsp70 in the retina of the Sprague-Dawley rat[19-21], only minor increases in hsp70 were seen in the OS and ONL of young Fischer rats. This outcome may have been a consequence of the fact that, on the average, the length of time that the rats' body temperatures were above 41° C was slightly less than the 15 minutes (Figure 4) that we found to be the minimum necessary for a robust induction of hsp70[20]. Perhaps the Fischer rats have a better thermoregulatory system or have a less vigorous hsp70 response to hyperthermia than do Sprague-Dawley rats. A direct comparison of the two strains would be necessary to determine which of these possibilities is the case.

Despite the fact that the hyperthermia did not produce the expected vigorous induction of hsp70 in these animals, it still revealed an intriguing distinction between the two ages. During the post-hyperthermia period, the 20 month old rats showed consistently lower levels of hsp70 immunoreactivity than the 3 month old rats in all of the retinal layers that include the various parts of the photoreceptor cells (Figures 6 and 7). Since hsp70 is known to protect cells from the lethal effects of metabolic stress[30,31], the decline observed in the older rats suggests that there is an impairment of the retinal response to metabolic stress which may render the photoreceptors more susceptible to light damage and destruction. A similar impairment of the heat shock response has been documented in cultured fibroblasts and lung tissue[32-34]. The only other *in vivo* studies of the hsp70 response in aged animals, done with Wistar rats, also showed a decreased response[25,35], but found it was, at least in part, a result of the fact that the average body temperature increase of the older rats was less than that for the young rats under the same conditions of heat stress[25]. Such a difference in body temperature increase with age could not account for our observations, since we found a decline in relative hsp70 level in old rats after hyperthermia, not just a smaller increase. Furthermore, under the heating conditions that we used, both young and old rats reached the same peak body temperatures for similar durations. Thus, our results suggest that one factor contributing to the age-related loss of photoreceptors may be an aberrant metabolic response to environmental stress, a hypothesis that is consistent with the observation by Blake *et al.*[35] that the hsp70 response in the adrenal cortex of aged rats was attenuated. Further work will be required to determine if this defect is a causative factor or an epiphenomenon of age-related retinal degeneration.

## ACKNOWLEDGEMENTS

This work was supported by the Section on Ophthalmology of the Department of Surgery of Bowman Gray School of Medicine and by NEI grant EY07616 to M.T. The authors wish to thank Carol R. Hollman and Kristie Reynolds for their contributions in conducting the experiments and the data analysis and to Ruoyu Xiao for the preparation of the photomicrographs.

## REFERENCES

1. Y.-L. Lai, R.O. Jacoby, and A.M. Jonas, Age-related and light-associated retinal changes in Fischer rats, *Invest. Ophthalmol. Vis. Sci.* 17:634 (1978).
2. N.L. Shinowara, E.D. London, and S.I. Rapoport, Changes in retinal morphology and glucose utilization in aging albino rats, *Exp. Eye Res.* 34:517 (1982).
3. W.K. O'Steen, A.J. Sweatt, and A. Brodish, Effects of acute and chronic stress on the neuronal retina of young, mid-age and aged Fischer-344 rats, *Brain Res.* 426:37 (1987).
4. W.K. O'Steen, A.J. Sweatt, J.C. Eldridge, and A. Brodish, Gender and chronic stress effects on the neural retina of young and mid-aged Fischer-344 rats, *Neurobiol. Aging* 8:449 (1987).
5. J. Marshall, J. Grindle, P.L. Ansell, and B. Borwein, Convolution in human rods: An ageing process, *Br. J. Ophthalmol.* 63:181 (1979).
6. P.E. Kilbride, L.P. Hutman, M. Fishman, and J.S. Read, Foveal cone pigment density difference in the aging eye, *Vision Res.* 26:321 (1986).
7. C.K. Dorey, G. Wu, D. Ebenstein, A. Garsd, and J.J. Weiter, Cell loss in the aging retina, *Invest. Ophthalmol. Vis. Sci.* 30:1691 (1989).
8. H. Gao and J.G. Hollyfield, Aging of the human retina, *Invest. Ophthalmol. Vis. Sci.* 33:1 (1992).
9. S. Gartner and P. Henkind, Aging and degeneration of the human macula. 1. Outer nuclear layer and photoreceptors, *Br. J. Ophthalmol.* 65:23 (1981).
10. L. Feeney-Burns, R.P. Burns, and C. Gao, Age-related macular changes in humans over 90 years old, *Am. J. Ophthalmol.* 109:265 (1990).
11. A.G. Balazsi, J. Rootman, S.M. Drance, M. Schulzer, and G.R. Douglas, The effect of age on the nerve fiber population of the human optic nerve, *Am. J. Ophthalmol.* 97:760 (1984).
12. D.T. Organisciak, P. Favreau, and H.M. Wang, The enzymatic estimation of organic hydroperoxides in the rat retina, *Exp. Eye Res.* 36:337 (1983).
13. R.D. Wiegand, N.M. Giusto, L.M. Rapp, and R.E. Anderson, Evidence for rod outer segment lipid peroxidation following constant illumination of the rat retina, *Invest. Ophthalmol. Vis. Sci.* 24:1433 (1983).
14. R.D. Wiegand, C.D. Joel, L.M. Rapp, J.C. Nielsen, M.B. Maude, and R.E. Anderson, Polyunsaturated fatty acids and vitamin E in rat rod outer segments during light damage, *Invest. Ophthalmol. Vis. Sci.* 27:727 (1986).

15. R.A. Linsenmeier, Effects of light and darkness on oxygen distribution and consumption in the cat retina, *J. Gen. Physiol.* 88:521 (1986).

16. R.H. Steinberg, Monitoring communications between photoreceptors and pigment epithelial cells: Effects of "mild" systemic hypoxia, *Invest. Ophthalmol. Vis. Sci.* 28:1888 (1987).

17. R.I. Morimoto, A. Tissières, and C. Georgopoulos, The stress response, function of the proteins, and perspectives, in: "Stress Proteins in Biology and Medicine," R.I. Morimoto, A. Tissières, and C. Georgopoulos, eds., Cold Spring Harbor Laboratory, NY (1990).

18. W.J. Welch, G.P. Garrels, G.P. Thomas, J.J. Lin, and J.R. Feramisco, Biochemical characterization of the mammalian stress proteins and identification of two stress proteins as glucose and Ca+2 ionophore regulated proteins, *J. Biol. Chem.* 258:7102 (1983).

19. M. Tytell and M.F. Barbe, Synthesis and axonal transport of heat shock proteins, in: "Neurology and Neurobiology vol. 25 - Axonal Transport," R.S. Smith and M.A. Bisby, eds., Alan R. Liss, Inc., New York (1987).

20. M.F. Barbe, M. Tytell, D.J. Gower, and W.J. Welch, Hyperthermia protects against light damage in the rat retina, *Science* 214:1817 (1988).

21. M. Tytell, M.F. Barbe, D.J. Gower, and I.R. Brown, Localization of heat shock protein and its mRNA correlates with protection of photoreceptors from light damage, *Invest. Ophthalmol. Vis. Sci.* 30:462 (1989).

22. I. Weisse and H. Stötzer, Age- and light-dependent changes in the rat eye, *Virchows Arch. A Path. Anat. Histol.* 362:145 (1974).

23. W.K. O'Steen and P.W. Landfield, Dietary restriction does not alter retinal aging in the Fischer 344 rat, *Neurobiol. Aging* (1992).

24. K. Yamaguchi, V.P. Gaur, M. Tytell, C.R. Hollman, and J.E. Turner, Ocular distribution of 70-kDa heat-shock protein in rats with normal and dystrophic retinas, *Cell Tissue Res.* 264:497 (1991).

25. A.D. Johnson, P.A. Berberian, and M.G. Bond, Effect of heat shock proteins on survival of isolated aortic cells from normal and atherosclerotic cynomolgus macaques, *Atherosclerosis* 84:111 (1990).

26. M.J. Blake, J. Fargnoli, D. Gershon, and N.J. Holbrook, Concomitant decline in heat-induced hyperthermia and HSP70 mRNA expression in aged rats, *Am. J. Physiol.* 260:R663 (1991).

27. M. Tytell, M.F. Barbe, and D.J. Gower, Photoreceptor protection from light damage by hyperthermia, in: "Inherited and Environmentally Induced Retinal Degenerations," M.M. LaVail, R.E. Anderson, and J.G. Hollyfield, eds., Alan R. Liss, Inc., New York (1989).

28. W.J. Welch and J.P. Suhan, Cellular and biochemical events in mammalian cells during and after recovery from physiological stress, *J. Cell Biol.* 103:2035 (1986).

29. R.J. Deshaies, B.D. Koch, and R. Schekman, The role of stress proteins in membrane biogenesis, *Trends Biochem.* 13:384 (1988).

30. R.N. Johnston and B.L. Kucey, Competitive inhibition of hsp70 gene expression causes thermosensitivity, *Science* 242:1551 (1988).

31. K.T. Riabowol, L.A. Mizzen, and W.J. Welch, Heat shock is lethal to fibroblasts microinjected with antibodies against hsp70, *Science* 242:433 (1988).

32. J. Fargnoli, T. Kunisada, A.J. Fornace, E.L. Schneider, and N.J. Holbrook, Decreased expression of heat shock protein 70 mRNA and protein after heat treatment in cells of aged rats, *Proc. Nat. Acad. Sci. (USA)* 87:846 (1990).

33. A.Y.-C. Liu, M.S. Bae-Lee, H.-S. Choi, and B. Li, Heat shock induction of HSP 89 is regulated in cellular aging, *Biochem. Biophys. Res. Comm.* 162:1302 (1989).

34. A.Y.-C. Liu, H.S. Choi, Y.K. Lee, and K.Y. Chen, Molecular events involved in transcriptional activation of heat shock genes become progressively refractory to heat stimulation during aging of human diploid fibroblasts, *J. Cell. Physiol.* 149:560 (1991).

35. M.J. Blake, R. Udelsman, G.J. Feulner, D.D. Norton, and N.J. Holbrook, Stress-induced heat shock protein 70 expression in adrenal cortex: An adrenocorticotropic hormone-sensitive, age-dependent response, *Proc. Nat. Acad. Sci. (USA)* 88:9873 (1991).

# A MORPHOMETRIC AND IMMUNOPATHOLOGIC STUDY OF RETINAL PHOTIC INJURY IN PRIMATE

Jun Fu, Tim T. Lam, Nancy J. Mangini and Mark O.M. Tso

From the Georgiana Dvorak Theobald Ophthalmic Pathology Laboratory, Department of Ophthalmology and Visual Sciences, University of Illinois at Chicago, College of Medicine, Chicago, Illinois

Reprint requests to Mark O. M. Tso, M.D., Department of Ophthalmology and Visual Sciences, University of Illinois at Chicago, 1855 West Taylor St., Chicago, IL, 60612, Telephone (312) 996-6505, Fax (312) 243-3937

## INTRODUCTION

Clinical, histopathological, and ultrastructural studies examining the pathogenesis of retinal photic injury in animal models were performed by methods and yielded important qualitative information[1,2]. Morphometric studies of retinal photic injury in rodents had provided us with additional understanding of the more subtle pathophysiologic factors. For example, with morphometric techniques, it was determined that superior and temporal quadrants of the rodent retina are more sensitive to retinal photic injury than the inferior and nasal quadrants[3]; aging rats are more susceptible to retinal photic injury[4]; light-adapted animals suffer less than the dark-adapted animals[5]; and different schedules of light exposure inflicted different severity of retinal lesions[6]. Hormonal influences on the severity of retinal lesions have also been defined[7]. Furthermore, with morphometric comparative methods, pharmacologic agents such as vitamin C[8], dimethylthiourea[9], flunarizine[10], dexamethasone[11], and methylprednisolone[12] were demonstrated to rescue photoreceptor cells from light exposure.

While retinal photic injury in humans and monkeys has been extensively examined by ultrastructural, clinical, and histopathological methods[13,14,15], there was a lack of parallel morphometric analyses of these retinal insults. Consequently, the many subtle pathophysiologic factors underlying retinal photic damage in primates had not been explored. In this series of papers, by using morphometry, immunocytochemical staining with a polyclonal anti-S-antigen anti-48K serum to recognize blue cones[16], as well as clinical and pathologic evaluation, we attempted to differentiate the susceptibility of rod cells, cone cells, and pigment epithelial cells to photic injury in monkeys, and to quantitatively evaluate retinal thickness changes in order to define the morphometric parameters. In the first paper of a series, we describe the morphometric methods and document the increased susceptibility of cone cells to white light photic injury when compared with rod cells. In the subsequent papers of this series we will report the differential susceptibility of rod and cone cells, injuries of photoreceptor cells exposed to different wavelengths, the effects of continuous versus intermittent light exposure to photoreceptor cells, and the effect of pharmacologic agents on the injury and repair of photoreceptor cells.

## MATERIALS AND METHODS

### Light Exposure

Six eyes of four healthy adult cynomolgus monkeys (2 to 3 kg each) were examined by indirect ophthalmoscopy, fundus photography, and fluorescein angiography and were found to be normal before light exposure experiments.

The animals were sedated with ketamine hydrochloride (20 mg/kg), intubated, and anesthetized with intravenous sodium thiamylal Surital (20 to 30 mg/kg). The heads were stabilized in a stereotaxic holder. The pupils were dilated with 10% phenylephrine and 1% tropicamide. Body temperature was monitored with a rectal thermometer and maintained at $36 \pm 1$ °C with an electric thermal blanket. The eyelids were retracted with a wire speculum, and the superior and inferior rectus muscles were fixed with a Bridle suture for positioning of the eyeball during light exposure. The light exposure system (Carl Zeiss Inc, Thornwood, NY), consisted of a high-pressure mercury light bulb (HBO 50/ac, Osram, Germany), a 100-W lamp housing from a Zeiss standard microscope 18, a three lens collector delivery system (Vertical illuminator IV FL), and a mirror. The light beam was adjusted by an aperture to produce a 2.0 mm diameter spot on the cornea and yielded an irradiance of $0.322 \pm 0.016$ W/cm$^2$, as measured with a radiometer (Model 200, Photodyne, Newbury Park, Calif). The homogeneity of the field was determined by measuring the intensity at multiple points in the field of the beam and was noted to be within 15%. A magnifying vitrectomy lens (OMVI Machemer) with an infusion handle (Storz, St. Louis, Mo) was placed on the cornea to keep the cornea moistened and to focus the exposure beam on the retina. The light beam was focused on the retina with a spot size of approximately 1.0 mm in diameter using the vertical diameter of the optic disc as a reference which was measured consistently in four eyes after enucleation as 1.1 mm. Three to six spots equidistant from the macula were exposed in the posterior retina of each eye. Exposure time varied from 15, 30, or 60 minutes. An operating microscope (Carl Zeiss Standard WL 4672 59-9901) was used to monitor the exact location of the light spots on the retina. At the end of the exposure session, the radiance of light on the cornea was checked again. Retinal irradiance was calculated to be $0.325 \pm 0.015$ W/cm$^2$ by the following formula[4]: $Er = Ec \times T \times a/b \times t$, where $\underline{Er}$ indicates retinal irradiance. $\underline{T}$ is the ocular transmittance (90%)[4], $\underline{Ec}$ is the corneal irradiance, $\underline{a}$ is the area of image at the corneal plane (in square millimeters), $\underline{b}$ is the area of image at the retinal plane (in square millimeters), and $\underline{t}$ is the transmittance of the magnifying vitrectomy lens (28%). The retinal irradiance after 15, 30, or 60 minutes of exposure was calculated to be $293 \pm 15$ J/cm$^2$, $585 \pm 30$ J/cm$^2$, or $1170 \pm 60$ J/cm$^2$, respectively.

All procedures involving animals were performed according to the guidelines established in the Association for Research in Vision and Ophthalmology Resolution for the Use of Animals in Research.

### Clinical and Histopathologic Studies

Clinically, the retinal lesions were monitored with ophthalmoscopy, fundus photography, and fluorescein angiography at 3 and 10 days, and then every 4 weeks after light exposure until the animals were killed. For histopathologic study, the eyes were enucleated at 3, 10, or 90 days after exposure. These three time points of observation were selected to represent the acute, reparative, and degenerative phases of the retinal injury. The total numbers of retinal lesions exposed to different energy levels and examined at various post-exposure periods are summarized in Table 1. For histopathologic study, the eyes were immediately opened equatorially after enucleation, and the retinas were fixed in a 1% glutaraldehyde-4% paraformaldehyde fixative (McDowel and Trump's fixative) for

**Table 1.** Number of Retinal Lesions Examined

| Follow-up Period | Total Retinal Energy ($J/cm^2$) | | |
|---|---|---|---|
| | 293 | 585 | 1170 |
| 3 d | 4 | 4 | 4 |
| 10 d | 3 | 4 | 4 |
| 3 mo | 3 | 3 | 3 |

3 hours. Retinal lesions were blocked out, the tissue was osmified, and embedded in epoxy resin. One micron thick sections were cut for light microscopy.

### Morphometric Measurement

Serial sectioning was performed on each of the 32 retinal lesions. The section which showed the maximum width of lesion area was selected as the center of the lesion and was used for morphometric measurement. At the center of these sections, cell counts were taken from a length of 0.25 mm under a light microscope (magnification x400) with an objective reticule. The numbers of nuclei in the outer nuclear layer (ONL), the numbers of cone inner segments (IS) observed along the outer limiting membrane, and numbers of nuclei of the retinal pigment epithelium (RPE) were counted. Rod nuclear count was taken from an area of 0.25 x 0.025 $mm^2$ in the inner portion of the lesion outer nuclear layer, so that the cone nuclei were not included in this cell count. Furthermore, cell counts were also made in retina of the same section which was adjacent to the lesion area, showed no morphologic change, and served as normal control. These control cell counts were further averaged from six monkey eyes.

In addition, the ONL thickness in the 3-month lesion and control areas were also measured. The image of a 1 um section was projected from a microscope on a digitizing pad coupled to a microcomputer with customized software. The measurements were obtained from the center of the lesions for 1 mm in length and from unexposed control areas.

### Immunocytochemistry

Sections were deplasticized in sodium methoxide and pretreated for 7 min with 1% sodium metaperiodate and then rinsed in deionized water[17]. A two step immunoperoxidase detection system using jars filled with diluted antiserum[19] was employed for immunolabeling. Polyclonal antibody against toad 48K was prepared at a dilution of 1:500, and the sections were incubated overnight at 4 °C with primary antibody[18].

## RESULTS

The retinal lesions were divided into three groups according to the length of the exposure time: mild (15-minute exposure, retinal radiance of 293 $J/cm^2$), moderate (30-minute exposure, retinal radiance of 585 $J/cm^2$), and severe (60-minute exposure, retinal radiance of 1170 $J/cm^2$). Each group of lesions showed distinct clinical and histopathologic features.

### Clinical Study

**Mild Photic Injury.** -- Three days after light exposure, the lesions could be seen by ophthalmoscopy as faint, yellowish spots with ill-defined margins (Fig 1A). On fluorescein angiography at the center of the lesions, leakage from the RPE was

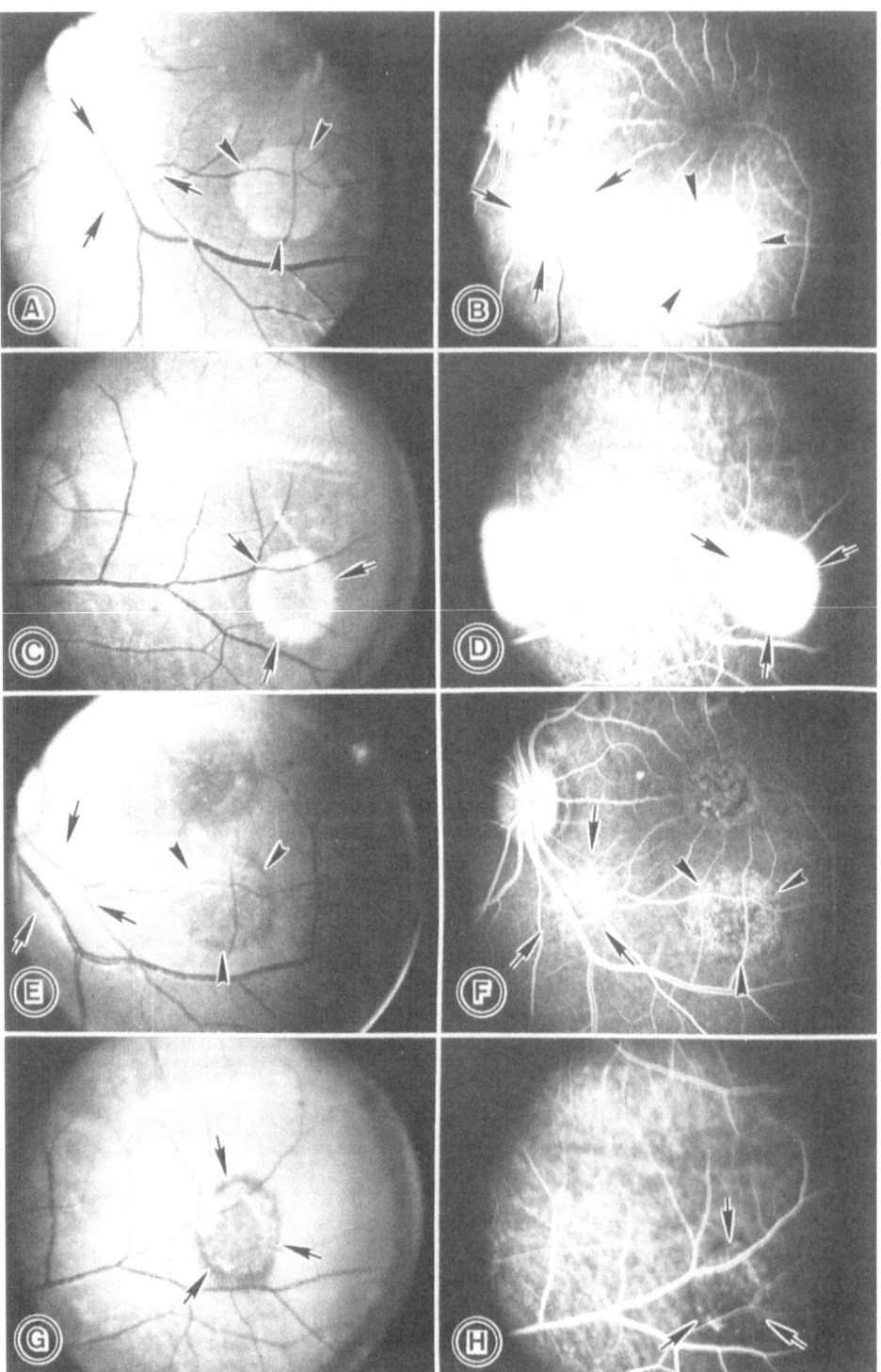

Figure 1. Retinal lesions 3 days and 3 months after light exposure. (A) Fundus photograph of lesions 3 days after exposure to 293 J/cm$^2$ (arrow) and 585 J/cm$^2$ (arrowhead) of energy. (B) Angiogram of the same lesions seen in Fig 1A. (C) Fundus photograph of lesions 3 days after exposure to 1170 J/cm$^2$ of energy (arrow). (D) Angiogram of the lesions seen in Fig 1C. (E) Fundus photograph of lesions 3 months after exposure to 293 J/cm$^2$ (arrow) and 585 J/cm$^2$ (arrowhead) of energy. (F) Angiogram of the same lesions seen in Fig 1E. (G) Fundus photograph of lesions 3 months after exposure to 1170 J/cm$^2$ of energy (arrow). (H) Angiogram of the lesions seen in Fig 1G.

noted spreading toward the periphery (Fig 1B). At 10 days, the lesions presented as faint yellowish areas without clear margin (Fig 2, top left). Angiograms showed minimal RPE staining and irregular hypofluorescence at the centers of the lesions (Fig 2B). At 3 months, the lesions were ill-defined and showed slightly irregular pigment mottling (Fig 1E). On fluorescein angiography, the lesions were noted as granular window defects (Fig 1F).

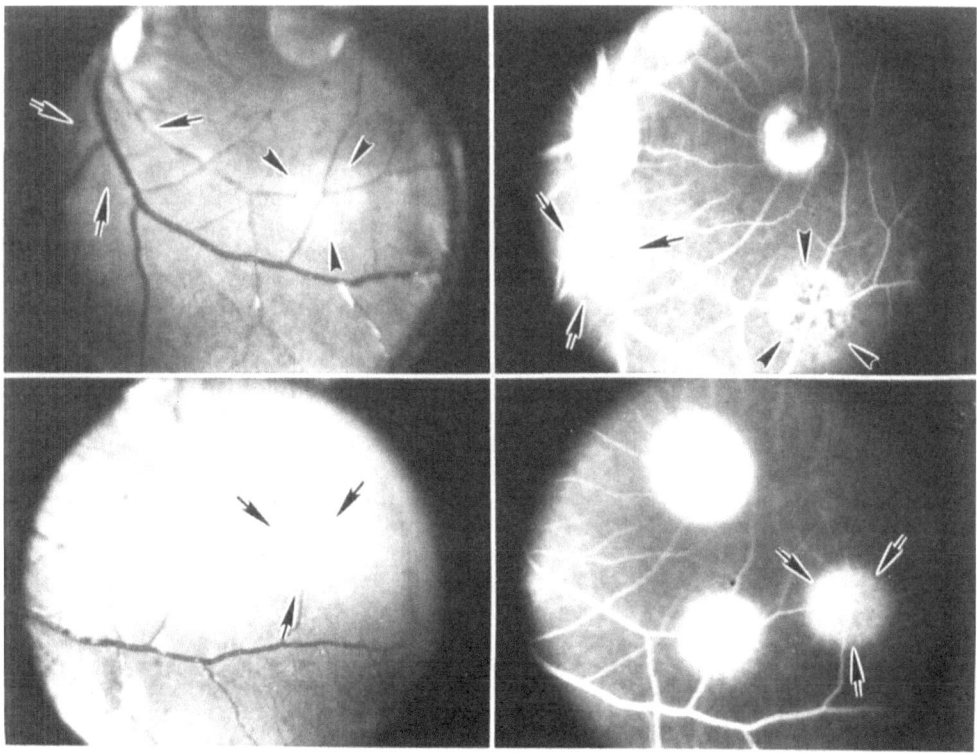

Figure 2. Retinal lesions 10 days after light exposure. Top left, Fundus photograph of lesions after exposure to 293 $J/cm^2$ (arrow) and 585 $J/cm^2$ (arrowhead) of energy. Top right, Angiogram of the same lesions  Bottom left, Fundus photograph of lesions 10 days after exposure to 1170 $J/cm^2$ of energy (arrow). Bottom right, Angiogram of the lesions.

**Moderate Photic Injury** -- Three days after light exposure, the retinal lesions appeared as whitish-yellow edematous spots with distinct margin on ophthalmoscopy (Fig 1A). Moderate leakage from the RPE with a hypofluorescent ring was observed on the early-phase of fluorescein angiography (Fig 1B). Ten days later, the edema subsided but an irregular margin could still be seen ophthalmologically (Fig 2, top left and bottom left). An angiogram showed leakage with irregular granular hypofluorescence (Fig 2, top right and bottom right). Three months later, the lesions appeared as irregularly pigmented areas ophthalmoscopically (Fig 1E) and as granular window defects on angiogram (Fig 1F).

**Severe Photic Injury** -- The severe retinal lesions showed granular grayish centers surrounded by a wide yellowish rim with distinct margins (Fig 1C). Fluorescein angiography showed diffuse and extensive leakage of the dye throughout the retinal lesions extending into the surrounding retina (Fig 1D). At 10 days, the granular grayish centers persisted, but the edema at the periphery of the lesions improved (Figs 1C and 1D, Fig 2, bottom). Three months after light exposure, irregular depigmented spots were observed ophthalmoscopically surrounded by irreg-

325

ular brownish pigmented margins (Fig 1G). Corresponding hypofluorescent dots with staining of the RPE on the angiogram were observed (Fig 1H).

### Histopathologic Study

All three groups of lesions showed histopathologic changes in the outer retinal layers with varying degrees of severity. The lesions that received the highest irradiance of exposure presented the most severe changes. No pathologic changes were identified in the inner retinal layers in all retinal lesions and at all times after light exposure.

**Mild Photic Injury** -- Three days after light exposure (Fig 3A), the RPE was focally necrotic. Few macrophages were noted at the margins of the lesions in the subretinal space. The outer segments (OS) of photoreceptors were fragmented and the inner segments (IS) were swollen and occasionally necrotic. The ONL was edematous exhibiting cytoplasmic swelling of Muller cells and photoreceptor cells, and a few photoreceptor nuclei, especially cone nuclei, were pyknotic. Densified axons of the photoreceptor cells were seen in the outer plexiform layer (OPL) which was vacuolated (Fig 3A). Ten days after light exposure (Fig 3B), the RPE became focally depigmented and vacuolated. A few macrophages were observed in the subretinal space. The OS of the photoreceptors were shortened, and edema in the IS partially subsided. The edema of the ONL also improved, but loss of a few cone cells was noted and the OPL was still edematous (Fig 3B). Three months later (Fig 3C), the RPE returned to a single cell layer but remained irregularly depigmented or hyperpigmented. Few macrophages were found in the subretinal space. The OS and IS appeared to return to normal, as did the ONL, except for mild loss of cone cells. The edema in the OPL subsided (Fig 3C).

**Moderate Photic Injury** -- Three days after light exposure (Fig 3D), the RPE was markedly necrotic, and a few cells were sloughed into the subretinal space. Macrophages were observed at the lateral margins of the retinal lesions. The OS of the photoreceptors were fragmented. The IS were edematous, and some were densified and necrotic. The edematous ONL showed loss of cone nuclei. Vacuolation was noted in the OPL (Fig 3D). At 10 days (Fig 3E), moderate proliferation of depigmented RPE with marked edema was seen. Macrophages were scattered in the subretinal space. The OS were shortened, and some of the IS were densified. Edema and densified axons of photoreceptors were still present in the ONL and OPL (Fig 3E). At 3 months (Fig 3F), some of the RPE cells were heavily laden with melanophagosomes and others were irregularly depigmented. Scattered macrophages remained in the subretinal space. The OS and IS of the rod cells appeared to be intact. The loss of cone nuclei and IS was seen along the external limiting membrane. Edema in the OPL largely subsided, but densified axons of the photoreceptor cells were still noted (Fig. 3F).

**Severe Photic Injury** -- The lesions were characterized by total necrosis of the RPE and extensive loss of cone IS three days after light exposure (Fig 3G). Also seen were macrophages at the margins of the lesions, fragmentation of the OS, edema of the IS, densification of axons of the cone cells, and edema of the ONL and OPL (Fig 3G). At ten days (Fig 3H), flattened and depigmented RPE slid over bare Bruch's membrane, and necrotic RPE cells sloughed into the subretinal space. Macrophages were scattered in the subretinal areas. The OS of the photoreceptor cells were almost absent, and some of the IS were densified. Edema could be observed in the ONL and OPL, and many cone nuclei had disappeared (Fig 3H). At three months (Fig 3I), the RPE showed irregular hyperpigmentation and depigmentation. Scattered pigment-laden macrophages were still noted in the subretinal space. The OS and IS had recovered. Extensive absence of cone nuclei was observed along the outer limiting membrane. Edema of the ONL and OPL mostly subsided, but the densified axons of the cone cells were still seen (Fig 3I).

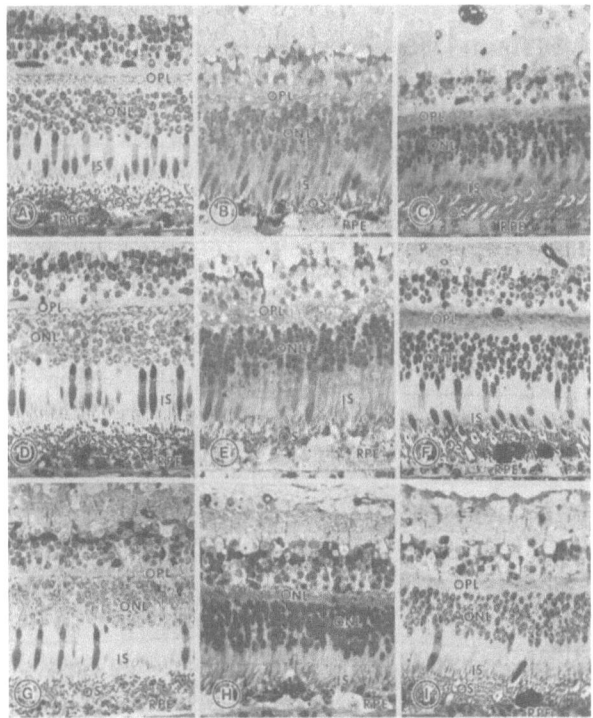

Figure 3. Pathologic changes in retinal lesions. (A) Retinal lesion 3 days after exposure to 293 J/cm$^2$ of energy. (B) Retinal lesion 10 days after exposure to 293 J/cm$^2$ energy. (C) Retinal lesion 3 months after exposure to 293 J/cm$^2$ of energy. (D) Retinal lesion 3 days after exposure to 585 J/cm$^2$ of energy. (E) Retinal lesion 10 days after exposure to 585 J/cm$^2$ of energy. (F) Retinal lesion 3 months after exposure to 585 J/cm$^2$ of energy. (G) Retinal lesion 3 days after exposure to 1170 J/cm$^2$ of energy. (H) Retinal lesion 10 days after exposure to 1170 J/cm$^2$ of energy. (I) Retinal lesion 3 months after exposure to 1170 J/cm$^2$ of energy.

## Immunocytochemical Study

The normal unexposed monkey retina immunoreacted with the polyclonal anti-48K serum, the outer segments of rod cells and the perikarya, the inner and outer segment, axon and pedicle of blue cone cells showed positive immunoreactivity. At the posterior pole of the normal retina, distribution of immunoreactive blue cone cells was regular with approximately one of every 10 cone cells (Fig 4A). After exposure to white light of 293 J/cm$^2$, an occasional blue cone cell could still be identified in the lesional area. No immunoreactive blue cone cells could be identified after light exposure to 585 and 1170 J/cm$^2$ (Fig 4B), even though blue cone cells were present in an adjacent unexposed area.

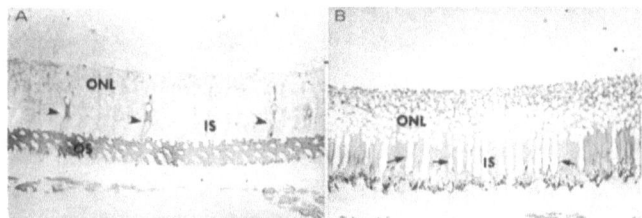

Figure 4. Immunocytochemical stain. (A) Unexposed monkey retina stained with anti-48K antibody. The outer segment (OS) of photoreceptor cells and whole blue cone cells (arrowhead) stained positive. (B) Retinal lesion 10 days after light exposure of 585 J/cm$^2$. Whole rod cells stained positive, no blue cone cells were present in lesion area. Red and green cone cells (arrow) still stained negative.

## Morphometric Study

**Total Number of Nuclei in the ONL.** -- Figure 5 shows the alteration of the number of photoreceptor nuclei in 0.25 mm of the ONL in the center of the lesions with time. All three groups of retinal lesions exposed to 293, 585 or 1170 $J/cm^2$ showed an initial drop in the number of photoreceptor nuclei in the edematous lesional areas at 3 days (P<.05) but returned to the approximate normal value at ten days (P>.05), and remained unchanged up to 90 days after exposure. While there was a tendency of increased photoreceptor cell loss with higher levels of retinal irradiances, no statistical significance (P>.05) among the loss of these 3 groups was observed.

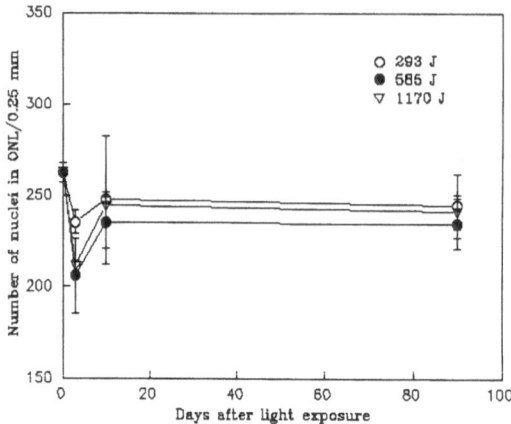

Figure 5. Number of nuclei in the outer nuclear layer (ONL) per 0.25 mm. Three days after exposure, there were significant differences between the unexposed normal area and exposed lesions (P<.05). No statistical change was noted at all other time points (P>.05). Vertical bars indicate standard error.

**Number of cone inner segments.** -- The inner segments of cone cells were counted as described in methods. All three groups of lesions showed loss of cone inner segments with time (Fig 6). There was an initial drop in the number of cones (P<.01) from normal values at three days. A further decrease was noted at ten days (P<.01) but no further significant change thereafter could be seen at 90 days, (P>.05). The cone loss at all time points examined showed a dependence on the retinal irradiance. At 3 months after exposure, the loss of cone cell was 19.48%, 46.91% and 65.72% at 293, 585 or 1170 $J/cm^2$ exposure respectively.

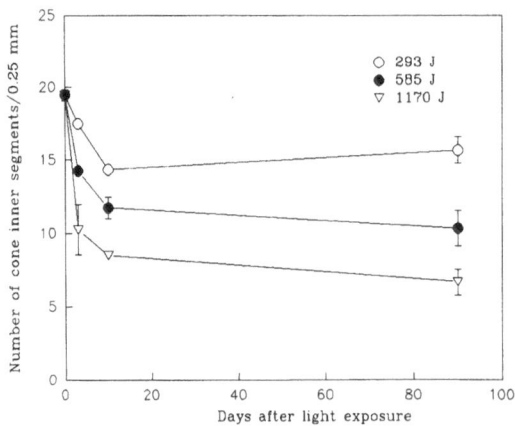

Figure 6. Number of cone inner segments (IS) per 0.25 mm. Three and 10 days after exposure, significant decreases of the cone IS were noted (P<.05). Three months after light exposure, further loss of cone IS was only noted in lesions exposed to 1170 $J/cm^2$ of energy (P<.05). Vertical bars indicate standard error.

**Rod Cell Density in the ONL** -- The density of rod cells was determined by nuclei counts per unit area at the inner portion of the ONL as the cone nuclei were present along the external limiting membrane at the posterior pole[20]. A marked decrease of rod nuclei per unit area was noted at three days after light exposure in all three grades of retinal lesions (P<.01, Fig 7). However, the rod cell density per unit area increased in ten days (P>.05). At 3 months after exposure, the rod cell density in moderate and severe lesions decreased significantly (P<.05); meanwhile, the cell density in the mild lesions returned to normal (P>.05).

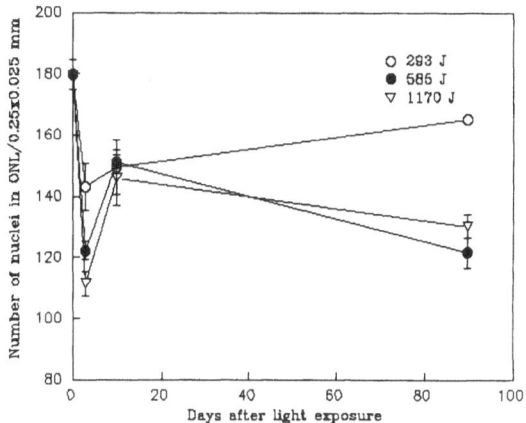

Figure 7. Number of nuclei in the outer nuclear layer per 0.25 x 0.025 mm$^2$. Compared with unexposed normal retina, numbers significantly decrease (P<.05). Vertical bars indicate standard error.

**RPE Cells** -- The numbers of the RPE nuclei in all three grades were markedly decreased three days after light exposure (P<.01, Fig 8), returned to normal range (P>.05) at ten days, and remained unchanged up to three months. No significant difference was noted among the three exposure groups (P>.05).

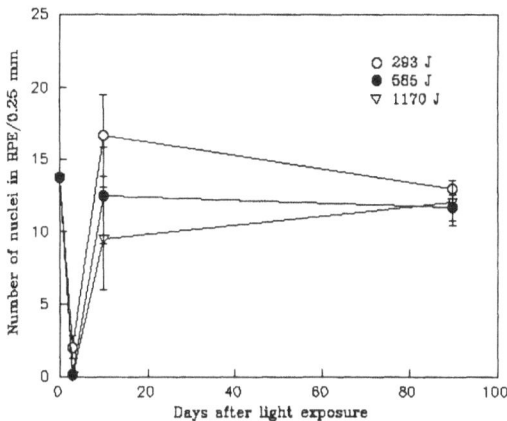

Figure 8. Number of nuclei in the retinal pigment epithelium per 0.25 mm in unexposed normal retina and retinal lesions. Three days after light exposure, a significant decrease of the numbers was noted (P<.05); ten days and 3 months, no dramatic difference was observed (P>.05). Vertical bars indicate standard error.

**ONL Thickness** -- Three months after light exposure, the ONL thickness in moderate and severe lesions remarkably increased compared with unexposed normal areas (P<.05, Fig 9), but no statistical changes in the thickness of the mild lesions were observed (P>.05).

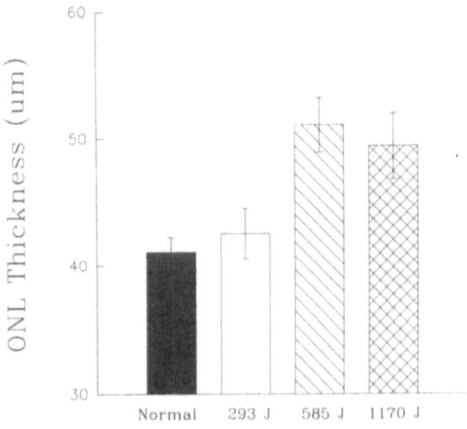

Figure 9. The outer nuclear layer thickness 3 months after light exposure. Compared with unexposed normal retina, the thickness in moderate and severe lesions showed a significant increase (P<.05). Vertical bars indicate standard error.

## DISCUSSION

In this study of retinal photic injury in the nonhuman primate, we studied the retinal lesions inflicted by three levels of retinal irradiance using clinical, histopathological, immunocytochemical and morphometrical methods. Our clinical and histopathological findings were comparable to previous studies[13,14,15,21,22]. Our morphometric study enabled us to dissect the cellular reactions of the initial injury and the recovery of cone cells, rod cells, retinal pigment epithelium, and Muller cells in three levels of retinal radiances. Furthermore, by immunocytochemical study with an anti-S-antigen antiserum, we could delineate blue cone injury.

Histopathologically, all retinal lesions were characterized by necrosis and proliferation of retinal pigment epithelium, infiltration of pigment laden macrophages in the subretinal space, loss of photoreceptor cells, and swelling in the outer nuclear layer and outer plexiform layer. Tissue damage was fairly uniform across the lesion with a definitive margin as the retinal exposure field was measured at different points in the field of the beam and was noted to be +7.5%. Thus the homogeneity of the field of exposure was insured. Yet, one of the most remarkable findings in this report was that the loss of cone cells was dependent on retinal radiances. In this experiment, the retinal radiances ranged from 293, 585 to 1170 J/cm², and the percentage loss of cone cells appeared to parallel the increase of retinal radiances and ranged from 19.48%, 46.91% to 65.72% at 3 months after light exposure. Furthermore, the loss of cone cells appeared most prominent in the first ten days, and the loss of cone cells between ten days and three months was not statistically significant in all retinal irradiances levels. As there is no previous reported morphometric study of retinal photic lesions in primates to the best of our knowledge, this is the first quantitative report dealing with cone cell susceptibility to white light.

As the spectrum of our light source consisted of peaks of 300, 380, 440, and 600 nm, we expected that blue, red, and green cones would be affected in our retinal lesions. Nork and Mangini reported that their anti-48K antibody immunoreacted positively with rod and blue cone cells but not red-green cone cells[16]; we used this antibody to recognize blue cone cells. The result showed few blue cone cells were left in the retinal lesions exposed to 293 J/cm² and no blue cone cells could be identified in retinal lesions induced by retinal radiance of 585 and 1170 J/cm², suggesting that the blue cone cells were most susceptible to our light source. However, the morphometric study showed that when retinal radiance was increased from 585 to 1170 J/cm², additional cone cells were lost. Since all blue cone cells were lost after exposure to 585 J/cm², this additional loss of cone cells was interpreted to be from the red-green cone cells.

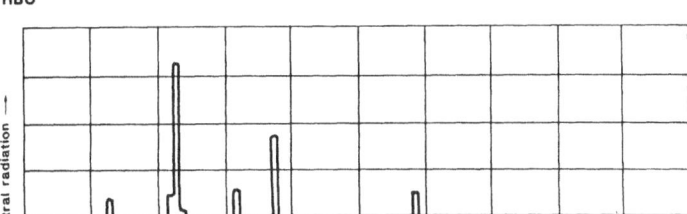

Figure 10. Spectrum of the light source.

Another observation was that total loss of the rod and cone photoreceptor nuclei in the outer nuclear layer was statistically insignificant in retinal lesions resulting from three levels of retinal irradiances at 10 days to 3 months after light exposure, even though cone cells were markedly injured. This discrepancy might be partially explained by the fact that there are 6 million cone cells and 120 million rod cells in the retina, and the cone cell population accounted for 5% of the photoreceptor cells. The loss of 19.48% to 65.72% of cone cells was not statistically significant in the total loss of photoreceptor nuclei.

Three days after light exposure acute swelling of the Muller fibers and the photoreceptor cytoplasm was noted histopathologically. This edema led to an initial drop of photoreceptor nuclear count in a fixed length of section, a decrease in rod nuclear density per unit area in the inner portion of the ONL, and an increase in the ONL thickness. At ten days, the cytoplasmic edema decreased resulting in a relative increase of nuclear counts in the ONL and rod nuclear density, and a decrease of the ONL thickness. At 3 months after light exposure, many of the cone nuclei and inner segments disappeared. However, the rod and cone nuclear counts in the ONL did not significantly decrease at all 3 levels of retinal radiances. In previous reports, threshold lesions of retinal photic injury were all defined by retinal pigment epithelial changes noted clinically or histopathologically[23]. No survival of rod cells with morphometric study has been previously reported. As such, this is the first report to illustrate that the rod cells in the nonhuman primate retina could tolerate retinal irradiance up to 1170 J/cm².

The differential sensitivity of rods and cones to light damage has been reported previously. Sykes et al estimated that in rhesus monkeys the threshold of cones to white light in photochemical injury was about half of that of rods[22]. Lawwill noted that mitochondria in rod cells seemed more resistant to light damage when compared to that of cones in the monkey[14]. Tso et al noted that rod cells survived in spite of severe cone cell loss in monkey after Argon laser injury[24]. Our morphometric study confirmed these observations with quantitative data.

The retinal pigment epithelium appeared to be most susceptible to retinal photic injury in our experiment. Clinically, early leakage from retinal pigment epithelium was observed. Histopathologically, extensive necrosis of retinal pigment epithelial cells was observed in retinal lesions of various severity. However, our morphometric study of counting the retinal pigment epithelial nuclei showed that the proliferative changes occurred most prominently in the first ten days. In the recovery phase at three months after exposure, the nuclei of the retinal pigment epithelium in the retinal lesions were not significantly different from the normal control. No placoid proliferation of retinal pigment epithelium was observed in this experiment. It was instructive to compare the cellular response of cone cells, rod cells and RPE cells to light injury, noting (1) the high sensitivity of cone cells without regeneration, (2) the great susceptibility of the retinal pigment epithelium with reactive proliferation and (3) the resistance of rod cells to damage. Our morphometric observation further highlighted that the determination of threshold retinal lesion using retinal pigment epithelium does not reflect the visual function deficit.

The response of Muller cells in the retinal photic injury in our lesions was also remarkable and was reflected in the increased thickness of the outer nuclear layer at the recovery phase. Combining this with significant decrease of rod cell density in moderate and severe retinal lesions, it appeared that the Muller cell fibers were hypertrophic and replaced the lost cone cells resulting in increased thickness of the outer nuclear layer.

Our experiment led to a nonhuman, primate model to examine differential susceptibility of rod and cone cells to light, varying degree of injury of photoreceptor cells with different wave lengths, and the effect of different pharmacological agents on the injury and repair of photoreceptor cells, because the experiment parameters of this animal model could be straightly controlled. Our light source consisted of a high pressure mercury light bulb (HBO 50/AC, Osram, Germany). The output of the light bulb was repeatedly measured during the experiment and was found to be very constant. The retinal lesions were placed equidistantly from the macula in order to minimize the variation of photoreceptor cells within different regions of the retina. Each retinal lesion was serially sectioned in order to find the center of the lesion so that comparison of the severity of retinal insult of different retinal lesions may be comparable. Twenty percent of the output of the mercury lamp consisted of blue light, and in future studies of this series we will report the comparison of the different wavelengths of light on different types of cone cells in the retina.

**Acknowledgments** - This investigation was supported in part by research grants RO1 EY06761, RO1 EY01903, and core grant EY01792 from the National Eye Institute, Bethesda, Md; an unrestricted research grant from Research to Prevent Blindness Inc., New York, NY; gifts from the Lions of Illinois Foundation, Maywood, Ill., the Clifford Sawyer Estate, Black Canyon, Ariz., and the McGraw Foundation, Arlington Heights, Ill.

## REFERENCES

1. W.K. Noell, V.S. Walker, B.S. Kang, and S. Berman, Retinal damage by light in rats. *Invest Ophthalmol.* 5:450 (1966).
2. M.O.M. Tso, Experiments on visual cells by nature and man: In search of treatment for photoreceptor degeneration. *Invest Ophthalmol Vis Sci.* 30:1032 (1989).
3. W.K. Noell, Possible mechanisms of photoreceptor damage by light in mammalian eyes. *Vision Res.* 20:1163 (1980).
4. Y.L. Lai, R.O. Jacoby, and A.M. Jonas, Age-related and light associated retinal changes in Fischer rats, *Invest Ophthalmol Vis Sci.* 17:634 (1978).
5. W.K. Noell, Effects of environmental lighting and dietary vitamin A on the vulnerability of the retina to light damage. *Photochem Photobiol.* 29:717 (1979).
6. F.L. Lee, D.Y. Yu, and M.O.M. Tso, Effect of continuous versus multiple intermittent light exposures on retina. *Cur Eye Res.* 9:11 (1990).
7. W.K. O'Steen, Hormonal and dim light effects in retinal photodamage. *Photochem Photobiol.* 29:745 (1979).
8. Z.Y. Li, M.O.M. Tso, H.M. Wang, et al, Amelioration of photic injury in rat retina by ascorbic acid. *Invest Ophthalmol Vis Sci.* 25:1980 (1985).
9. S. Lam, M.O.M. Tso, and D.H. Gurne, Amelioration of retinal photic injury in albino rats by dimethylthiourea, *Arch Ophthalmol.* 108:1751 (1990).
10. J. Li, D.P. Edward, T.T. Lam, and M.O.M. Tso, The ameliorative effect of flunarizine and dimethylthiourea in light-induced retinal degeneration in albino rats. *Invest Ophthalmol Vis Sci.* 32:1096 (1991).
11. J. Fu, T.T. Lam, and M.O.M. Tso, Dexamethasone ameliorates retinal photic injury in albino rats, *Exp Eye Res.* 54:583 (1992).
12. M. Rosner, T.T. Lam, J Fu, and M.O.M. Tso, Methylprednisolone ameliorates retinal photic injury in rats. *Arch Ophthalmol.* 110:857 (1992).
13. J. Lanum, The damaging effects of light on the retina. Empirical findings, theoretical and practical implications. *Surv Ophthalmol.* 22:221 (1978).

14. T.L. Lawwill, Three major pathologic processes caused by light in the primate retina: a search for mechanisms. *Trans Am Ophthalmol Soc.* 80:517 (1982).

15. M.O.M. Tso, Retinal photic injury in normal and scorbutic monkeys. *Trans Am Ophthalmol Soc.* 85:498 (1987).

16. T.M. Nork, N.J. Mangini, and L.L. Millecchia, Localization of S-antigen in rods and blue cones of the human retina and in photoreceptors of the cat retina. *Invest Ophthalmol Vis Sci.* 32:1150 (1991).

17. R.E. Marc, W.S. Lui, M. Kalloniatis, S.F. Raiguel, and E.V. Haesendonck, Patterns of glutamate immunoreactivity in goldfish retina. *J Neuroscience.* 10:4006 (1990).

18. D. Tacha, M. Brown, and W.T. Galey, A two step immunoperoxidase detection system with repeated use of diluted antibodies for bulk staining of slides. *J Histotechnol.* 13:15 (1990).

19. N.J. Mangini, and D.R. Pepperberg, Immunolocalization of 48K in rod photoreceptors: Light and ATP increase of labeling. *Invest Ophthalmol Vis Sci.* 29:1221 (1988).

20. S.S Duke-Elder, Ocular tissue, *in:* "The Anatomy of the Visual System." S.S Duke-Elder, ed., The C.V. Mosby Co., St. Louis (1961).

21. H.G. Sperling, and C. Johnson, Histological findings in the receptor layer of primate retina associated with light-induced dichromacy. *Mod Probl Ophthalmol.* 13:291 (1974).

22. S.M. Sykes, W.G. Robison Jr, M. Waxler, and T. Kuwabara, Damage to the monkey retina by broad-spectrum fluorescent light. *Invest Ophthalmol Vis Sci.* 20:425 (1981).

23. W.T. Ham Jr, and H.A. Mueller, Photopathology and nature of the blue light and near-UV retinal lesions produced by lasers and other optical sources, *in:* "Laser Applications in Medicine and Biology", M.L. Wolbarsht, ed., Plenum Publishing Corp, New York, NY (1991).

24. M.O.M. Tso, I.H.L. Wallow, and J.O. Powell, Differential susceptibility of rod and cone cells to argon laser. *Arch Ophthalmol.* 89:228 (1973).

# CUTANEOUS MELANOMA-ASSOCIATED RETINOPATHY

Ann H. Milam,[1] Kenneth R. Alexander,[2] Samuel G. Jacobson,[3] John C. Saari,[1] Wojciech P. Lubinski,[3] Lynn G. Feun,[3] and Kirk E. Winward[4]

[1]University of Washington, Seattle, WA 98195; [2]University of Illinois at Chicago, Chicago, IL 60612; [3]University of Miami, Miami, FL 33136; [4]University of Utah, Salt Lake City, UT 84108

## INTRODUCTION

In a number of paraneoplastic syndromes, patients develop autoantibodies to tumor cells that are also reactive with normal host tissues, including neurons of the central and peripheral nervous system.[1] Two forms of paraneoplastic retinopathy are now recognized: cancer-associated retinopathy (CAR),[2,3] and cutaneous malignant melanoma-associated retinopathy (MAR).[4-7] CAR occurs in a small percentage of patients having small cell carcinoma of the lung and other cancers, and the patients are thought to develop autoantibodies against a tumor cell antigen that cross react with their retinal rods and cones. It was recently demonstrated that sera of some CAR patients contain antibodies that react with recoverin, a calcium-binding protein found in photoreceptors and some types of cone bipolar cells.[8-10] These antibodies may cross the blood retinal barrier and interfere with normal photoreceptor function. In CAR, the patient's visual symptoms may precede or occur at about the time of tumor identification; loss of vision typically occurs over weeks to months. There is clinical, electroretinographic (ERG) and histopathologic evidence of progressive dysfunction and death of rod and cone photoreceptors in this form of retinopathy.[2,3]

The clinical course of MAR patients differs from those with CAR in several features. MAR is found in patients with malignant cutaneous melanoma, and their visual symptoms typically occur months to years after diagnosis of metastatic disease. These patients experience relatively acute onset of night blindness and persistent photopsias, and have elevated dark-adapted thresholds and an ERG resembling that of patients with congenital stationary night blindness (CSNB).[4-7] Photoreceptor function is intact but there appears to be a defect in signal transmission between photoreceptors and second-order retinal neurons.[4] Based on ERG test results, it was hypothesized that the transmission defect is specific for the retinal ON system.[6] Although MAR was suggested to represent a paraneoplastic retinopathy,[5,6] autoantibodies reactive with retinal cells were only recently identified.[7] These findings are described below.

## METHODS

### Patients

Two MAR patients were the main subjects of the study. As described previously,[6,7] the clinical course and visual function abnormalities of both patients were similar. MAR-1, a 36-year-old man, had a cutaneous malignant melanoma removed from his left leg, and MAR-2, a 58-year-old man, had a malignant melanoma excised from his scalp. Both patients had surgery for metastatic disease and MAR-2 but not MAR-1 also received chemotherapy following the onset of his visual symptoms. A number of years after removal of the primary melanoma (MAR-1; 4 yrs; MAR-2, 15 yrs), each patient noticed visual sensations variously described as "moving lights", "looking through water" or "shimmering lights". Soon thereafter, both patients experienced loss of night vision that had a sudden onset and became progressively worse. For MAR-2, there was progressive reduction in the amplitude of the b-wave of the rod ERG over this same time period. Both MAR patients had initial visual symptoms in one eye, with bilateral involvement after several weeks.

The best corrected VA of MAR-1 was: 20/20 (+0.50) OD and 20/40 (+0.50) OS; and of MAR-2: 20/25 OU. Goldmann kinetic perimetry with the V-4e target showed full visual fields for both patients. For MAR-1, the I-4e target was detectable only within the 10-20° isopter. For MAR-2, the right eye had an enlarged blind spot and a small parafoveal relative scotoma with the II-4e target, while the left eye showed no scotomas with this target. A slit lamp exam of MAR-1 revealed fine cells in each vitreous and a fundus exam was normal except for slight mottling of the retinal pigment epithelium (RPE) in the midperiphery and attenuation of the retinal arteries. Similarly, a fundus exam of MAR-2 showed nonspecific pigment mottling in the posterior pole with a slight degree of vascular attenuation. A trial of oral corticosteroids was undertaken in each patient. Neither MAR patient showed a change in night vision or in ERG responses, but MAR-1 reported decreased severity of the photopsias.

We evaluated sera from our two MAR patients and 38 other subjects: 28 patients with cutaneous malignant melanoma (ages 26-85) who had no visual symptoms (four patients at Stage I; one at Stage II; six at Stage III; and 17 at Stage IV); one patient with autosomal recessive CSNB (age 73); one with bird shot chorioretinopathy (age 38); and eight normal subjects (ages 21-59).

### Visual Function Tests

Visual function tests included: two-color (500 nm and 650 nm) dark-adapted static threshold perimetry (MAR-1 and -2); short wavelength (S, blue) cone perimetry (MAR-1); dark- and light-adapted spectral sensitivity measurements (MAR-1 and -2); midspectral cone ERG (MAR-1 and -2) and S cone ERG (MAR-1) measurements.

### Immunocytochemistry

Normal human retinas 3 hr or less *post mortem* were processed unfixed or fixed (1 to 4% paraformaldehyde; 15 min to 6 hr). Cryostat sections were incubated overnight at 4°C in serum or IgG (1:100) from each patient, followed by incubation in secondary antibody (goat anti-human IgG labeled with fluorescein isothiocyanate [FITC]), and photographed with a fluorescence microscope. To demonstrate rod bipolar cells,[11] retinal sections were incubated in a mixture of patient IgG (1:100) and a monoclonal antibody made in mouse against protein kinase C (1:50; Seikagaku Kogyo Co., Tokyo, Japan). The sections were rinsed and treated with a mixture of secondary antibodies (goat anti-human and anti-mouse IgG) labeled with FITC and rhodamine, respectively. The sections were photographed as consecutive exposures using alternate filters for FITC and rhodamine.

## Biochemistry

Samples of human retinal homogenate were analyzed by SDS-PAGE. For Western blots, separated proteins were transferred to PVDF membranes and probed with patient sera or IgG fractions at 1:100 dilution using the ProtoBlot alkaline phosphatase based system. For ELISA, microtiter plates were coated with human retinal supernatant (homogenized in buffer or buffered detergent; diluted to a concentration that produced a color response ~10% above background with control antisera, ~38 µg/ml) and serial dilutions of control and patient sera and IgG fractions were analyzed.

## RESULTS

### Visual Function Tests

The rod ERGs of MAR-1, MAR-2 and the CSNB patients showed decreased b-wave amplitudes at lower stimulus intensities and reduced b- to a-wave ratios at higher intensities. The midspectral cone ERGs of these subjects all showed a characteristic reduction in the amplitude of the initial portion of the cone b-wave. The cone flicker ERGs of both MAR patients differed from those of normal subjects and CSNB patients, in that the implicit times were prolonged, while the amplitudes were either normal (MAR-1) or reduced (MAR-2). The oscillatory potentials of both MAR patients were markedly reduced compared to normal, while those of CSNB patients were only slightly reduced but had a less complex waveform than normal. By dark adapted perimetry, neither MAR patient had measurable rod sensitivity in the central field, and rod sensitivity was very reduced in the peripheral field. Cone sensitivity for a long wavelength test stimulus was also reduced at most test loci in both patients.

S cone ERGs were not detectable in MAR-1 (they were not measured in MAR-2). By S cone perimetry, MAR-1 had no detectable S cone function except at fixation, where sensitivity was reduced by ~10 dB. In contrast, two CSNB patients had normal S cone sensitivity at fixation and at a few paracentral loci, and there was reduced but detectable function at about 60% of the test loci. MAR-1 thus showed a greater degree of S cone sensitivity loss by perimetry than the CSNB patients. S cone sensitivity was only measured in the fovea for MAR-2, and there he showed a ~7.5 dB sensitivity loss, which was comparable to his sensitivity loss for the middle and long wavelength cone systems. MAR-2 also showed a tritan axis on the FM 100-hue test, consistent with a loss of S cone sensitivity.

### Immunocytochemistry

Indirect immunofluorescence using sera or IgG from most control patients produced no specific labeling of retina sections. RPE and photoreceptor cells contained autofluorescent lipofuscin granules and the retina showed a dim autofluorescence throughout (Figure 1, left). Sera and IgG from some normal subjects (n = 3) and non-MAR melanoma patients (n = 7) produced higher non-specific background labeling of all parts of the retina and specific staining of filamentous structures in the nerve fiber layer.

Sera and IgG from both MAR patients produced strong labeling of bipolar cells whose somata lay midway in the inner nuclear layer (Figure 1, middle) and whose dendrites formed a row of bright dots in the outer plexiform layer. Both MAR sera also produced weak labeling of rod outer segments and strong labeling of filaments in the nerve fiber layer. The inner plexiform layer (IPL) of the unfixed retinas was poorly preserved; only hints of axons could be traced from the labeled bipolar cells and their axons stratified deep in the IPL (Figure 1, right). None of the sera or IgG fractions from the normal subjects, the non-MAR melanoma patients, or the patients with other retinal diseases produced bipolar cell labeling.

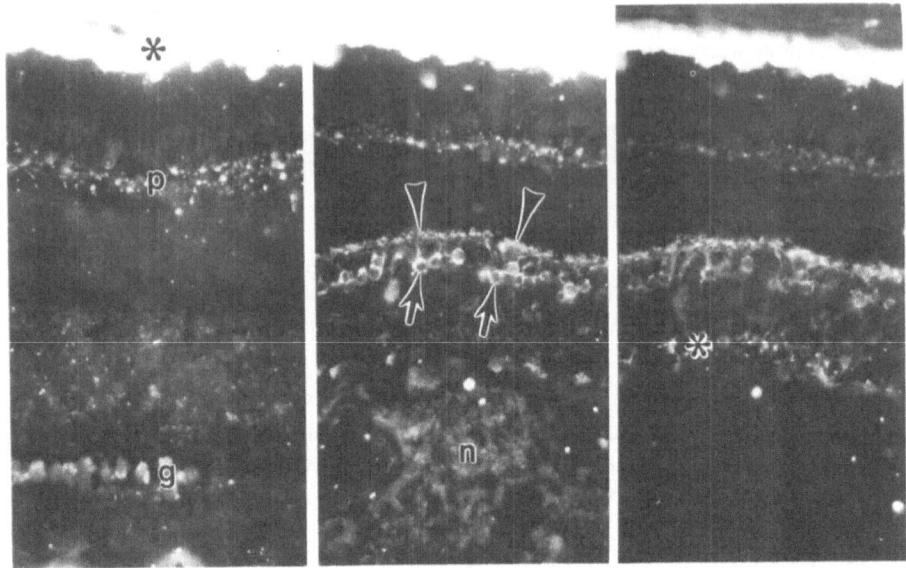

**Figure 1.** Cryostat sections of normal unfixed human retina processed for indirect immuno-fluorescence using IgG (1:100) from a normal subject or MAR-1, followed by anti-human IgG labeled with FITC. Left: Section processed with IgG from a normal subject. Note autofluorescent lipofuscin granules (*) in the RPE and photoreceptor cells (p). The remainder of the neural retina shows dull autofluorescence, including the ganglion cells (g). X100. Middle: Section of same retina processed using a mixture of IgG (1:100) from MAR-1 and anti-protein kinase C (1:50) made in mouse, followed by a mixture of FITC-labeled anti-human IgG and rhodamine-labeled anti-mouse IgG. This photograph was made using filters for detection of FITC fluorescence. In addition to autofluorescent structures as found in Figure 1, left, note brightly stained cells (arrows) in the inner nuclear layer and a row of bright dots (arrowheads) corresponding to bipolar dendrites in the outer plexiform layer. The reactivity of the nerve fiber layer (n) is also found in some normal subjects and non-MAR melanoma patients. X100. Right: The same area of the retina shown in the middle panel was photographed using filters for detection of rhodamine fluorescence. Note that many of the cells labeled with MAR IgG are also reactive with anti-protein kinase C, a marker for rod bipolar cells. The bipolar axons (*) are faintly labeled. X100.

Serum samples from MAR-1 and -2 taken before and after steroid treatment produced equivalent labeling of the bipolar cells. Double labeling experiments were performed using a mixture of each MAR serum and anti-protein kinase C, a specific marker for rod bipolar cells.[11] Many of the cells labeled with the two MAR sera were also labeled with anti-protein kinase C (Figure 1, right). These results indicate that many of the retinal neurons labeled with MAR sera are rod bipolar cells. Similar staining of rod bipolar cells was found with the two MAR sera using unfixed sections of normal rat retina, where the rod bipolar cells are more numerous (Figures 2 and 3).

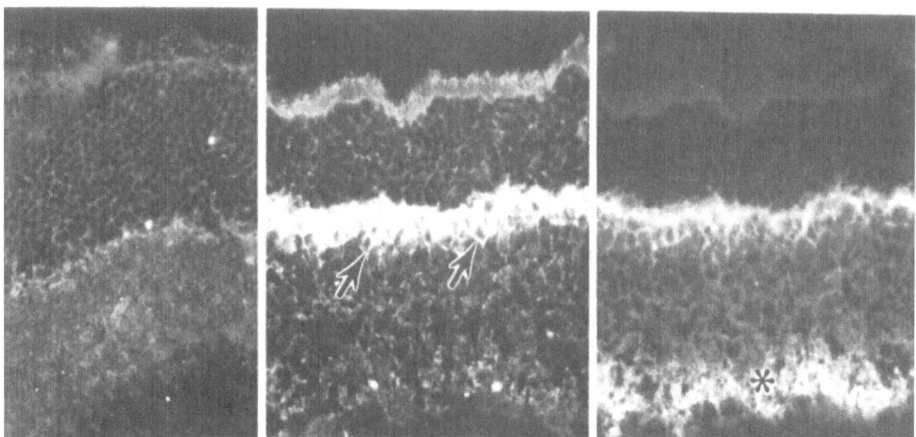

**Figure 2.** Cryostat sections of rat retina processed as in Figure 1. Left: Section processed with IgG (1:100) from another normal subject. Note low background autofluorescence. X100. Middle: Section processed with IgG (1:100) from MAR-2, illustrating brightly stained rod bipolar cells in the inner nuclear layer (arrows). Rod outer segment labeling is above background. X100. Right: Same section processed with anti-protein kinase C, illustrating labeling of the rod bipolar cells and their axon terminals deep in the IPL (*). X100.

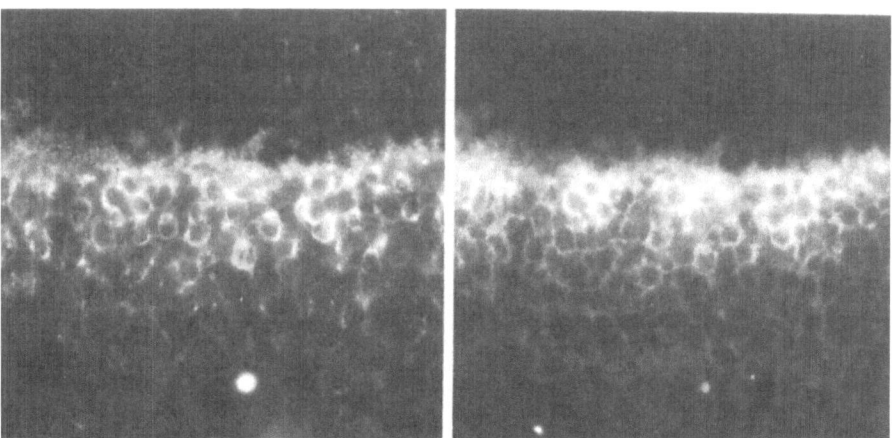

**Figure 3.** Cryostat sections of rat retina processed with IgG (1:100) from MAR-2 mixed with anti-protein kinase C (1:50), as in Figure 1. Left: Labeled bipolar cells are prominent in this section photographed using FITC filters. X250. Right: Same area photographed using rhodamine filters, illustrating that the same population of bipolar cells is labeled with anti-protein kinase C. X250.

The experiments were repeated using MAR sera on sections of human and rat retinas that had been fixed with varying concentrations of paraformaldehyde. Bipolar cells were very weakly labeled in retinas that had been fixed for up to 1 hour in 1% paraformaldehyde, but no specific labeling was found in retinas fixed with 4% paraformaldehyde. This result indicates that the bipolar cell antigen that is recognized by MAR sera is sensitive to fixation and would not be recognized in sections processed by conventional fixation methods.

**Biochemistry**

To detect the antigen responsible for bipolar cell staining, human retinal homogenates were analyzed by SDS-PAGE and Western blotting. Control and patient sera or IgG appeared to recognize numerous retinal components; however, several proteins were labeled in the absence of primary antibody. Sera and IgG from MAR-1 stained a component(s) of ~75 kDa (doublet) in retinal homogenates; however, a co-migrating component was also stained by IgG from one normal subject, from one non-MAR melanoma patient, and not by IgG from MAR-2. Thus, staining of the ~75 kDa band did not correlate with the staining of bipolar cells found with the two MAR sera. Further analysis of homogenate, pellet, and supernatant retinal fractions did not reveal a component that was recognized uniquely by the two MAR sera.

ELISA plates were coated with supernatants from human retinal homogenates and tested for reactivity with IgG from MAR patients and controls. No differences were noted.

**DISCUSSION**

At least two mechanisms have been suggested to cause the retinopathy found in CAR and MAR patients: a.) damage to the retina by a circulating toxic tumor cell product, and b.) production of an autoantibody directed initially against a tumor specific epitope that cross reacts with a retinal antigen. Specific labeling of retinal bipolar cells by sera from two MAR patients supports hypothesis (b). However, cross reactivity of our patients' sera with their melanoma cells has not yet been tested because these cells are not available for study.

Our results suggest that circulating antibodies to rod bipolar cells may be related to altered function of the rod system in MAR. A defect at the level of the rod bipolar cell leading to impaired signal transmission from rod photoreceptors would be consistent with the reduced b-wave amplitude and preserved a-wave in the dark-adapted ERG, the normal rhodopsin levels by fundus reflectometry,[4] and the marked rod-mediated sensitivity losses across the visual field by dark-adapted perimetry that have been observed in MAR patients. However, we emphasize that although retinal function tests and immunologic staining both implicate components of the same pathway, there is still no evidence in CAR or MAR that directly links the presence of circulating autoantibodies with retinal dysfunction. In addition, clarification is required on the pathophysiology that underlies abnormal cone-mediated functions in MAR, including an ERG ON-response defect, cone oscillatory potential abnormalities, middle/long wavelength cone sensitivity losses across the visual field, and S cone function abnormalities.

We were unable to detect a protein responsible for the immunostaining of rod bipolar cells by MAR sera using biochemical techniques. In contrast, the sera of some but not all CAR patients produce specific staining of a 23 kDa retinal component recently identified as recoverin.[8,9,14] Bipolar staining with MAR sera was sensitive to paraformaldehyde fixation, suggesting that antigenicity is abolished by amino group modification and perhaps by conformational alteration. Denaturation by SDS may have also disrupted a conformational epitope, accounting for the absence of specific staining by the two MAR sera on Western blots. A component stained specifically by MAR sera could also co-migrate with a non-specifically stained component and be hidden. Melanoma specific antigens include cell surface gangliosides and proteoglycans; thus, the antigen responsible for bipolar staining may not be a protein. Biochemical studies are ongoing to identify the retinal bipolar antigen that is recognized by MAR sera.

Very recently, a vaccine produced from three melanoma cell lines was administered to patients with metastatic melanoma and found to produce specific immune responses.[12] The vaccine contained three gangliosides and three proteins that are melanoma specific, and antibodies to these antigens were found to bind complement and kill melanoma cells *in vitro*. Patients who received the vaccine and developed titers against the melanoma antigens had statistically longer survival times than non-vaccinated controls. The presence of antibodies to these melanoma specific antigens in non-vaccinated melanoma patients was also found to correlate with their survival. These data suggest that autoantibodies against melanoma specific antigens may strengthen patients' immune surveillance of their tumor cells. It is possible that such antibodies may also be involved in MAR.

In contrast to a recently reported CAR patient,[2] we found no definite beneficial effects of steroid therapy on the vision of MAR-1 or -2. Serial monitoring by immunocytochemistry and measurements of retinal function provided no strong evidence to warrant this therapy.

Since initiating our studies of MAR-1 and -2, we have received serum samples from additional patients with metastatic melanoma and visual symptoms characteristic of MAR. Our preliminary immunocytochemistry analyses indicate that sera and IgG from some of these patients produce the same bipolar labeling pattern observed with sera and IgG from MAR-1 and -2, while sera and IgG from other patients produce diffuse labeling of the entire inner retina, including the inner nuclear, the inner plexiform, the ganglion cell, and the nerve fiber layers. The latter sera did not label the photoreceptor layer, and specific bipolar labeling was not detectable against the high background labeling of the inner retina. It is possible that the latter patients had retinal inflammation sufficient to allow release of antigens from the inner retina into the circulation and elicit a secondary autoimmune response, although details of the role of inflammation in MAR remain to be determined. As emphasized in recent reviews, the neurologic paraneoplastic syndromes, including retinopathy, are complex and the underlying pathophysiological events are still poorly understood.[13,14]

Additional research is required to identify the MAR retinal antigen, to determine if rod bipolar cells share an epitope with malignant melanoma cells, and to document the mechanism by which MAR antibodies penetrate the blood retinal barrier and cause bipolar cell dysfunction. Histopathologic examination of the retina from a MAR patient is also needed to determine if rod bipolar cells are abnormal or absent, and if immune complexes are present in the tissue.

## ACKNOWLEDGMENTS

Supported by the National Retinitis Pigmentosa Foundation, Inc.; NIH grants EY0-1311 (AHM), -2317 (JCS), -1730, -5627 (SGJ), and -8301 (KRA); by departmental awards from Research to Prevent Blindness, Inc.; and by the Chatlos Foundation. AHM is a Senior Scholar and SGJ is a Dolly Green Scholar of Research to Prevent Blindness, Inc.

# REFERENCES

1. M. Matsui, D.A. Hafler, and H.L. Weiner, Neurologic aspects of autoimmunity, *in:* "Molecular Autoimmunity", N. Talal, ed., Academic Press Ltd., London (1991).

2. C.E. Thirkill, P. FitzGerald, R.C. Sergott, A.M. Roth, N. K. Tyler, and J.L. Keltner, Cancer-associated retinopathy (CAR syndrome) with antibodies reacting with retinal, optic-nerve, and cancer cells, *N Engl J Med* 321:1589-1594 (1989).

3. J.L. Keltner, C.E. Thirkill, N.K. Tyler, and A.M. Roth, Management and monitoring of cancer-associated retinopathy, *Arch Ophthalmol* 110:48-53 (1992).

4. H. Ripps, R.E. Carr, I.M. Siegel, and V.C. Greenstein, Functional abnormalities in vincristine-induced night blindness, *Invest Ophthalmol Vis Sci* 25:787-794 (1984).

5. E.L. Berson and S. Lessell, Paraneoplastic night blindness with malignant melanoma, *Am J Ophthalmol* 106:307-311 (1988).

6. K.R. Alexander, G.A. Fishman, N.S. Peachey, A.L. Marchese, and M.O.M. Tso, "On" response defect in paraneoplastic night blindness with cutaneous malignant melanoma, *Invest Ophthalmol Vis Sci* 33:477-483 (1992).

7. A.H. Milam, J.C. Saari, S.G. Jacobson, W.P. Lubinski, L.G.Feun, and K.R. Alexander, Autoantibodies against retinal bipolar cells in cutaneous melanoma-associated retinopathy, *Invest Ophthalmol Vis Sci.* (In Press).

8. A.S. Polans, J. Buczylko, J. Crabb, and K. Palczewski, A photoreceptor calcium-binding protein is recognized by autoantibodies obtained from patients with cancer-associated retinopathy, *J Cell Biol* 112:981-989 (1991).

9. C.E. Thirkill, R.C. Tait, N.K. Tyler, A.M. Roth, and J.L. Keltner, The cancer-associated retinopathy antigen is a recoverin-like protein, *Invest Ophthalmol Vis Sci* 33:2768-2772 (1992).

10. A.H. Milam, D.M. Dacey, and A.M. Dizhoor, Recoverin immunoreactivity in mammalian cone bipolar cells, *Visual Neuroscience* (In Press).

11. H. Wässle, M. Yamashita, U. Greferath, Y. Grünert, and F. Müller, The rod bipolar cell of the mammalian retina, *Visual Neuroscience* 7:99-112 (1991).

12. D.L. Morton, L.J. Foshag, D.S.B. Hoon, J.A. Nizze, L.A. Wanek, C. Chang, D.G. Davtyan, R.K. Gupta, R. Elashoff, and R.F. Irie, Prolongation of survival in metastatic melanoma after active specific immunotherapy with a new polyvalent melanoma vaccine, *Ann. Surgery* 216:463-482 (1992).

13. N.E. Anderson, Anti-neuronal autoantibodies and neurological paraneoplastic syndromes, *Aust NZ J Med* 19:379-387 (1989).

14. J.F. Rizzo III, and J.W. Gittinger, Jr., Selective immunohistochemical staining in the paraneoplastic retinopathy syndrome, *Ophthalmology* 99:1286-1295 (1992).

# THE EFFECT OF NAPHTHALENE ON THE RETINA OF RABBIT

Nicola Orzalesi, Luca Migliavacca, Stefano Miglior

Eye Clinic of the University of Milan
Institute of Biomedical Sciences
S. Paolo Hospital
Milan, Italy

## INTRODUCTION

Naphthalene has been known to be a cataractogenic agent for more than a century. The first report concerning its effect on the lens was presented at the French Academy of Medicine by Bouchard and Charrin (1886), and was subsequently experimentally proved in animals such as rabbits, cats, rats, guinea-pigs and chickens. Ocular toxicity in man has also been reported following accidental ingestion of the poison (for a review of the literature see Duke Elder, 1966).

The mechanism of action of naphthalene on the lens has been the subject of extensive biochemical investigations. Early hypotheses on the pathogenesis of naphthalene cataract, such as those based on the dramatic drop in lenticular cystine concentration or the loss of ascorbic acid in the aqueous humour, were not conclusive, and it is only recently that lenticular metabolic changes have been elucidated by van Heyningen and Pirie (1967), and Rees and Pirie (1967). The primary toxic agent is 1,2-dihydroxynaphthalene which is formed in the lens and ciliary body from the

*Retinal Degeneration*, Edited by J.G. Hollyfield *et al.*
Plenum Press, New York, 1993

liver metabolite, 1,2-dihydro-1,2-dihydroxynaphthalene. The self-oxidation of 1,2-dihydroxinaphthalene produces hydrogen peroxide and 1,2-naphthoquinone. Hydrogen peroxide and the quinone induce the drop in ascorbic acid which has always been observed during the development of naphthalene cataract (ascorbic acid was found to protect rabbits from naphthalene-induced cataract by Yoritaka et al (1958)). Quinone also interacts with alpha, beta and gamma crystallins to give products with a reduced SH content, reacts with amino acids and the SH group of glutathione, and in its semi-quinone form, can also generate free radicals through redox cycling (Brunmark and Cadenas, 1989). Thus, the development of naphthalene cataract has been explained as a consequence of oxidative stress at the level of a number of metabolic targets. .

Interest in naphthalene has been recently revived by the suggestion that the use of aldose reductase inhibitors may prevent cataract formation (Tao et al, 1991). Two recent studies by Xu et al (1992 a, b) have reported that the administration of the aldose reductase inhibitor, Alcon AL01576, prevented in vivo and in vitro experimental cataract in both albino and pigmented rats as a consequence of preserved glutathione and mixed protein-thiol disulfide levels. These results offer further evidence as to the possible oxidative damage that naphthalene may cause to the lens.

Treatment with naphthalene also causes a series of other ocular and systemic changes (signs of general toxaemia, particularly of a gastro-intestinal nature), most of which were reported in the old literature. Hyperaemia and haemorrhages were described in the ciliary body (Hess, 1887; Klingmann, 1897; Sala, 1903; Lindberg, 1922); and retinal changes resembling diabetic retinopathy, such as exudates and sometimes haemorrhages, were also reported as initially arising in the periphery and later spreading to the central retina. These lesions were histologically described as edematous (Adams, 1930), sometimes associated with albuminous exudates located between different retinal layers, between the rod and cones, and between the RPE and the choroid (Panas, 1887; Igersheimer and Ruben, 1910; Takamura, 1912). A peculiar clinical picture of the retina was that of a "clear starry heaven" consisting of most probably calcium oxalate crystallin deposits within the retina and the vitreous (Adams, 1930).

Retinal tissue is highly exposed to oxidative insult because of the high concentration of polyunsaturated fatty acids (PUFAs) in the photoreceptor membranes, which are potential targets for lipid peroxidation; because of the high degree of oxygen turnover; because of the continuous production of free radicals derived from exposure to light. However, like the lens, the retina is protected against oxidative stress by antioxidant agents such as ascorbic acid, glutathione (GSH), superoxide dismutase, catalase, etc. (Berman, 1991). Therefore, the integrity of retinal tissue may depend upon the balance between oxidant and antioxidant agents, and it is conceivable that any agent altering this system may induce retinal degeneration.

The aim of the present study was to evaluate the clinical and morphological effect of naphthalene, a highly oxidative agent, on the retina of rabbit.

## MATERIAL AND METHODS

Thirty-five young adult male pigmented rabbits weighing 1.2 Kgs, all treated according to the ARVO Resolution on the Use of Animals in Research and with food and water ad libitum, were housed in laboratory conditions under a 12 hours light (100/200 Lux) - 12 hours dark cycle. Thirty animals were assigned to treatment, and 5 used as controls. A 10% (wt/vol) solution of naphthalene (99%, SIGMA Chemicals S. Louis, MO, U.S.A.) dissolved in paraffin oil (SIGMA Chemicals S. Louis, MO, U.S.A.) was administered by gavage to all treated animals at a dose of 1 g/Kg body weight on alternate days for five weeks; the control animals received only paraffin oil. The treatment was well tolerated by all of the animals, who looked healthy, showed no change in their feeding and behavioural habits, and continued to increase their weight regularly. Using a slit lamp and indirect ophthalmoscopy their eyes were examined every week for five weeks, and then at monthly intervals. In the case of significant changes in the picture of the fundus, photography and fluorescein angiography (FAG) were performed. All of the animals were manipulated under a general anesthesia obtained using an intramuscular injection of ketamine hydrochloride and xylazine. The pupils were dilated by means of topical 1% tropicamide and 10% phenylephrine hydrochloride. The rabbits were killed by intravenous embolization at different stages of intoxication (i.e. 10, 15, 21, 30, 40 and 60 days after the beginning of treatment), the controls being killed at day 30. The eyes were immediately enucleated and immersed in 2.5% glutaraldehyde and 2% paraformaldehyde in 0.1 M cacodylate buffer (pH 7.4). They were then bisected at the ora serrata and the posterior cup kept in the fixing fluid at room temperature. Three hours later, they were divided into small pieces which were fixed for a further hour. After fixation they were immersed in 1% osmium tetroxide, washed and dehydrated in graded ethanol. The pieces were then embedded in SPURR resin or directly processed for embedding in hydrosoluble JB4 resin. Semi-thin sections (1 micrometer) were obtained at the posterior pole, the mid periphery and the periphery, and whenever possible, at the site of the ophthalmoscopical lesions. The sections were stained with toluidine blue for examination by light microscopy. In the case of significant pathology, the pieces embedded in SPURR resin were ultra-thin sectioned (50-60 nm), stained with uranil acetate and lead cytrate or bismuth subnitrate (Riva, 1974), and then examined by means of a Jeol transmission electron microscope.

## RESULTS

### Clinical Picture

About three weeks after the beginning of treatment, 22 animals (70%) developed initial fundus lesions represented by white spots ("snow flakes" aspect) scattered throughout the retinal periphery (Fig. 1a). At the subsequent examinations, the spots spread toward the posterior pole; at the periphery, they tended towards confluence and scarring and gave rise to white pigment-edged plaques occasionally accompanied by small retinal hemorrhages (Fig. 1b).

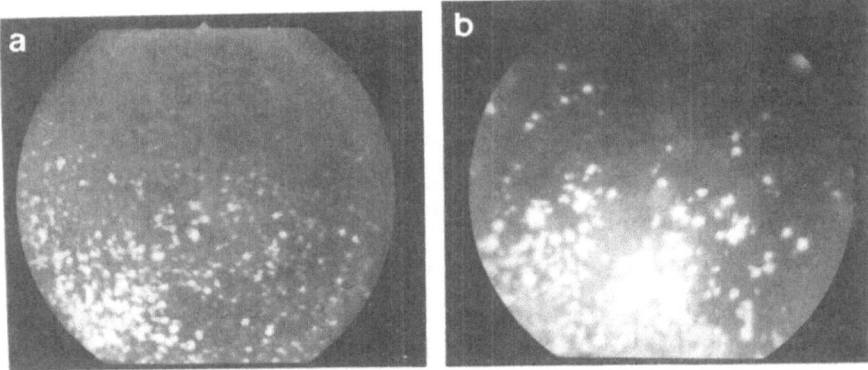

Fig. 1. a): The fundus in the early stages of the degeneration with "snow flakes" aspect. b): confluent peripheral plaques.

At FAG, the white lesions showed masked choroidal fluorescence bordered, seldom centered, by a faint hyperfluorescence (Fig. 2a). After two months, the periphery of the fundus of the majority of the animals showed irregular atrophic changes associated with retinal scarring whereas the posterior pole showed fine and diffuse pigmentary disturbance. In this latter area FAG revealed only diffuse background hyperfluorescence interspersed with granular pigmentation (Fig. 2b). No case showed any clear FAG signs of subretinal neovascularization (SRN) and/or dye pooling. In these animals,

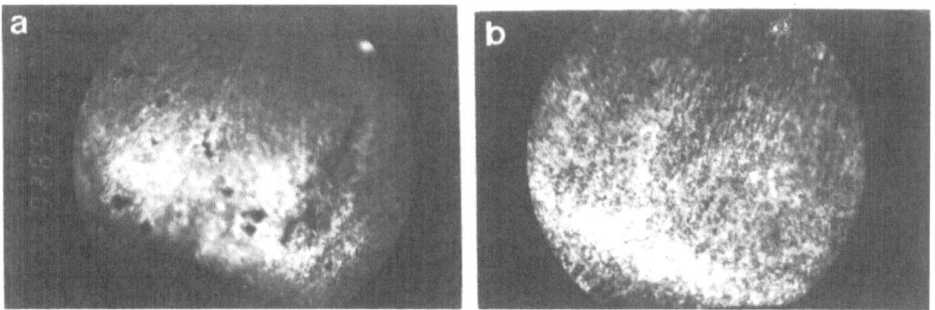

Fig. 2. FAG. a): confluent peripheral fundus lesion. b): end-stage retinal atrophy with pigmentary disturbance.

the lens remained clear or showed faint white-ringed perinuclear opacity; no inflammatory reaction was present.

Three animals (10%) developed an early complete cataract accompanied by severe uveitis after about two weeks. Like the controls, another five treated animals (16%) did not develop any clinically relevant ocular change at any time during the experimental period.

## Morphological Observations

The first histological lesions appeared in the same place and at the same time as the fundus changes (i.e. in the retinal periphery, about two weeks after the beginning of treatment) and were represented by the swelling and derangement of photoreceptor outer segments (Fig. 3a-b).

After three weeks, constant changes were seen at the level of the outer retinal layers, which showed various degrees of photoreceptor degeneration and RPE reaction. Both the outer and inner segments of photoreceptors were shortened and disorganized, some nuclei were pycnotic and some prolapsed into the subretinal space beyond the external limiting membrane, coming in contact with the RPE (Fig. 4a). In TEM, well preserved inner segments and nuclei were often seen in contact with the apical border of RPE cells after the disappearance of outer segments (Fig. 4b). Marked phagocytosis of degenerated parts of photoreceptors by the RPE cells was also apparent (Fig. 4b). RPE cells facing the photoreceptor lesions showed reactive changes represented by cell enlargement and bilayering (Fig. 5a).

Enlarged R P E cells still showed polar organization in their cytoplasm as long as they were lying on Bruch's membrane, whereas

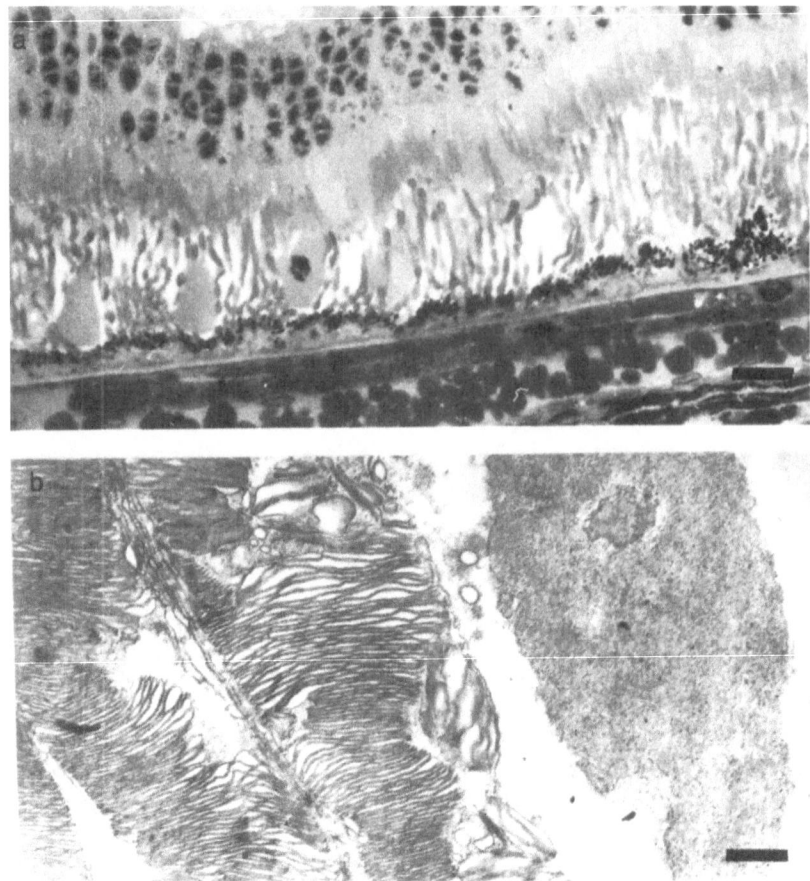

Fig. 3. a): Early changes of photoreptors' outer segments two weeks after the beginning of treatment. b): TEM shows disorganization of disk membrane and homogeneization of outer segment. Calibration bars: a) = 16 µm. b) = 1.25 µm.

polarity was lost when the cells migrated into subretinal space and/or formed multiple layers. These cells showed the usual features of so-called "RPE derived subretinal macrophages" with inclusion bodies and microvilli and, even in bilayers often maintained junctional apparatus between adjacent cells (Fig. 5b). The detachment of RPE cells from their basement membrane was always marked by a change in the pattern of basal infoldings, which were flattened or transformed into slender pseudopodia (Fig. 6). These changes were focally distributed and separated by healthy or less damaged retina (Fig. 5a). In each focal lesion, the extent of the damage was greatest at the center (often involving less than 10 photoreceptors per section in the earliest stages), and progressively decreased towards the borders. Muller's cells were hypertrophic with bumps projecting into subretinal space, but swelling

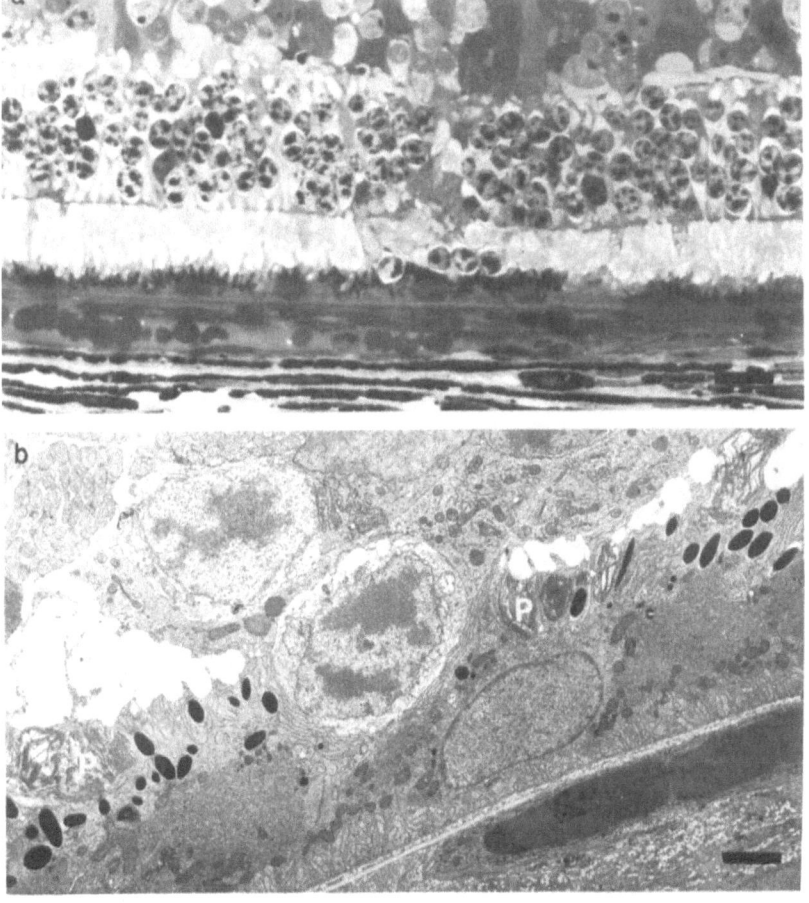

Fig. 4. Histological a) and TEM b) aspect of photoreceptor damage three weeks after the beginning of treatment. Outer segments are shortened or absent (for comparison with normal height, see Fig. 3a). Prolapse of viable nuclei and phagocytosis are also shown. P: phagosomes. Calibration bars: a) = 16 μm. b) = 4 μm.

and disorganization of the inner retinal layers was less constant. Inflammatory signs were absent at both the retinal and choroidal level.

After 30 - 40 days, more advanced stages were characterized by the complete degeneration of photoreceptors accompanied by glial reaction and the obliteration of subretinal space (Fig. 5a). The reaction of RPE cells was more evident at this stage, with marked pleomorphism and the formation of pseudoacinar cavities. TEM showed marked changes at the level of the Bruch's membrane-RPE junction, represented by irregularities of the basement membrane, interruptions in the elastic layers and cytoplasmic projections of RPE cells into the Bruch's layer associated with overproduction of basement membrane material (Fig. 6). While some of these projections may be considered residual bodies discharged by the RPE, others are more suggestive of an invasion of the Bruch's layer by RPE pseudopodia. In this respect, it must be considered that no RPE-derived cells completely migrated outside the basement membrane, and the abnormal cells found in the Bruch's layer at this stage were more likely to be fibroblasts, pericytes and blood-borne macrophages (Fig. 6).

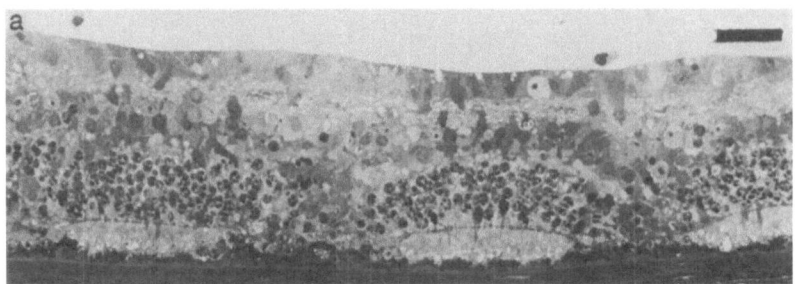

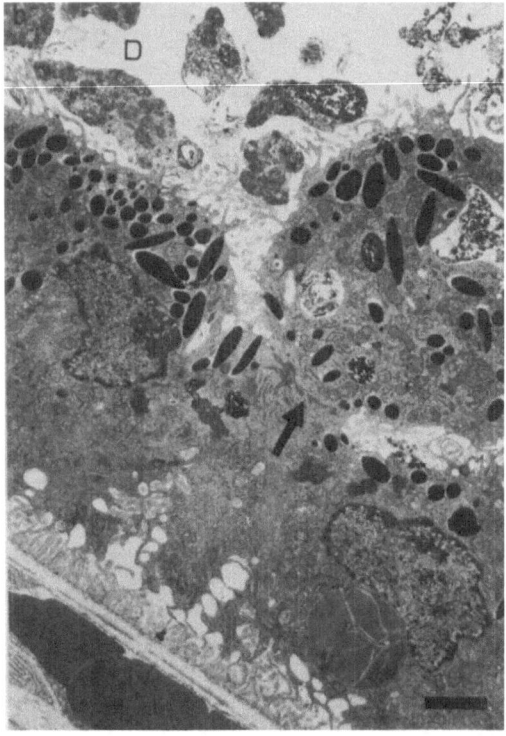

Fig. 5. a) Focal degeneration of photoreceptors and reaction of RPE. b) TEM shows loss of polarity and pleomorfism of the migrated RPE cells, still connected by junctional apparatus (arrow). D: degenerated photoreceptors. Calibration bars: a) = 40 μm. b) = 4 μm.

In some cases the relationship between RPE and the degenerated retina was more complex: this was true of RPE mounds and retinal folds, and in the case of multilayered pigmented cells contacting remnants of retinal tissue. In these stages, subretinal neovascularization (SRN), always associated with pigmented cells, was also evident. The above description refers to the majority of treated animals in which retinal degeneration was associated with mild lens opacity. Minor photoreceptor damage was also present in the few animals with no clinical signs, and an early and severe retinal necrosis followed by extensive inflammatory infiltration and scarring characterized the animals developing early and complete cataract with uveitis.

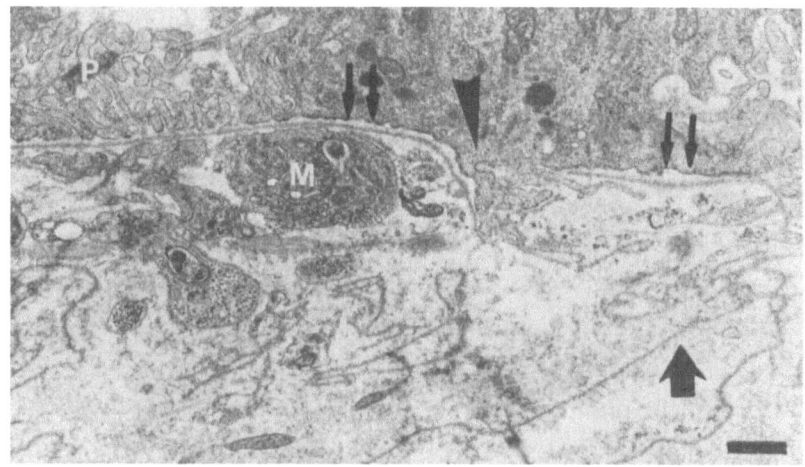

Fig. 6. An example of cytoplasmic process (arrow-head) of RPE cell extending into the Bruch's membrane associated with overproduction of basement membrane material (bold arrow). Flattening of RPE cell membrane (narrow arrow) or formation of pseudopodia (P) are also evident. M: macrophage. Calibration bar = 1.5 μm.

## DISCUSSION

The results of the present investigation clearly show that naphthalene may severely damage rabbit retina. The first target of the naphthalene effect seems to be the photoreceptor, whose degeneration is accompanied by the reaction of RPE cells. The main features of such a degeneration closely resemble the retinal damage induced by visible light (Hoppeler et al, 1988): an initial disruption, followed by the disappearance of outer and inner photoreceptor segments, swelling and the proliferation of RPE cells and scarring.

The mechanism of photoreceptor degeneration may be related to the potential oxidative action of naphthalene, which has been clearly demonstrated to be the cause of its cataractogenic effect. Furthermore, it is has been suggested that not only 1,2-naphthoquinone, but also its precursors (such as 1,2 diol 1,2-dihydroxynaphthalene, 1,2-dihydro-1,2-diol and 1,2-naphthalene oxide) affect the redox system in the lens, just as GSH levels and GSH S-transferase activity decrease in the lens prior to the appearance of 1,2-naphthoquinone in the aqueous humor and vitreous body (Iwata and Maesato, 1988). In the case of the retina, it is possible that oxidative metabolites such as naphthoquinone (which has been found in rabbit serum even after a single administration of naphthalene (Iwata and Maesato, 1988)) may reach the photoreceptors and RPE cells through choroidal circulation. Retinal tissue is particularly exposed to oxidative damage, just as the high concentration of PUFAs makes photoreceptor membranes potential targets for lipid peroxidation (Berman, 1991). Moreover, its environment is particularly favourable to the production of free radicals with the potential of initiating peroxidative chain reactions in the PUFAs of the disk membranes (Ham et al, 1984).

The lesion of photoreceptors is constantly associated with RPE reaction, which initially aims at scavenging the remnants of degenerated photoreceptors, and only later takes on a reparative function probably associated with scarring and SRN. RPE reaction as a consequence of the degeneration of the photoreceptors has also been observed in experimentally induced light damage to the retina (Hoppeler et al, 1988). It is well known that RPE cells produce a series of diffusible factors which influence various types of cells (Glaser, 1991), but the release of factors from photoreceptors which promote changes in RPE cells has not yet been demonstrated. The possibility that damage to photoreceptors may be part of an inflammatory reaction (of which RPE changes may be an aspect) cannot be ruled out, just as immunogenic uveoretinitis experimentally induced by retinal antigens (Broekhuyse et al, 1991; Dua et al, 1991) closely resemble the reactive changes following the photoreceptor degeneration induced by naphthalene. Moreover, severe uveitis accompanies the early complete cataract observed in a small number of rabbits (but in this case the possibility of facoanafilactic reaction should also be considered).

In conclusion the naphthalene effect on the rabbit retina offers a simple and reliable experimental model to study oxidative damage to photoreceptors in an early stage, as well as SRN following degeneration of photoreceptors and RPE reaction in the advanced ones. Further investigations aimed at elucidating the sequential morphological features of the development of SRN in the advanced stages of naphthalene degeneration are on the way.

# REFERENCES

Adams, 1930. Quoted by Duke-Elder.

Berman ER, 1991. Biochemistry of the eye. Plenum Press, New York.

Broekhuyse RM, Schuurmans Stekhoven JH and Arends M, 1991: Experimental autoimmune retinitis in the rat induced by immunization with rodopsin: an ultrastructural study. Exp Eye Res, 53:141-9.

Brunmark A and Cadenas E, 1989: Redox and addition chemistry of quinone compounds and its biological implications. Free Radical Biol Med, 7:435-77.

Dua HS, Mc Kinnon A, Mc Menamin PG and Forrester JV, 1991: Ultrastructural pathology of the "barrier sites" in experimental autoimmune uveitis and experimental autoimmune pinealitis. Br J Ophthalmol, 75:391-7.

Duke-Elder S: System of Ophthalmology. Henry Kimpton Publisher London (1969).

Glaser B, 1988: Extracellular modulating factors and the control of intraocular neovascularization. Arch Ophthalmol, 106:603-7.

Hess, 1887. Quoted by Duke-Elder.

Hoppeler T, Hendrickson P, Dietrich C and Reme' C, 1988: Morphology and time-course of defined photochemical lesions in the rabbit retina. Curr Eye Res, 7:849-60.

Igersheimer and Ruben, 1910. Quoted by Duke-Elder.

Iwata S and Maesato T, 1988: Studies on the mercapturic acid pathway in the rabbit lens. Exp Eye Res, 47:479-88.

Klingmann, 1897. Quoted by Duke-Elder.

Lindberg, 1922. Quoted by Duke-Elder.

Panas, 1887. Quoted by Duke-Elder.

Rees JR and Pirie A, 1967: Possible reactions of 1,2-naphthoquinone in the eye. Biochem J, 102:853-63.

Riva A, 1974: A sample and rapid staining method for enhancing the contrast of tissues previously treated with uranyl acetate. J Microsc (Paris), 19:105-8.

Sala, 1903. Quoted by Duke-Elder.

Takamura, 1912. Quoted by Duke-Elder.

Tao RV, Takahashi Y and Kador PF, 1991: Effect of aldose reductase inhibitors on naphthalene cataract formation in the rat. Invest Ophthalmol Vis Sci, 32:1630-7.

Van Heyningen R and Pirie A, 1967:The metabolism of naphthalene and its toxic effect on the eye. Biochem J, 102:842-52.

Yoritaka, Tominaga, Semba and Hiwaki, 1958. Quoted by Duke-Elder.

Xu GT, Zigler SJr and Lou MF, 1992: The possible mechanism of naphthalene cataract in rat and its prevention by an aldose reductase inhibitor (AL01576). Exp Eye Res, 54:63-72.

Xu GT, Zigler SJr and Lou MF, 1992: Establishment of a naphthalene cataract model in vitro. Exp Eye Res, 54:73-81.

# BIOLOGICAL EFFECTS OF RETINOIDS AND RETINOID METABOLISM IN CULTURES OF CHICK EMBRYO RETINA NEURONS AND PHOTORECEPTORS

Deborah L. Stenkamp[1] and Ruben Adler[2]

[1]Department of Neuroscience and
[2]Retinal Degenerations Research Center
The Wilmer Institute
The Johns Hopkins University
School of Medicine
Baltimore, MD 21287-9257

## I.    INTRODUCTION: RETINOIDS IN THE RETINA

The retinoids comprise a family of small hydrophobic molecules which are chemically similar to retinol; those that share also some of its biological properties are additionally considered part of the vitamin A family.[1]  The pleiotropic effects of vitamin A family members are best exemplified by those of retinoic acid, which include activities as a developmental signal, a teratogen, and as a regulator of tumor cell behavior, among others.[2]  Recent progress in the cloning and characterization of several different retinoic acid receptors, belonging to the glucocorticoid family of ligand activated transcription factors, has provided new insights into the mechanisms supporting the diversity of retinoic acid biological effects.[2]  Another vitamin A derivative with well established function is 11-cis retinaldehyde, the chromophore of rod and cone visual pigments.  This compound binds covalently to the visual pigment apoproteins, and its isomerization to the all-trans isomer in response to photon absorption by the pigment initiates the process of visual transduction.[1,3-5]

A possible trophic function for vitamin A in the retina was disclosed by the finding of Dowling and Wald[6] that vitamin A-deprived rats undergo degenerative changes in their photoreceptors, resulting eventually in the death of these cells.  The initial abnormalities in the photoreceptors (including outer segment degeneration) were reversible by dietary administration of retinol.  Retinoic acid, however, failed to prevent the degenerative changes in the retina, although it did prevent pathologies in other vitamin A-dependent tissues.

These findings raised questions regarding both the identity and the mechanism of action of the retinoid(s) responsible for the trophic effects of vitamin A in the retina.  An initial suggestion[7,8] was that visual pigment molecules are unstable in the absence of the 11-cis retinaldehyde chromophore, and that their alteration leads to abnormalities in the visual pigment-rich outer segment, and eventually to the death of the photoreceptor cell.  An important corollary of this hypothesis (which is relevant for the interpretation of experimental results presented below), is that the retinal pigment epithelium (RPE) would be essential for the trophic effect of retinol to occur through this mechanism, because the RPE is the site where 11-cis retinaldehyde is synthesized by metabolic transformation of retinol.[3]  It must also be noted that, despite the ineffectiveness of dietary retinoic acid supplementation, a possible involvement of retinoic acid cannot be excluded.  One reason for this is that retinoic acid is apparently not transported across the RPE,[9] but it is known to be synthesized within the retina,[10] which also contains retinoic acid binding proteins in Müller glia and in some types of neurons.[11] There are also reports of the presence of retinoic acid receptors in the retina,[12] but more information is needed in this area. Theoretically, however, retinol could also have trophic effects of its own, without requiring transformation into any other type of retinoid.

We have recently investigated some of these questions using a cell culture system in which retinal neurons and photoreceptors differentiate in the absence of glia and RPE.[13] Retinol, 11-cis retinaldehyde, and retinoic acid were studied regarding their effects upon these cells and their metabolic transformations in culture. Our findings indicate that these retinoids can affect the survival of both retinal neurons and photoreceptors in culture, even in the absence of RPE and glial cells. They also provide some evidence that retinol itself could potentially function as a trophic factor, without being converted into either retinaldehyde or retinoic acid.

## II. THE CULTURE SYSTEM

The cultures are prepared from an embryonic day 8 chick, a stage at which the retina is essentially unlaminated and undifferentiated. The retina is dissected free of RPE and other eye tissues, dissociated, and seeded at low density on polyornithine-coated dishes.[14] Under these conditions, isolated, identifiable photoreceptors and nonphotoreceptor neurons differentiate *in vitro*, in the absence of glia, epithelial, and endothelial cells.[15] Special precautions were adopted for the experiments reported here in order to prevent retinoid degradation, including manipulation of solutions under dim red illumination, and growth of the cultures in darkness in an atmosphere with reduced $O_2$ concentrations.[13]

The neurons and photoreceptors that differentiate in these cultures can be distinguished by morphological, immunocytochemical and biochemical criteria. As shown in Figure 1, photoreceptor cells develop a highly polarized structure *in vitro*, with one short

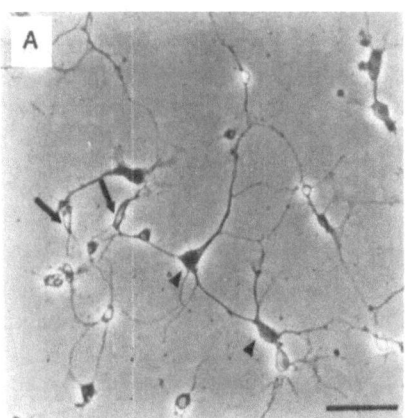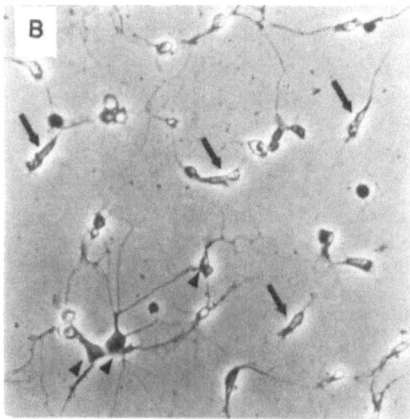

**Figure 1.** Effect of retinol on cultured embryonic chick retinal cells. Cells were cultured for 6 days in the presence of A) ethanol or B) 10 μM retinol. Photoreceptors are indicated by arrows, neurons by arrowheads. Scale Bar = 20 μm. From Stenkamp *et al.*, (1993), *Invest. Ophthalmol. Vis. Sci.*, in press.

neurite, a cell body (occupied almost exclusively by the nucleus), a lipid droplet-containing inner segment, and an "outer segment region" which is immunoreactive with visual pigment antibodies.[15] We have recently shown that these photoreceptors are capable of photomechanical length changes in response to light.[16] Additionally, N-acetyl transferase activity has been associated with the cultured photoreceptor cells,[17] indicating that this cell type may produce melatonin, a neuromodulator known to regulate cyclic behaviors in the retina. The multipolar neurons are morphologically very different from the photoreceptors, showing a larger cell body and longer (and usually more numerous) neurites. Several neurotransmitter-related activities have been associated with the presence of multipolar neurons in the cultures, including acetylcholine synthesis and high-affinity uptake systems for gamma-amino butyric acid (GABA).[15]

## III.    EFFECTS OF RETINOIDS ON EMBRYONIC CHICK RETINA CELLS IN VITRO

Retinol, retinoic acid, and 11-cis retinaldehyde were investigated for their possible effects on the survival and differentiation of embryonic retinal cells *in vitro*. When nontoxic concentrations of retinoids were added to the cultures at the time of seeding, the number of photoreceptors observed after 6 days *in vitro* (div) was substantially increased, and there were smaller but consistent increases in the number of multipolar neurons as well. This is illustrated in Figure 1B; Figure 2 quantitatively compares the numbers of neurons (Fig. 2A) and photoreceptors (Fig. 2B) present on each day *in vitro* following treatment with retinol

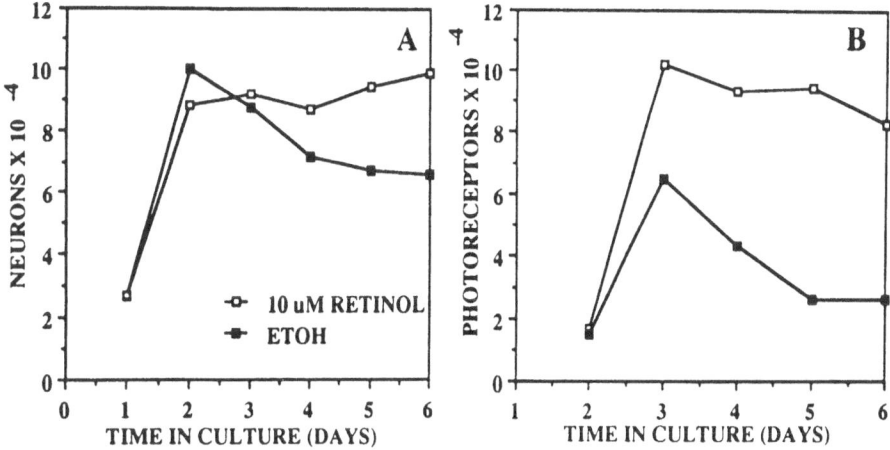

Figure 2. Effect of retinol on numbers of A) neurons, and B) photoreceptors, as observed over time in culture. Cultures were treated at seeding with 10 nM retinol or the ethanol vehicle, and fixed at daily intervals. Each point represents the mean ($\pm$ S.D.) of triplicate cultures from a typical experiment. From Stenkamp *et al.*, (1993), *Invest. Ophthalmol. Vis. Sci.*, in press.

or the vehicle, ethanol. Overall, the numbers of both photoreceptors and neurons are higher in retinol-treated cultures, with a concomitant decrease in the number of morphologically undifferentiated cells.[13] The higher numbers of both photoreceptors and neurons appear to reflect a survival-promoting effect of the retinoid, but it is noteworthy that there was also an increase in the number of photoreceptors appearing in retinol-treated cultures at early *in vitro* stages, when morphologically differentiated cells can first be identified. Taken together, these findings indicate that embryonic retina neurons and photoreceptors (and/or their precursors) are responsive to retinoids without need for the presence of RPE or glial cells. While the survival-promoting effects were much more pronounced for the photoreceptor cells [as previously reported by Dowling in the adult rat *in vivo*],[7] there was also a more modest but nonetheless reproducible stimulation of neuronal survival by the retinoids, which had not been previously observed.

Somewhat surprisingly, retinol was the most effective of the three retinoids tested based on its effects on photoreceptor survival *in vitro* (Figure 3). Subtoxic concentrations of exogenous 11-cis retinaldehyde and retinoic acid elicited lesser increases in numbers of differentiated cells, with retinoic acid being 10-fold less potent than retinol at the same dosage. The interpretation of these results, however, required information regarding the fate of the exogenous retinoids upon exposure to the cultured cells. We therefore used high pressure liquid chromatography to analyze retinoid-supplemented culture medium both before ("starting medium") and after ("conditioned medium") its exposure to the cultured cells, which were also extracted and similarly analyzed.[13] The exogenous retinoids were 5-10 fold higher in concentration than the small amounts of retinol and retinyl acetate that are present in our basal culture medium. Figure 4 shows the retinoids detected in *starting medium* upon addition of exogenous retinaldehyde at culture onset, as well as in *conditioned*

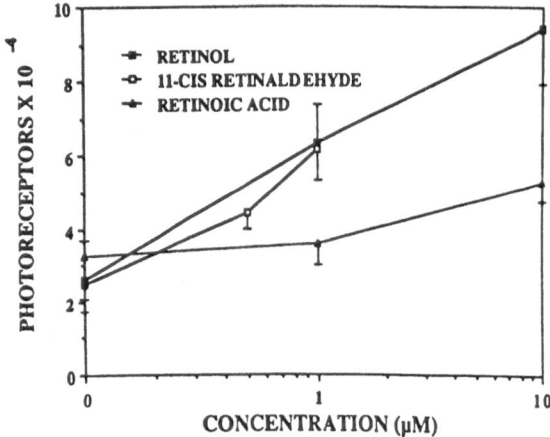

**Figure 3.** Comparative dose-dependent effects of retinol, retinoic acid, and 11-cis retinaldehyde on photoreceptor number. Cultures were treated at seeding with subtoxic concentrations of retinoids and fixed 6 days later. Each point represents the mean (± S.D.) of triplicate cultures from a typical experiment. From Stenkamp et al., (1993), *Invest. Ophthalmol. Vis. Sci.,* in press.

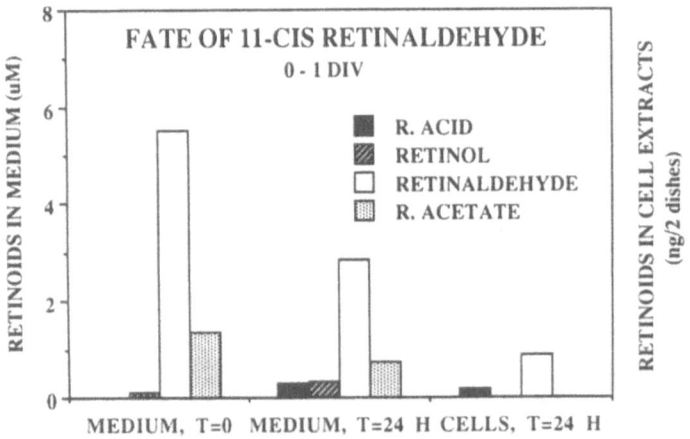

**Figure 4.** Fate of 11-cis retinaldehyde incubated for 24 hours in the presence of cultured neural retina cells. 11-cis retinaldehyde was added at seeding, and medium was removed immediately (T=O; starting medium), and following a 24 hour incubation (T=24H; conditioned medium), at which time cells were harvested from 2, 35 mm dishes. Results are expressed as μM retinoid present in medium (means of duplicates ± S.D. from a typical experiment), or ng retinoid present in cells. From Stenkamp *et al.,* (1993), *Invest. Ophthalmol. Vis. Sci.,* in press.

medium and in the cultured cells 24 hours later. Although retinaldehyde is less stable in culture than other retinoids, it can enter cells (as can other retinoids), and is metabolized to produce retinol and retinoic acid. Interestingly, the capacity to synthesize retinoic acid from retinaldehyde disappears at later in vitro stages, when retinol production increases.[13] Separate determinations showed that retinol is also taken up by the cells, that it is much more stable than retinaldehyde, and that it can be esterified into retinyl acetate at a very low rate at early *in vitro* stages, but is recovered unmodified thereafter.[13] Retinoic acid is also highly stable, can enter the cells, and undergoes no detectable metabolic conversions.[13]

These metabolic results, when taken together with the above mentioned findings regarding trophic activity, suggest that 11-cis retinaldehyde could exert trophic effects either directly (since a fraction remains unchanged in culture), or as a consequence of its conversion into retinoic acid or retinol. Our observations are more conclusive regarding both retinol and retinoic acid, and suggest that these retinoids are likely to function directly, without need for conversion into other metabolites. Since the presence of 11-cis retinaldehyde does not appear necessary for the trophic effects of vitamin A in culture, it is unlikely that a mechanism based on chromophore stabilization of visual pigment[7,8] plays a significant role in the *in vitro* situation, especially considering that retinoids are also trophic for non-photoreceptor neurons, which lack visual pigments. Our results, however, do not exclude the possibility that visual pigment stabilization may play some role in the *in vivo* situation. Our results also suggest that the effects of retinol do not appear to depend upon the production of retinoic acid which (although capable of supporting *per se* the survival of neurons and photoreceptors), is not detectable in retinol-treated cultures. It is not possible to rule out at this time the possible formation and biological activity of other retinol metabolites which would escape detection with our methods (especially if these hypothetical metabolites are active at $< 10^{-8}$M concentrations). If the effects are indeed exerted directly by retinol, however, it would be tempting to speculate that there may be nuclear receptors for retinol, equivalent to those described in recent years for retinoic acid and for other small hydrophobic ligands.

## IV. SUMMARY AND FUTURE DIRECTIONS

Retinoid research has virtually exploded in recent years, and many of the mechanisms responsible for the wide variety of retinoid effects on many different cell types are becoming less mysterious. It is likely that the unexplained aspects of the trophic functions of retinoids in the retina will follow suit. Together with previous observations on the retinoid-induced upregulation of opsin expression in cultured mouse photoreceptors,[18] these *in vitro* studies are consistent with a possible role of vitamin A in the survival and differentiation of retinal cells, suggesting that retinoids may have therapeutic potential for intervention in the course of retinal degenerations.

## ACKNOWLEDGEMENTS

Original studies described in this article were supported by NIH grant EY04859. RA is a Senior Investigator from Research to Prevent Blindness, Inc. This material is based upon work supported under a National Science Foundation Graduate Fellowship (DLS).

## REFERENCES

1. J.C. Saari, Enzymes and proteins of the mammalian visual cycle, *Prog. Ret. Res.* 9:363 (1990).

2. P. Chambon, A. Zelent, M. Petkovich, C. Mendelsohn, P. Leroy, A. Krust, P. Kastner, and N. Brand. "Retinoids : 10 Years on," Basel Karger, Geneva (1991).

3. R.R. Rando, T.S. Bernstein, R. J. Barry, New insights into the visual cycle, *Prog. Ret. Res.* 10:161 (1992).

4. D. Bok, Processing and transport of retinoids by the retinal pigment epithelium, *Eye* 4:376 (1990).

5. G.J. Chader. Retinoids in ocular tissues: Binding proteins, transport, and mechanisms of action, *in*: "Cell Biology of the Eye," D.S. McDevitt, ed., Academic Press, New York (1982).

6. J.E. Dowling, G. Wald, Vitamin A deficiency and night blindness, *Proc. Natl. Acad. Sci. USA* 44:648 (1958).

7. J.E. Dowling, Nutritional and inherited blindness in the rat, *Exp. Eye Res.* 3:348 (1964).

8. L.D. Carter-Dawson, T. Kuwabara, P. O'Brien, J.G. Bier, Structural and biochemical changes in vitamin A deficient rats, *Inv. Ophtalmol. Vis. Sci.* 18:437 (1979).

9. C.D.B. Bridges, S-L. Fong, G.I. Liou, R.A. Alvarez, and R.A. Landers, Transport utilization and metabolism of visual cycle retinoids in the retina and pigment epithelium. *Prog. Ret. Res.* 1:137 (1983).

10. R.B. Edwards, A.J. Adler, S. Dev, R.C. Claycomb, Synthesis of retinoic acid from retinol by cultured rabbit Müller cells, *Exp. Eye Res.* 54(4):481 (1992).

11. A.H. Milam, A.M. Deleeuw, V.P. Gaur, J.C. Saari, Immunolocalization of cellular retinoic acid binding protein to Müller cells and or a subpopulation of GABA-positive amacrine cells in retinas of different species, *J. Comp. Neur.* 296(1):123 (1990).

12. P. Dolle, E. Ruberte, P. Leroy, G. Morrisskay, P. Chambon, Retinoic acid receptors and cellular retinoid binding proteins .1. A systematic study of their differential pattern of transcription during mouse organogenesis, *Development* 110(4):1133 (1990).

13. D.L. Stenkamp, J.K. Gregory, and R. Adler, Retinoid effects in purified cultures of chick embryo retina neurons and photoreceptors, *Invest. Ophthalmol. Vis. Sci.* In press (1993).

14. R. Adler, Preparation, enrichment and growth of purified cultures of neurons and photoreceptors from chick embryos and from normal and mutant mice, *in*: "Methods in Neurosciences," P.M. Conn, ed., Academic Press, Orlando (1990).

15. R. Adler, The differentiation of retinal photoreceptors and neurons *in vitro*, *Prog. Ret. Res.* 6:1 (1986).

16. D.L. Stenkamp, and R. Adler, Photoreceptor differentiation of isolated retinal precursor cells includes the capacity for photomechanical responses. *Proc. Natl. Acad. Sci. USA.* 90:1982(1993).

17. P.M. Iuvone, G. Avendano, B.J. Butler, and R. Adler, Cyclic AMP-dependent induction of serotonin N-acetyltransferase activity in photoreceptor-enriched chick retinal cell cultures - characterization and inhibition by dopamine. *J. Neurochem.* 55(2):673 (1990).

18. R. Adler and L.E. Politi, Effects of 11-cis retinal and other retinoids on opsin expression by isolated mouse and chick photoreceptor cells in culture. *Invest. Ophthalmol. Vis. Sci.* (Suppl.) 30:157 (1989).

# INDEX